Rhetoric and Incommensurability

Rhetoric and Incommensurability

Edited and Introduced by
Randy Allen Harris

Parlor Press
West Lafayette, Indiana
www.parlorpress.com

Parlor Press LLC, West Lafayette, Indiana 47906

S A N: 2 5 4 - 8 8 7 9

Library of Congress Cataloging-in-Publication Data

Rhetoric and incommensurability / edited and introduced by Randy Allen Harris.
p. cm.
Includes bibliographical references and index.
ISBN 1-932559-49-3 (pbk. : alk. paper) -- ISBN 1-932559-50-7 (hardcover : alk. paper) -- ISBN 1-932559-51-5 (adobe ebook)
1. Science--Methodology. 2. Comparison (Philosophy) I. Harris, Randy Allen.
Q175.R434 2005
501--dc22
2005010812

Printed on acid-free paper.

Cover art: "Tango," oil on canvas, by Mark Forth, used courtesy of the Artist and The Tory Folliard Gallery, Milwaukee, Wisconsin.

Parlor Press, LLC is an independent publisher of scholarly and trade titles in print and multimedia formats. This book is available in paper, cloth and Adobe eBook formats from Parlor Press on the World Wide Web at http://www.parlorpress.com or through online and brick-and mortar bookstores. For submission information or to find out about Parlor Press publications, write to Parlor Press, 816 Robinson St., West Lafayette, Indiana, 47906, or e-mail editor@parlorpress.com.

This book is for

Indira

Incommensurability is a difficulty for philosophers, not for scientists.

—Paul K. Feyerabend, *Farewell to Reason*

Properly understood—something I've by no means always managed myself—incommensurability is far from being the threat to rational evaluation of truth claims that it has frequently seemed.

—Thomas S. Kuhn, *The Road since Structure*

To divide humanity into irreconcilable groups with irreconcilable attitudes, having no common language of truth and morality, is, ultimately, to rob both groups of their humanity.

—Stephen Spender, *World within World*

No incommensurability [is] absolute

—Barbara Herrnstein Smith, *Belief and Resistance*

The English term 'incommensurable' is somewhat unfortunate.

—Sir Karl Popper, *The Open Society and Its Enemies*

Contents

Preface

> Within the comprehensibility chasm lies the condition of incommensurability.
>
> —Carolyn R. Miller, "Rhetoric and Community"

The incommensurability thesis represents the most profound problem facing rhetoric—of science, surely, but of any symbolic encounter, any attempt to cooperate, find common ground, get along, make better knowledge and build better societies. It's too big and too deep for me. So I invited the smartest, most clear-eyed rhetoricians I know—of science and of any symbolic encounter—and an equally gifted philosopher, to help me wrestle with it. The result is this book, which, hand-to-my-heart, you will find seriously illuminating about the way scientists and other value-holders achieve or fail to achieve shared understandings.

I would like to thank, first of all, these brilliant and good-hearted professors. Even the customary dog-work of copy editing and proof reading has been a joy on this project, as I got to read and re-read the paradigms they crafted, learning something new on every pass; and watching those essays come together in the first place was a lesson in scholarly collegiality I hope never to forget.

In addition to help from these model scholars, I have been very lucky in the range of support and feedback I have had in working through the problems of incommensurability, starting with my one unfailing source of insights and challenges, the students at the University of Waterloo. My contact with nearly all of them over the course of this project has been tremendously rewarding, but a small group of them deserve an extra measure of gratitude for specific help with various aspects of this book: Jim Brookes, Jacqueline Chioreanu, Paul Clifford, Ryan Devitt, Zarsheesh Divecha, Olga Gladkova, David Hoff, Kim Honeyford, Laura Knudsen, Sheila Hannon, Christopher Hutton, Shirley Lichti, Karen Menard, Sarah Mohr, Tiffany Murray, Stephen Noel, Joel Pearce, Jeff Stacey, Rachel Stuckey, Y-Dang Troeung, Lara Varpio, Karl Wierzbicki, and Robert Jing Zhu;Y-Dang deserves a yet greater helping of thanks for her hard work on the volume, as does Ryan for his incredible generosity, stamina and dedication.

I have also benefited from the opportunities, advice, opinions, and direct feedback provided by a diverse network of generous colleagues, including Kelvin Booth, G. Thomas Goodnight, Brian Hendley, Andrew Jewett, Tim Kenyon, John Lyne, Michael MacDonald, Andrew McMurry, Kathryn Northcut, Brian Orend, Trevor Pearce, Howard Sankey, Paul Thagard, Christopher Tindale, Jonathan Tsou, James Van Evra, and Charles Willard; from the scholarly and editorial direction of the tireless David Blakesley; from the support of Kevin McGuirk, Neil Randall, and Norma Snyder; from the eunoia and phronesis of Pauline McAughey; and from the financial assistance of the Social Sciences and Humanities Research Council of Canada.

I am especially grateful for the thoughtful and encouraging commentaries of Michael Leff and Harvey Siegel on the overall scope and shape of the project.

This book would have been much poorer without the industry, intelligence, and intellectual integrity of Michael Truscello.

And personally, I would have been much poorer without the daily engagement of my closest rhetorical collective, Oriana, Galen, and Indira. They give me all the argumentation I can handle, and more reasons than I deserve to strive constantly for commensuration.

—*Randy Allen Harris*

I

Incommensurability, Rhetoric

There are two possibilities. Perhaps incommensurability just does not obtain of scientific programs (theories, paradigms, . . .). Or perhaps there are ways around it, remedies—rhetoric.

—Randy Allen Harris, "Introduction"

Incommensurability had two fathers, unusual even for philosophical terms, and the joint paternity of Thomas S. Kuhn and Paul K. Feyerabend has contributed to much subsequent confusion. While the signifier is the same for both of them, each father engendered a different signified, taking the mathematical metaphor in largely overlapping, but subtly dissimilar directions.

—Paul Hoyningen-Huene, "Three Biographies"

1

Introduction

Randy Allen Harris

> Arguments that seem powerful to one side seem unimportant to the other. What looks like striking insight to one side looks like perverse illusion to the other. Often, the parties simply see the world differently, in some way that is not directly observable. [. . .] What I take to be essentially this phenomenon has been most clearly identified and articulated by Kuhn, and his term is the one I will use: incommensurability.
>
> —Howard Margolis, *Paradigms and Barriers*

Incommensurability—the lack of a common standard for taking the measure of two systems with respect to each other—has crippling implications for science, the domain in which it was first raised to widespread contemporary attention.* It disables progress. If one can't mea-

* I thank, very gratefully, Bruce Dadey, Alan Gross, Paul Hoyningen-Huene, Michael Leff, Carolyn Miller, Kathryn Northcut, Harvey Siegel, Christopher Tindale, and Michael Truscello, for helpful and perceptive comments on earlier drafts of this introduction, and Charles Willard for a trenchant response to some of its arguments in another forum. Jim Brookes, Paul Clifford, Ryan Devitt, Olga Gladkova, Kim Honeyford, Sarah Mohr, Tiffany Murray, Joel Pearce, and Jeff Stacey also kicked its tires and drove it around the block. Needless to say, not all of them agree with everything I say, but these few people, and a good many others, have had such a deep and polylogic effect on this essay that all errors, omissions, and lapses of style or taste can safely be traced to one or another of them. I am innocent.

sure theories with respect to each other, how can one choose which is best? If one can't choose, how do new theories arise? What would be the point?

There is a hitch, though. Science isn't crippled. It's not even limping.

That's not to say there aren't problems with science. The areas that get attentive research and those that don't, the flow of capital, the authority science has in the public sphere—to sample narrowly from a welter of political, ethical, and sociological issues impinging on science, and vice versa—all require steady vigilance, even steady suspicion, from the citizenry of the twenty-first century. Science and its cousin, technology, permeate almost everything we do, certainly every breath we take. But in terms of its daily duties with respect to its primary job—making knowledge about the phenomenal world—there is no sign of disability. Incommensurability should destroy science. It doesn't. Why not?

There are two possibilities. Perhaps incommensurability just does not obtain of scientific programs (theories, paradigms, . . .). Or perhaps there are ways around it, remedies—rhetoric. Both positions are reasonable, depending of course on how one construes *incommensurability.* It also matters how one construes *rhetoric.* We will need to fix both of these construals in the course of this introduction, one broadly, the other much more narrowly. (Other consequential construals are of course implicated here and there as well:what science is; what knowledge is; what theories are; what standards, values, methods, and meanings are. Sometimes they are explored briefly and interdependently in what follows, but often they are just assumed or stipulated, or even ignored. The meeting of rhetoric and science is a complex web, and investigating a couple of strands at a time is a pretty good day's work.)

Rhetoric, the term we will construe broadly, we need to fix only enough to distinguish it from the ordinary language sense of specious, empty, or shallowly florid talk, and from the widespread scholarly sense in which it is a synonym for *irrational.* To that end, I will anchor it in a long and fairly continuous tradition of suasion studies that stretches to ancient Greece, briefly outlining its themes and practices; in the short term, we can get by with a stipulation that it is a discipline which investigates the ways symbols induce action and belief. That discipline governs the primary methodologies of this book.

Incommensurability, the focal term of the book, requires much more detailed analysis. It, too, stretches to ancient Greece, but in a quite restricted sense that ultimately bears little resemblance to the principal range of usages we are concerned with, the ones that arose out of mid-twentieth century philosophy of science. That range is shifty and myriad. *Incommensurability* is hard to pin down, and the more one examines it, the more facets it has.

In what follows, I chart the origins of incommensurability, the notion, and rhetoric, the field. I then outline both a taxonomy and a scale of the ways in which *incommensurability* is deployed in various literatures—chiefly philosophy of science, but also ethics and postmodernist cultural critiques—before returning to the question of why such a potentially debilitating concept does not in fact debilitate the production of knowledge. I explore the two chief possibilities: (1) it does not obtain, and (2) there are remedies. Each accounts for part of the picture, depending on how rigidly *incommensurability* is construed. There's more.

Not only does incommensurability not hamstring science, it proves epistemologically wholesome in several respects. After investigating those respects, this introduction concludes with an outline of the impressive contributions to this volume by rhetoricians looking at the interplay of incommensurability and argumentation.

Incommensurability

> Those magnitudes are said to be *commensurable* which are measured by the same measure, and those *incommensurable* which cannot have any common measure.
>
> —Euclid, *The Elements*

The story goes like this. A disciple of Pythagoras—one Hippasus of Metapontum—approached the master with an unsettling discovery one day while a colligation of Pythagoreans was out on the bounding main in the sixth century BCE. The universe for Pythagoras, as you surely recall, was profoundly geometrical—composed of perfect shapes allied with perfect numbers. He was especially devout in two articles of faith, that (1) everything could be measured; and (2) all measurements were either whole numbers or ratios of whole numbers. The first Pythagorean article meant that the structure of reality was wholly amenable to numeration (geo-metry is, after all, "earth-measuring"). The second meant that this structure was therefore unutterably glorious, a reflection of the compelling intellectual beauty of mathematics.

Pythagoras was not, therefore, amused by Hippasus's revelation: that the diagonal of a square whose side was one unit could be expressed neither as a whole number nor as a ratio of a whole number: the two whole numbers needed to satisfy Pythagoras's second article of faith did not exist in this case. There is no common divisor (except one and zero). One needs to use $\sqrt{2}$, a decidedly unwhole number, whose value could never be exhaustively computed—the sort of number that we now call *irrational* but that the Pythagoreans called *arrhētos*, 'unspeakable.'

Hippasus had discovered there were pairs of numbers that were *asummetra:* without a (rational, speakable) common measure. The discovery was fatal for the Pythagorean cosmos: "The point-to-point correspondence between arithmetic and geometry had broken down—and with it the universe of number-shapes" (Koestler 1959, 39–40).

It also proved fatal for Hippasus. Pythagoras did the only thing a rational man could do in such circumstances, throwing him overboard and swearing all of his other followers to secrecy, about the murder, but especially about the ghastly idea of asummetric number pairs.[1] Or, that's the story. If the universe is poetic enough for the story actually to be true, however, someone squealed. The notion of number pairs without a common measure made its way into Euclid's *Elements,* where it rested comfortably for two and a half millennia (somewhere in the Latin Middle Ages the label becoming *incommensurabilis,* a term nearly isomorphic with *asummetra*),[2] until the middle of the twentieth century, when two scholars interested in theory change—Paul Karl Feyerabend, a historically minded philosopher of science, and Thomas Samuel Kuhn, a philosophically minded historian of science—applied it to pairs of successive theories (and, in Kuhn's case, to the overlapping but looser notion of successive paradigms).

Their use of the word was metaphorical, with a twist. The metaphor is the type Aristotle called proportional (*Rhetoric* 1411[a]), running in this case something like "neutral algorithm" is to "theory" as "common measure" is to "number" (Kuhn 1996, 200).[3] Some theories are incommensurable because there is no neutral algorithm by which to plumb them, just as some numbers are incommensurable because there is no common measure by which to size them both up. The twist is the introduction of a component irrelevant to geometry: critical agents, scientists. In Euclid, no one chooses one number over another to champion as true and to invest his career in.

Though many philosophers apparently wished someone would throw them overboard, Feyerabend and Kuhn survived; incommensurability (the nominalization of the anglicized adjective, *incommensurable*) proliferated in its new habitats.[4] It is fashionable in language circles to talk of the global ubiquity of the language rooted in the noises and notions that came along with the Angles, Jutes, and Saxons in their fifth-century CE invasion of Britain, not so much as the spread of English, a presumed monolithic language, but as the creation of Englishes, a diverse range of locally and situationally determined languages. A similar observation holds for the key word rooted in the noises and notions of Feyerabend and Kuhn. There are multiple incommensurabilities, none of them tidy and precise and walled off from the others, but all of them with distinct, identifiable features.

We will get to them in due course. First, though, our other key term has an origin myth as well, rhetoric.

Rhetoric

> Rhetoric, I shall urge, should be a study of misunderstanding and its remedies.
>
> —I. A. Richards, *The Philosophy of Rhetoric*

The contributors to this volume have been working out their analyses and arguments in various forums for several years. One of the most recurrent responses when they have come before non-rhetoricians has been, "ah, yes, well, that's a darn good analysis of the data—I especially like what you said about P, and your suggestion that it implicates Q, in the conclusion, is intriguing—but what's rhetorical about it?" The answer to that question depends, first, foremost, and initially, on knowing *rhetorical* beyond its use in ordinary discourse as an adjective signaling empty decoration or dangerous manipulation, and even more on knowing it beyond its synonymy with *irrational* in much (putatively) scholarly discourse.

Rhetoric has a genealogy as ancient as that of incommensurability, and philosophy, and science. While Pythagoras was busy mathematicizing nature and drowning heretics, other scholars were investigating the symbolic means of influencing belief and behaviour.

The story goes like this (B. Smith 1921, Hinks 1940). A customer of the legendary Corax, one Tisias, studied with him, refused to pay, and—therefore—ended up in court. Corax was a Sicilian master of argumentation who coached others in his strategies and techniques, so they could defend themselves and pursue their interests in court, and influence the development of the polis—in other words, he was a sophist. He is the reputed author of a treatise on the use of probability in argumentation called the *Art of Rhetoric.* His common sobriquet is "the inventor of rhetoric." But what *inventor* means here is not that after Corax people were rhetorical, but not before; that after Corax people arranged their arguments, but not before; considered their audience, but not before; appealed to emotion, but not before; used figures of thought and speech, but not before. Arrangement, audience adaption, emotive appeals, figuration, . . . everything one might want to identify as central to suasive language, is present in our oldest texts—*Gilgamesh,* the *Old Testament,* the Homeric epics. The *Iliad,* especially, is one oration after another, all of them highly crafted.

Inventor of rhetoric for Corax means simply that he was the first to codify some of the long-known principles of effective suasion, and maybe the first to theorize about probability in argumentation, and, in any case, the first to peddle principles of suasion as an art or a craft—in fact, as both, the relevant Greek word being *technē,* a term which bridges artiness and craftiness. This was the man across the court from Tisias.

Undaunted, Tisias argued that the jury could not possibly settle in Corax's favor. Either, he said, I win the case, it is settled to my advantage, and that is the end of it; Corax goes home without his fee. Or I lose the case, which would prove my case, that the lessons were worthless—I can't even win a simple case before such discerning citizens as yourselves—and *that* is the end of it; Corax goes home without his fee. One way or the other, I must win. The jurors scratched their heads and turned to Corax. *Au contraire,* he said. Either I win the case, it is settled to my advantage, and that is an end of it; I go home with my fee. Or, I lose the case, which would prove how valuable the lessons were, cheap at half the price—Tisias will have defeated the inventor of the art of rhetoric before such discerning citizens as yourselves—and *that* will be the end of it; I go home with my fee. One way or the other, I must win.

The arguments, that is, come down to:

Tisias	Corax
If I win, well then I win.	If I win, well then I win.
If I lose, that proves I should win (since Corax didn't teach me well enough to be able to win).	If I lose, that proves I should win (since I taught Tisias so well he is able to win against me).

Aside from some muttering about a bad egg coming from a bad crow (*corax* means 'crow,' *tisias* 'eggs'), history does not record what the jury did, though surely drowning crossed a few of their minds.[5] Tisias went on to become a speechwriter (logographer), and the next move in the story is that he traveled to the Greek mainland in the company of a sophist who dramatically eclipsed him, Gorgias of Leontini—Tisias reportedly taking rhetoric-as-*technē* with him, Gorgias taking his renowned oratorical skills, relativistic epistemology, and general-purpose brilliance—setting up the contentious turf-war with philosophy that was soon to emerge in Athens.

Rhetoric, as a field of study, originates with the speculations into (and, necessarily, out of) language: language as filter, as tool, as medium; first of knowing, and then of distributing what is made known. Corax is not the only sophist credited with the invention of rhetoric. Aristotle conferred that honor on Empedocles (Diogenes Laërtius viii.57), a sophist on the vanguard of a widespread pre-Socratic program to use language as an instrument for understanding the buzz and flux of experience—at that pivotal moment in the history of the West when, as Ortega y Gasset phrases it, "the gods were downgraded into causes" (1967, 104). Drawing on the resources of poetry for shaping perception and attitude and response; assimilating and discriminating, balancing and opposing, cutting and splicing linguistic formulations for pleasure and profit; using paradoxes, puns and polyptotons to probe the senses and referents of everyday terms like *river* and *god* and *stuff*; manipulating the variables of argumentation that induce belief and certainty and action, that discover and build and propagate values; zealously pushing a technique that would later be called *reductio ad absurdum*; piling up examples, counter-examples, and counter-counter-examples, until nothing seemed inevitable, but everything seemed possible; the sophists were a breathtakingly exuberant force.

The god (goddess, actually) that sophists downgraded into the most powerful social and epistemological cause was *Peitho*, persuasion. They were preoccupied with the social and cultural power of monologic suasion, and the epistemological power of reciprocal suasion. On the one hand, this fed a bustling industry of handbooks, schools, private lessons, and speech-writing, as well as a culture of public oratorical displays, recreational jury-sitting, and massive argumentative involvement in the polis—to a culture of suasion. On the other hand, it fed dialectic, the relentless pursuit of knowledge and the grounds for knowledge. It was Plato's genius (though not his only one) to appropriate dialectic for his program, pin monologic suasion on the sophists, under the label, *rhetoric,* and hound most of them right out of Athens.

The origins of rhetoric as a discipline grow out of the linguistic investigations of the sophists, the methodologies and codifications they engendered, and a definitional opposition to philosophy. Rhetoric as a *practice*—the use of suasive argumentation in social reasoning—has no origin we can recover. It has been with us not just since the earliest textual productions, like *Gilgamesh* and the *Iliad,* but surely as long as there has been an us, a species capable of deploying language. Gorgias's student, Isocrates, observes that

> in the other powers which we possess we are in no respect superior to other living creatures; nay, we are inferior to many in swiftness and in strength and in other resources; but, because there has been implanted in us the power to persuade each other and to make clear to each other whatever we desire, not only have we escaped the life of wild beasts, but we have come together and founded cities and made laws and invented arts; and, generally speaking, there is no institution devised by man which the power of speech has not helped us to establish. (*Nicocles,* 5–6)

From an Isocratean perspective, as well as from the perspective of current rhetorical theory, the crucial phrase here is "the power *to persuade each other*"—*tou peithein allêlous,* more strictly rendered as "persuasiveness of/to one another"; that is, reciprocal suasion, negotiation. Reciprocal suasion, however, was not the dominant Greek perspective on *rhetoric,* a word Plato appears to have coined in order to nominalize the main professional concern of the sophists he loathed.[6]

The dominant Greek perspective on rhetoric, to the extent we can recover it, given the massive loss of primary sophistic texts,[7] saw it as a monologic enterprise, in which an orator attempts to work his will on an audience. That is the activity Plato characterizes under the label *rhetoric.* More importantly, that is largely the activity Aristotle systematizes under that label. Aristotle is the historical anchor point for the principles and practices of the rhetorical tradition, not their beginning but their first, best extant formulation. Plato saw a bunch of thought-and-language quacks flogging their manuals on how to lie, and attacked them brilliantly enough to almost singlehandedly turn the word *sophist* into a synonym for *charlatan* (at the same time detaching it from people he venerated, most notably Socrates). Aristotle didn't. Where Plato saw cheats and recipes for cheating, Aristotle saw suasion merchants and a diffuse body of maxims and taxonomies for symbolic inducement. He collected, synthesized, and rationalized those materials, and then he peddled them under his own brand. Somehow, possibly through student notes, the Aristotelian brand of rhetoric made its way into the magnificent compendium we now call *The Rhetoric.* That book systematically treats rhetoric as the study of monologic public argumentation, in the assembly (*symboleutic,* or *deliberative*), in the court (*dikanic,* or *forensic*), and at ceremonies (*epideictic,* or *ceremonial*). But the elemental nature of suasion sometimes shows through that system. Aristotle's formal definition of *rhetoric,* for instance, shows the fundamental nature of suasion that Isocrates celebrates—"the faculty [or power; *dunamis*] of observing in any given case the available means of persuasion," he calls it (*Rhetoric* 1355[b]), and reciprocal suasion likewise also shows through, often in connection with Plato's god-term, *dialectic*:

> we must be able to employ persuasion, just as strict reasoning can be employed, on opposite sides of a question [. . .] No other of the arts draws opposite conclusions: dialectic and rhetoric alone do this. Both these arts draw opposite conclusions impartially. (*Rhetoric* 1355[a])

Still, the driving concerns of the *Rhetoric* are neither the elemental nature of a suasive faculty nor the marshalling of oppositional arguments. They are: the categories of rhetorical proof (*ethos*, *pathos*, and *logos*; or character, emotion, and reason); the strategies and techniques deployed within those categories; and their application to particular cases (that is, *heuresis*, the invention of arguments). Prominent among those strategies are the enthymeme (a species of syllogism tuned to the demands of probability), and the *topoi* (schemata for constructing arguments). The three types of proof provide reasons to believe something is true. You believe something because it comes from a credible source (*ethos*). You believe something because of the way you feel about it (*pathos*). And/or you believe something because of the form and structure of the symbols that encode it (*logos*). These three categories of proof correlate with the three critical elements of any rhetorical situation. There must be a speaker (manifesting character), an audience (experiencing emotion), and a speech (exhibiting form and structure). These are correlations, not encapsulated groupings. The character of the speaker, for instance, might be exhibited through a display of emotion, which in turn influences the emotional state of the audience. The form and structure of the speech, too, might influence the audience's perception of the character of the speaker—as rational, or imaginative, or careful, or incisive—and thereby influence their response to his argument. Rhetoric generates in audiences (or fails to generate in audiences) motivations to believe, and belief is not a simple or unitary phenomenon.

While Aristotle's emphasis is largely on productive strategies for accomplishing suasion, there is also a powerful current in the *Rhetoric* of appraising arguments and adjudicating cases. "The hearer" of rhetorical discourse, Aristotle says, "must be either a judge, with a decision to make about things past or future, or an observer [in the case of *epideictic* discourse]" (1358^{b}). The point of rhetoric for Aristotle is to direct or condition a decision, and the nexus is always the argument—its construction, its reception, its logical integrity.

The Roman period, our next major stop in the history of rhetoric, is known as the Age of Codification (Murphy *et al.* 2003), chiefly for its taxonomic attention to *stasis* theory and figuration, though the mature work of its most renowned rhetorician, Cicero, reveals him to be less than a fan of systems. *Stasis* (also status) theory dates at least to Gorgias, and is charted out by (who else?) Aristotle, in book three of the *Rhetoric* (1417b). A later Greek rhetorician teaching in Rome, however, is associated with its methodological elaboration, Hermagoras of Temnos. Figuration goes back much further yet, of course, to the origins of thought, constituting the earliest texts we have and receiving various levels of attention in the first manuals of civil discourse, the *logon technē*.

The word *stasis* is traceable to the Proto-IndoEuropean verb root, **sta*, which meant (and gives us the English verb) *to stand. Stasis* is richly polysemous in Greek, as most **sta*-decendants are, but orbits around the stability of things inherently motile—a *hippostasis,* for instance, is a stable, the place where horses stand—and had a wide range of technical applications. "The further one reads Aristotle's physical science," Otto Alvin Loeb Dieter wrote, "the better one understands stasis" (1994, 216):

> It is immobility, or station, which disrupts continuity, divides motion into two movements, and separates the two from one another; it is both an end and a beginning of motion, both a stop and a start, the turning, or the transitional standing at the moment of reversal of movement, single in number but dual in function and in definition. (Dieter 1994, 218)

All of that sounds rather mystical, a bit like one hand clapping, until the concrete physics are brought in: it is the point in the behavior of a pendulum, for instance, where the energy that pushed it in one direction is exhausted, but just before the force that will pull it back down in the opposite direction has begun to do so: not the equilibrium position, but the suspended mass, the very transient place/moment you occupy at the acme of the swing on a playground, before gravity takes over to pull you back. Or, if you haven't been on a swing lately, think of the furthest point your golf club, tennis racket, or baseball bat reaches just before your muscles start to bring it back towards the ball.

In rhetoric, then, the *stasis* is the 'place' between the two directions a controversy can take, and *stasis* theory identifies a series of four such canonical places: fact, definition, quality, and jurisdiction.[8] A dispute might hinge on whether or not something happened (*fact*), for instance (did Miss Scarlet fatally wallop Colonel Mustard with a candlestick, or not?). Granting the facts, a dispute might turn on what to call them (*definition*), on how those facts should be interpreted linguistically (was it first-degree murder? second-degree? manslaughter?). Granting the facts and the label, a dispute might revolve around the nature of the act (*quality*), on the values it instantiates (was it an act of self defence? the result of battered woman syndrome? a case of diminished capacity?). The fourth *stasis* (*jurisdiction*) addresses the legitimacy of the arguers, forum, or proceedings (maybe Miss Scarlet has diplomatic immunity, or is a juvenile, or the prosecutor has a conflict of interest).

These four stases don't detach quite so clinically in argumentation, of course, since words are involved at all levels; therefore, so are definitions and values. But the categories are effective inventional strategies for building and refining a case, equally effective diagnostics for taking the measure of another's argument, and, as Lawrence Prelli argues in his essay in this volume, because of both those inventional and diagnostic powers, stases also provide commensurability points for bringing divergent perspectives to a common place for negotiation and understanding.

Cicero's earliest work, *De Inventione,* falls easily under the Age-of-Codification rubric, plotting out the four-way *stasis* system, the five elements of rhetoric, the three genres of oratory, the six parts of an oration, and so on. But his later works reveal a decidedly Isocratean cast, pursuing a more dialogic sense of rhetoric, in which it is less of a methodical machine of persuasion and more of a fluid instrument of understanding. *De Oratore,* in particular, builds a vision of communicative action in which well-schooled and responsible rhetors collaboratively pursue common judgments, interpretations, and knowledge. Plato's *Phaedrus* is a striking contrast, in which Socrates takes a disciple aside, beneath a plane tree, and leads him by the nose through several views of rhetoric, each displacing the other, before arriving at a picture in which it serves as handmaid to philosophy. *De Oratore* brings a small group of highly able orators together (notably on the first day, beneath a plane tree), and they present each other with various interlocking and complementary views of rhetoric, subtly disputing and agreeing about a range of factors and themes, ultimately leaving the reader with a collective vision of a powerful civic vehicle that melds wisdom with eloquence (see Ceccarelli, this volume, 275-278, for further discussion).

The history of rhetoric after Cicero is a complicated affair, but for our purposes—that is, for the way rhetoric stands in relation to science and incommensurability—one theme is especially important: rhetoric's increasing association with aesthetics at the expense of argumentation. The disciplinary birth of rhetoric had involved applying the instruments of oral poetry to the problems of knowing and saying; aesthetic concerns have always been integral to rhetoric. Some sophists—most notably Empedocles's student and Tisias's travelling companion, Gorgias—had come to be known for their extravagant figuration, and even the it's-all-about-argument Aristotle had found time to explore style, but aesthetics had mostly been peripheral in the *logon technē.* In the Roman period, rhetoric became strongly linked to "nonliteral language" (figures, tropes, schemes); the *Rhetorica ad Herennium,* the most obsessive of the Roman codifications and the most influential for the Middle Ages, has a laundry list of sixty-four figurative maneuvers. In the Middle Ages, especially in the branch of rhetorical studies known as the *ars poetica,* this linkage bred a mania for cataloguing figuration in what retrospectively looks like sterile and superficial ways. With the rise of scientific rationalism and the empiricist craze of the early modern period, aesthetic elements of discourse came to be seen as worse than merely pretty; they came to be seen as misleading, corrosive to truth and knowledge.

It was not a difficult task for the language reformers associated with the rise of science to scapegoat rhetoric generally, figuration specifically, as part of the old epistemic order that needed to be swept aside, along with superstition, alchemy, and scholasticism. The works of Boyle and Newton, not to say Bacon and Hobbes and Locke and Hume and Descartes, include strategic screeds against the contaminating influence of fancy-Dan word-mongering, the lasciviousness of metaphor, the wheedling dishonesty of rhetoric. Thomas Sprat, in his *History of the Royal-Society of London, for the Improving of Natural Knowledge,* is representative:

> Who can behold, without indignation, how many mists and uncertainties, these specious Tropes and Figures have brought to our Knowledge? How many rewards, which are due to more profitable, and difficult arts, have been still snatch'd away by the easie vanity of fine speaking? For now I am warmed with this just anger, I cannot withhold my self, from betraying the shallowness of all these seeming Mysteries; upon which, we writers, and speakers, look so bigg. And, in few words, I dare say; that of all the studies of men, nothing may be sooner obtained,

> than this vicious abundance of phrase, this trick of metaphors, this volubility of tongue, which makes so great a noise in the world. (Sprat 1958 [1667], 112)

Modern science grew out of these beginnings with a view of itself as a nonrhetorical, even an antirhetorical, enterprise, and a corresponding view of rhetoric as a tumid excrescence on true, real, literal-minded language. That antipathy was all the more natural since science grew out of philosophy, many strains of which (remember Plato), had long regarded rhetoric as its opposite. Rhetoric entered a long stretch of ignominy and marginalization, and the word gained the connotations it wears in ordinary language today, evoking the specious, the florid, and the vacuous.

Rhetorical theory was rehabilitated in the twentieth century, at least for some scholars.[9] This rehabilitation was marked by two significant moves: first, a return to Aristotle, with his emphasis on argumentation and reason, and a concomitant diminution of an interest in style; second, an expansion of the Aristotelian synthesis back into the more embracing sophistic view of rhetoric as elemental to all symbolic exchanges.

The shape of this rehabilitation can be characterized by a shift away from the monologic, strategic-rhetor-addressing-pliant-sheep model of rhetoric, toward a mutual, give-and-take, negotiative, Isocratean model. The key word for ancient theory is *persuasion;* the key word for modern theory is *identification.* The shift is away from a body of principles for working your will upon others, and toward a cluster of practices for building collective arguments, for making ideas and institutions that we can share and which help define us into groups of Philosophers and neoKantians, Rhetoricians and neoAristotelians, Humans and Americans and Canadians. We identify each other, and identify with each other.

We exchange reasons for beliefs and actions. I accept some of your reasons, you accept some of mine, we both accept some from various third parties; in swapping our reasons, some get discarded, others get forged in the crucible of exchange—perhaps you develop a new way to counter some reason I give you, or perhaps you see a way to augment it. Other reasons to believe simply come into or fall out of fashion, like gods and miniskirts. And many reasons get their pull because of an oppositional push. Identification includes dissociation in the bargain: *we* are like each other in large part through our difference from *them.*

Kenneth Burke's view of identification, in particular, is suffused with ostracism and violence: "an imagery of slaying (slaying of either the self or other) is to be considered merely as a special case of identification in general" (Burke 1969b, 19–20).

This framework, where suasion is subsumed under the larger notion of identification, is most closely associated with Burke, and with I. A. Richards, Chaim Perelman, and Lucie Olbrechts-Tyteca. Burke was concerned with ideology and community, Richards with using rhetoric to heal ruptures of communication and belief, Perelman and Olbrechts-Tyteca with the contingencies of real argumentation in real contexts. This constellation of interests defined rhetoric for most of the last century. As a scholarly pursuit, rhetoric retained much of the Aristotelian methodological kit bag—*ethos*, *pathos*, *logos*; *topoi*; the structure of argumentation—but rhetoricians became less concerned with the preparation of effective public speakers than Aristotle was. The focus became critical, concerned now with understanding how it is humans come to hold and advocate beliefs—and, increasingly over the last few decades of the last century, and on into this one, with how it is we come to knowledge.

The recent focus on epistemology has rekindled interest in a range of intellectual and civic values associated with the Isocratean tradition, which Herbert Simons, in his essay in this volume, characterizes as "a process of coming to judgment, and of bringing others to that judgment" (239). Among the consequences of this tradition are "that what gets called 'knowledge,' even in the hallowed realms of science, is far less rule-based or data-derived and far more a consequence of interpretation, social construction and rhetorical choice than had previously been imagined."

Science, that is, becomes a prime concern of rhetoric—it is the locus of our most revered and successful knowledge-production activities—and rhetorical critics have been especially attracted to moments when reciprocal suasion fails most dramatically.

We return then to the problem of scientific ruptures, and the crippling implications that lacking a common standard has for the progress of science.

Incommensurabilities

> You cannot talk with a person whose basic premises are completely incompatible with your own. The words that are exchanged are without meaning, so that in the real sense there has not been discourse.
>
> —Richard Weaver, *The Ethics of Rhetoric*

"A new word," Ludwig Wittgenstein said, "is like a fresh seed sewn on the ground of the discussion" (1998, entry 1929). The ground for *incommensurability* (a borrowed word, but new to the context) was well tilled by the time Feyerabend and Kuhn sowed it in philosophy of science. Karl Popper's *Logik der Forschung* (1934) was translated and updated as *The Logic of Scientific Discovery* (1959), a signal work stressing the theory-driven nature of observation and its attendant terminology. Norwood Hanson, in *Patterns of Discovery* (1958), had used gestalt imagery to investigate the effect of prior commitments on perception, and had noted the semantic shifts among the terminology of successive theories. He also coined the indispensable term, *theory-laden* (1958, 19), which nicely summarizes in a scientific idiom the notion that words carry their dominant contexts with them, that Democritus's *atom,* for instance, is not Bohr's *atom,* because they each get their meanings from different theoretical contexts.[10] Michael Polanyi had characterized inter-framework disputes by saying that the opponents "speak a different language, live in a different world," that they are separated by a logical gap that can only be negotiated by processes of persuasion, and that switching frameworks is akin to "a conversion" (Polanyi 1958, 151).[11] A collection of essays by Benjamin Lee Whorf (1956) in anthropological linguistics had been published, many of which developed the line of thought that his teacher, Edward Sapir, put this way: "The worlds in which different societies live are distinct worlds, not merely the same world with different labels attached" (1949, 162). There, that's enough name-dropping for one paragraph. We can go back to the one with whom we began, and whose thought touches on all these matters, and influenced most of these thinkers, certainly Feyerabend and Kuhn: "The limits of my language are the limits of my world" (Wittgenstein 1922, 148; *Tractatus* 5.6).

These elements (and more), in various configurations, fed the notion of incommensurability that struck both Feyerabend and Kuhn at roughly the time, in the same place, as they worked through issues of theory change in the history of science, both of them publishing the landmark works that introduced the word in its contemporary non-mathematical sense in 1962 (as Paul Hoyningen-Huene chronicles in his essay in this volume).[12] They had very different aims. Feyerabend, rather modestly, wanted largely to disinfect empiricism of the principle of meaning invariance; Kuhn wanted to carry off a paradigm shift of his own, in epistemology.

After the sowing, the soil for *incommensurability* proved fertile indeed, and it has propagated far and wide,—undergoing, like any word at the core of heated argumentation, a range of semantic shifts and distensions. Our two principals took the concept in somewhat different directions. Early in Kuhn's career incommensurability seemed to be both a startlingly nonrational proposal and a cornerstone of his program. But he constrained it in his later writings, distancing himself from his more extreme earlier suggestions, repeatedly arguing that it was not as radical as everyone thought. At the end of his life he was working on a book devoted to incommensurability that argued it was "far from being the threat to rational evaluation of truth claims that it has frequently seemed to be" (Kuhn 2000, 91). By the 1980s Kuhn was talking of a constrained variant he called *local incommensurability*. Meanwhile, Feyerabend was going more and more global.

It is dicey to ascribe a trajectory to Feyerabend's career because of what John Preston calls "the consistently malleable nature of Feyerabend's views" (Preston 2002) and because Feyerabend himself routinely inoculated his work against such attempts ("When writing a paper I have usually forgotten what I wrote before and application of earlier arguments is done at the applier's own risk"—Feyerabend 1978a, 114). Still, incommensurability seemed early in Feyerabend's career mostly to be a fairly small but pointed stick with which to jab his erstwhile thesis supervisor, Karl Popper, a lexically-based notion derived from the convergence of theory-ladenness and meaning variance. In Feyerabend's hands this convergence compromised Popper's rationalist program, because it implied that there were at least a few theories that could not be compared by a philosophically coherent methodology. Feyerabend never tired of prodding Popper, but the stick grew and grew, becoming one of the pillars of his account of epistemology, tied closely to his advocacy of theoretical pluralism. Since incommensurability looks prototypically irrational, he was all the more glad to embrace it. As Preston has characterized the later course of Kuhn's and Feyerabend's work, "at a time when Kuhn was downplaying the 'irrationalist' implications of his own book, Feyerabend was perceived to be casting himself in the role others already saw as his for the taking" (Preston 1997).

"The phenomenon of incommensurability," Feyerabend crowed in *Against Method,* "creates problems for all theories of rationality" (Feyerabend 1978a, 214). In many ways that book represents one possible culmination of the program introduced in *The Structure of Scientific Revolutions,* a book Feyerabend called "Kuhn's masterpiece" (Feyerabend 1993, ix).

The problems Feyerabend was crowing about are terrifying for rationalists, going right to the core of what it means to be a social animal. It's not just that incommensurability makes discussion hard or evaluation problematic; lots of things do, things as trivial as not getting a decent cup of coffee, or as difficult as not knowing calculus, but all of them seem remediable. The problem is that incommensurability seems to be fundamentally irremediable, ruling out agreement and evaluation in principle. The rhetorical message is, "Why bother?" If we can't agree, or even decide on criteria by which it is conceivable to agree, we might continue talking *at* each other for our own expressive needs, but there would be no point in talking *with* each other.

While Feyerabend and Kuhn were developing incommensurability, others were reacting to, resisting, and reconfiguring our key term. If, as Alan Musgrave says, "a technical term is introduced to 'cut a long story short'" (Musgrave 1978, 344), then *incommensurability* fails as a technical term. The story gets longer every day. It kicked up a storm in philosophy of science, implicating history of science in the bargain and helping to initiate a tremendously fruitful alliance that still continues, and it sponsored new positions in sociology of science as that field entered a growth phase. It has spiraled out into other discourses as well—most notably suffusing schools of moral, political, and aesthetic philosophers in a body of scholarship associated with the slogan "incommensurability of values," and taking root in postmodern discourses, where it emblemizes fractionation of knowledge and language practices.

In 1993, John Earman said "issues about incommensurability present amorphous and shifting targets" (Earman 1993, 17), and they certainly haven't solidified or settled down in the time since. *Incommensurability* is, in Richard Weaver's terminology, a *charismatic* word, one of the aspects of which is to fall into uses that have "broken loose somehow and [. . .] operate independently of referential connections" (Weaver 1953, 227). There are also somewhat eccentric uses, in which the word has fairly tight referential connections, but they are embedded deeply within a specific approach, and those connections don't reach beyond that approach significantly enough to require any attention in a broad survey.[13] The word has begun to pick up the vogue that *paradigm* assumed in the 1990s. I fully expect to be informed by a waiter some evening that a *beaujolais nouveau* is incommensurable with rack of lamb.

We can begin to chart *incommensurability,* however, with an account that compactly includes most of the issues we will be sifting through. Here is what (mid-career) Feyerabend says about the consequences that follow when two theories, A and B, are incommensurable:

> [W]e cannot compare the contents of A and B. A-facts and B-facts cannot be put side-by-side, not even in memory: presenting B-facts means suspending principles assumed in the construction of A-facts. All we can do is draw B-pictures of A-facts in B, or introduce B-statements of A-facts into B. We cannot use A-statements of A-facts in B. Nor is it possible to *translate* language A into language B. This does not mean that we cannot *discuss* the two views—but the discussion will lead to sizeable changes of both views (and of the languages in which they are expressed). (Feyerabend 1978a, 206-7)

With this epitome as a platform, we can begin looking in two different ways at the manner the word is deployed in the relevant discourses influenced (and, in some cases, spawned) by Feyerabend's and Kuhn's introduction of the term—one categorical, one graduated.

For the categorical approach, we can invoke biology, counting up four species of incommensurability (by my count; others tote differently, usually stopping at two or three, and there is certainly room to subdivide these categories further, into upwards of four categories[14])—*brick-wall* incommensurability, *cosmic* incommensurability, *semantic* incommensurability, and *pragmatic* incommensurability. I explicate these in some detail as we move through the introduction, tracing lineages and applications, but as a cursory approximation, to sketch out the range of ways *incommensurability* gets used, the categories roughly plot as follows:

Brick-wall incommensurability labels situations in which communication is hopelessly stymied, where each party can only hear gibberish when the other speaks.

Cosmic incommensurability labels a situation in which communication is severely hindered because of different perceptions of the "same" phenomena, where the parties can't communicate coherently because they "live in different worlds."

Semantic incommensurability labels a situation in which communication is significantly complicated because clusters of meanings used by the parties are out of synch. It includes a noteworthy variety, the weakest form of incommensurability posited, *local incommensurability.*

Pragmatic incommensurability labels a situation in which argumentation is rendered very difficult because themes and practices of the contending parties are out of synch because the parties appeal, often covertly and vaguely, to different values; inevitably, there are substantial semantic components to such situations as well. It also includes a noteworthy variety, *meta-incommensurability,* which labels incommensurability about the presence/possibility of incommensurability.

There are also two related usages—*value* incommensurability, and *postmodernist* incommensurability—both of which align fairly closely to pragmatic incommensurability. With value incommensurability the genealogy is somewhat more diffuse than with the other classes, since it is largely outside the domain of science, and since there are clear pre-Kuhn-and-Feyerabend bloodlines. Postmodernist incommensurability, too, is substantially outside of science, not because postmodern scholars are uninterested in scientific discourse; just because they are interested in so much else as well. But there are relatively few complications with the lineage of postmodernist incommensurability. It comes quite directly from Feyerabend and Kuhn. All of these categories (varieties, usages) implicate various configurations of semantic, thematic, practical (or methodological), and perceptual features.

The second model, the graduated approach to the uses of *incommensurability*, is simpler. We can invoke the thermometer, identifying degrees, rather than categories, of incommensurability—a scale that runs from total homogeneity of paired symbolic networks (theories, languages, worldviews) to total heterogeneity.

The first model, the biological one, is attractive because it highlights different groupings of features and traits in various accounts of incommensurability, and because it captures the sense that when people talk about incommensurability, they do not always seem to be talking about the same thing. It also suggests cross-pollinations, bloodlines of usage, seminal texts, and other organic metaphors of influence that are always significant in the propagation of words and concepts. The second model, the scalar one, is attractive because it highlights the relative magnitudes of incommensurability—from zero to total semiotic mismatch—that might hold between two systems, and because it suggests that the 'type' of incommensurability may often be much less important than the 'amount' of incommensurability.[15]

Fortunately, these two approaches are commensurable, so long as we don't demand an unreasonably tight mapping. In particular, brick-wall incommensurability is effectively total incommensurability: a situation in which the two relevant systems are so completely alien to each other that communication isn't even possible; local incommensurability is not zero, or even negligible, or there wouldn't be enough incommensurability to constitute a category, but it is weak (in the sense of Thagard and Zhu 2002), the lowest categorical reading on our incommmensurometer. The commensurability of our two models is doubly fortunate because neither model works sufficiently on its own; one misses the partially graduated nature that most commentators observe in matters of incommensurability, while the other misses the range of features that accompany various instances of incommensurability.

In addition to our four categories and our zero-to-total continuum, there is one more critical facet to Feyerabend's and Kuhn's treatment of incommensurability: temporal proximity. Ferdinand de Saussure's terms for linguistics, *diachronic* and *synchronic,* capture this dimension nicely. Diachronic linguistics investigates language across time, usually charting historical 'stages' against each other. Synchronic linguistics investigates language at one particular 'stage.' Both Kuhn and Feyerabend were strongly motivated by the synchronic matter of communicative blockage during theory contestation (the moment when two programs are on the disciplinary table at the same time),[16] a phenomenon routinely called *theory succession* in the incommensurability literature, emphasizing a concern with dominance and displacement. But they both were also concerned with the diachronic matter of a modern scholar (a philosopher or a historian more than a scientist) confronting a long obsolete program; Kuhn more so than Feyerabend.

Incommensurability, the diachronic phenomenon, does not look like the same creature as incommensurability, the synchronic phenomenon. The historian's confrontation with an archaic program is exclusively hermeneutic, wholly unidirectional, with no possibility for exchange, and involves a relatively fixed set of ontological allegiances. In particular, the historian is far more likely to be sympathetic to an earlier ontology than a contemporaneous scientist is to be sympathetic to a rival ontology. More dramatically, the historian juxtaposes two highly distinct and encapsulated research programs. The scientist in a context of radical theory comparison usually has access to a range of proposals, from either 'side', but also from people occupying various intermediate positions, with more or less allegiance to the polarized camps. "The conceptual gaps that must be closed in typical cases of scientific theory choice," as Harold Brown notes, "are often much smaller than the gaps that an historian must cross," painting the picture of historical theory change as "a series of theory replacements running from T_1 to T_n, where T_1 and T_n have little in common although there is a great deal in common between any two adjacent members of the series" (Brown 2005, 157). This picture is still a bit too neat and pretty for the options available in contexts of theory change, but the diachronic cases of theory comparison clearly fit the T_1 / T_n situation, while the synchronic cases are much closer to a T_x / T_{x+1} situation.

The question that remains is whether there is sufficient overlap between the diachronic cases and the synchronic cases of communicative blockage that one would want to put them in the same basket—are they both really instances of incommensurability? Maybe not. But (1) since the literature assigns both phenomena that label regularly, we will continue to discuss them together, where appropriate, and, fortunately (2) they are compatible with our four-category and graduated treatments. Diachronic cases of incommensurability tend to correlate with high readings on our incommensurometer, synchronic cases with somewhat lower readings. Therefore, diachronic cases tend to fall into the high-degree cosmic category, synchronic cases tend to fall into the low-degree semantic cases.

These three ways of talking about incommensurability, please note, are not in competition. Their applications are different. The four categories of incommensurability concern etiology—what is behind the communicative breakdowns, what causes them? The graduated scale concerns severity—how substantial are such breakdowns? The temporal dimension concerns the populations affected—incommensurability between whom? And, as in medicine, while there is logical independence among etiology, severity, and populations, there are also significant correlations among the three—some strains of a virus trigger severe symptomologies, some trigger mild symptomologies, or none; and populations respond differently, some resisting, some succumbing utterly.

But all of these elements of the modern discursive notion, incommensurability, should however make one thing inescapably clear: we have left the no-common-divisor vehicle of our original mathematical metaphor very far behind indeed. It makes little sense to talk of the degree or the type or the temporal dimension of incommensurability with respect to a pair of numbers. Incommensurability is binary in Euclid. The antonymical structure of the word, with its negating prefix, still suggests simple opposition, even in science studies. But it has a far more multifaceted deployment in that domain, and in the other relevant discourses, than it does in its germinating field, mathematics. And we now move to examine the more prominent of these facets—brick-wall, cosmic, semantic, and pragmatic incommensurabilities—as well as the related usages, before considering why this rich and threatening configuration does not cripple science.

Brick-Wall Incommensurability

> One which my father saw in a hexagon on circuit fifteen ninety-four was made up of the letters MCV, perversely repeated from the first line to the last. Another (very much consulted in this area) is a mere labyrinth of letters, but the next-to-last page says *Oh time thy pyramids.*
>
> —Jorge Luis Borges, "The Library of Babel"

Brick-wall incommensurability is the situation where no sensible communication is possible across some imagined impassable communicative divide, where two programs are so mutually impenetrable that the participants of each see their rivals to be spouting gibberish. With respect to the straight philosophy-of-science literature, it might better be called *straw-man incommensurability,* since no one genuinely advocates it, although opponents of incommensurability often allude to such brick-wall states of affairs, as a kind of background *reductio,* and occasionally the allusion becomes explicit. Hilary Putnam (1981b), for instance, has an argument we will look at briefly later which is pitched effectively like this: if there really was such a thing as incommensurability, then we wouldn't even be able to see people on the other side of an incommensurable divide as people, just as animals making indecipherable noises—ergo, there is no incommensurability. Donald Davidson's "The very idea of a conceptual scheme" runs similarly: "Kuhn," he says, "is brilliant at saying what things were like" on the other side of a presumably incommensurable revolutionary divide, "using—what else?—our post-revolutionary idiom" (Davidson 1984, 184). Again, runs the argument, since we can figure out what is meant by/in other schemes, there is no incommensurability; indeed, for Davidson, there isn't even an "other" scheme. Feyerabend (whom Putnam has in his crosshairs with the indecipherable-noises argument) comes much closer than Kuhn to advocating total, brick-wall incommensurability, but he never forecloses either linkages or comparison—"incommensurable languages (theories, points of view) are not completely disconnected—there exists a subtle and interesting relation between their conditions of meaningfulness" (Feyerabend 1987, 272).

Jaakko Hintikka's treatment of total incommensurability probably comes closest to describing our brick wall,—a pair of theories such that "their respective consequences are somehow so different that they do not enable a scientist to compare them with each other" (Hintikka 1988, 28).[17] That definition on its own wouldn't look entirely out of place in Kuhn (even the hedging "somehow" could come straight from the pages of *Structure*), but Hintikka formalizes it to the point where it becomes obvious that such theories must be very, very different. They are not theory pairs like, say, Newtonian and Einsteinian physics, or phlogiston and oxygen chemistry. They are pairs like (to pick two that Ian Hacking suggests—1983, 345) genetics and astronomy, or (to pick two out of a hat) Pasteur's germ theory and Aristotelian mechanics, or (to pick two treated in different essays of this volume) Gilbert Ling's structured lattice theory of the cell and Murray Straus's Social Conflict theory of social groups. They are, that is, theories so different that one cannot find a single question about their domains that both of them can answer (Hintikka 1998, 29). Frankly, human ingenuity being what it is, especially among readers hunting for counter-evidence, I'm not sure even *these* theory pairs satisfy Hintikka's definition. They seem to meet his criterion from where I'm sitting, but I don't want to start an argument about it. You're welcome to find your own, and you're equally welcome to deny that in practice there ever could be brick-wall incommensurability, so long as you grant the meaning of this category—no conceivable points of comparison. As William James points out, such a state of affairs is at least conceivable, which is all this category calls for:

> We can easily conceive of things that shall have no connection whatever with each other. We may assume them to inhabit different times and spaces, as the dreams of different persons do even now. They may be so unlike and incommensurable, and so inert towards one another, as never to jostle or interfere. Even now there may actually be whole universes so disparate from ours that we who know ours have no means of perceiving that they exist. (W. James 1911, 125)

This brick-wall (total) version of incommensurability does the most violence to our base-line mathematical metaphor. The diagonal and side of a square are incommensurable precisely *because* they share something, the property of length, something rendered in numbers. In Euclid, the incommensurability of the diagonal of a square and, say, a ripe olive, does not come in for discussion.

That is why, practically speaking, brick-wall incommensurability is of little interest beyond its role as a rhetorical ploy. It just doesn't apply to things people care about in relation to each other. For theory-pairs that satisfy Hintikka's definition, whatever they might be, the chances are low in the extreme that anyone would *want* to compare them, let alone that the two sets of theorists would have any motivation to communicate. A minimal criterion for incommensurability to be an interesting concept, in science studies especially, is that the phenomena to which it applies have substantially overlapping domains.[18] If two theories are "totally incommensurable, we don't even know whether they are rivals, for rival points of view must be shown to disagree somewhere" (Laudan 1990, 123).

Brick-wall incommensurability, in short, is either trivial and banal, or it is fictitious. It refers to cases where situations are so different that one can't even imagine reasons to communicate—in which case, why call it anything, let alone "incommensurability"? Or it refers to situations where communication is completely and wholly stymied by radically different communicative modalities—in which case, there are really no such cases outside of fantasy (the novel *Native Tongue,* for instance, includes nonhumanoid languages so unutterably different that they have lethal consequences for human brains, literally—"no human mind can view the universe as it is perceived [through the language of] a non-humanoid extraterrestrial," the linguist Chornyak intones, "and not self-destruct," shortly after being informed of forty-three horrible deaths, and counting, in attempts to acquire the alien language, Beta-2—Elgin 1984, 66). Or, I suppose (but how would we know), it might be the case that humans and dolphins, or humans and ants for that matter, have an ongoing case of brick-wall incommensurability. Certainly humans try to communicate with dolphins, but it clearly hasn't gone anywhere; and I've seen my son earnestly talking to ants, with even less results. The ants may be talking too, or pheromoning, and dolphins could be clicking epic poems to us, or superstring theorems, but, if so, they aren't getting through, and it's doubtful that we're getting through either beyond (in the case of dolphins) "jump through that hoop and I'll give you a fish."

When the brick-wall species (or the total-incommensurability reading) enters into argumentation about the notion of incommensurability, it is fictive, marking either an ungenerous or—one can always find ironies in this discourse—uncomprehending version of someone else's position (usually Kuhn's or Feyerabend's).

Cosmic Incommensurability

> After discovering oxygen Lavoisier lived in a different world.
>
> —Thomas S. Kuhn, *The Structure of Scientific Revolutions*

There is no direct support for brick-wall incommensurability in the work of Kuhn and Feyerabend, despite accusations and assumptions to the contrary. Both of them articulate versions of a related notion, in which there is a severe-to-total breakdown of communicative possibilities, but where (unlike the brick-wall variant) the breakdown occurs largely within the same domain (mechanics, chemistry, astronomy). This, too, is a favorite target of critics, with somewhat more (but usually hyperbolized) justification. Stephen Toulmin, for instance, "refutes" incommensurability by defining it as "a situation in which there [is], inevitably, complete incomprehension [. . .] incomprehension [is] inescapable [. . . throwing] communication so completely out of joint that incomprehension [is] *guaranteed*" and then going on to illustrate points of contact in one of Kuhn's paradigm incommensurability cases (Toulmin 1970, 43ff).

Here is Kuhn's most famous formulation of cosmic incommensurability (my term, not his): "The proponents of [incommensurable] paradigms practice their trades in different worlds [. . .] Practicing in different worlds, the two groups of scientists see different things when they look from the same point in the same direction" (Kuhn 1996, 150). The consequences of this situation for science—for progressing, for making real knowledge, for the presumed rationality governing its very conduct—are dire: "Just because it is a transition between incommensurables, the transition between competing paradigms cannot be made a step at a time, forced by logic and neutral experience. Like the gestalt switch, it must occur all at once (though not necessarily in an instant) or not at all" (Kuhn 1996, 150). It is possible, of course, that a wholesale shift to another program, unmediated by logic or neutral experience, could result in a growth of knowledge, but how would one know? One day you're a phlogiston chemist, then wham-bam, thank-you Kuhn, the next day you're an oxygen chemist. "It is impossible to say that [the shift has] led to something *better,*" as Feyerabend annotates Kuhn's position (1981b, 136).[19]

What is most intriguing about the gestalt simile is that it appeals directly to perception.[20] There are two images latent in the prototypical gestalt picture, but an observer can only take them in one at a time, exclusively. She looks at two symmetrical lines (as in Figure 1) and sees either a goblet or a pair of faces, one percept displacing the other wholly, and the two of them alternating so completely and instantaneously as to suggest the flick of a switch. Crucially, the goblet-percept and the faces-percept can't be sustained simultaneously, and almost as crucially there is no fuzzy transition period where the goblet-percept fades and the faces-percept emerges. (Elsewhere, Kuhn gives us "scales falling from the eyes," and "lightning flash[es]," and "inundations" that completely sweep scientists into new and revealing visions—Kuhn 1996, 122.)

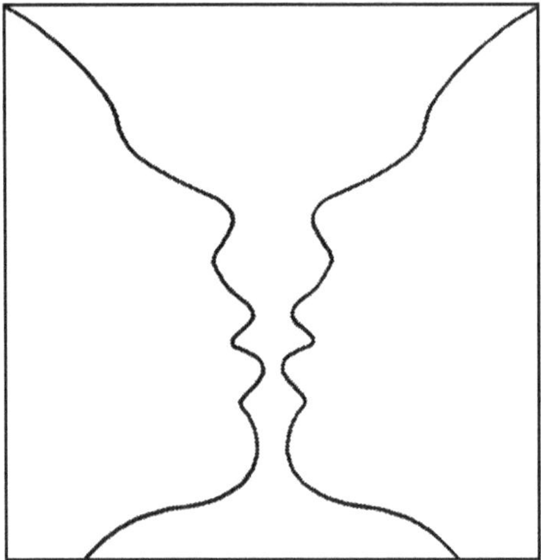

Figure 1.

It is significant, I think, that Kuhn's gestalt-shift imagery was sponsored largely by diachrony (specifically, his own Eureka moment with Aristotelian physics—1977a, xi). But he applied that imagery equally to synchrony: to revolution and theory succession. In the diachronic case, it fits rather well. The further one looks back in time—where attendant shifts in culture, language, and (to a surprisingly underestimated degree) technology effect very substantial obstacles, and where, whatever the preliminary time and effort, cognitive obstacles often fall in a rush, the position on the other side of the obstacles emerging rapidly, even instantaneously, as they drop—the more gestalt imagery seems to apply. But in the synchronic case—contemporaneous theory comparison and successive change—the gestalt imagery fits very uneasily. The diachronic realization moment, for one thing, is personal, just as a gestalt switch is. But theory succession, while it involves numerous individuals, is a distinctly social phenomenon. One individual's change of adherence is not a theory succession. Nor are stories of Eureka moments especially widespread. They exist, usually featuring the leader of a new program, but most adherents credit the power of argumentation, the beauty or simplicity or reach of a new theory, or the impact of personal research—processes that sound both more prolonged and more conventionally rational than gestalt switches. As Kuhn puts it, in one of the places where he confesses the inadequacy of this image, rather than a single sweeping change, "what occurs is an increasing shift in the distribution of professional allegiances" (Kuhn 1996, 158; see Barker 2001 for a cognitive-science account of this process in Kuhnian terms).

Unlike gestalt switches, too, putative incommensurability divides are not crossed easily. Deep commitments are forged to the truth of one of the pictures (let's say the picture of chemistry with oxygen), which requires that the other picture (chemistry with phlogiston) be discarded as false. Meanwhile, many scientists do not make the transition at all, remaining committed to the truth of the earlier (phlogiston) picture, and rejecting the proposed (oxygen) alternative as false. Kuhn offers an impressive roll call of scientists whose now-venerated work was resisted for decades by others, or conversely who resisted unto death work we now accept—Copernicus, Newton, Darwin, Priestly, Kelvin—culminating in his use of this telling quotation from Max Planck: "A new scientific truth does not triumph by convincing its opponents and making them see the light, but rather because its opponents eventually die, and a new generation grows up that is familiar with it" (Planck 1950, 33–34; Kuhn 1996,151; see R. A. Harris 1998). There is little sense in this situation of being able to flick the switch back and forth arbitrarily between two co-equal images.[21] This inability to apprehend co-equal (even if exclusive) pictures, this commitment to truth and falsity, is probably why Kuhn—moving now from simile to metaphor—draws religious conversion into his account: "[. . .] neither proof nor error is at issue. The transfer of allegiances from paradigm to paradigm is a conversion experience that can not be forced" (Kuhn 1996, 151). Conversion, too, has limitations as an analogy. It catches the evangelical character of many new scientific programs, and broaches the social dimension more than the gestalt image, but it also misses aspects of deliberation.

Feyerabend's different-world cast to incommensurability bloomed somewhat later than Kuhn's. It is implicit his early papers. "Scientific theories are ways of looking at the world," he said in 1962, "and their adoption affects our general beliefs and expectations, and thereby also our experiences and our conception of reality" (1981a, 45),[22] but this language is tentative indeed when set cheek-by-jowl with Kuhn's rhetoric of gestalts and conversions. Within a decade, however, Feyerabend was using Kuhnian perceptual terminology, arguing that incommensurable theories "deal with different worlds and that the change (from one world to another) [is] brought about by a switch from one theory to another" (1978a, 193), and bringing in perceptual diagrams—including a Necker cube, which Hanson had used, and a network of which festoon the second-edition cover of *Structure*. Feyerabend's discussion of the cube (and related images) stresses both the switch-like nature of the change in percepts, and their mutual exclusivity. In Figure 2, we can perceive two cubes (one whose front is projected to the top left, and one whose front is projected to the bottom right), which two "we may compare [. . .] in our *memory*, but *not* while attending to the *same* picture"(1978a, 226), and "there is no way of 'catching' the transition from one to the other" (1978a, 227). He goes even further, calling

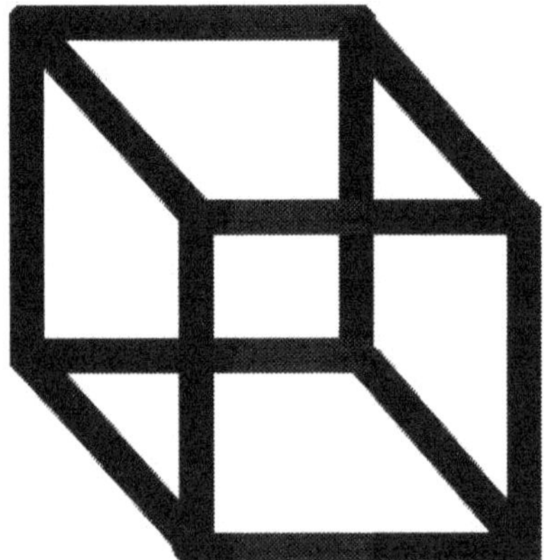

Figure 2. Necker cube.

different maturational phases of (a Piagetian theory of) perceptual development incommensurable, and noting that in passing to another stage the child's "earlier visual world [. . .] literally *disappears*" (1978a, 228).[23] Feyerabend's epistemological interests also spread considerably further than standard-issue scientific theories, taking incommensurability with them (or vice versa), and his prized example became the successive incommensurable worldviews of archaic Greece (as realized in the cosmological stories of Homeric myth) and Classical Greece (as realized in the cosmological stories of the Eleatics and their successors), charting the transitional downgrading of gods into causes.[24]

The only attempt I know of that applies the gestalt perceptual model of incommensurability seriously to a scientific episode is Marcello Pera's account of the Galvani-Volta controversy in physiological electricity. Pera characterizes the protagonists (Luigi Galvani and Alessandro Volta) as holding "rigid, exclusive 'all or nothing' position[s . . . which] clashed without any margin for compromise, exactly as the two ambiguous figures [of a gestalt image] exclude each other." He even drafts a pseudo-gestalt version of a central visual image from the dispute (Pera 1992, 168–169). On Pera's largely Feyerabendian story (the section which delivers the verdict is entitled "Frogs against method"), the controversy falls in Volta's favor from a combination of propaganda and solid argument (Pera 1992, 170ff).

Ian Hacking's exemplum of cosmic incommensurability[25] is a diachronic contrast, between (1) the thought patterns manifest in the writings of sixteenth-century German physician-alchemist, Theophrastus Bombastus von Hohenheim, known as *Paracelsus,* and (2) the thought patterns of you. "If you try to read him," Hacking says, "you will find someone utterly different from us" (1983, 70).[26] Even managing to hack your way through the mongrel Latin and Renaissance Low German to understand the terms and the propositions, you will be "left in a fog," because the concepts and purposes and registers commingle into a textual web that is markedly alien to our own: "Paracelsus's discourse is incommensurable with ours, because there is no way to match what he wanted to say against anything we want to say [. . .] We do not strain a metaphor if we say that Paracelsus lived in a different world from ours" (Hacking 1983, 71; see also Hacking 1993, 296f).

For example, an internal administration of mercury should be used to treat syphilis, runs a Paracelsian line of reasoning, because the metal is the sign of the planet and the planet is the sign of the market place, and, as we all know, you contract syphilis in the market place. The links of cause and effect and association are so exotic that we have no purchase on the style of reasoning. We don't get it. (This situation is not, however, brick-wall incommensurability. We *can* get it with lots of hard work and charity, and it *does* answer questions in the domain of medicine. See Hacking 1983, 71.)

Hacking's linkage between different worlds and different discourses is not accidental. Both Kuhn and Feyerabend invoke Benjamin Lee Whorf's research in connection with their different-worlds talk, insinuating linguistic relativity deeply into cosmic incommensurability. A brief excursus is called for.

Linguistic Relativity

> Proponents of different theories [. . .] are [. . .] like native speakers of different languages.
>
> —Thomas S. Kuhn, *The Essential Tension*

Whorf got this idea from anthropological linguistics (via that legendary triple-play, von-Humboldt-to-Boas-to-Sapir); or, perhaps he had the idea already—it is not uncommon—and that is what attracted him to study under Sapir, who had written in his very popular *Language* that "language and our thought-grooves are inextricably interrelated, are, in a sense, one and the same" (Sapir 1921, 217–218). In either case, he fell into orbit around the position and championed it in several popular forums. The idea is that languages can embody entirely unique concepts and dispositions (or, if not entirely unique, then at least strongly alien to other classes of languages); that these concepts and dispositions cannot be adequately inter-translated between at least certain languages; that there is an inescapable cultural relativity deeply engrained in the primary tool of culture, language.[27]

The contrast between Amerindian languages and European languages especially seemed to support this notion-cum-hypothesis; Whorf's favorite example was Hopi (a language Boas had investigated). Before he died (in his, and the century's, early forties), Whorf had published several articles in general, high brow periodicals, most notably *Main Currents in Modern Thought,* which had put this notion into general circulation, but it was a collection of his essays published fifteen years after his death with the evocative title, *Language, Thought, and Reality,* that most directly influenced the milieu of Kuhn and Feyerabend (and, for that matter, Quine). Among other maneuvers, Whorf gave the hypothesis its name (actually, both of its chief names, since one he sponsored eponymously, the Whorfian hypothesis), and the allusion to Einstein is not an accident: "We are thus introduced to a *new principle of relativity,* which holds that all observers are not led by the same physical evidence to the same picture of the universe, unless their linguistic backgrounds are similar, or can in some way be calibrated" (Whorf 1956, 214).

His principal examples came from Amerindian/European contrasts. For instance, one theme that Whorf mined in his relativity data is the preponderance of verbs and verbal morphology in Hopi, in contrast to English. Hopi has a prototypical eventness to its assertions in contrast to the prototypical thingness of English (Whorf 1956, 57–64).[28] Words corresponding to our nouns *lightning, wave, flame, meteor,* and *pulsation,* and to phrases like *puff of smoke,* are verbs in Hopi (Whorf 1956, 51–56), so that they convey something on the order of 'lightning occurs' or 'it lightnings,' though these glosses include substantial and inevitable distortions because English can't do without tense in our verbs; Hopi, on the other hand, deploys not temporal morphemes but evidential morphemes (what Whorf called *validity markers*). A closer approximation to the functional meaning might be I-report-as-fact-that-lightning-occurs(-though-I-make-no-temporal-commitment-about-said-occurrence), but, to understate the case massively, some nuance is lost. As Native American linguist Dan Moonhawk Alford (2000) has articulated the difficulties of incorporating Amerindian perspectives into English, "the bitch of it is, even *verb* is a noun!"

Linguistic relativity is a romantic notion, linked both to an egalitarian cultural pluralism and to the myth of the profound savage, as well as a general affection people apparently have for superficial philosophy of language and mind. What else accounts for the incredible cultural currency of the "Eskimos have N words for snow"? (Where *N* is some number the speaker assumes to be significantly large without being outlandish, or, in a few rare cases a number the speaker has actually read somewhere or heard someone else use.) The statement is apparently only true for an N of 2 (Martin 1986). Or, take this meme: "If we spoke a different language, we would perceive a somewhat different world." It is very widespread on the Web, in collections of quotations, always attributed to Wittgenstein, but never linked to a specific text. As it turns out, Wittgenstein never wrote it.[29] But people like what it expresses, and, what the hell, Wittgenstein is famous for being really smart and deep, so it sounds a lot better coming from him. Whorf was a subtle thinker, but certainly not immune to romanticism, and celebrates the Hopi for having a worldview remarkably consonant with the grand and (to Western minds) counter-intuitive claims of Einstein ("The Hopi actually have a language better equipped to deal with such vibratile phenomena than is our latest scientific terminology." [Whorf 1956, 55]).

It is difficult to overestimate the ease with which this thesis could be assimilated to late nineteen-fifties philosophy of science, at the tail-end of logical positivism, a philosophy full of technical expressions like *thing-language,* and *system-language,* and *observation-sentences,* and *protocol-sentences.* Books with titles like *The Language of Physics* (Hutten 1956) and *The Grammar of Science* (Pearson 1957) were commonplace in philosophy. There was certainly a sense in which the languages of science were "much more tightly knit," as Feyerabend described them in 1962 (1981a, 66) than daily-affairs languages, that their designations were more rigid, their syntax more constrained, but these were differences in quantity, not quality. Science was overwhelmingly seen as a matter of linking linguistic descriptions with linguistic explanations, aided and abetted by the language of mathematics. Somewhat ironically, in retrospect, one of the strongest motivations for this linguistic turn, was that it "carries the discussion into a domain where both parties are better agreed on the objects [. . .] Words, or their inscriptions, unlike points, miles or classes and the rest, are tangible objects of the size so popular in the marketplace, where men of unlike conceptual schemes communicate at their best" (Quine 1960, 272). But add to this background conception of the centrality of the theory-as-language metaphor, an increasing conviction that the meaning of theoretical terms came not by their anchorage in observation but by their place in the semiotic web of the theory, and you come very naturally to the sense that different theories are different languages, that "competing theories within the same domain 'speak different languages'" (Hacking 1983, 68).

As Hacking's not-quite-literal, quotation-mark distancers suggest, the use of the word *language* in these contexts carried some metonymy or metaphoricity—standing either for some constrained, structured subset of natural language, or for some symbol-system that was partially analogous to natural language. But the figurative nature of *language* frequently blurred in the literature. By the time we get to Kuhn's *Structure* and Feyerabend's "Explanation," the theories-are-languages claim is so routine that it looks literal, and in *Against Method,* Feyerabend explicitly erases all figurative aspects of the theory-as-language commonplace: "Scientific theories, such as Aristotle's theory of motion, the theory of relativity, the quantum theory, classical and modern cosmology are sufficiently general, sufficiently 'deep,' and have developed in sufficiently complex ways to be considered along the same lines as natural languages" (Feyerabend 1978a, 224–225).

The different-world talk of cosmic incommensurability that attends the Whorfian (and gestalt) analogies is one of the most prominent targets for critics of Kuhn and Feyerabend, but perceptual issues are not a necessary component of incommensurability. Kuhn includes a lot of that talk in his introduction of incommensurability, but Feyerabend does not. On the other hand, though Feyerabend later pursues them, Kuhn leaves the perceptual issues behind in his later advocacy of incommensurability. Moreover, even in *Structure*—treated routinely in the literature as profligate in this regard—the perceptual issues are brought in more reluctantly than many people seem to be aware, certainly than many of Kuhn's critics care to notice.[30] "According to Kuhn," Donald Davidson says, "scientists operating in different scientific traditions (within different 'paradigms') 'work in different worlds'" (Davidson 1984, 186–187, citing Kuhn 1996, 135). Well, no, that sort of flat assertion about working in different worlds is *not* "according to Kuhn." He was certainly wrangling with the perceptual implications of incommensurability, and circles around such locutions repeatedly in *Structure,* but he never once makes the unequivocal claim that scientists operating out of different paradigms work in different worlds. What actually *is* according to Kuhn, in precisely the location of *Structure* that Davidson cites, is that "we may want to say that after a revolution scientists work in a different world" (Kuhn 1996, 135). *May want to say* is much more heavily modalized than Davidson's representation of Kuhn's claim. Earlier in the book, Kuhn had said "living in a different world" was "a strange locution" which he was trying to avoid (1996, 118). Other phrases in which Kuhn frames his cosmic incommensurability speculations include:

> We may want to say that after a revolution scientists are responding to a different world. (111) [. . .] The very ease with which astronomers saw new things when looking at old objects with old instruments may make us wish to say that, after Copernicus, astronomers lived in a different world (117) [. . .] Schools guided by different paradigms are always slightly at cross purposes. (112) [. . .] The proponents of competing paradigms fail to make complete contact with

> each other's viewpoints (147) [. . .] The proponents of competing paradigms are always at least slightly at cross-purposes. (148) [. . .] The inevitable result is what we must call, though the term is not quite right, a misunderstanding between the two competing schools. (148) [. . .] I am, for instance, acutely aware of the difficulties created by saying that when Aristotle and Galileo looked at swinging stones, the first saw constrained fall, the second a pendulum [. . .] Nevertheless, I am convinced that we must learn to make sense of statements that at least resembles these. (121)

Or, in perhaps his most tortuous attempt to wrestle with these conflicting positions:

> In a sense that I am unable to explicate further, the proponents of competing paradigms practice their trades in different worlds. One contains constrained bodies that fall slowly, the other pendulums that repeat their motions again and again. In one, solutions are compounds, in the other mixtures. One is embedded in a flat, the other in a curved, matrix of space. Practicing in two different worlds, the two groups of scientists see different things when they look from the same point in the same direction. Again, that is not to say that they can see anything they please. Both are looking at the world, and what they look at has not changed. But in some areas they see different things and they see them in different relations one to the other. That is why a law that cannot even be demonstrated to one group of scientists may occasionally seem intuitively obvious to another. (150)

These are not the modalities of a different-world zealot—"We may want to say," "may make us wish to say," "acutely aware of the difficulties," "unable to explicate further."—especially in combination with systemic realist assurances, like "outside the laboratory everyday affairs usually continue as before" (Kuhn 1996, 111), and "Whatever he may then see [after a revolution], the scientist is still looking at the same world" (Kuhn 1996, 129). Wes Sharrock and Rupert Read see

Kuhn as a bit woolly-minded here, as "using the expression 'different worlds' in a routinely idiomatic fashion" that confused even himself (Sharrock and Read 2002, 172). Kuhn's usage *does* draw more from ordinary language than from philosophic usage (not that they are always easy to tease apart, even in critically influential works). But there is nothing routine about his deployment of the phrase. He struggles with it, knows it doesn't quite fit, but can't quite find anything better for his sense that perceptions shift as a function of linguistic apparatus. And he is honest enough to expose the struggles. Even the gestalt simile, for which he was much abused, is introduced with its inadequacies on display ("[paradigm transformations are like gestalt switches, ...] though usually more gradual and almost always irreversible"—Kuhn 1996, 111). Herbert Simons, in this volume, characterizes the different-world imagery that Kuhn deployed as part of his tendency "toward the melodramatic" (245). Kuhn was highly aware of this tendency, recurrently saying in effect, "look, I know this talk is melodramatic, but it seems like the only way to broach these issues."[31] He acknowledged the aptness of John Earman's label for stretches of such talk in *Structure*—the "purple passages" on incommensurability—saying that even before the book came out he realized his "attempts to describe [incommensurability in it . . .] were extremely crude" (Kuhn 2000, 228; see Earman 1993).

Yet even if we ignore the overtly metaphorical character of the discussion in *Structure,* and strip away the may-be-required-to-wish-to-say sort of hedges—if we fall, like Davidson, for the melodrama—and take Kuhn to be making bald X-and-Y-live-in-different-worlds claims, the degree of difference is still not clangingly large: "Slightly at cross-purposes," he says, and "fail to make complete contact" and "a misunderstanding." These are not descriptions of total, cosmic incommensurability, nor even of the binary notion that many critics take incommensurability to be; these are descriptions of partial communicative slippages, not of irremediable communicative failure.

In short, cosmic incommensurability is not the straw position that brick-wall incommensurability is. There are very clear claims, especially in the most influential document in the field, Kuhn's *Structure,* that incommensurability involves significant perceptual mismatches on either side of paradigm fissures. But neither is it necessarily a debilitating condition that inevitably leads to mutual incomprehension. In terms of our graded approach, it may be at the more severe end of the scale, but it is never total, and therefore leaves open possibilities of resolution and understanding, of reconfiguring the mismatches in pragmatic and semantic terms, where the ground is comparatively firmer.

Semantic Incommensurability

> The men who called Copernicus mad because he proclaimed that the earth moved [. . .] were not either just wrong or quite wrong. Part of what they meant by 'earth' was fixed position. Their earth, at least, could not be moved. Correspondingly, Copernicus's innovation was not simply to move the earth [. . . He] necessarily changed the meaning of both 'earth' and 'motion.'
>
> —Thomas S. Kuhn, *The Structure of Scientific Revolutions*

For cases where incommensurability is diagnosed, there are certainly linguistic complications beyond theoretical terminology. With Paracelsus and us, for instance—presuming we get past the referential obstacles of an archaic, commingled-Latin-German text, as Hacking assures us we can (Hacking 1983, 70)—there are overwhelming differences in register and purpose and theme. But many cases can be localized to technical terminology, which sponsors almost all of the concrete examples in Feyerabend's and Kuhn's writings. Misalignments of the meanings of theoretically pivotal terms, in fact, is the area of incommensurability that has most exercised philosophers of science; Howard Sankey's (1994) philosophically definitive *The Incommensurability Thesis,* for instance, is almost entirely semantic.

The textbook case, one that Kuhn was especially fond of (his first book circled around it), is the word *planet* before and after Copernicus.[32] From the Greek verb, *planasthai,* 'to wander,' the term picked out the *asteres planetai,* the wandering stars, the ones that moved across the heavens in ways that cut across the more uniform motion of the rest of stars, the non-wandering ones. In the heliocentric model of Claudius Ptolemy, those objects included points of light named *Hermes*, *Aphrodite*, *Ares*, *Zeus*, and *Kronos* (in the Latin, *Mercurii*, *Veneris*, *Martis*, *Jovis*, and *Saturni*). But the wandering stars did not include Earth. How could they? Earth was not in the heavens at all; it was the stable site of the observer about which the celestial bodies moved. (Nor did they include Neptune, Uranus, or Pluto, of course, none of which can be seen unaided.) There were two other planets for Ptolemy as well, Selene and Helios, the bodies we now designate not with proper nouns but with names built out of common nouns and definite determiners: *the moon* and *the sun.*

So, a Ptolemaist and a Copernican might look up, indicate the same point of light in the night sky (say, the one we call *Venus*), utter an observation sentence, "lo, planet," and be making a true proposition within their different cosmologies. But the Ptolemaist could also indicate a different point (the moon) and, with precisely the same utterance, say something true in his system, false in the Copernican; he could also utter it at midday (pointing at the sun), with the same results. Meanwhile, the Copernican could utter the phrase while looking down at the ground and hopping in the dirt (of Earth), with the opposite consequences—true for him, false for the Ptolemaist. Lo, incommensurability.

All of this comes down, in Feyerabend and in Kuhn, to meaning variance. Sometimes called a "thesis," meaning variance is better recognized as a "phenomenon"—ubiquitous and endemic. Languages change, generation to generation, region to region, interest cluster to interest cluster. In philosophy of science, this fact of human symbol-use runs thusly, "The definition of meaningfulness must be relative to a theory T, because the same term may be meaningful with respect to one theory but meaningless with respect to another" (Carnap 1956, 4833), but exactly the same is true of ordinary language—the meaningfulness of *dude* or *gay* or *babe,* for instance, is relative to age and region and interest (and context of utterance).

Linguistic variation is, in fact, one of the most startling differences between animal communication systems, which are highly stable, and human symbol-use (prototypically, natural languages). Language sounds change, syntactic patterns change, lexicalizations change, meanings change. The reference of words expands, shrinks, reverses, or just hops over to some other category. Once upon a time in English, *meat* (*mete*) was virtually any solid food, now it is only comestible animal flesh; *dog* (*dogge*) was a specific breed, now it includes all canis domesticus; *hound* (*hund*) was all *canis domesticus,* now it is a small set of breeds; *bead* (*gebed*) was a prayer, now it is a little piece of glass with a hole through it that you can string together to make decorative artifacts. Once upon a time in chemistry, *element* was earth, air, fire, and water, now there are 109 of them, none of which is earth, air, fire, or water. Once upon a time in astronomy, *planet* (*planete*) was a collection of seven celestial objects, now it's a larger, partially overlapping set of objects, some gone, some added.

Meaning variance is not only a diachronic issue. Two languages at the same moment in time can synchronically exhibit meaning variance. English and French have these cognates, for instance: *veal* and *veau, beef* and *boeuf, pork* and *porc.* The English terms have smaller ranges of extension, referring only to comestible animal flesh; the French terms also refer to the flesh donors themselves, alive or dead, in the field, barn, or freezer. Indeed, two dialects, or two registers, or two jargons can exhibit meaning variance.

Meaning variance, diachronic or synchronic, only becomes an issue when communication across the divide is hindered, when some untutored modern encounters *mete* in Chaucer and the referent is bread, or some English parent encounters *porc* in a French picture book he's reading with his daughter and has to explain why there is no plate; or, like Feyerabend and Kuhn in 1962, when someone investigates theory comparison, or tries to plumb a superseded theoretical picture.

What Carnap has in mind when he says "meaningful with respect to one theory" is 'meaningful with respect to all the other terms in that theory.' The terminology of a theory is a network of interrelated, mutually delineating words, a semiotic web. Again, this situation is standard linguistic fare. *River* gets its meaning in English by way of its place in a semiotic web of water-body terms (*lake, puddle, stream, rivulet, canal, inlet,* and the like), and more generally, by way of its place in the overall semiotic web of English (*water, ice,* H^2O, . . . but also, *fountain, rink, hose,* . . . and even *green, Hopi, incommensurability,* . . .). To see the way meanings can traverse semiotic webs, think of Paracelsus and the words *metal, planet, marketplace,* strands from which run to *mercury* and *Mercury* in the web of meanings implicated by his medical program. It is diachronic alterations to these webs—deletions, insertions, amalgamations, and transpositions of its nodes—and synchronic discrepancies in parallel webs, that give us meaning variance. In science studies, it was recognition of these webs that led philosophers to the significant web-altering term, *theory-laden.* As Feyerabend puts it, "Words do not 'mean' something in isolation; they obtain their meanings by being part of a theoretical system" (1965b, 180). And here is Kuhn, putting terminological interrelation in learning-theory terms, by way of a very tidy two-term case:

> In learning Newtonian mechanics, the terms 'mass' and 'force' must be learned together, and Newton's second law must play a role in their acquisition. One cannot, that is, learn 'mass' and 'force' independently and then empirically discover that force equals mass times acceleration. Nor can one first learn 'mass' (or 'force') and then use it to define 'force' (or 'mass') with the aid of the second law [Force = mass × acceleration; F=ma]. Instead, all three must be learned together, parts of a new whole (but not wholly new) way of doing mechanics. (2000, 44)

Newtonian *force,* by this story, is psychologically unattainable without Newtonian *mass,* and vice versa; neither is attainable without the second law, and vice versa. All are dependent on each other, and together they constitute a cornerstone of the overall Newtonian framework. *Force* is a theory-laden term, ditto *mass,* both getting their positions in the local semiotic web through their relations to other terms implicated by the second law, specifically *acceleration,* but also *inertia* and *gravity;* but, of course, Kuhn understates the case dramatically, since a good many other terms relevant to the problem space, such as *body, rest, motion, resistance,* and so on, are also necessary for getting a fix on Newtonian *force* (or *mass*). *Force,* like *river* and *incommensurability,* is a node in a network, a cell in a matrix, whose meaning derives from its position and its neighbors.

Newtonian semantics, that is, function like natural-language semantics, more or less—as do Aristotelian semantics, Einsteinian semantics, and so on—a point that is axiomatic in the literature on incommensurability. If it is not immediately apparent what this has to do with the governing metaphor, a lack of common measure, here's the point: when each program has its own semiotic web there is no neutral place for an adjudicator to stand in order to evaluate those programs. The Newtonian is caught in his web, the Aristotelian in his, the Einsteinian in his or hers. Moreover, any other system that might be brought in or built from scratch, will function semantically as a system, not by reference outside itself. There is no still Archimedian web. For any two semantic systems, there is no third (fourth, fifth, . . .) system that has some set of privileged terms for adjudicating between them. It's semiotic webs and meaning variance all the way down.

All incommensurability, in short, has a semantic component. Brick-wall incommensurability, for instance, is the case where the relevant semantic networks are so far out of alignment that communication is impossible; there are no shared referents. Cosmic incommensurability is when the networks are sufficiently out of phase that communication is seriously confused; there are shared referents—a common domain—but the fundamental conceptualizations of that domain are discrepant. Compare Paracelsus's mercury/Mercury story with more recent theory-laden counsel: take penicillin because the pathogen, *Treponema pallidum* causes syphilis, and penicillin destroys *Treponema pallidum.* The domain is a specific symptomology, but the way it is linguistically (and materially) conceived offers few if any points of contact. All incommensurability is semantic, to a degree.

The type I am specifically calling *semantic incommensurability* here (adopting Howard Sankey's term)[34] is one of the crucial focal points of the general term, *incommensurability,* the type that most closely satisfies the no-common-measure, mathematical metaphor. Words are bigger, lumpier, more permeable concepts than numbers, but they lend themselves more readily to talk of measurement than do other components associated with incommensurability—percepts, methodologies, or values (all of which, of course, are themselves lexically enwrapped).

There are two major interrelated themes in the philosophical debates over semantic incommensurability—reference and translation. The extensive literature on the relevance of reference theories for incommensurability boils down to "Yes, there is incommensurability, because the X-version of the Y-theory of reference is the best theory for describing the relevant aspects of scientific reference, and it mandates incommensurability" vs. "No, there is no incommensurability, because the Y-version of X-theory of reference is the best theory for describing the relevant aspects of scientific reference, and it dissolves incommensurability quite nicely, thank you." A dominant leitmotif in these arguments is the notion of comparability. So, for instance, arguments regularly sum up with "Yes, there is (semantic) incommensurability, but because of the X-Y approach to reference it doesn't get in the way of comparison" (or, in Alexander Bird's formulation, "incommensurability [. . .] exists, but it is no big deal"—2000, 188).

This caricature of an elaborate, even baroque, literature is sure to alarm some readers, who might wish for a little more responsibility here about the lines of the debate—particularly by way of detaching causal theories of reference from the more traditional variants, by mapping out ascribed differences between reference and meaning, perhaps by attending to the ramifications of the extension/intension distinctions, maybe even evoking a Ramsey sentence or two. But the theory-of-reference debates are mostly in-principle exercises about whether there is such a phenomenon as incommensurability, or not, and the implications of that putative phenomenon for the possibility of theory comparison. I think we can bite the bullet and say there are some phenomena that correlate with Feyerabend and Kuhn's use of the word *incommensurability,* in various ways, and get on with addressing the phenomena more directly, which is precisely what the second main theme in the investigation of semantic incommensurability does, translation.

Translation issues have attended incommensurability from the beginning. There's no other way to make sense of talk like "Is it then possible to define a concept such as impetus in terms of the theoretical primitives of Newton's theory?" (Feyerabend 1981a, 66—the question is of course rhetorical, and receives a negative answer). If you cannot define, say, *derrière,* in English, then you cannot translate it into English. (That's not to say English can't accommodate it—by, say, swallowing the signifier and signified together—but that you can't render it into the existing lexical structure of English.) Kuhn was more explicit. In his second-edition postscript, marking his first withdrawal from the strong cosmic implications of his 1962 formulations, he prescribes translation for scientists in different paradigms who still want to talk to each other, taking a lesson directly from history of science:

> What the participants in a communication breakdown can do is recognize each other as members of different language communities and then become translators. Taking the differences between their own intra and inter-group discourse as itself a subject for study, they can first attempt to discover the terms and locutions that, used unproblematically within each community, are nevertheless foci of trouble for inter-group discussions. [. . .] Each will have learned to translate the other's theory and its consequences into his own language and simultaneously to describe in his language the world to which that theory applies. That is what the historian of science regularly

> does (or should [do]) when dealing with out-of-date theories. (Kuhn 1996, 202; see also 1970b, 277)

Kuhn continued his taming project throughout his career, leading in the 1980s to the re-configuration he termed "local incommensurability," to which we will turn shortly, but he found himself in one of those tortuous messes scholars get themselves into when they want to be consistent and innovative at the same time, and ended up (1) endorsing translation success, but (2) problematizing translation as well, in a defense of incommensurability.

Kuhn's problematizing of translation proceeds by way of Willard van Orman Quine's indeterminacy of (radical) translation arguments (Kuhn 1970b, 268–9). Quine's case hangs on ambiguity of a radical sort. Take this sentence, uttered by Ford Prefect in *Hitchiker's Guide to the Galaxy* after he has warned Arthur Dent to prepare for a jump into hyperspace:

> Ford: It's unpleasantly like being drunk.

The follow-up reveals an ambiguity:

> Arthur: What's so unpleasant about being drunk?
> Ford: You ask a glass of water. (Adams 1979, 59)

The example is a simple one, the complication resting (pretty much) on the sequence "being drunk." Ford intends this sequence to be a passive verb phrase, but Arthur receives it as a copular verb phrase—*being* is an auxiliary verb for Ford, *drunk* a main verb; *being* is a main verb for Arthur, *drunk* an adjective. The ambiguity, however, dissolves, as it almost always does when two people are trying to understand one another, on the next turn, Ford's "ask a glass of water." (Notice while we're here, that ambiguity is another face of meaning variance—*drunk* is "theory-laden" differently here by the respective contexts of inebriation and liquid ingestion.)

Quine (1960), positing an imaginary language, and an imaginary linguist, trying to translate an imaginary utterance of that imaginary language, *gavagai,* exploits ambiguity extravagantly to set up his indeterminacy argument: in principle, ambiguity presents an unbridgeable gap between minds. Let's say Arthur, more concerned about the impending hyperspace jump than about Ford's conception of "being drunk," doesn't ask his question, or let's say Ford doesn't respond, or that the jump happens before they get to converse any longer, and they come out the other end too discombobulated to pursue the topic further—whatever—but that they retain their own encapsulated meanings for "being drunk" in the context of Ford's utterance. Ford meant one thing, but communicated something else to Arthur, yet (in this scenario) they both assume communication was flawless—Ford, that Arthur got what he intended; Arthur, that Ford intended what he got.

Now multiply this situation by some Whorfian coefficient—different languages, different cultures, different perceptions—and you come out of hyperspace with Quinean indeterminacy of (radical) translation.[35] Presuming no follow-up exchange in our *being-drunk* example above, Ford and Arthur wouldn't know that anything went wrong. The actual utterance-chunk "being drunk" is indeterminate (and its reference is fixed by the interlocutors in different ways). Well, we communicate with each other all the time, never *really* knowing how reference is being fixed by our interlocutors, only knowing that external behaviors suggest intended reference has matched construed reference; meaning is indeterminate. In translating from an alien language (especially an imaginary one, where its inventor is maliciously trying to thwart full translation), the referential indeterminacy can only be greater.[36] Quine elaborates his example into multiple linguists working on parallel translations (translation manuals), one on the assumption that *gavagai* can be fixed at "undetached rabbit part," another that it can be fixed at "rabbit occurrence," and so on. Each can construct translations that achieve full predictive accuracy (*gavagai* is uttered in all and only those circumstances in which both "undetached rabbit part" and "rabbit occurrence" and so on are potential referents). Indeterminacy of translation, as both the being-drunk and the *gavagai* examples show clearly, is indeterminacy of reference.

The point of Kuhn's deployment of the indeterminacy of translation thesis (like meaning variance, it's also more of a phenomenon than a thesis: language is always indeterminate, at least when it is used for communication, because there are always at least two distinct minds at work in the use of any given utterance), the point for Kuhn is to assimilate incommensurability partially to Quine's arguments (Sankey 1994, 17), and thereby to begin localizing the notion to semantic concerns. Kuhn does not go all the way with Quine, who wants to make the notion of reference "inscrutable." (Kuhn 1976, 191). Kuhn just wants to demonstrate the profound difficulties of translation, something Quine's arguments do. In part, this assimilation is probably just a gesture for philosophical legitimacy, always a concern for Kuhn, but in part it was a serious response to the challenge of Popper and others to the Whorfian analogy that helped to sponsor incommensurability. "It's just a dogma—a dangerous dogma—that the different frameworks are like mutually untranslatable languages," Popper had said. "The fact is that even totally different languages (like English and Hopi, or Chinese) are not untranslatable" (Popper 1970, 56); at the least, Quine's argument compromises Popper's fact.

Hilary Putnam prosecuted Popper's case more thoroughly. If the incommensurability thesis "were really true," Putnam argued, in a move meant to dissolve the thesis once and forever,

> then we could not translate other languages—or even past stages of our own language—at all. And if we cannot interpret organisms' noises at all, then we have no grounds for regarding them as thinkers, speakers, or even persons. In short, if Feyerabend (and Kuhn at his most incommensurable) were right, then members of other cultures, including seventeenth-century scientists, would be conceptualizable by us only as animals producing responses to stimuli (including noises that curiously resemble English or Italian). To tell us that Galileo had 'incommensurable' notions and then to go on to describe them at length is totally incoherent. (Putnam 1981b, 115)

Putnam is surely right here, if "incommensurability thesis" means brick-wall or perhaps even cosmic incommensurability.[37] The history of mankind has no shortage of examples of one group dehumanizing some poorly understood others, seeing them as animals; the word *barbarian* derives onomatopoeically from the Greek view that non-Greeks spoke crude gibberish, that their language was just the sheep-like noises "bar-bar." But for readings a little lower on our incommensurometer, where we find semantic incommensurability, things are not so dire. We can fail utterly to get Newtonian force and mass and not be compelled to regard Newton as an animal producing responses to stimuli (including marks on the page that correspond to early-modern English or Latin). More to the point, we needn't fail. Kuhn sees inter-paradigmatic translation to be difficult, but not impossible.[38] Neither he nor Feyerabend had proposed anything like complete translation failure of the sort Putnam says that incommensurability entails. Or, more strictly, Kuhn continues to abjure the word *translation* and siphons off the interpretive component of all translation, as well as the pedagogical component of technical translation, to frame the process this way:[39] "Interpretation is the process by which the use of those terms is discovered [. . .] Once it has been completed and the words acquired, the historian uses them in his own work and teaches them to others. The question of translation does not arise" (Kuhn 2000, 45). This move allows Kuhn to maintain translation failure but allow communicative success, though any robust notion of translation cannot get along without interpretation. Kuhn's move is a shell game, but the emphasis he places on interpretation as a remedy for incommensurability is important.

For his part, Feyerabend has two primary responses to Putnam. The first is a whopper. "We can learn a language or a culture from scratch," Feyerabend claims, "as a child learns them, without detour through our native tongue" (1987, 76). The claim is remarkably inconsistent for someone who endorses Whorfian perception-through-language notions, though we should probably not let that worry us; consistency was not a prime concern for Feyerabend. But to suppose that one's native language somehow can shut off when learning another language, when it doesn't even shut off to learn a new dance step, is gratuitous at best (though Feyerabend apparently believed that "linguists, historians, and anthropologists" do this regularly). In principle, one guesses, such learning is possible, but the circumstances would have to be extreme, like someone learning to walk again, from scratch, as a child would, if the ability had somehow been erased while leaving the nerves and muscles intact. It might be possible to learn new languages without any mediation from a native language, but no one does, not even linguists or anthropologists, let alone historians. Indeed, all three scholarly communities have richly augmented their native languages for just the purpose of acquiring new languages and new understandings.

Feyerabend's second response, however, is devastating, cutting right to the heart of the translation analogy. Putnam's argument rests on the assumption that the translation is precluded from affecting the target language, which is not the way translations work at all, especially when novel phenomena are involved. When Anglo-Americans, for some reason, wanted to start consuming a spongy pap of coagulated bean juice, they had a number of lexical options for the signs in their supermarkets—assimilating it to an approximating substance ("soy cheese"), for instance, or using a metaphor from a more distant semantic field ("bean foam"), or using a calque ("rotten bean"), or, what we in the event actually did, importing the source-language signifier as best we could, *tofu,* along with the signified substance. In short, "we can change our native tongue," Feyerabend points out, "so that it becomes capable of expressing alien notions" (Feyerabend 1987, 266); Kuhn, too, points to target-language enrichment (Kuhn 2000, 45).

One of Whorf's two caveats on his hypothesis, remember, is: "unless their linguistic backgrounds [. . .] can in some way be calibrated" (Whorf 1956, 214). Calibration, both Kuhn and Feyerabend argue, is always possible. Kuhn's specific case for why calibration is always possible is linked to his move toward local incommensurability.

Local Incommensurability

> Only for a small subgroup of (usually interdefined) terms and for sentences containing them do problems of translatability arise. The claim that two theories are incommensurable is more modest than many of its critics have supposed.
>
> I shall call this modest version of incommensurability, 'local incommensurability.'
>
> —Thomas S. Kuhn, "Commensurability, Comparability, Communicability"

Kuhn's changing claims about incommensurability illustrate clearly the shaping power of dialogic rhetoric. "How can a conceptual scheme," he puzzled of the Ptolemaic cosmos in *Copernican Revolution,* "that one generation admiringly describes as subtle, flexible, and complex become for a later generation merely obscure, ambiguous, and cumbersome?" (Kuhn 1957, 76). His answer five years later, of course, was "incommensurability," but that answer came in a lumpy stew of assertions, concerns, and hunches—about communicative breakdowns between proponents of successive programs, and about the reduced accessibility of long-superceded programs. *Incommensurability* in *Structure* is at least as difficult to pin down as *paradigm.* But three clusters of definitional features stand out in the book: different, possibly contrary perceptions of scientists operating out of different programs (e.g., 1996, 150); conflicting themes and practices (e.g., 1996, 148); and semantic noise interfering in the arguments between programs, or interfering between past programs and subsequent historians (e.g., 1996, 149).

But this undifferentiated package—representing cosmic, pragmatic, and semantic incommensurability in my scheme, and usually called *paradigm* or *framework incommensurability* in the associated literature, when it is not just called *incommensurability*—was not very well received. In particular the cosmic, different-world perceptual talk was greeted with fear and loathing, along with its attendant gestalt-switch and conversion analogies. The themes-and-practices issues were treated as obvious, even tired, but certainly not supporting the incomparability claims Kuhn was seen to be advancing. The semantic-interference issues, too, were either rejected or downgraded by the community: but they were also seen to present the most immediately tractable ground of the problem field, the firmest place to stand on which to battle incommensurability; the ground was seized, the flag was planted. It helped, too, that this was the species of incommensurability that Feyerabend was advancing, someone with better philosophical credentials in 1962, and with considerably more rigorous, more audience-aware argumentation.

As interlocutors, philosophers made it pretty clear what they wanted to talk about; as audience, they made it equally clear what they wanted to hear. In both cases, as much was excluded as was included. A good rhetor (one who wants not just to communicate, nor simply to persuade, but to reach understandings and agreements), observes the affordances an audience opens up, and the rhetorical constraints it exercises. Kuhn was a good rhetor, constantly working with others' arguments and constantly refining his own positions (most famously, reanalyzing *paradigm* into *disciplinary matrix* plus *exemplar*).[40] The debates that followed *Structure* led Kuhn to narrow his incommensurability thesis exclusively to matters of semantic interference, and specifically to local clusters of words that encode the taxonomy of the theory (the categorization of its defining entities, into terms such as *force* and *mass*); that is, to what he called local incommensurability.[41]

What's local about it is that the source of incommensurability is confined to a small district of the semiotic web that structures the claims of the theory. Kuhn's *force*/*mass*/second-law example occupies such a district in Newtonian theory. Here is another district, with Kuhn explaining the implications of its lexical occupants for local incommensurability:

> Among the phrases which describe how the referents of the term 'phlogiston' are picked out are a number that include other untranslatable terms like 'principle' and 'element.' Together with 'phlogiston,' they constitute an interrelated or interdefined set that must be acquired together, as a whole, before any of them can be used, applied to natural phenomena. Only after they have been thus acquired can one recognize eighteenth-century chemistry for what it was, a discipline that differed from its twentieth-century successor not simply in what it had to say about individual substances and processes but in the way it structured and parceled out a large part of the chemical world. (Kuhn 2000, 43-44)

The lexical keystone in this passage for the notion of incommensurability, of course, is *untranslatable.* The members of this interdefined set have left only faint and misleading impressions on the current vocabulary of chemistry. *Phlogiston* has disappeared altogether, and it disappeared because a partial antonym that offered a more compelling account of the data, *oxygen,* was introduced to chemistry. But there is no point-to-point mapping between *phlogiston* and *oxygen—phlogiston* as *negative oxygen,* for instance. As Philip Kitcher surveys out the referential commingle: *dephlogisticated air* is sometimes equivalent to *oxygen,* sometimes to *oxygen-enriched atmosphere* (i.e., to regular air with a higher than normal oxygen content); *phlogisticated air* is usually *oxygen-depleted atmosphere,* or *oxygenless atmosphere; phlogiston* itself is sometimes equivalent to *hydrogen,* but sometimes (in a phrase like "phlogiston is emitted during combustion") it does not refer at all, in a modern chemical parceling of the world.[42] If that were the whole of it, translation would be awkward indeed, but not foreclosed. The real communicative barriers arise, under Kuhn's account of local incommensurability, with the interdependence of *phlogiston* (and its morphological relatives) with *element* and *principle. Principle* lives on, but in an unrecognizable form—it denotes a methodological heuristic, not a theoretical entity—while its original referent has evaporated. And *element* has (roughly) bifurcated into the contemporary *element* and *state of aggregation*. Moreover this cluster of words/concepts/phenomena fundamentally determines the taxonomic structure of phlogistic chemical theory—the way it categorizes the phenomena in its domain. So, local incommensurability is a variety of semantic incommensurability localized to a small area of the overall semiotic web, but one that is fundamental to the relevant theory.[43]

The increasing centrality of vocabulary in Kuhn's work is obvious from the successor term to *paradigm* (and *disciplinary matrix*), *lexicon*. *Paradigm*, too, was a word from linguistics, but it focuses on representative morphological alternation; the metaphor implicates form and pattern. For linguists, lexicons are much grander concepts than paradigms (which frequently don't even enter linguistic theory these days, being confined to grammar books in second-language acquisition), though *paradigm* had got so bloated from Kuhn's wide-ranging use of it, and even more from its adoption hither and yon, that his shift to *lexicon* was a restriction of the claim-and-practice-defining elements characteristic of scientific programs. Kuhnian lexicons are structured vocabularies learned and deployed in specific disciplinary contexts—like phlogiston-, or oxygen-based, chemical theory—that link words, concepts, and practices. They are "inextricable configuration[s] of words and things; mastering the lexicon means acquiring the skill to recognize its appropriate application in various settings, and to encounter the world in those terms" (Rouse 1998, 45). Rival lexicons (in theory succession) are the locus of synchronic incommensurability; archaic lexicons (and their discrepancies with respect to modern lexicons) are the locus of diachronic incommensurability.

While translation (of a narrowly construed sort) remains closed to the lexicons of rival theories, interpretation is not only possible but desirable, and interpretation takes place on the level of the taxonomies. "It is taxonomies," Kuhn told Giovanna Borradori in the late 1980s, "that are commensurable with one another, the things one can really talk about." (Borradori 1994, 166). Moreover, Kuhn felt that this was the notion motivating him from the beginning, just that it took years of debate to bring it out. "Insofar as incommensurability was a claim about language, about meaning change," he said, "its local form is my original version" (Kuhn 2000, 36). Perhaps. The linguistic analogy had always been lexical in nature, and further, was always confined to lexical reference; except for a few scattered and vague allusions, syntax and morphology and even lexical class had never entered into accounts of incommensurability. But what is most significant about the local form is that it no longer concerns paradigms, except incidentally, nor even disciplinary matrices, nor even theories, but "lexicons," a much tidier notion, one that localizes the translation difficulties closely enough, encapsulates them even, to make the overall problem of incommensurability more pliable.

I won't explicate this category of incommensurability any further here—Alan Gross's chapter takes it up in some detail (albeit with somewhat different conclusions)—except to note that it is the weakest type in our scheme, certainly not degree-zero incommensurability (which would be absolute commensurability), but fairly low on our incommensurometer compared to the other varieties we have explored. Paul Thagard and Jing Zhu call it, in contrast to the cosmic implications of Khun's early formulations, "an unthreatening observation on difficulties of translation and communication" (Thagard and Zhu 2002, 81).

If local incommensurability between any old A and B is unthreatening, and if local incommensurability satisfies Kuhn's attempt to show that semantic incommensurability in general is practically tractable; if brick-wall incommensurability is beyond all reckoning, where the points of contact between A and B are so negligible and accidental that there is no motivation to bring them into comparison, evaluation, or even communicative abutment; if cosmic incommensurability—the different-world conflict of perceptions—is a matter of metaphysics (I don't even know what you perceive, let alone what A and B holders respectively perceive); if, that is, on the one hand, incommensurability is "no big deal" (Bird 2000, 188), and on the other, "incoherent" (Putnam 1981b, 115), why all the ink spilled on this topic? Because there really are rhetorical problems among rational, earnest scientists (and other people) that amount to very different argument appraisals, so that what looks powerful to one side looks unimportant or perverse to the other. Which brings us to pragmatic incommensurability.[44]

Pragmatic Incommensurability

> An Athenian addressing a Spartan—assuming that weapons can be laid aside long enough for some talk—has a categori-

> cally different problem from the one he faces when talking to the Athenians.
>
> —Wayne Booth, *Modern Dogma and the Rhetoric of Assent*

Incommensurability is a phenomenon of misaligned meanings. But meaning is not a simple function (nor even a complex function) of semantics alone. The lines of separation between semantics and pragmatics are notoriously blurry. But, whatever those lines might be, as almost everyone who has looked at the problem field of incommensurability agrees, it always implicates more than "just" semantics. Context counts.

For matters incommensurable, context is almost all that counts. It was, of course, the admission of context that led to the whole rumpus in the first place. The doctrine of theory-ladenness introduced local context to terminological meaning. So, for instance, when one says, "Mercury is a planet," in Ptolemaese or in Copernicanese, *planet* draws its meaning from the home theory; therefore, the predication draws its truth-conditions from the home theory. The sentence is true in both theories, but its meanings (while overlapping) are different,—ho hum: once more around the block with our old friend, semantic incommensurability.

But the Ptolemaists and the Copernicans knew all this as clearly as we do. Most of them surely knew it an awful lot more clearly. Their disputes were more subtle and far-ranging, and to the extent that they genuinely talked at cross-purposes, failing to see the power in each other's arguments that they saw in those of their kith and kin, seeing rather perversity, the misalignments were more subtle and far ranging. The semantic incompatibility of *planet* in two rival programs arguing about (among other things) what planets are, what a fruitful definition of *planet* is, would have been lost on no one. The trouble comes from incompatibilities that rest further below the surface of the dispute.

Tacit theory-ladenness is one of the strongest sources of both success and failure in communication, depending on whether the interlocutors are operating out of the same or different theories. We agree about vast tracts of things when we exchange words—that *river* implicates water, banks, motion, possibilities of travel, shipping, and recreation, for instance, and a division of sides that might require a bridge, or a ferry, and termination in a lake or sea, and on and on and on—the sorts of things that linguistic pragmatics charts. These very extensive agreements and alignments—on which language depends for its daily, moment-to-moment successes—are submerged, the bulk that keeps the tip of the communicative iceberg afloat. Also under the surface, however, there can equally be disagreements and misalignments, sometimes extensive ones, leading interlocutors to talk past each other, potentially escalating to confusion and hostility. As Rupert Crawshay-Williams notes, in a probing study of controversy, "It is because such unrecognized agreements are so common that unrecognized disagreements are also common; in so far as we fail to realise what it is that makes our discussions go smoothly, we fail also to realise what it is that makes them go wrong" (Crawshay-Williams 1957, 4).

Semantic calibration, getting the denotative ducks in a row, is a relatively simple matter compared to pragmatic calibration, getting the presuppositional ducks in a row. "Really, universally," Henry James wrote, "relations stop nowhere" (H. James 1948, 5), but in questions of incommensurability, two classes of relations are especially important—both rather broad—thematic and methodological.

Pragmatic incommensurability, that is, concerns the (non-referential, non lexical) components of a program (theory, paradigm) that lead proponents on either side of a communicative divide to resist and misunderstand each other, the patterns of talk and practice that pull disputants in opposite directions and retard argumentation—what philosopher Wolfgang Stegmüller calls "paradigmatic dispositions" (Stegmüller 1979, 56), and rhetorician Edwin Black calls "stylistic proclivities" (Black 1976, 85; see Halloran 1984). The patterns of talk are largely themes that inform appeals: what values one can base an argument on (simplicity, scope, consonance . . . ; see Kuhn 1996, 184–5), what authorities one can invoke, what sorts of solutions to what sorts of problems count as support for one side of the debate or the other (Kuhn 1996, 184). Patterns of practice inform these patterns of talk: what sorts of data feeds one side, what sort compromises the other, what even counts as data, how do you get it, with what conceptual or mechanical instruments? And the process is mutual. The thematic availability of one type of solution over another informs the lab or field practices; those practices make available the solutions one can then talk about. Pragmatic incommensurability follows from, and reciprocally feeds, the materials of argumentation.

In theory-succession terms, Kuhn puts it thus:

> the reception of a new paradigm often necessitates a redefinition of the corresponding science. Some old problems may be relegated to another science or declared entirely "unscientific." Others that were previously non-existent or trivial may, with a new paradigm, become the very archetypes of significant scientific achievement. And as the problems change, so, often does the standard that distinguishes a real scientific solution from a mere metaphysical speculation, word game, or mathematical play. The normal-scientific tradition that emerges from a scientific revolution is not only incompatible but often actually incommensurable with that which has gone before. (Kuhn 1996, 103)

Feyerabend's initial formulations were largely lexical in nature—concerning semantic, possibly even local, incommensurability—but his later descriptions take the increasingly broader tones of pragmatic incommensurability. Here he is in *Against Method:*

> We have a point of view (theory, framework, cosmos, mode of representation) whose elements (concepts, 'facts,' pictures) are built up in accordance with certain principles of construction. [. . .] Now take those constructive principles that underlie every element of the cosmos (of the theory), every fact (every concept). Let us call such principles *universal principles* of the theory in question. Suspending universal principles means suspending all facts and concepts. Finally, let us call a discovery, or a statement, or an attitude *incommensurable* with the cosmos (the theory, the framework) if it suspends some of its universal principles. (Feyerabend 1978a, 269)

Together asynchronously—Kuhn early in his career, Feyerabend later on—the architects of incommensurability drew attention not (just) to the communicatively obstructive nature of misaligned words, but of misaligned problems, and standards, and principles, as well—themes and practices.

Take the antagonism in biology at the turn of the century between the two strains that Ernst Mayr (1988, 25) calls *functional* and *evolu-*

tionary. Functionalists sought explanation in immediate circumstances, evolutionists in the tradeoffs between mutation and natural selection. Here is a functionalist, Jacques Loeb, in *The Organism as a Whole from a Physicochemical Viewpoint,* rebutting an evolutionary account of amphibian legs:

> The [evolutionary] writers explained the growth of the legs in the tadpole of the frog or toad as a case of adaptation to life on land. [...But] we know through Gudernatsch that the growth of the legs can be produced at any time even in the youngest tadpole, which is unable to live on land, by feeding the animal with the thyroid gland. (Loeb 1916; qtd. by Mayr 1988, 28)

The evolutionary account makes no sense to Loeb because causation for him is biochemical and individually based, not environmental and population based, because his concern is with individual development, not species development, and because time is eliminated as an explanatory principle. He is looking at the matter, as he declares in his title, from a physicochemical viewpoint; evolutionists look from a temporal, natural-selectionist viewpoint.

The incommensurability here is not semantic (except to the extent that words like *cause* and *explain* have different referents in the functionalist program than in the evolutionary program, but that's putting the referent before the horse). It's not cosmic. There are no perceptual issues of note at stake. The differences in "seeing," such as they are, derive from differences of theatre (individual vs. species; a few days in the lab vs. a few epochs in the distant prehistoric wild), not from differences in theoretical lenses looking in the same direction. It's certainly not brick-wall. There's no question that the domains strongly overlap (explaining the growth of legs in the life-cycle of frogs and toads). But it is incommensurability, if anything is; the passage exemplifies, as well as anything might, Kuhn's recurrent dictum about opposing frameworks satisfying his descriptor, *incommensurable*, "communication is inevitably only partial" (Kuhn 1996, 198).

The blockages—the reasons that this passage can only partially communicate to evolutionists, and that whatever Loeb had heard or read before writing this passage must have only communicated partially with him—are thematic and methodological. The appeals to biochemistry, to the inability of tadpoles with thyroid-gland-induced legs to live on land, and to the whole apparatus of experimentation are presented as devastating counter-evidence to evolutionary claims. Loeb sees the problem field as so localized, it is as if the only evidence he would take to support the adaptation-to-life-on-land hypothesis would be an experiment which dropped a tadpole on the sand and observed it sprout legs, possibly some lungs, and go about its terrestrial business. Loeb has no trouble at all recognizing what the evolutionists were claiming—that the growth of legs in tadpoles exemplifies adaptation—only in recognizing that (within his own set of assumptions) those claims are coherent. It is not a failure to build some lexical interface between the two theories, not a failure of translation. It is a failure of connection that follows from an underlying conflict in paradigmatic dispositions.

We know, of course, that this dispute was resolved—not exactly amicably, but without the intransigence-unto-death of the Max Planck Effect—with the neoDarwinian blending of biochemistry and evolution in the 1930s, a case less of theory succession than of theory amalgamation. (Loeb, alas, died in 1924, before the great synthesis.)

Especially important among the patterns of talk are values, which ground themes and shape practices.[45] A constellation of shared values between Ptolemaists and Copernicans, for instance, retarded the widespread acceptance of the latter, which only came when there was a reconfiguration of values more than eighty years later. Both Ptolemaists and early Copernicans valued perfection in celestial mechanics, holding circular motion to be perfect. Objects in both of their respective heavens inscribed perfect circles. Copernicus offered a few savings in elegance, a reduction of the number of epicycles, and a more natural account of retrograde motion (one that derived from the model and did not have to be simply stipulated). But neither elegance nor deriveable accounts were enough to tip the geo- to heliocentric balance. Both programs also valued predictive accuracy very highly, but neither had a notable edge. That changed with Kepler, whose elliptical orbits brought significantly greater mathematical elegance (no more epicycles), and, most importantly, whose Rudolphine tables demonstrated impressive gains in predictive accuracy. But the shift in models could not be accomplished without restructuring the value system. Perfect circles could no longer be maintained, and celestial perfection overall was seriously compromised (if Earth was a planet, then all the wandering bodies in the heavens might be made of the same mutable, corruptible materials). For Ptolemaists with greater allegiance to circular orbits or superlunary perfection than to predictive accuracy or mathematical elegance, there could be no conversion. Values can become so entrenched and entangled with terminological meaning, with practices, even with intuitions and perceptual possibilities, certainly with premises and argumentative structures, that they can (1) provide rapid passage to solutions and equally rapid acceptance of another's solution, fostering tight cooperative engagement, but also that they can (2) block solution paths which violate those values, and the acceptance of arguments from others who have rejected or downgraded or never held those values, fostering hostile competitive engagement.[46]

All value-structures are personal—different scientists having different rankings at different times, deploying them differently in different circumstances, for different audiences. The researcher who has investigated value-clusters most thoroughly in science studies is Gerald Holton, who calls them *themata* (and *anti-themata*), and uses them most extensively to explore the individual psychology of prominent scientists, like Kepler and Bohr, especially as they contribute to the development of those scientists' major contributions (Holton 1988, 1998).[47] But, like the idiolects (personal speech patterns) that comprise dialects, personal value clusters overlap into defining social clusters, giving programs their distinctive ethoi, their distinctive arguments, their stylistic proclivities; and, reciprocally, social value clusters, like dialects shaping idiolects, feed and constrain and configure the personal value systems of individual scientists (R. A. Harris 1993b). Sometimes these collective ethoi are willfully dissimilar, especially when a movement flamboyantly rejects its established predecessor, or an individual repudiates a leading figure, perhaps even a mentor. The generative-semantics movement in linguistics, for instance, valued politically charged, whimsical example sentences, and generated data with the aim of showing that "grammaticality" was determined by context, not by abstract rules of grammar (and therefore that pragmatics subsumed syntax). Early transformational grammar built its arguments around data sentences like:

> John is easy to please. (Chomsky 1965)

Generative semanticists, on the other hand, preferred data like

> Spiro conjectures Ex-Lax. (Morgan 1973)

Traditional and earlier transformational grammars treat this word string as ungrammatical because *conjecture* was held to require sentential complements, but it is perfectly fine as an answer to "Does anyone know what Pat Nixon frosts her cakes with?" Of course, the same grammatical point might be made with "John conjectures vanilla," answering a similarly mundane question, but that would miss a definitional opportunity by, among other effects, thumbing a nose at the seriously prosaic Chomsky.

Astronomers venerated sphericity until Kepler, and superlunary immutability until Galileo. Pythagoreans valued certain numbers, like multiples of ten, and found the appeal of geometry irresistible in accounts of reality (Mason 1962, 28–35), a "number mysticism" that also pervaded much of twentieth-century microphysics in Planck's wake (Heisenberg 1971, 34–35). Twentieth-century microphysicists also shared notions of isolation and existentialism that gave rise to themes of "disintegration, violence, and derangement" in their work (Holton 1988, 79). Molecular biologists after Watson and Crick had an entrepreneurial, Mr. Fix-it spirit, valuing commercial tie-ins and breezy terminology (Halloran 1984).

When rival programs clash, their allegiance to different values, and their differing levels of allegiances to the same values, contribute much to the dissonance, usually in quite subterranean ways. "Neither side," as Kuhn said, "will grant all the non-empirical assumptions that the other needs in order to make its case. [. . . They] will often disagree about the list of problems that any candidate for paradigm must resolve. Their standards or their definitions of science are not the same" (Kuhn 1996, 148). These value clusters are interdependent, which makes them mutually reinforcing, so that (1) they are individually hard to overcome, but that (2) when one *does* fall in the hierarchy—or, like sphericity, out of the hierarchy altogether—there is a reordering, sometimes substantial.[48] This rapid restructuring might be a partial explanation for the scales-from-the-eyes personal Eureka moments Kuhn sought to explain with gestalts and conversions.

Related Usages of the *Incommensurability* Suite

> Worthy Hellenes, here is one of the things of which we said that ignorance is a disgrace and knowledge on a point so necessary. [. . .] The real relation of

> commensurability and incommensurability to one another. A man must be able to distinguish them on examination, or else must be a very poor creature.
>
> —Plato, *Laws*

One of the projects in this introduction is to chart some of the meaning variance and theory-ladenness of the *incommensurability* suite of terms. In some usages, it collocates strongly with totalizing terminology (*absolute, complete, total*); in others, with perceptual terminology (*gestalt, different-world, relativity*); in others, with semantic terminology (*meaning, reference*); in others, with pragmatic terminology (*practice, problem, value*). I don't mean to suggest that this exhausts the lexical traveling companions of *incommensurability* in science-studies discourse (lest we forget *translation, comparison,* and *theory-laden* itself); nor that there aren't patterns of usage which show that my tidy little categories leak. But (cross-cut with diachronic/synchronic topics, and significantly intersecting with a continuum model) those are the ways the meanings of *incommensurability* vary in the philosophy (history, sociology, rhetoric) of science.

But incommensurability as a non-mathematical scholarly concern is not confined to philosophy (etcetera) of science. It participates in other patterns; one of them a considerably older pattern, which assimilates naturally to pragmatic incommensurability, because it zeros in closely on thematic material that is critical to many incommensurable oppositions: values. Truthfully, *all* incommensurability is value-driven, even more fundamentally than it is semantic. Differences in meaning, perception, themes, and practices all correlate very strongly with values, and if one of the bodies of discourse we are about to look at hadn't cornered the market on the phrase "value incommensurability," I would have drafted it as one of my main category labels for scientific discourse (as, in fact, Lawrence J. Prelli does, in his contribution to this volume—329), as well as for ethics discourse. But precedence is very difficult to fight in lexical matters; I have followed prior usage. And the ethics usage is an old one—for the adjective *incommensurable*—going back at least to the 1600s.

Value incommensurability marks a distinct set of technical uses of the *incommensurability* suite in a range of fields that were once called the *moral sciences,* though I will only look at it in broad outlines, not in specifics—looking chiefly for the way the suite participates in issues that overlap with those of pragmatic incommensurability. The other main usage of the suite in scholarly discourse, the postmodernist usage, is also related to pragmatic incommensurability, but the suite is not part of the technical terminology of that discourse. *Incommensurability* is a loan word from science studies and (subordinately) from value studies that has brought with it some general baggage of disconnection and severe incompatibility, with allusions to some related issues (particularly comparison and translation), but little in the way of technical specifics—nor has it grown any technical specifics of its own in postmodernist deployments, though it has accrued a characteristic plural usage, *incommensurabilities.*

Value Incommensurability

> To claim that goods are incommensurable is not to rank them. It is to say that they cannot be ranked.
>
> —John Gray, *Two Faces of Liberalism*

In this literature, the suite collocates with *value,* of course, but also with *comparison, rank, choice, pluralism,* and *tragic,* as well as a cluster of words treated as near-synonyms for *incommensurable—incompatible, irreconcilable,* and *incomparable.* The usage is unquestionably an old one, but it gained new currency in the wake of Feyerabend's and especially Kuhn's widespread cultural influence. The nonmathematical use of the word *incommensurability* before Feyerabend and Kuhn was restricted and arcane; they brought fresh attention to the concept, creating a problem field, and popularizing the nominalization. But the nonmathematical use of *incommensurable* dates at least to the seventeenth century.[49] Long before 1962, that is, there was an adjective, a quality that might hold of some pairs of phenomena, but there was little presence of the corresponding noun, to mark a locus of attention. noun, a locus of attention.

What is most interesting about the adjective is that it very often implicates values (not, say, meanings or percepts). The second edition of Oxford English Dictionary (OED2) defines what it calls the "general" use of *incommensurable,* as "having no common standard of measurement; not comparable in respect of magnitude or value," exemplifying that definition with this collection of paradigmatic sentences:

> **1660** *R. Coke Justice Vind.* 12 Whether such things so apprehended by the Senses, be pleasant, profitable,

> just or unjust [. . .] commensurable, or incommensurable. **1664** ***H. More*** *Myst. Iniq.* Apol. 539 Will not this Position prove as incommensurable to humane affairs and be laden with as great inconveniences? **1796** ***Burke*** *Let. Noble Ld.* Wks. 1842 II. 260 Between money and such services [. . .] there is no common principle of comparison: they are quantities incommensurable. **1845** ***De Quincey*** *Nat. Temperance Movem.* Wks. XII. 167 The two states are incommensurable on any plan of direct comparison. **1881** ***Westcott*** & ***Hort*** *Grk. N.T.* II. 46 The rival probabilities represented by relative number of attesting documents must be treated as incommensurable.

Some of the uses of *incommensurable* catalogued here have technical connotations to them, and some are fairly indeterminate from what we can see in these brief snatches. But overall these five sentences, sampling from three centuries of use, illustrate a lengthy, if not dense or widespread, deployment of *incommensurable* in the realm of ethical values. This usage has clear precedent over science studies (and one wonders how much these "general" usages may have influenced Kuhn or Feyerabend). The usage also clearly penetrated moral philosophy, perhaps even originating with it. Roger Coke's (1660) treatise, the earliest OED2 citation, falls into that genre—the full title is *Justice Vindicated from the False Fucus put upon it by Thos. White, Gent., Mr. Thomas Hobbs, and Hugo Grotius*—but his use of the term, while embedded in a discussion of moral values, clearly belongs to geometry.[50] We can say, in any case, that by at least the turn of the last century the theme of value-(in)commensurability had noteworthy currency in ethics.

Hastings Rashdall, for instance, in his two-volume *Theory of Good and Evil,* argues that any

> two kinds of value are not absolutely incommensurable. However much superior the value of a good act may be to that of a transitory pleasure, we still use the term 'value' of both, and we use it in the same sense: the two kinds of value differ as being at the top and the bottom of the same scale, not as representing two totally incommensurable scales. There can be only

> one ultimate scale of values, however heterogeneous the objects which we appraise by that scale. (Rashdall 1907, 1:174)

Rashdall implicitly treats incommensurability as a graduated notion (as the inverse of the graduated notion that he elevates, commensurability); what he repudiates is the far end of the scale, absolute incommensurability.[51] A cornerstone chapter in the second volume, "The Commensurability of All Values," epitomizes his "scientific treatment" of ethics. Every pleasurable experience for Rashdall (1907, 2:15) is "a sum of other pleasures" (with *pleasure* construed in a utilitarian or even Socratic sense—knowledge, virtue, and so on, are pleasures) which can be calculated via a Benthamite "hedonistic calculus." Commensurability is not always a concern; it becomes an issue only when we have to choose between differently ranked goods, perhaps in differing quantities, "*when we cannot have both.*" In that case, the calculus can be brought in: "we can compare them, and pronounce that one possesses more value than the other" (Rashdall 1907, 2:39).

Rashdall addresses a variety of opponents, who reject the measurement and/or calculation of pleasures (values, goods), but it is not clear there was a distinct position that might be connected with the slogan "incommensurability of values"[52] until work by Sterling Lamprecht on ethical pluralism and prudential values in the early 1920s, which flatly contradicts Rashdall:

> Many times men are faced with situations in which the potential goods are woefully incompatible, in which the choice of one good involves the abandonment of another; and sometimes men are faced with still more trying situations in which the potential goods are unknown and can not be brought to light except on the basis of a daring decision, a decision which is frankly a hazard and will not be proved true or false until the outcome has made investigation of other expedients forever impossible. The goods of life are utterly incommensurable. Health, beauty, courtesy, knowledge, friendship, all these cannot be measured by a common scale and tabulated in a common calculus. (Lamprecht 1920, 564)

> [. . .] The prolonged clash of rival codes and standards of right is so grave an evil that at all costs we must seek to limit and prevent its occurrence. But in so doing, we are more likely to meet success if we remember that we are not always judging between a right and a wrong, but often between two irreconcilable rights, two irreconcilable choices of incommensurable goods. (Lamprecht 1920, 570)

Lamprecht is especially focused on using value incommensurability to counter ethical monism, not just of Rashdall's utilitarian sort, but any universal-solvent project for ethical issues, and to promote its converse: pluralism. Arguing that there are no available "eternal principles" by which to measure potential actions that we find valuable, Lamprecht endorses an approach in which we make largely personal choices and do not require others to act according to our constellation of values. There are noteworthy parallels here with Feyerabend and Kuhn—parallels that play out through all the incommensurability-of-values literature—starting with the entire line of thought from logical empiricism right through to the early 1960s philosophy of science milieu that worked against a compelling earlier monism known as "*the* scientific method"— in Feyerabend's case, ending up with an advocacy of *proliferation*, a strong version of methodological pluralism. And the incomparability of goods resembles the incomparability-of-theories (paradigms) readings that many scholars give *Structure* in particular. There is also a significant theme in Lamprecht and subsequent scholars in this tradition that the existence of rival-but-incommensurable values compromises rational decision making. Lamprecht's "daring decision" even sounds like the sort of move that would be required of a Kuhnian revolutionary.[53]

Lamprecht situates his essay within "The general tendency in American philosophy during the last two decades [. . .] towards pluralism" (Lamprecht 1920, 561), and some while later we find Henry W. Stuart, a student of leading American pluralist John Dewey, pursuing the theme of incommensurability in a survey of Dewey's ethical views. Summing up the moral he draws from a series of situations that lead to choice among values, a process Dewey called *valuation*,[54] Stuart writes:

> The ends in an ethical situation are, then, variously described in the above as incompatible, discrepant, heterogeneous, opposed. They get in each other's way; they cannot readily be measured and chosen, one as against the others, because no common denominator can be found in terms of which to express their relative worth. In ethical situations, that is to say, the rival ends toward which the individual finds himself attracted are found to be incommensurable. (Stuart 1939, 298)

Where Lamprecht clearly uses a general sense of *incommensurable,* Stuart makes the mathematical linkage explicit, and he not only uses the nominalized variant, he uses it in a way that suggests he is circumscribing a problem field—"ethical incommensurability" (Stuart 1939, 301, 311). For him, *incommensurable* and *incommensurability* are technical terms.

It isn't clear what contact, if any, Lamprecht and Stuart had with each other. They surely knew each other on some level; both were active professional philosophers in an overlapping period, at established U.S. schools, at a time when the philosophical community was not large.[55] Nor is it clear how common talk was about value incommensurability among the American pragmatic pluralists; there are no adequate corpora available for electronic searches. Nor can I say what direct influence pragmatists generally, Lamprecht or Stuart specifically, and/or such talk might have had on subsequent ethical philosophy. There is no citation presence of this work to speak of in the current literature of value incommensurabilists, and, in particular, the most substantial figure associated with this line of thought later in the century, Sir Isaiah Berlin, does not cite earlier pluralists, aside from occasional notice of William James; indeed, in the article "My Intellectual Path" he says rather disarmingly of value incommensurability, "I do not know who else may have thought this" (1998, 60).[56] Finally, it is not clear what influence, if any, Feyerabend and/or Kuhn may have had on Berlin (whose earliest major statement of pluralism, *sans incommensurability,* comes in 1958, with "Two Concepts of Liberty").

That's a good deal of murk; my apologies. But two things *are* clear: the fortunes of discourse on incommensurability of values were improved substantially by Berlin's work, especially his *Four Essays on Liberty* (1969), from which some commentators date the movement;[57] and the charisma of the word *incommensurability* was raised immeasurably by the work of Feyerabend, and especially Kuhn. (Feyerabend's *Against Method* went through three editions, achieving substantial notoriety. Kuhn's *Structure of Scientific Revolutions* also went through three editions—the very widely sold second appearing contemporaneously with Berlin's *Four Essays*—making it onto the list of the one hundred most influential books since the Second World War by the *Times Literary Supplement* [1995]. It has sold well over a million copies. Feyerabend's book and Kuhn's book have been—mild irony here—respectively translated into nineteen and twenty-five languages.) It is in any case noteworthy, I think, that this strain of discourse is not known, for instance, under the slogan "the irreconcilability of values," or "the incomparability of values," or "the incompatibility of values."[58]

The incommensurability-of-values literature over the last quarter of the last century, that is, and on into this one, manifests the convergence of two currents of philosophical usage, the one Berlin represents, and the Feyerabend-and-Kuhn semantically infused incommensurability-of-scientific-programs usage, though the latter is distinctly less important for the topics and instruments of values literature. Value incommensurabilists almost never mention Feyerabend, and mention Kuhn largely to exorcize his massive ghost. Ruth Chang's landmark collection of essays, for instance, *Incommensurability, Incomparability, and Practical Reason,* introduces Kuhn on the very first page solely to banish him (or, rather, a confused-but-popular version of him):

> We can reject one notion straight off as inapplicable for [incommensurability of values discourse]. This is the idea, spawned by the writings of Thomas Kuhn, that evaluation across different conceptual schemata, ways of life, or cultures is impossible. (Chang 1997, 1)

The authors in Chang's book observe her banishment, with the lone exception of James Griffin, who invokes Kuhn in kind—as responsible for an "extreme position," one that is "implausibly strong" (Griffin 1997, 39). Indeed, despite the title of her book, Chang attempts to dispose entirely of the word, *incommensurability,* early on, in favor of *incomparability,*[59] this time far less successfully. All of the authors feature the word *incommensurability* very prominently in their articles, often in their titles.

What value incommensurabilists mean by the term is, like science-studies incommensurabilists, various, but it trucks much more explicitly with comparison—moral, political, and aesthetic. In the science-studies literature, the relation between comparison and incommensurability is controversial. Feyerabend, Kuhn, and their more sympathetic expositors deny that incommensurability forecloses theory comparison;[60] many of their opponents insist that it *does* foreclose comparison (and, further, that since it does, it must be wrong, because theories are compared all the time). Often, comparability is either absent altogether from science studies discussions, or simmers quietly on the back burner, present only through allusion. For value incommensurabilists, there is little or no controversy on this core: "What nearly all of us [. . .] mean by the 'incommensurability' of values," James Griffin volunteers for this discourse community, "is their 'incomparability'—that there are values that cannot be got on any scale, that they cannot even be compared as to 'greater,' 'less,' or 'equal'" (Griffin 1997, 35); this focus on (in)comparability, is perhaps the clearest sign of the influence of philosophy-of-science incommensurability literature. Comparability in value discourse is sometimes linked to value systems of a sort that correspond to frameworks in science studies, as in Berlin's juxtaposition of Classical and Christian values:

> [The Greek and Roman heroic system was based on the virtues] of courage, vigor, fortitude in adversity, public achievement, order, discipline, happiness, strength, justice, above all assertion of one's proper claims and the knowledge and power needed to secure their satisfaction [... which contrasts with the Christian virtues of] charity, mercy, sacrifice, love of God, forgiveness of enemies, contempt for the goods of this world, faith in the life hereafter, belief in the salvation of the individual soul as being of incomparable value—higher than, indeed, wholly incommensurable with, any social or political or other terrestrial goal. (Berlin 1979, 45)

This wholesale way-of-living contrast looks very much like the framework incommensurability that occupies much of *Structure,* and it clearly echoes Lamprecht's concern for the "rival codes and standards of right" (1920, 570). It is not a subject much treated in the literature;

however, it will no doubt escalate in the work of value incommensurabilists as religious and economic frameworks clash politically in the wake of the terrorism and warfare that has defined the early years of the twenty-first century.

Far more often, value incommensurabilists concern themselves with intramural clashes, when incommensurability makes choosing among system-internal values arduous.[61] The claim is not that all values are incommensurable. The choice between a cheeseburger, even a really good cheeseburger, and the education of a child, is not one that many people would have trouble making. Further, there are some values that may well be incomparable but are not worth worrying about—choosing between a cheeseburger and a rerun on TV—the sorts of things we have whims for. But there are some values—"ultimate values," Berlin calls them, "sacred values" in another theorist's terms (Lukes 1997)—values that are very much worth worrying about, and some of them just cannot be commensurated these theorists argue.

These sacred values are the familiar, compellingly large abstract notions that have preoccupied moral and political philosophy from the days when it was all anima and gods. Berlin argues that some of them are, in consummate form, antithetical:

> Liberty [. . .] is an eternal human ideal, whether individual or social. So is equality. But perfect liberty (as it must be in the perfect world) is not compatible with perfect equality. If man is free to do anything he chooses, then the strong will crush the weak, the wolves will eat the sheep, and this puts an end to equality. If perfect equality is to be attained, then men must be prevented from outdistancing each other, whether in material or intellectual or spiritual achievement, otherwise inequalities will result. [. . .] Similarly, a world of perfect justice—and who can deny that this is one of the noblest of human values?—is not compatible with perfect mercy. I need not labor the point: either the law takes its toll, or men forgive, but the two values cannot both be realised. (Berlin 1998, 60)

While the notions of comparability and ranking drive much of the value-incommensurability literature (especially a concern for the two relations, better-than and equal-to, which are said not to hold in cases of incommensurability), they are not compelling interests of that movement's most immediate sponsor, Berlin. There is no overt concern for the lack of a common measure in Berlin, no express worry about comparing liberty and equality, justice and mercy. One can certainly *imagine* a calculus for both liberty and equality or justice and mercy (if *you* can't, Rashdall can); indeed, truly antithetical notions are *necessarily* measured by the same metric (heat and cold, light and dark, strong and weak—pick your antonyms, and you will find them at either end of the same scale). Moreover, if the values are both ultimate they must satisfy a comparative relation. They must be equal. Like infinite quantities, one would expect ultimate values to be equal; if not, one must be less ultimate than the other (in which case, another comparative relation kicks in, better-than).

What concerns Berlin is not measurement but repulsion. The values push in opposite directions in some tightly coupled way—more of one means less of the other—which in fact evokes a common metric. Berlin customarily plays ultimate values to the full. So the issue is not really that "the two values cannot be both realized," even though Berlin puts it that baldly in the excerpt above—one might easily have a half-measure of justice and a half-measure of mercy, for instance—but that they cannot both be *fully* realized at the same moment. Consummate liberty, on Berlin's story, can only be got by eliminating equality as a value; consummate equality, by eliminating liberty.[62] Moreover, life will sometimes throw us—personally, but especially politically—into situations where there is no option but to choose between conflicting ultimate values, which means inevitable tragic loss. We are required by circumstances to choose between sacred and mutually exclusive values: one of them has to give. There's no way out.

What to do?

The solution for Berlin, and for many value-incommensurabilists (though, like all scholarly literatures, this one is far from homogeneous), is liberalism, which can only be got via value-pluralism-plus-incommensurability. What is relevant is not "the mere plurality of values," James Griffin argues. One might have values galore, with varying allegiances to all of them (as Rashdall's lexical essentialism implies, we wouldn't call them "values" if they didn't at least have in common that they all have value to us) and still be content (as Rashdall is) with a one-size-fits-all calculus; if the values are incommensurable, however, there can be no such calculus. The situation calls not for a method, Griffin says, but an attitude, an attitude that brings together multiple worth and incommensurability:

> a certain important picture of how [values] are related—that they clash, that they all matter, that they all have their day, that there are no permanent orderings or rankings among them, that life depressingly often ties gain in one value to terrible loss in another, that persons may go in very different directions and still lead equally valuable lives—call this picture 'liberalism.' (Griffin 1986, 91)

All of these factors—the clashing, the distributed significance, the unavailability of absolute rankings, the negation, the inverse relationship of gain and loss, the susceptibility to the tragic, and, somewhat perversely, the notion of "equally valuable life"—follow from the incommensurability fly in the pluralistic ointment, and collectively they constitute the ground for liberalism.[63] The enemy of Berlin's liberalism—as it is with John Dewey, William James, Sterling Lamprecht, Henry W. Stuart, and the whole plurality of American pluralists—is monism.[64] For James, monism is a philosophical disease, pluralism the cure. For value-incommensurabilitists, monism is a political disease, liberalism the cure. The cure cannot be a strict methodological one, which would be another monism. It must be a stance towards values, and especially towards other value-holders, that counsels tolerance and respect. The catch, of course, is the same with all ethical solutions to moral and political dilemmas—the stance must be shared, indeed, universal. *Everyone* needs to realize that values are incommensurable so that they don't attack or coerce other value-holders, or follow some other hegemonic program for their own slate of values.

Berlinian liberalism is a philosophy of the particular—of the individual, context-embedded decision. Much of his case about the incommensurability of ultimate values is conducted as a prosecution against the possibility of utopias. "The very notion of the ideal society," Berlin says, "presupposes the conception of a perfect world in which all the great values in the light of which men have lived for so long can be realised together, at least in principle" (Berlin 1998). Such a world is not only unrealistic, or utopian, it is logically impossible. X is ultimate. Y is ultimate. Not X and Y. Therefore, there is no logically possible world in which all ultimate values are realized fully. But the fall-out of this argument from abstract values is a required attitude one must bring to the particular case.

While the driving theme of this literature is prudential values, political and aesthetic values also concern various incommensurabilists, especially the lack of ranking possibilities in those domains. Take "Aeschylus, Shakespeare and Samuel Beckett," for example. They "are supremely great dramatists," offers John Gray, "but we cannot rank their work in value" (Gray 2000, 37). He doesn't mean, of course, any more than does anyone in the prudential domain, that we don't rank dramatists (Aeschylus is better than Beckett, Shakespeare better than the other two combined—there, I just did it; and, while I'm at it, liberty is better than equality). Rather, it's that these values (or, in aesthetic domains, the representatives of value) properly understood, cannot be compared in a rational way, *should not* be compared. There is no common metric.

Value pluralism has tremendous appeal for rhetoricians, of course, especially rhetoricians of a sophistic stripe. The history of rhetorical theory, aside from brief exceptions—the sophists, the Italian humanists, Giambattista Vico—is a handmaid's tale. Truth and value are obtained elsewhere, through other resources, and then rhetoric is brought in to market that truth or distribute (or enforce) those values. We can thank Plato for the earliest articulation of this position, in the one dialogue in which he lets his foot off the throat of rhetoric long enough to allow it a mildly wholesome role in the life of men. Having secured the truth, Socrates tells little Phaedrus, "The dialectician selects a soul of the right type, and in it he plants and sows his words founded on knowledge" (*Phaedrus* 277[a]). If truth and value, however, are not universal, immutable Forms, if they are contingent products arising from the exchange of argument, if they emerge from reciprocal suasion, if they grow in the medium Wayne Booth calls "symbolic communion" (Booth 1974, 194), then rhetoric has a constitutive role, not an auxiliary role, in human affairs. And if, as Aristotle argued, rhetoric is allied inextricably with ethics (*Rhetoric* 1356[a], 1359[b]), there is also duty associated with this constitutive role.

The dangers of pluralism (alienation, tribalism, persecution) are at least as great as the dangers of monism (zealotry, totalitarianism, persecution), and a constitutive rhetoric is the strongest weapon to combat them. Herbert W. Simons, for instance, sketches out a program of reconstructive rhetoric in his contribution to this volume, for exactly this purpose, as something of an antidote to the deconstructive, always-another-story-to-tell elements that drive interlocutors apart and endlessly defer resolution. Another of our contributors, Carolyn R. Miller, argues in an earlier paper for "a rhetoric of pluralism" that (in keeping with Aristotle) addresses itself to community values in preference to an exaggerated sense of the individual, a sense which encourages "anomie and disaffection and ultimately the conviction that reasoned argument is not possible because each individual is entitled to his or her conception of the good incommensurable by definition with everybody else's" (Miller 1993, 87). She advocates a rhetoric addressed "not only to the diversity within any community but also to the diversity of communities that co-exist and overlap each other" (Miller 1993, 91). In part, this call is very much of a kind with Simons's reconstructive rhetoric. It is motivated by the liabilities Miller, like Simons, sees in postmodernism, to whose use of *incommensurability* we now turn.

Postmodernist Incommensurability

> Postmodern knowledge [. . .] refines our sensitivity to differences and reinforces our ability to tolerate the incommensurable.
>
> —Jean-François Lyotard, *The Postmodern Condition*

Postmodernism is a notoriously amorphous body of discourse that partially overlaps Berlin-school liberalism, with its emphasis on value incommensurability, pluralism, and tolerance, but that draws its sense of incommensurability more directly from Kuhn and Feyerabend. *Incommensurability* is a not a technical term in postmodernist writing, which in fact does have a fairly sharp technical term that captures much of the notion, *différend* (Lyotard 1988). But the word and its lexical cohort are broadly used in postmodernism, in most of its widespread varieties, frequently showing up in the plural, as in

> What The Antibody Rhetoric dreams of, then, in its struggle against author/ity, is to enhance our abilities to tolerate the incommensurabilities that make up what cultural critics are calling post-modern knowledge. (Vitanza 1987, 49)

Indeed, the postmodern usage appears more common than the specifically Kuhn-and-Feyerabendian usage in nonspecialist language. In the massive British National Corpus of Modern British English, the suite of *incommensurability* terms collocates with recognizably postmodernist words (*Lacanian, Lyotard, Foucault, Deleuze, narrativization*) more often than it collocates with science-study words (*Kuhn, Feyerabend, science, paradigm*); the values usage is somewhat further back (*moral, pleasures*), while the mathematical usage is attested just once in the corpus's one hundred million words, a collocation with *numbers.*

The roots of *incommensurability*'s postmodern usage reach into the value literatures, into Kuhn and Feyerabend, and into Richard Rorty's influential arguments, but they are massed most fully in Jean-François Lyotard's *Postmodern Condition* (1984), which diagnoses a number of late-twentieth-century social maladies, chiefly the politically enforced fractionation of knowledge and of language, institutions which keep us apart, but also which give us reasons to value each other. An active awareness of this fractionation—much like the liberalist stance toward values (knowing that they clash, that they all matter, that they all have their day)—is the defining emulsion of postmodern knowledge. The virtue of such knowledge is that "it refines our sensitivity to differences and reinforces our ability to tolerate the incommensurable" (Lyotard 1984, xxv). Also like pluralism, the postmodern condition is resentful of monisms (totalizing discourses).[65] But it has a much deeper distrust of explanatory matrices (meta-narratives), which are necessarily focal points of power and authority. Lyotard's recurrent example is science, and he extrapolates the Kuhnian revolutionary moment into the genesis of postmodernism—a "crisis of narratives" (Lyotard 1984, xxiii)[66]. Where value pluralism is largely concerned with stances toward other value-holders, and is thereby epiphenomenally political, postmodernism is thoroughly and inherently distrustful of political structures—which is very nearly debilitating, since it sees all structures as political:

> following a logic which implies that their elements are commensurable and that the whole is determinable [. . . decision makers] allocate our lives for the growth of power. In matters of social justice and of scientific truth alike, the legitimation of that power is based on its optimizing the system's performance efficiency. The application of this criterion to all of

> our games necessarily entails a certain level of terror, whether soft or hard: be operational (that is, commensurable) or disappear. (Lyotard 1984, xxiv)

Commensurability, on this story, is the greater evil; fractionation into a plethora of incommensurabilities is not only inevitable, but preferable. There is no single postmodernist story, of course—how could there be?—so the opposite terror that many of us would see in the isolation, alienation, and anti-social impulses implicit in this strain are not universally pursued by everyone participating in postmodernist discoursing. Pauline Marie Rosenau, for instance, distinguishes between two strains in a convenient (and of course—ho hum—necessarily artificial) binary between "skeptical" and "affirmative" postmodernists (Rosenau 1992, 15–16). And in the current volume Herbert Simons turns his energies to articulating an affirmative, reconstructive rhetoric, one that counters the negative proclivities of postmodernist deconstruction.

Despite the incredible diversity, reach, and diffuseness of the postmodern sensibility, and the pride it takes in avoiding categorization, the presence and purpose of the incommensurability suite in that discourse is relatively easy to outline. It evokes non-trivial divisions, almost always in clusters (rather than between two isolated systems). It evokes multiple subjectivities, and competition among standards. It evokes, in different circumstances, but sometimes simultaneously, both tolerance and alienation. It evokes alternate rationalities (not irrationality, but decidedly not the Parmenides-to-Plato-to-Aristotle foundational sense of Rationality that defines much of the Western tradition, and was especially highly venerated in modernist discourse). *Incommensurability* is, in fact, something of a shorthand for the failure of that rationality—a word that works here by alluding to bodies of evidence for the breakdown of modernist rationality (in short, by summoning the ghosts of Kuhn and Feyerabend). By exactly the same token, it is effectively a citation to the deprivileging of scientific authority.

This collection of implications might seem like an awful lot of freight for one word to carry, even the big, gangly, octosyllabic nominalization we are dealing with here (along with its lexical relatives). It's not. *Incommensurability* is not a technical term in this discourse, just an important mood setter, so it doesn't require precision. Ordinary language words are routinely polysemous—just to pick a word from the sentence we are in, the OED2 gives twenty-six distinct definitions for *ordinary,* not counting various subordinate senses, and the OED2 is a conservative dictionary. Technical terms, in any case, are susceptible to luxuriant semantic growth as well, whatever the intentions of their coiners and users, and the technical sources postmodernism draws on—Kuhn and Feyerabend, with a side order of value pluralism—are hardly sparse. And, while *incommensurability* is non-technical in postmodernism, it is nonetheless central. The postmodern condition, Lyotard says, is a wholesome "incredulity toward metanarratives" (Lyotard 1984, xxiv), which makes incommensurability almost exactly as attractive to postmodernist sensibilities as it was repellent to modernist sensibilities. Widespread incommensurability, just as with pluralists critiquing monism, is key to metanarrative skepticism.

And incommensurability is *very* widespread on this story. Richard Rorty, one of the few Anglo-American philosophers to have any influence in this discourse, put incommensurability at the centre of his celebrated critique of modernist philosophy, *Philosophy and the Mirror of Nature* (though the affection wasn't entirely requited, and his influence has therefore waned somewhat). Rorty's *incommensurability* in that book is mundane and endemic. "It is [a] commonplace fact that people may develop doubts about what they are doing," he says, "and thereupon begin to discourse in ways incommensurable with those they used previously" (Rorty 1979, 386). This sort of quotidian incommensurability, coupled with a daily sense of profuse isolations, defines the postmodernist outlook.

The incommensurability suite plays a prominent role in postmodernist catechisms of heterogeneity, being duoed, trioed, quarteted—multiplexed—to other terms associated with profusion, division, and alienation:

> [. . .] essential heterogeneity of art by reference to the fundamental divergences and incommensurabilities. (Eldridge 1987, 256)

> Is this the sense in which we are not modern? Incommensurability, heterogeneity, the différend, the persistence of proper names, the absence of a supreme tribunal? Or, on the other hand, is this the continuation of romanticism, the nostalgia that accompanies

> the retreat of ... etc. (Lyotard 1988, 135; his faux-elision)
>
> [. . .] plurality, relativity and incommensurability—rather than consensus, objectivity and rationality. (Reed 1994, 164)
>
> [. . .] genuine radical differences and incommensurabilities between 'knowledges,' conceptual repertoires, languages or whatever. (Blake *et al.* 1998, 10)

The plural usage—*incommensurabilities*—is especially interesting, because it instantiates morphologically the key notion of value pluralism, and an important sublimated notion of science studies; not that there are two theories (frameworks, ways of life, whatever) that effect some kind of communicative logjam, but that there are passels of them. This sense of multiplicity was present from the beginning for Kuhn and Feyerabend. Kuhn wrote of a "proliferation of versions of a theory" at times of theory succession (Kuhn 1996, 71); Feyerabend, of "a whole set of partly overlapping, factually adequate, but mutually inconsistent theories" (Feyerabend 1981a, 72). Methodologically, confining discussion of incommensurability to theory pairs, rather than expressing it with respect to theory clusters, was always a convenient fiction, and their respective allegiances to the fiction is one of the defining characteristics of their subsequent career trajectories. By the time of *Against Method* Feyerabend had shed the fiction, but it continued to tighten its grip on Kuhn's work throughout his career.

The *incommensurability* suite is deployed in characteristic postmodern ways, but it doesn't do any real work beyond establishing or reinforcing a tone.[67] Postmodernism is, however, in need of technical terms for phenomena so crucial to its perspective. So critical aspects of the notion of incommensurability are realized in other terms, principally the "différend"—called "incommensurability politicized" by Lloyd Spencer. "The notion of incommensurability captured the idea of opacity between languages," Spencer says. "The notion of a 'différend' suggests an opacity, and a deformation, created by the action of one language [. . .] on another." (Spencer 1998). Here it is from the horse's mouth: "A *différend* would be a case of conflict, between (at least) two parties, that cannot be equitably resolved for lack of a rule of judgment applicable to both arguments. One side's legitimacy does not imply the other's lack of legitimacy" (Lyotard 1988, xi). This sounds a bit like a corax (if I win, I win; if I lose, I win; me too): equal legitimacy on the formal level with no procedure to adjudicate them. But there is a major difference. Lyotard does not end here. The *différend* is not a formal matter. Even though there is no applicable rule of judgment in such cases, one gets applied anyway, just as in all the practical cases labeled *incommensurable* in the history of science. One wins. One loses. The principal reason there is no rule of judgment, is that the parties are locked inside distinct language systems (phrase regimes), and cannot therefore agree on such a rule: Party A wants the rule to be drawn from her regime; Party B from his. In the event, one of them is unfortunate, a victim. The other supplies the regulatory language. "The 'regulation' of the conflict" in cases of *différend*, Lyotard says, is carried out "in the idiom of one of the parties while the wrong suffered by the other is not signified in that idiom" (Lyotard 1988, 9).

The *différend* is a heavily ramified notion—*very* theory laden, within a dense refraction of Wittgensteinian philosophy of language—and my sketch here is brutally terse. In several ways, the *différend* is a much more finely honed instrument for dealing with the diachronic discontinuities we label with *incommensurability,* a word that contains a late-modernist ideology—especially in its sense that there is always a resolution which excludes and renders one of the parties partially unintelligible or mute. There is even a nicely symmetrical sense of the irrational to Lyotard's concept, in a linkage to the Greek term for irrational numbers, which date to the Pythagorean discovery of incommensurability—*arrhētos,* "unspeakable." The victim in a *différend,* because the regulation occurs in an unaccommodating idiom, is left without a way to speak his case, perhaps being forced entirely into silence But for all that, *différend* remains peripheral to the current project, and we can only glance at a word that overlaps in fascinating ways with our cornerstone term, in a discourse preoccupied with the sorts of communicative misalignments that drove Feyerabend's and Kuhn's work, and that feeds value incommensurabilists.

Meta-Incommensurability

> What's in a word? Often, an ideology.
>
> —Ian Hacking, "The Disunities of the Sciences"

There is even—not that you would have doubted it by this point—incommensurability among incommensurabilities.[68] Feyerabend noticed, in his contribution to the 1965 symposium centered on Kuhn's work, that the same theory pairs might be commensurable for one observer, incommensurable for another. In particular, an instrumentalist and a realist can look at the same theories, right down to the last datum and equation, and they will be commensurable for the former, incommensurable for the latter (Feyerabend 1970, 220); again, in *Against Method,* he uses a similar example, and the identical phrase, "commensurable in some interpretations, incommensurable in others" (Feyerabend 1988, 221). Paul Hoyningen-Huene, Eric Oberheim, and Hanne Andersen make the same observation for versions of incommensurability. Appealing to semantic incommensurability, they say "there are several terms that change meaning when one crosses the line from realism to non-realism: namely, 'reality,' 'world,' 'theory comparison,' 'fact,' and even 'reference' itself" (Hoyningen-Huene, Oberheim and Andersen 1996, 138). When theorists of differing metaphysical commitments use *fact* or *world* or *theory comparison,* therefore, in arguments about whether A and B are incommensurable, or whether incommensurability is a coherent notion, they are always somewhat at cross-purposes. There is, they say, a meta-incommensurability among people arguing about incommensurability.

Alan Gross and Herbert Simons take up this notion in various ways in their contributions to this volume.

The Two Possibilities

> Rhetorically, incommensurability appears as mutual suspicion.
>
> —Eugene Garver, "Rhetorical Arguments and Scientific Arguments"

We started by noting that incommensurability does not debilitate scientists. The observation is not news. Feyerabend insisted on it from the beginning, and Kuhn noted it regularly. Kuhn, a historian who spent much of his career courting philosophers, contrasts significantly with Feyerabend, who spent the bulk of his career galling other philosophers, but they were both clear on the point that—whatever incommensurability implied in principle about the ways and means of theory comparison—it is not a practical obstacle for scientists comparing theories. Kuhn tried to show that incommensurability could be tamed in various ways, downgrading it to a lexically local notion that was amenable to an interpretive repertoire, and suggesting a set of scientific values for theory appraisal. Feyerabend simply put his faith in the pragmatically flexible argumentation of scientists:

> Incommensurability is a difficulty for philosophers, not for scientists. Philosophers insist on stability of meaning throughout an argument while scientists, being aware that "speaking a language or explaining a situation means both following rules and changing them' [. . .scientists] are experts in the art of arguing across lines which philosophers regard as insuperable boundaries of discourse. (Feyerabend 1987, 272)

Many others have subsequently noted the same thing. "The ferocity [of the resistance to incommensurability]," Rorty noted, "was found mainly among professional philosophers. Kuhn's description of how science works was no shock to the scientists whose rationality the philosophers were concerned to protect" (Rorty 1979, 333).

There are two possibilities. Kuhn and Feyerabend (and Rorty) may be right that there are ways around incommensurability (ways that strongly implicate rhetoric for both of them, and for Rorty), or some of their critics may be right, that it just doesn't obtain in science (and the problem was only illusory to begin with). Both of these possibilities have led several contributors to this volume—Lawrence Prelli, Thomas Lessl, and John Angus Campbell—to drop the term *incommensurable,* with its connotations of finality and its legacy of irrationality, in favor of the more plastic *incommensurate.*[69]

Incommensurability Does Not Obtain

> The doctrine of extreme incommensurability remains flawed; for it is fundamentally incoherent. How can *any* two things be *completely* incomparable?
>
> —Dudley Shapere, "Evolution and Continuity in Scientific Change"

Nobody denies there is incommensurability. As a mathematical concept, it is ancient, stable, and has massive consensus. The question is, does the concept extrapolate reliably to other domains? I won't pursue the ethical or postmodernist discourses any further. The emphases in that work on comparison, opposition, and eliminative relations, as well as the driving themes of pluralism and monism, give me pause about extrapolations into those domains. But there is certainly widespread scholarly agreement something is going on in the realm of values that sponsors phrasing like Lamprecht's "utterly incommensurable" and Rashdall's "not absolutely incommensurable."

With respect to science, however, there is substantial controversy. Some theorists, like Putnam and Davidson, argue that the claim 'scientific theories are incommensurable' is incoherent, that incommensurability is a figment of Kuhn's untutored and Feyerabend's overactive imaginations. A return to the ground-floor metaphors may help with this controversy. There are two of them, coming from opposite directions, from the kempt realm of mathematics and the unkempt realm of language.

The basic, etymological metaphor equates theories and comparison-criteria to numbers. The numbers 4 and 6 are commensurable because there is a number that divides into them both, 2, and can be used to measure them with respect to each other (letting us know, for instance, that 4 is 2/3s of 6); π and 6 are not commensurable, they are incommensurable, because no proportion of π can go cleanly into 6. Theories are bigger, messier, richer and more malleable things than numbers—even unspeakable numbers like π—which substantially compromises the analogy. A number is simplex and final (or, in any case, infinitely simpler than a theory, and infinitely more stable). A theory is not simplex, not final. Nor is a word (which is, at various times by various theorists, including Kuhn and Feyerabend, said to be the locus of incommensurability). Possibly mitigating this compromise is the most primitive etymological sense of *incommensurability,* which reaches to geometry and thereby to abstract spatial notions that have numeric relationships with each other. What caused the Pythagoreans so much grief, after all, was not the discovery of incommensurability or irrationals *per se,* but the implications those coincident discoveries had for the numeric amenability of the structure of reality—the sundering of arithmetic and geometry, the separation of the counting art from earth-measuring. Theories are like numbers in this rudimentary sense, and like lines—abstract concepts or shapes that model reality.

But theories of the sort that anyone asserts incommensurability about are reality-models on a massively larger scale than numbers and lines, with massively more resources for accommodating change. If the number 2 changes—is enlarged, contracted, distributed, any transformation at all—it is no longer the number 2; ditto π. A line that grows angles or curves is no longer the same line. They are simplex notions. If a complex notion changes, however—say, heliocentricity grows elliptical orbits—it can remain the same notion. It is heliocentricity still. Indeed, it is Copernicanism still. Theories can't change limitlessly. At some point, they become other theories (effectively, when their theorists declare them so). But theories have access to what Imre Lakatos called a *protective belt* of claims, hypotheses, and data configurations maintained by "a partially articulated set of suggestions or hints on how to change, develop, [. . .] modify, sophisticate" them, as required, so that they can accommodate challenges without a collapse of the whole structure (Lakatos 1970, 135).

Words, too, if we localize incommensurability to them (rather than involving pragmatic incommensurability), have this flexibility, this capacity to roll with the punches, that marks them as much less simple and complete than numbers; they can change sound, structure, meaning, syntactic function, or the whole caboodle. *Knight* no longer sounds very much at all like it used to, no longer refers primarily to a not-yet-mature human male, has a social designation rather than a biological one, can function as a verb sometimes, and so on. Again, words can't change without limit. At some point, we have a new word (effectively, when the users no longer recognize any part of an earlier sense, identifying only with the later one). The verb *park,* for vehicles, no longer has any linkage to the noun *park,* for designated green spaces, from which it emanated; *holiday* no longer has any religious implications for most people, though it was fairly recently *holy-day;* for that matter, *holy* has little to do with *halig,* which sponsored it but was closer in sense to our *whole;* and on, and on, and on. Flexibility begets new words, as old associations fade, but people remain very comfortable with statements like "*knight* is the same word, but its meaning has changed" or "*meat* used to mean any kind of food, not just meat" or, of course, "people used to call the sun and the moon *planets.*" Numbers don't have that sort of elasticity. There is no corresponding "2 used to be the square root of 6 as well as 4" or "2 used to be greater than 3" or "2 used to be irrational."

Theories are a lot like languages in these ways, with their protective belts, elasticity, and political definitions. But—here's where the second critical metaphor comes in—are they enough like natural languages for the Whorfian elements of the analogy to kick in? It is clear that the original sense of *incommensurability* is too sharply pointed for *that* sense of the word to obtain very directly of scientific theories. The metaphor is too sloppy. It misleadingly suggests a precision that just does not apply, occupying a ground that might otherwise be taken by somewhat more literal predicates, like *incompatible* or *inconsistent* or *intractable* (or, diachronically, *discontinuous*). And, of course, the entire literature on Kuhnian and Feyerabendian incommensurability stretches the geometrical metaphor to the breaking point. The various types and amounts of incommensurability that recurrently enter the debate, explicitly or implicitly, and the diachronic elements of a great many of the examples, and the active presence of comparing agents, all are without parallel in mathematical incommensurability, which has neither categories nor degrees, which is entirely synchronic, and which does not involve agents choosing between numbers to believe in and reject.

Moreover (as Lawrence Prelli notes in his chapter—298f) there is a critical morphological obstacle with the key term of this book, linked to its binary roots in geometry—an obstacle that strongly implicates a closed-door finality of the sort that is always avoidable with a determined and charitable application of language. The central *in . . . able* structure of the word (compounded by the last suffix,*–ity*) suggests a situation beyond remediation. The words in this class—*inexplicable, inevitable, inescapable*—are adjectives which leave no hope that the root verb can be realized. Even words in which the root verb has evaporated into time, like *ineffable* and *inscrutable,* carry that sense of finality, as do morphophonologically related words like *impossible* and *immutable, irremediable* and *irretrievable, illegible* and *illimitable.* It is the sense of preclusion built into *incommensurability* which fuels the incommensurability-of-values project; its favored descriptive synonyms for values are *incompatible* and *incomparable* and *irreconcilable,* not *inconsistent* or *uneven* or *misaligned,* and Ruth Chang seeks to trade in the guiding term for another foreclosing one, *incomparability* (Chang 1997).

The root verb in our case concerns measurement, which implies precision, a deeply embedded material value in science; undermining it is disquieting. And the implication that no resolution is possible inevitably suggests the response, "why bother?" Kuhn was especially uncomfortable with that suggestion, and worked much of his career to offer the hope of rational progress in science. Feyerabend was certainly warmer to the possibility of foreclosing reconciliation between rival theories, but he would have been the last one to draw from it the conclusion that critical discourse should stop.

But—moving now toward the linguistic metaphor—languages change. We know that. We just cut loose all the strings to the geometrical vehicle except some vague sense of things whose stock can't be taken with respect to each other. How, then, about the different-theories-are-not-intertranslatable-therefore-their-stock-can't-be-taken-with-respect-to-each-other sort of incommensurability? It retains the foreclosure-of-hope implications that attend the mathematical metaphor, but the relevant vehicle, language, is much more capacious and theory-like than numbers. The metaphor here is different-theories-are-significantly-different-languages (or different paradigms-, frameworks-, etc.-, are-significantly-different-languages). Are they? Not for Whorf. Take the answers on the right to the questions on the left:

Why does water rise in a pump?	Because Nature abhors a vacuum.
Why does water quench fire?	Because [. . .] the fiery principle and the watery principle are antithetical.
Why do flames rise?	Because of the lightness of the element fire.

The examples (among others) are Whorf's, in a context which implicitly contrasts these answers to modern accounts of the same phenomena:

> If once these sentences seemed satisfying to logic, but today seem idiosyncrasies of a particular jargon, the change did not come about because science has discovered new facts. Science has adopted new linguistic formulations of the old facts, and now that we have become at home in the new dialect, certain traits of the old one are no longer binding on us. (Whorf 1956, 222)[70]

Whorf knew language. He knew science pretty well too. His first degree was in Chemical Engineering, from MIT. He even had a decent grasp on philosophy of science. He died before theory-ladenness became a commonplace (which would reject the neat dualism of facts and formulations he employs here), and only partially overlapped the time when philosophy of science became a distinct profession, but his work makes regular informed allusions to philosopher-scientists like Duhem and Einstein. So it is worth taking his analogy seriously, which is linguistically more responsible than those of many philosophers in the context of incommensurability: Aristotelian physics is to modern physics not as Hopi is to English, but as Texas English is to Bronx English. The difference is dialectal. Whorf understood the incredible linguistic naiveté it would take to confuse two variant theories of motion for distinct natural languages, let alone for languages as different as an Amerindian one and a European one—that is, for languages of the sort that empirically undergird linguistic relativity. Indeed, the principal caveat to Whorf's linguistic relativity formulation is "[. . .] unless their linguistic backgrounds are similar" (Whorf 1956, 214), a situation he repeatedly says holds of (1) modern Western languages, which he calls "a few recent dialects of the Indo-European family," and (2) the "rationalizing techniques" of Western science (Whorf 1956, 218), because of their fundamental reliance on those languages.

For Whorf, the most crucial factor about differences in theories of motion versus differences in Amerindian and European conceptual schemes might be that the former are overwhelmingly terminological, not grammatical. Recall that his chief examples for linguistic relativity concerned morphology and word class, not the sense and reference of specific words. The senses and references of the Hopi and English words for "lightning" overlap a great deal, probably more than the senses and references of *planet* for Ptolemaic and Copernican astronomical theories. It is the overall eventness of Hopi contrasting with the overall thingness of English, manifest in words like these, which preoccupies Whorf and exemplifies linguistic relativity for him. But the basis of the theory-as-language commonplace in philosophy of science, back to its inception, is denotation and assertion. It is language-according-to-Frege, not to, say, von Humboldt or Boas. On rare occasions one sees a more developed analogy: "Just as [. . .] one who is not a native speaker can almost never acquire the Frenchman's feel for the use of the imperfect tense," Daniel Goldman Cedarbaum writes, illustrating the intuition-shaping nature of paradigms, "the physicist trained in the nineteenth century would almost certainly not have the same facility with problems involving discontinuous motion as the researcher committed to the quantum mechanical paradigm" (Cedarbaum 1983, 205). But this sort of analogy is rare indeed. The theory-as-language claims follow very directly from the positivist interest in a strategically impoverished version of language—rigid lexicalization and a narrow, largely propositional syntax—filtering off aspect, voice, often tense, most word classes, virtually everything that might help an appeal to Whorf go through. Whorf's extended example, contrasting Hopi with English, exclusively focuses on "the background linguistic system [. . .] in other words, the grammar" (Whorf 1956, 212). As Paul Thagard notes, after charting the relevant elements of several languages and several theories, "grammatically, acquiring a scientific conceptual system is unproblematic" (Thagard 1992, 115). Grammatically, successive theories are all pretty much the same, certainly within the boundary conditions Whorf uses to qualify linguistic relativity. Even in the realm of lexicalization, the analogy is questionable; Kuhn, for instance, who attended to only that dimension in his work, had come to feel by the end of his life that "the language metaphor [is] far too inclusive" to represent the character of scientific theories (Kuhn 2000, 92).

Applying linguistic terminology to science yields extremely rich analogies, but they aren't the sorts of analogies that have been traditionally pursued. It makes more sense, for instance, to speak of different fields, rather than rival theories within the same field, as different languages, particularly if one wants to discuss heterogeneity of a Whorfian sort. Competing programs (frameworks, paradigms, theories) within the same field are, as Whorf suggests, much closer to different dialects, with each scientist speaking his own idiolectal version of one or more dialects. A field in science—refracting one of linguists' most favored aphorisms—is a dialect with an army and a navy, a dialect willing and able to fight for its borders.[71] There are even strong parallels between interdisciplinary work (such as Charles Bazerman and René Agustín De los Santos take up in this volume) and pidgins, and between second-generation syntheses (such as molecular biology) and creoles.[72]

Returning to the two metaphors—theories as numbers and theories as significantly different languages—and to the central question of whether incommensurability obtains in science, between rival theories in the same domain: the verdict is strongly against incommensurability. The plasticity of theories, in the first place, and the linguistic commonality of rival theories, in the second place, vitiates claims of *incommensurability* in the context of theory succession or theory comparison. For similar reasons, the word's use in value discussions is similarly unwarranted. And, to boot, the phantom obstacles reared by the *in . . . able* morphology compromise the accuracy and usefulness and the word further yet.

But, listen, I'm a rhetorician, and I've looked at the incommensurability debates: I know both professionally and empirically that declaring (as I now hereby do) "incommensurability does not obtain for rival theories in the same domain" is not about to make everyone nod in agreement, murmur, "there, that settles that" and knock off for a latte. Nor will explicating the unwholesome morphosemantics of the *incommensurability* suite cause very many people to abandon those words everywhere but in math.

Although Davidson's no-conceptual-schemes argument, as devastating as he (along with many others) believes it to be, is widely cited and deeply respected, it has not managed to get *incommensurability* struck from the science-studies lexicon. My brief is not likely to do any better. I'm not even going to stop using the word myself, since it would be troublesome in the extreme, for you and for me, to abandon it at this point, deep into the introduction of a book which investigates arguments associated with that word, though I fully support some of our contributors' (Campbell, Lessl, and Prelli) choice to use *incommensurate,* a less-absolute term without the same history of exaggerated fears, without the implied technical precision of the geometrical metaphor or the different-worlds implications of the linguistic metaphor, without the abandon-all-hope morphology of *incommensurable*—a word which just says things are out of whack, but not irredeemably so, and have the potential to become commensurate under the right conditions.

In the end, of course, it depends on what we say incommensurability *is,* how we define it, and how we use it. As Rupert Crawshay-Williams argues, though hardly him alone, meaning depends on context. As he further argues, in somewhat smaller company, argumentation itself is about selecting and building the context that constrains and affords meaning. The appropriately chosen and assembled context can assimilate virtually anything to anything else (think of quantum theory: if the corpuscular and the undular can be assimilated to the same phenomenal story, what can't?). Artful context assembling, on the other hand, can equally distinguish virtually anything from anything else (energy from force, weight from mass, properties from substances, style from content); or, indeed, have anything pass anything else by in the night. The point can be trivially made with set theory: with the right descriptors, any two things can find themselves in the same Venn circle, or in complementary ones, or just in different, unrelated and mutually irrelevant ones. Within the incommensurability debates, Feyerabend's now-you-see-it (if you're a realist), now-you-don't (if you're an instrumentalist) account of incommensurability is a case of assimilation and distinction (Feyerabend 1970, 220). And, if we were required to pick one thing to hang all of incommensurability on, it would have to be context. Meaning variance is semantic variation as a function of context. In the cosmic realm, Hacking notes the importance of context in his remarks about the possibility of adjusting himself to the ideas of Paracelsus: "Either I can't project Paracelus's terms, or I do and drop out of my community" (Hacking 1993, 296). Projection is important, especially in light of cosmic incommensurability. The best sense one can probably make of this category might be to see cosmic incommensurables as perspectives, in the Jamesian pluralist sense: "A Beethoven string-quartet is truly, as some one has said, a scraping of horses' tails on cats' bowels, and may be exhaustively described in such terms; but the application of this description in no way precludes the simultaneous applicability of an entirely different description" (W. James 1984, 30–31). If A is [horses' tails on cats' bowels], and B is [two violins, a viola, and a cello, in musical conversation]—or, say, A is [corpuscular matter] and B is [undular matter]—then perhaps we have a different-world sort of incommensurability. The perspectives metaphor isn't exactly what Kuhn had in mind (the different-worlds situation is diagnosed when scientists "look from the same point in the same direction" but "see different things"—Kuhn 1996, 150), but it is consonant with the linguistic underpinnings he invokes; Whorf said that "anyone who really knows two or more tongues realises that even that small enlargement [. . .] gives him new perspectives" (Kunz and Whorf 1941, 14).

As contemporary microphysics shows very clearly, chickens looking from different directions at the "same" event can certainly be brought to roost commensurably with each other. Two perspectives can be assimilated to the same overall framework.

These procedures—assimilation and distinction on the basis of context—are not new with Crawshay-Williams. "I am myself a lover of these divisions and collections," Socrates lectures Phaedrus on the central instruments of dialectic; "whenever I deem another man able to discern an objective unity and plurality, I follow in his footsteps where he leadeth as a god" (*Phaedrus* 266[b]). What *is* different with Crawshay-Williams, and with others who have explored these maneuvers in the spirit of contingency rather than absolutism (Protagoras, for instance, and Burke, who calls them *merger* and *division* [1969a, 402ff], and Perelman and Olbrechts-Tyteca, who call them *association* and *dissociation* [1969, 190–191 *et passim*]), is the deletion of Socrates's *objective* before *unity and plurality.* Contexts are built for these thinkers, not found. And Crawshay-Williams's treatment also has the great virtue for our purposes of exploring assimilation and distinction at play in domain of controversy.

Crawshay-Williams's instruments have two implications for our project, one at the ground level, one at the meta-level. First, at the ground level of theories and values, any two of them, or any two component parts of them (versions of mass, notions of good), but not only two, might be assimilated to the same context, and therefore made commensurable. This view, effectively, informs the brunt of Shapere's rhetorical question, "How can *any* two things be *completely* incomparable?" (Shapere 1989, 422); one can always, with enough ingenuity and assimilationist motivation, build a context in which some metric applies. But one can also always cut the pie so fine in context-building that any two things can be distinguished, and therefore declared incommensurable on some metric—fingerprints, snowflakes, utterances of the vocable *incommensurability,* even buttons produced from the same machine of the same materials on the same day, are hugely similar, but one can, with enough ingenuity and distinguishionist motivation, find grounds to declare them incommensurable. If nothing else, one can just fall back on a metaphysic that every element of creation is unutterably unique. (In)commensurability, taking this highly Protagorean line, is in the eye of the arguer.[73]

Second, at the meta-level, the context one builds for the notion of incommensurability determines whether it obtains or not; that is, it's not just the "incommensurability of A and B" which is in the eye of the arguer, but the theoretical construct, "incommensurability," itself. Given some definitions, it does not obtain of scientific argumentation.

Given others, it does, but not in any interesting ways. Given yet others, it does, and it is debilitating, though we know those ones don't accord with what we see. Science is not debilitated. It's not even limping.

Crawshay-Williams, no less than us, wants some governors on context construction, one collection of which is empirical. Because we can see that science still operates, we can put aside the incommensurability-is-debilitating-to-science option, leaving "does not obtain" and "obtains with remedies." Crawshay-Williams was a distinctly modernist philosopher, affectionate toward traditional rationality and irritable toward relativism. He was looking for criteria—aren't we all?—which could help decide the legitimate argument/context match-ups. His general tack was to regard virtually all statements as methodological, providing information about the appropriate method of treating something for a given purpose:

> Thus, "It is wasteful to use a steamhammer for cracking pea-nuts" is an explicit methodological judgement in the sense that it is a sort of shorthand for "For the combined purposes of saving fuel and altering the shape of things it is not expedient to treat pea-nuts in the same way as we treat a sheet of steel." [. . .] the statements which constitute the methodology of a particular science are judgements about the methods of research, classification, etc., which are appropriate to that science. (Crawshay-Williams 1957, 5)

Returning to our ground-level implications for a moment, Crawshay-Williams offers the prescription that "when theoretical and philosophical statements are disputed, they should usually be thought of, not as stating facts, but as themselves embodying disguised methodological judgements" (Crawshay-Williams 1957, 4). That is, getting to the root of communicative impediments and uncrossing Kuhnian cross-purposes requires uncovering the tacit contexts of assimilation or distinction behind propositions. For Copernican purposes, one assimilates Earth with Venus and distinguishes both from the sun; for Ptolemaic purposes, one assimilates Venus with the sun and distinguishes both from Earth.

Back up at the meta-level, we need to ask what a responsible application of *incommensurability* is to scientific issues. What purposes are we trying to serve? We know *something* is going on, and it is convenient to have a label for somethings—whether the label is *incommensurability, incommensurateness,* or *macaronics.* The main problem with the Kuhn-and-Feyerabendian word, its metaphors, and its traditional treatment is that they collectively tend to obscure rather than reveal that something. The word, rooting the metaphors in measurement, focuses inevitably on *things.* The geometrical metaphor focuses on shapes and objects, the linguistic metaphor on systems. But the real issues are misunderstanding and disagreement, which concern not things directly, but people's attitudes to things (and to each other), and, socially, the arguments they build around those attitudes. The most useful application of *incommensurability,* in the end, is a name for a condition not of theories (things), but of theorists (people). Focusing on misunderstanding and disagreement does not mean that all species of confusion or belligerence fall into the category of incommensurability, nor does it mean that there are no incompatibilities among theories, that differences in understanding are only a matter of scientists' thick heads or thick tongues.[74] Theories differ, propositions clash, and wrestling with alternatives is the guiding preoccupation of science. But they clash in relatively straightforward ways. A proposition from A might contradict a proposition from B (or, indeed, another proposition from A), it might compete with a proposition from B (ditto), or it might do both (ditto redux). (Propositions might also pass in the night, in which case they are mutually irrelevant and the word *clash* isn't appropriate, though someone in the back row might murmur something about brick-wall incommensurability.) What else is there?

These clashes are usually resolved, even when there are serious semantic and/or pragmatic misalignments. The classic study of such resolution is Martin J. S. Rudwick's *Great Devonian Controversy,* in which he meticulously charts out the positions, usages, and allegiances among a handful of Victorian scientists shaping a new and enduring picture of the geological record "on the anvil of heated argumentative debate" (Rudwick 1985, 455). As Rudwick is at pains to point out, his study engages a rather typical scientific episode—conflicting interpretations that resolve into a single interpretation (or set of related interpretations). It is not a battle between, say, I^A and I^B, in which one emerges triumphant, but among I^1, I^2, I^3, . . . I^N, some of which form part of the initial conditions, some of which arise in the course of debate, all of which overlap substantially in mechanisms, meanings, and values, and none of which emerges whole from the agonistic comparative struggle. The weaker interpretations fall away, sacrificing any virtues and any associated data they might have that can be assimilated to the general framework (which in turn is altered, sometimes slightly, sometimes substantially, in the Devonian case moderately, in the course of those assimilations). The stronger interpretations amalgamate, their vices and any recalcitrant data getting distinguished away, and while the emergent, consensual interpretation often wears the terminology of only one of the candidate interpretations, its components always come in part from the others. These typical, quickly forgotten episodes have little or no whiff of the incommensurable about them, yet they have all the ingredients found in the standard diagnoses of incommensurability clashes, except perhaps for the scale and publicity of the disagreements.

Competitions and contradictions, that is—even when they are subtle, partial, or dependant on long entailment chains, and have to be dug out of larger matrices—offer sites of comparison, negotiation, and resolution. They can be dropped, modified, augmented by other propositions, or held to steadfastly as the theorists go their separate ways, but it is not the propositions, not the theories, that determine which of those options are followed; it is the comparing, negotiating, and resolving agents—factors that Feyerabend and especially Kuhn introduced along with the incommensurability problem—the theorists. When they disagree in certain ways, we get crises, revolutions, diagnoses of incommensurability. When they don't, we don't. It is important to notice that not all disagreements, not even all violent disagreements over rival theories, are incommensurable in this way. For one thing, the notion of incommensurability requires a modicum of goodwill and practical reason to be sensible. To say that hostile, unreasonable, or just plain stubborn people disagree unto death is trivial. Meanings, values, or practices—that is, semantic and/or pragmatic misalignments, in various degrees—are at the heart of those incommensurable disagreements and their attendant misunderstandings.

If we choose to identify those sorts of disagreements with the use of *incommensurability* (rather than reserving that word for theory-laden perceptual discontinuities, for instance, or for total conceptual misalignment, or for the claim that scientists can't compare theories), then we have a problem field where arguments matter. We needn't throw our hands up in the horrified belief that incommensurability negates rationality, requiring us to abandon one or the other. If incommensurability is a syndrome, then rhetoric—the search for ways to adjust ideas to people, people to ideas—is the cure.

There are Remedies, Rhetoric

> Rhetorical theory and experience both show that arguments over incommensurable values are not impossible or pointless; experience also shows that they are a difficult sort of argument that most people work hard to avoid.
>
> —Eugene Garver, "Arguing over Incommensurable Values"

Both Kuhn and Feyerabend implicate rhetoric in incommensurability, to different degrees, though neither of them in any informed way. Kuhn occasionally uses words like *rhetoric* and *persuasion* to describe the sorts of things scientists are up to (e.g., 1996, 198–204), but never gets much beyond the ordinary language sense of those terms. And he often says things that dovetail very closely with the interests of rhetoricians. "One can deny," for instance, "as Feyerabend and I do, the existence of an observation language shared in its entirety by two theories and still hope to preserve good reasons for choosing between them" (2000, 127); one of the defining themes of contemporary rhetorical theory, developing out of an important paper by Karl Wallace, is the concern for good reasons, what they are, how they work, and why (K. Wallace 1963). Of course, Kuhn's use of "good reasons" is entirely coincidental with Wallace's use, but the overlap in concerns is difficult for a rhetorician to miss.

Feyerabend is a much more interesting case. He frequently talks in ways that make most rhetoricians' hearts race appreciatively:

> Learning to argue is part of learning how to get along with people. One does not study the 'nature of man'; one studies individual people and learns the many ways of living with them, arguments included. One learns how to adapt one's persuasion to the idiosyncrasies of the person one confronts rather than to an abstract creature 'rational man.' There is no distinction between logic and rhetoric. (Feyerabend 1981b, 6)

Or, here he is, in an explicit disavowal of a monotonic positivist view of language that leads to the sort of alien-organisms counter-argument against incommensurability that Putnam offered:

> If discourse is defined as a sequence of clear and distinct propositions (actions, plans, etc.) which are constructed according to precise and merciless rules, then discourse has a very short breath indeed. Such a discourse would be often interrupted by "irrational" events and soon be replaced by a new discourse for which its predecessor is nonsense pure and simple. If the history of thought depended on a discourse of this kind, then it would consist of an ocean of irrationality interrupted, briefly, by mutually incommensurable islands of sense. (Feyerabend 1999, 32–33)

The richness and plasticity of discourse that this view endorses is exactly what rhetoric has always operated under. But Feyerabend goes further yet, explicitly pledging allegiance, in strategic places, to rhetoric—as in the index entry for *rhetoric* in the first edition of *Against Method* ("1–309;" the book is 309 pages long, excluding the index) and in his invocation of Protagoras as an intellectual antecedent (Feyerabend 1981a, 1981b, xiv). He even says that the principal function of scientists is the one that got the sophists into so much trouble, making the weaker case the stronger (Feyerabend 1978a, 30). These various moves, however, appear directed more at the kidneys of philosophers than the hearts of rhetoricians; their design is to antagonize rather than to endear.[75]

But, whatever their knowledge of, or allegiance to, traditions of rhetorical research, both of them were critically important sponsors (unwilling in Kuhn's case) for the science-studies movement that Richard Rorty labeled "the rhetorical turn" (Simons 1990, vii), a diverse movement that includes philosophers, historians, and sociologists.[76] The work all of these folks do is incredibly varied, as are their instruments, commitments, and goals, but it comes together in a focus on discourse from the perspective of suasion, aesthetics, and argumentation outside the varieties traditionally sanctioned by philosophy. Within the field of rhetoric the corresponding turn might better be called, "the scientific turn," but rhetoricians are overly fond of one particular preposition, so the label is "rhetoric of science."[77] No one single concept can be charged for the general movement or the specific sub-field, but if we wanted to choose only one, as a metonym for all of the issues and interests feeding into the rhetorical investigation of science, the hands-down winner would be incommensurability.[78]

Incommensurability is the one idea in post-positivist philosophy of science that is seen as most corrosive to rationality and to progress. Incommensurability is the biggest intoxicant in Feyerabend's brew, leading him into declarations like:

> Reason, at last, joins all those other abstract monsters, such as Obligation, Duty, Morality, Truth and their more concrete predecessors, the Gods, which were once used to intimidate man and restrict his free and happy development: it withers away. . . . (Feyerabend 1978a, 180; Feyerabend's faux-elision)

Incommensurability is the notion that led Kuhn, effectively, to put out his call for "a sort of study that has not previously been undertaken," one which investigated science in terms of "techniques of persuasion [. . .] argument and counterargument in a situation in which there can be no proof" (Kuhn 1996, 152). He did not mean for this new study to be a general rhetorical turn, which is clear from his reaction especially to the sociology of scientific knowledge (e.g., "absurd [. . .] deconstruction gone mad"—Kuhn 2000, 110), nor a specific rhetoric of science, which is clear from his reaction to me (we corresponded briefly by e-mail in the early 1990s; he was courteous, even encouraging, but couldn't see what rhetoric could contribute to understanding science, or what it had to do with him). But he got them anyway. Incommensurability was an invitation for the sort of analysis that looks to arguers and the processes of arguing far more than to the products of their arguments—sociologists and rhetoricians were natural candidates to take up the invitation, and many did, the first seeing incommensurability in terms of interest cohorts, the second in terms of suasive argumentation; a noteworthy contingent of historians, a discipline of practiced hermeneuts, also responded.

Since it does not cause the practical problems one would predict for science; since communication is hindered and confounded, perhaps, but not blocked; since evaluation and comparison are difficult, perhaps, but not impossible—incommensurability is an invitation to explore how scientists get their jobs done in the face of misalignments and discontinuities in meanings, themes, and practices. The most interesting cases, the ones that seem to fit the situations where Feyerabend and Kuhn's term, *incommensurability,* applies, occur when there is no significant perception of those misalignments and discontinuities, where the disputants think they agree about terms and referents but don't—along the lines of what Charles Arthur Willard calls "incommensurability by spurious agreement" (Willard 1989, 7). Certainly very few scientists in the midst of a dispute will go very far with "I don't understand what Jones is saying." Rather, it is more on the order of "Jones can't really believe what he is saying" or "Jones has missed the point," or "Jones is a chowderhead."

Scientists do in fact say, "I don't understand Jones," of course, and they say it rather frequently in disputes of the incommensurable sort. But it rarely means exactly that. It is almost always a polemical maneuver. Scientists usually believe they understand other scientists quite well, thank you very much, and claim not to understand one another only when it reinforces an assault for incoherence or absurdity or irrelevance. In effect, the claimant pretends to take the onus on himself (coupling it with an often covert ethotic appeal that he is rational, knowledgeable, well intentioned) while launching an *ad hominem,* so that what looks like "I don't understand Jones" is really "Jones is not understandable (even by rational, knowledgeable, well-intentioned listeners)."[79] But, polemical maneuvers aside, for the moment, there *is* misunderstanding among scientists, about core issues; misaligned meanings, themes, and practices could not but breed it.

The pursuit of ways to align (or realign) meanings, themes, and practices is a major focus of contemporary rhetoric. One of the more influential definitions of *rhetoric* (they are legion) in the last century's academic revival of suasion studies is Donald C. Bryant's "the *function of adjusting ideas to people and people to ideas*" (Bryant 1953, 423). Exploring that function is the drive behind Wayne Booth's rhetoric of assent (Booth 1974), behind Chaim Perelman and Lucie Olbrechts-Tyteca's search for grounds of adherence (Perelman and Olbrechts-Tyteca 1969), behind I. A. (Ivor Armstrong) Richards's preoccupation with misunderstanding (Richards 1936), behind Karl Wallace's program of good reasons (K. Wallace 1963).

(Do me a favor, will you? Don't ask "What is a rhetorician?" That way I can avoid confessing that I.A. Richards primarily studied philosophy, that Chaim Perelman was a philosopher and jurisprudent, that Lucie Olbrechts-Tyeteca's training was in literature, social sciences, and economics, and that Wayne Booth is professionally a literary critic. I also wouldn't have to answer uncomfortable questions about Aristotle, or why even Saint Isocrates called his field *philosophy,* and penned an oration entitled "Against the Sophists." Donald Bryant and Karl Wallace, I would have a chance to note proudly, however, were Professors of Speech.)

Much of most of these efforts has an unfortunately negative cast to it, diagnosing aliments rather than prescribing cures, but that cast, much like the critical reasoning focus on fallacies, is inevitable; remedies are rare when there is no theory of causes. The way to proceed with communicative breakdowns of this sort is to chart the misalignments, to catalogue the misunderstandings, to find the leaks so you can gum them up. Booth's and Richards's diagnostic heuristics also have the virtue that one often just needs to flip them on their heads to find the remedies. On the level of values, for instance, Booth's One Rhetorical Purpose Dogma is the belief that argumentation has only one function, to win (Booth 1974, 77). People under the spell of this dogma—and it is a rare scientist, just as it is a rare human, who does not follow it at least some of the time—view every rhetorical event as an opportunity to crush, lest they be crushed themselves. The remedy? Revoke the dogma, recognize the importance of reciprocal suasion. Cooperate. Negotiate. Get along.

The same procedure works for Richards's One Proper Meaning Superstition, which also operates on the pragmatic level of values but strongly implicates the semantic level: the belief that "a word has a meaning of its own (ideally, only one) independent of and controlling its use and the purpose for which it should be uttered" (Richards 1936, 11). People who hold to this semantic idolatry—and again, whatever we may profess, who doesn't often hold this one rather deeply, at least about cherished words?—look on deviations from the One Proper Meaning of a word not as differences in usage or interpretation or tone or purpose, but as error. If Jones uses *planet* for something you don't use *planet* for, that doesn't manifest a difference to be negotiated with Jones, but an error to be eradicated. "The word," Sapir says, "is not only a key; it is also a fetter" (Sapir 1921, 17).

We hear once again, that is, that meanings are theory laden; they can severely hinder our capacity to see over the fence of our theory (system, framework). Most users of technical language operate under the assumption of univocality, the One Proper Meaning Superstition. But technical terms, no less than ordinary-language terms, "summarize an aggregate of acquired knowledge, rules, and conventions" (Perelman and Olbrechts-Tyteca 1969, 99). In Crawshay-Williams's version of this superstition (he was influenced by Ogden and Richards), the focus falls on context and purpose in a way that makes its relevance both to science and Kuhn's "cross-purposes" clearer:

> The [interlocutors] fail entirely to notice that they have got their purposes crossed, and that in consequence they fail to realise that context and purpose have any relevance to the discussion at all. What then happens is that they both assume that there is one so to speak universal context. And, naturally, they each take for granted that, since their own statement is current in what they assume to be the universal context, the other's statement must be incorrect—so to speak 'universally.' (Crawshay-Williams 1957, 39)

The remedy? Suspend the superstition. Cooperate. Negotiate. Get along. Recognize that the way you use a term may not be precisely and only the way it is ever used; indeed, may not be entirely precise at all. The move here comes down to (1) acknowledging meaning variance and theory-ladenness (that is, context and purpose), and (2) denying privilege (if not affection) to the terms laden by your own particular theory.

As the recurrence of *One* in Booth's and Richards' ailments should indicate, large-scale communicative breakdowns frequently involve a lack of perspective, an inability to see out of your own back yard, which shows perhaps most closely that cosmic and pragmatic incommensurability differ almost entirely in degree. But very frequently this incapacity to see is chosen, or at least settled for. With a little motivation, almost anyone can peer over the fence, if they genuinely want to share someone else's perspective, and are willing to put in the work. As rhetor, one needs to strive for clarity, coherence, relevance, truth; as audience, one needs to strive for charity, empathy, generosity, truth. In short, attitude is more important, by far, than maxims.[80]

Historians have typically worked in terms of blocks—pre-war and post-war, decades and centuries, eras and ages—a tendency that has strongly influenced history of science, and in the process given Feyerabend and Kuhn sites for much of their incommensurability data, discontinuous blocks with labels like "Copernican" and "Ptolemaic" astronomy, "Aristotelian" and "Newtonian" mechanics, "classical" and "quantum" physics. Of course, history *does* chunk like that, conveniently, for generalizations about big political or social or belief structures. But the transitions are often both longer and more piecemeal than these blanketing labels tend to suggest. In science, these longer, lumpier changes make fictions of the "Aristotelian"-talking-to-a-"Newtonian" sort substantially misleading. Moreover (returning expressly to our remedy theme; sorry for the seeming detour), these extended transitions show more evidence of people striving for clarity, coherence, and relevance, for charity, empathy, and generosity—for consensual, rather than evangelical, truth—than the block-periodization incommensurability accounts would indicate.

Peter Galison, a historian who especially stresses the piecemeal nature of transitional development, prefers an anti-block model he calls "intercalated periodization," with theoretical scientists, experimental scientists, and instrument-manufacturers, all proceeding at different paces, with various levels of reciprocal engagement, mutual support, internal consensus, and so on, helping to determine those paces. (This account might be elaborated somewhat with the addition of further interest groups, such as policy makers, journalists, and exogenous scientists.) The image that comes along with intercalated periodization is

> an irregular stone fence or rough brick wall rather than [in contrast to the relevant block periodization image] adjacent columns of stacked bricks [. . .] just as the offsets between joints in a brick wall give the wall much of its strength, it is this intercalation of diverse sets of practices (instrument making, experimenting, and theorizing) that accords physics [Galison's chosen exemplum] its sense of continuity as a whole, even while deep breaks occur in each subculture separately considered. (Galison 1997, 19)

Looking at incommensurability issues in light of this image, far removed from individualist gestalt switches or blinding conversions, brings two matters quickly into sight: since the various groups operate with overlapping but not identical vocabularies and themes, and with possibly quite disparate practices, we have ready sources for the growth of misaligned meanings, themes, and practices; and since the groups have interdependencies that could lead naturally to competitive marketing for specific audiences ("our hypothesis is richer than theirs, because . . ."), we have ready sources of strategies for theory comparison. Whenever there are rival proposals, we have a situation of comparative marketing already, for audiences of grad students, post-docs, and junior faculty, and for granting agencies, sometimes for editors and referees. But Galison's picture—developed through a very detailed case study of high-energy particle physics—shows how the rhetorical demands of multiple, allied but distinct, audiences and interlocutors are daily present in the lab. Given that allied-but-distinct groups manifest the sorts of misalignments characteristic of incommensurability, it is worth asking how they meet those demands. At the pragmatic level, they meet them by dropping the One Rhetorical Purpose Dogma; at the semantic level by, if not dropping, then certainly confronting the Proper Meaning Superstition, by facing up to theory-ladenness.

Galison recurrently describes situations in which various groups use terms "in such heterogeneous ways that on the face of it carry such different meanings that we might expect to locate them in different and incommensurable conceptual schemes," yet, surprised, he finds passionate cooperative engagement. "The actors worked furiously," as he describes one situation, "to coordinate and adjudicate among alternatives" (Galison 1997, 814). Galison's circumstances, in fact, do not fit the profile of classic incommensurability encounters. His encounters bring together groups of experimenters, instrument-makers, and theorists. They are shared-interest, separate-sphere encounters. Rival-theorist disputes, on the other hand, are competing-interest, same-sphere encounters. But, whether we reckon the camps classically incommensurable or not, the sort of aggressively reciprocal suasion Galison's cohorts engage in: (1) is in principle equally available to rival theorists; (2) can only operate at least partially outside the argue-exclusively-to-win dogma; and (3) satisfies what Feyerabend and Kuhn had said from the beginning one finds by looking at practicing scientists, that incommensurability is not a problem. Indeed, Galison, though he is anxious to refute Kuhn and especially Feyerabend (see, in particular, Galison 1997, 812–816), provides a useful addendum to their diachronic accounts, accounting for disciplinary continuity in the face of theoretical discontinuity. Rival theorists do, in fact, engage in reciprocal suasion, though its effects are frequently covert (which we turn to in the next section, "The Persistence of Incommensurability"). Even working outside an attitude that argues only to win, how does one get round the hampering effects of vocabularies laden by different theories?

Thomas Russman's response to theory-ladenness is helpful; wrong, but helpful. "There is always a neutral observation language available," he says, "lying under the biased one" (Russman 1987, 2). Russman is confused over how deep the value saturation of language runs (all the way down), but he points in the right direction for counter-balancing and calibrating the semantic effects of those values, especially as manifest in specific terminology: beneath the layers of discernible bias. The only way to overcome semantic bias is to detect it, to suspend the One Proper Meaning Superstition, and to call on the constrained resources of the very language that instantiates that bias. That, in the event, is what Galison finds, and puts into the terms of another revealing analogy about the communicative activities in his interstitial trading zones. When groups with different languages (and conceptual schemes) share interests—forced interests because they are thrown together in circumstances of slavery or indentured labor, or natural interests because they each have something the other wants—they develop stripped-down communicative instruments called *pidgins.* Pidgins are what you get when you pare languages of most of the grammatical subtleties from which Whorf draws his linguistic relativity arguments, ending up with small vocabularies, minimal morphology, and a highly restricted, nuance-constrained, information-driven syntax (statements, questions, and commands).

In a partial but effective use of this vehicle, Galison develops the concept of an interlanguage—effectively, it's an attenuated lexicon, formed at the nexus of the users' shared linguistic resources.[81] It has one crucial similarity to a pidgin, but is not used in the manner or the circumstances in which pidgins are used. With a pidgin, that is pretty much all the relevant groups have for communication (augmented, as in most communicative encounters, with gestures, any germane physical objects that might be present, and so on). But the groups that concern Galison (*subcultures,* he calls them) share a meta-language or two (English and German in his case study). So their syntactic and morphological resources have all the riches of a full-fledged language, or two. They also share technical vocabularies and syntaxes—weights, measures, mathematical operations, that sort of thing. The unique interlanguage component of the communicative medium that Galison identifies is a specific vocabulary assembled out of terms drawn from the registers and jargons of the various groups, but in a semantically stripped down form, in order to reduce ambiguity and optimize agreement.

This vocabulary reduces the implications those words have for the respective conceptual schemes of the negotiators, and therefore simplifies one party's beliefs and practices for presentation to the other—which is the critical similarity between an interlanguage and a pidgin. That is, the restricted vocabulary works by ignoring the One Proper Meaning Superstition and building new, reduced, common meanings out of the resources of the reciprocal rhetors in order to get their business done,[82] perhaps by focusing on just those words that Kuhn identifies as epitomizing the taxonomic structure of a theory. For instance, "a trading language in terms of particles, their energies and motions" was deployed in the late 1940s, Galison tells us, that "functioned by employing neither the high theory of Dirac or Yukawa equations nor the experimental details of expansion chambers and optics [. . . binding] together the various subcultures of microphysics" and allowing them to build the field cooperatively, despite markedly different commitments and different understandings of key concepts. Take electrons, for instance:

> Electrons, for the cloud chamber experimenter, were inseparably bound to characteristic visual appearances—of the tight spirals of delta rays, of the striking offset helices associated with synchrotron radiation, of the minimum ionizing tracks, of a particle with specific mass, spin and charge. [. . .] For the theorist [. . .] an individual electron was an entity with no fixed and recoverable identity as it passed some of its life as virtual versions of other objects, with no fixed charge and mass because those quantities were momentum dependent; the whole representational scheme was based formally on the new calculus of renormalization, not a whiff of which could be scented in the atlases of cloud chamber photographs. (Galison 1997, 134)

Galison's subcultures may or may not qualify as incommensurable. Certainly they don't fit the classic profile of mutual hostility, even with the conceptual misalignments Galison highlights. But look at how different their concepts of the 'same' crucial theoretical constituent are. Given conceptual misalignments on this scale, it does not seem a stretch to suggest their negotiative strategies would be available to parties that *do* fit the classic profile. *Neutral* languages are unavailable, granted. But that doesn't mean carefully *balanced* and strategically *attenuated* linguistic resources cannot be fashioned as needed, by reducing the involvement of obstructive values. In Lyotard's *différend,* for instance, the unsatisfactory resolution is that the dispute gets articulated in one of the disputing phrase regimes, sublimating the other. In a sense, Galison's notion of interlanguage depends on mutually sublimating the conflicting elements of *both* phrase regimes, and building a resolution on what is left. (Lyotard would call this a *litigation*—one side may well lose, but it is in terms that both phrase regimes can articulate, both parties comprehend.)

In their essay in this volume, Charles Bazerman and René Agustín De los Santos explore a case with very interesting parallels to the ones Galison and Rudwick chart. They chronicle the 'revolutionary' emergence of ecotoxicology out of the social and intellectual resources of toxicology, a case where Kuhn would lead us to expect tell-tale symptoms of incommensurability, but where signs of incomprehension and hostility are significantly muted, and the order of the day is inverse. The parties engage in mutual communication responsibly and successfully. They employ each other's respective entities and measures as needed. They get along. Bazerman and De los Santos use a quite different set of theoretical tools than Galison or Rudwick (they deploy a sociocultural, interpersonalist, pragmatist rhetorical approach influenced by thinkers such as Volosinov, Vygotsky and Mead). But they arrive at much the same point.

Looking at only one scientist shaping his work out of the materials he inherited, John Angus Campbell (in this volume) arrives at that point as well, or one close to it. Campbell traces Charles Darwin's development, from a deeply Lyellian geologist, one who took the anti-evolutionist *Principles of Geology* as the equivalent of scientific scripture, to the (co-)inventor and overwhelmingly most powerful advocate of natural selection; from anti-evolution to evolution. From almost any conventional vantage, *Principles of Geology* and *Origin of*

Species are incommensurable texts. But Darwin's inventional path—a swerve here, a completion there, an overall hermeneutic reconfiguration—commensurates them. Again, the tools are different. Campbell works out of the insights of scholars in three different fields (literary criticism, rhetorical theory, philosophy of science). But the moral is the same. Incommensurability is surmountable.

The cooperative phenomena that Rudwick, Galison, and Bazerman and De los Santos discover may be more common than the hostility, incomprehension, and unshakeable obstinacy that characterizes other apparent incompatibilities. The accommodation of new developments to prior concepts and frameworks that Campbell finds in Darwin may be more common than the tear-it-down-and-start-again rhetoric often associated with revolutions. Moreover, this cooperation and accommodation is largely based on . . . a common measure. Scientists, as Edwin Hubble put it, judge theories "by a single criterion—do they work?" (Hubble 1954, 12). *Work*, of course, is a deeply value-saturated word, with different implications for different scientists, shaped in various ways by different conceptual, empirical, disciplinary, and personal allegiances, but it is a commensurability nexus for all that, the solidest point of negotiation in the sciences.

We do, nonetheless, see evidence of incomprehension, obstinacy, rejection, and hostility in science. Incommensurability is surmountable, but it's not always surmounted. Why?

The Persistence of Incommensurability

> In Italy for 30 years under the Borgias they had warfare, terror, murder, bloodshed—they produced Michelangelo, Leonardo da Vinci and the Renaissance. In Switzerland, they had brotherly love, 500 years of democracy and peace and what did they produce? The cuckoo clock.
>
> —Harry Lime, *The Third Man* (Orson Welles)

Why incommensurability is sometimes not surmounted has a range of possible answers. Leah Ceccarelli 's essay in this volume, for instance, outlines a Ciceronian view of civil argumentation (reciprocal suasion) that is up for the challenge of paradigmatic conflict, but then charts a scientist (E. O. Wilson) who steadfastly resists the possibilities of such argumentation, operating under an agonistic, all-or-nothing model of scientific argument (that is, he follows Booth's One Rhetorical Purpose Dogma). Some scientists simply want to conquer, at the expense of achieving the sort of consensus that might advance their theories more effectively. The failure here can be located in the personal.

Lawrence Prelli's contribution finds a number of commensurability points in a controversy between family/spousal violence researchers, but finds that nonetheless there have been only marginal and unsuccessful attempts to bring the sides into alignment. The failure here is social and institutional, in the competition for the same resources (grants, adherents, prestige, employment). As he also notes, moving from that specific dispute to the general level of rhetorical strategy:

> once controversy becomes polarized, the disputing sides argue as though the issues dividing them are ultimately resolvable by proving the preferred perspective conclusively right and the adversary's perspective conclusively wrong. Such argumentation presumes that there are standards—neutral or otherwise—which, once applied, would definitively settle the issues. Those presumed standards usually turn out to be the preferred values and assumptions of one or the other clashing perspective. That sort of argumentation is not without rhetorical advantages, but any advantages are purchased at the cost of foreclosing prospects for *commensurate deliberation* about divisive issues (Prelli 1996, 424). As Mann put it, strategies that "polarize," "de-legitimate," and "discredit" fail to address or "resolve issues" (Mann 2000, 13). (Prelli this volume, 326)

The failure in these circumstances can be located in the centrifugal pressures of dialectical exchange that force disputants apart rather than bringing them together.

Alan Gross's contribution tells the story (in brief) of David Brewster, a physicist of considerable talent who assimilated wave theory to selection theory, while the rest of the community abandoned selectionist accounts altogether. Conceptually, in terms of theory construction, the assimilation was very successful, but "the price" Gross argues, was "prohibitively high" (189)—accommodation, effectively, for the sake of accommodation, which was necessarily ad hoc and not at all generative of future research. The failure here can be located in the respective Kuhnian lexicons of selection theory and wave theory. They can be brought together, but not fruitfully.

To say that A and B can be made commensurable, that is, does not mean they should, or even can, work together. The bringing together might well just precipitate a choice between them, a bet placed on the one that scientists feel most likely to be epistemically (and socially, and professionally) most successful. But different scientists will make different bets. There are broad social-epistemic reasons for failing to resolve conflicts partaking of incommensurability. Feyerabend, in any case, certainly thought so.

There is an irony, that is, in looking for remedies to incommensurability that should not be missed; for Feyerabend, incommensurability itself is a remedy. Ever the philosopher, he eagerly dispensed prescriptions for how to think and act. As early as "Explanation" he saw "a whole class of mutually incompatible and factually adequate theories" as "a most potent antidote" against hegemonic dogmatism (Feyerabend 1962, 69; 1981a, 74), and much of his later career develops this theme relentlessly. He came increasingly to view (traditional notions of) rationality, progress, and the putative growth of knowledge as misfortunes, incommensurability as central to the cure, and advocated an agonistic, anarchic epistemology, a sophistic ecosystem of knowledges in an explicit counter-hegemonic position to a Platonic edifice of Knowledge:

> [Epistemology] so conceived is not a series of self-consistent theories that converges towards an ideal view; it is not a gradual approach to the truth. It is rather an ever increasing *ocean of mutually incompatible (perhaps even incommensurable) alternatives,* each single theory, each fairy tale, each myth that is part of the collection forcing the others into greater articulation and all of them contributing via this process of competition, to the development of our consciousness. Nothing is ever settled, no view can ever be omitted from a comprehensive account. [. . .] Experts and laymen, professionals and dilettanti, truth-freaks and liars—they are all invited to participate in the contest and to make their contribution to the enrichment of our culture. The task of the scientist, however, is no

> longer 'to search for truth' or 'to praise god' or 'to systematize observations,' or 'to improve predictions.' These are but the side effects of an activity to which his attention is now mainly directed and which is '*to make the weaker case the stronger*' as the sophists said, *and thereby to sustain the motion of the whole.* (Feyerabend 1978a, 30)

Incommensurability may not be required by this picture—which, aside from the prescriptive optimism of "forcing the others into greater articulation" and "*sustain the motion of the whole,*" is a blueprint for postmodernism—but it clearly nourishes Feyerabend's anarchic epistemology, a source of wholesome friction against the consistency and homogenization he sees as inherent curses and postmodernism sees as the Satanic Mills of the Meta-narratives. In Kenneth Burke's terms, it is dialectical in his "general" sense, a mutually reinforcing unconscious collective effort, manifesting now "the competition of coöperation," now "the coöperation of competition" (Burke 1969a, 403); an enterprise in which "the coöperative competition of *all* the voices [. . .] modify one another's assertions, so that the whole transcends the partiality of its parts" (Burke 1969a, 89). "It is a deep-seated part of intellectual structures," Randall Collins notes in the *Sociology of Philosophies,* "that questions are asked, debates take place; polemics and denunciations also often occur, in a circulating structure that resembles equally the kula ring, the potlatch, and the vendetta" (Collins 1998, 28).

If we calibrate Feyerabend's picture with Burke's insistence on cooperation (which is not missing from Feyerabend, just sublimated), we get a picture of scientific knowledge-making as personally and collectively a competitive enterprise, in which the competition helps ensure overall robustness by requiring that rival proposals bolster themselves to weather each other's hostile attacks—precisely what close studies of science usually reveal. In his meticulous study of microphysics, for instance, a study that stresses negotiation and common ground, Galison remarks on "the incessant drive on each side alternately to defeat and encompass the other" (Galison 1997, 26). In his similarly fastidious examination of nineteenth-century geology, studying an episode chosen especially because of a successful resolution which left virtually no trace of the agonistic contest it resolved, Rudwick draws this moral: "scientific growth may derive from, and even require, intense conflict and controversy" (Rudwick 1985, 27). David Hull's close charting of the specific play of "cooperation and competition, secrecy and openness, rewards and punishments" (Hull 1988, 3) in evolutionary biology and systematics returns a number of general findings.

> For instance, one thing that is true of science is that it takes time. Any scientist who is not willing to put in the hours formerly reserved for factory workers in Victorian England is not too likely to succeed. Strong motivations are needed to induce such protracted labours. Scientists acknowledge that among their motivations are natural curiosity, the love of truth, and the desire to help humanity, but other inducements exist as well, and one of them is to "get that son of a bitch." Time and again, the scientists whom I have been studying have told stories of confrontations with other scientists that roused them from routine work to massive effort. No matter what the cost, they were going to get even. (Hull 1988, 160)

On a somewhat broader canvas, Herbert Butterfield's narrative of *The Origins of Modern Science,* a fairly conservative set of lectures delivered in 1948, chronicles that an atmosphere of "healthy friction" among rival Aristotelian positions set the stage for the emergence of modern scientific practices, themselves a mélange of "variations, uncertainties, controversies, and developments," a multi-party contest of "rival explanations" and "rival systems," replete not only with truth-freaks like Descartes and liars like Priestly and fairy-tale collectors like Boyle, but with the catalytic participation of men who were not the direct source of new hypotheses or procedures but the irritant for them, like Marin Mersenne, "not himself a great discoverer, but a central depot of information and a general channel of communication—a man who provoked enquiries, collected results, set one scientist against another, and incited his colleagues to controversy."[83]

In short, science is a rollicking Feyerabendian, Protagorean epistemic carnival, in which incommensurability—misaligned meanings, practices, and values in situations of reciprocal hostility—helps to keep the various groups agitated. It is also dull, dogged, drudgery much of the time, long taxi-driver hours with taxi-driver tedium, which might be enough on its own to account for the punctuation that bursts of passion introduce. But it is certainly both cooperative and competitive, and incommensurability is instrumental in both.

I don't really propose to answer in any depth the question I casually bridged into this section with—Why is incommensurability, which is surmountable, not always surmounted?—a question that is I suppose, ultimately for social psychology to answer. But the shallow answer is simply that it works. Open hostility helps generate knowledge, so can covert hostility, and wallowing in incommensurability sustains hostility. It begins, often, in the intensely cooperative flip side of incommensurability, the remarkably effortless intra-group communication that can grow in collectives, especially in small, tight collectives with a mission—extremely high agreement quotients, based on strongly shared meanings, practices, and values. As Jerry Fodor recalls about early transformational-grammar research in Building 20 at MIT,

> We shared a general picture of the methodology for doing, not just linguistics, but behavioral science research. We were all more or less nativist, and all more or less mentalist. There was a lot of methodological conversation that one didn't need to have. One could get right to the substantive issues. (qtd. in R. A. Harris 1993a, 68)

Such groups depend on a balance, "the right kind of insularity with the right kind of homogeneity," as Malcolm Gladwell puts it (Gladwell 2002, 106), generating "conversation, validation, the intimacy of proximity, and the look in your listener's eye that tells you you're on to something" (Gladwell 2002, 105)[84]. It is a commonplace of communication studies generally and rhetorical studies specifically that the more commitments you share with someone, the easier it is to get your message across, and get it accepted, what Grandy (1983, 18) calls the "relative ease of agreement."[85] A great deal can be taken for granted in such circumstances. The members of such groups are on the same general wave length, and many of the same specific wave lengths, but not all, and meanings-out-of-phase can be incredibly synergistic (which might be part of what Feyerabend meant by "without a constant misuse of language there cannot be any discovery, any progress"—Feyerabend 1978a, 27).

Geoffrey Pullum—a linguist from a different, somewhat later, collective than Fodor's, and somewhat antagonistic to it—illustrates this meaning-out-of-phase synergy extremely well in a *conte a clef* of his collaborations with Gerald Gazdar and Ivan Sag in the development of Generalized Phrase Structure Grammar:

> At some point Pillock said something about PP [prepositional phrase] distributions and subscripts that he immediately realised was embarrassingly incorrect and anyway irrelevant, but he was misunderstood by Gander, who said, apparently about a different idea, "Hey, that's a thought, they only clash when . . .", and straightaway Sachs was on to something. [. . .] Pillock didn't get it. Gander and Sachs explained to Pillock what they were thinking of doing, though this was made difficult by the fact that each had a different idea in mind. It never became fully clear who had thought of what, or whether Pillock's remark had stimulated Gander to notice something that was the germ of what Sachs had incorrectly based on what he interpreted Gander to be assuming. But as they sat and worked together, collaboration worked its odd magic, and an idea did take shape. (Pullum 1991, 207)

This parable manifests what Oppenheimer described (of the incessant arguing-in-the-dark characteristic of early twentieth century physics) as "explaining to each other what we don't know" (Cline 1965, 126; see also Hull 1988, 7). Pullum's glimpse of intellectual slapstick is meant, of course, to raise a chuckle. But it's also meant to suggest that something mysterious happens when people work together to solve problems they only partially understand. And it does. But the mystery of apparent communicative bumbling is not quite as deep as it looks. The relevant ingredient is what Polanyi calls *conviviality,* a kind of instinctual cooperative generosity (Polanyi 1958, 203ff). Misconstruals in an atmosphere of hostility—when each side dons the "mental blinkers [that] make us take another man's words in the ways in which we can down him with the least trouble" (Richards 1936, 25)—move in the direction of ascribing error. Misconstruals in an atmosphere of conviviality—when a very different set of blinkers lead us to take another person's words in the ways we can build on them as indulgently as possible—move in the direction of ascribing insight.

But Pullum's narrative point of view also misrepresents the process somewhat. The centre of consciousness in the story is his alter-ego, Pillock, who "realises" he has made an embarrassing and irrelevant error. He hasn't. What Pillock has done is produce an indeterminate utterance, one that Gander happens to determine in a way Pillock didn't foresee. Meaning is not intention; meaning is collaboration; dialogic, not monologic. "Science is a conversation with nature," David Hull says, "but it is also a conversation with other scientists," and it is "not until scientists publish their own views and discover the reactions of other scientists [that they] can [. . .] possibly appreciate what they have actually said." Hull's model here is unduly literate, but if we take *publish* in its older, broader sense of 'make public,' it is spot on. "It is impossible," Hull goes on, now that we've given him a wider canvas, "to avoid all possible misunderstandings, and not all such misunderstandings are plainly 'misunderstandings.' Frequently scientists do not know what they intended to say until they discover what it is that other scientists have taken them to be saying" (Hull 1988, 7).

When meanings-out-of-phase meet values-out-of-phase, in an atmosphere of hostility, we have a recipe for the events prototypically described as satisfying Feyerabend and Kuhn's incommensurability thesis. Rival groups become infused with what George Campbell calls "party spirit" in his *Philosophy of Rhetoric,* becoming "inflexible and [. . .] unjust, [. . .] arrogant, uncharitable, and malevolent" toward other parties (Campbell 1873, 97, 110); that is, not disposed to negotiate, in the sense of reciprocal suasion. But there is a sort of negotiation going on, in the sense of Burke's general dialectic. Here is Rudwick again on the protagonists of great Devonian controversy: "Their rhetorical strategy was of course to deride their opponents' opinions and to impugn their competence. But they could not afford to ignore those opinions, and they were forced to meet them with every weapon they could muster" (Rudwick 1985, 422). There is little in the way of overt attempts at détente, but at the level of the ideas, there is constant adjustment, data-borrowing, principle echoing. Each theory maintains its nominal independence as long as the theorists have the will and ingenuity to do so, but the theories cannot neglect the other's successes, striving to match or surpass them, nor the other's failures, striving to pick up speed precisely when and where the other stumbles, and a higher-order negotiation transpires as they drive each other to greater levels of articulation.

William Whewell, outlining what happens to the losing theory in such a clash, the one that is superseded and goes down in the historical record in the error column, describes its efforts this way, not as quiet, resigned deference to the successor:

> And thus, when different and rival explanations of the same phenomena are held, till one of them, though long defended by ingenious [theorists], is at last driven out of the field by the pressure of facts, the defeated hypothesis is transformed before it is extinguished. Before it has disappeared, it has been modified so as to have all palpable falsities squeezed out of it, and subsidiary provisions added, in order to reconcile it with the phenomena. It has, in short, been penetrated, infiltrated, and metamorphosed by the surrounding medium of the truth, before the merely arbitrary and erroneous residuum has been finally ejected out of the body of permanent and certain knowledge. (Whewell 1984, 392)

We are not so sanguine these days, most of us, about the body of permanent and certain knowledge as was Whewell (the essay is from 1851), and most of us would also view the penetration, infiltration, and metamorphic pressures as coming from the successful theory in the clash—that is, from another collection of ingenious and motivated theorists—not from some omnipresent neutral light of truth. The transformation, in short, even if we see it with Whewell as a desperate and cynical attempt by a doomed theory to retain its market hold, is forged in the crucible of debate. The thing is, if we now look at the other side of the debate, we find palpable falsities being squeezed out, subsidiary provisions being added, phenomena-reconciliation wrought, metamorphosis. Take one of the most successful theories of the twentieth century, which took shape largely under pressure from antagonistic theorists:

> Bohr never yielded; his entire epistemological edifice was constructed in dialogical response to Einstein's ceaseless challenges [. . .]
>
> The Copenhagen interpretation was erected not as a consistent philosophical framework, but as a collec-

> tion of local responses to changing challenges from the opposition. (Beller 1999, 135, 167)

Such examples could be multiplied at will: critical debate changes theories, almost always for the better, in the sense of becoming more resilient and more responsible. And the few examples I have offered of this process, just now and scattered around in this introduction, are all misleadingly narrow, two-position confrontations. Feyerabend's as-many-theories-as-there-are-bunsen-burners account is misleadingly broad (and, in truth, he offers it as a *pre*scription not a *de*scription), but it is much closer to what goes on in the cooperative competition of alternatives—Butterfield's "variations, uncertainties, controversies, and developments"—than the artificially tidy two-position accounts available by vignette.[86]

What does incommensurability—misaligned meanings, practices, and values, among rival theorists—buy us on this story? In a sense, everything; but, on its own, nothing. There is at least a low-grade incommensurability always at work in science (and, *a la* postmodernism, in symbolic life generally): even lab mates are rivals sometimes, and meanings are always indeterminate, values always ultimately personal, practices always ultimately individual. But meanings come into phase, values converge, practices synchronize. Scientists cooperate. Dark-matter incommensurability may always be present, but it needn't always be an irritant. The rapid acceptance of Watson and Crick's paradigm-altering structure-of-DNA paper, which certainly came in a milieu of aggression, competitive cooperation, and a good deal of cross-purposes talk, is sufficient evidence that incommensurability not only can be overcome naturally, it can be ignored.[87] On other occasions (like the Devonian controversy), it can be consciously bargained away. On other occasions, it might never be overcome; indeed it escalates until we get triumphant winners (like the Copenhagen interpretation) and extinguished losers (Cartesian vortices, which is Whewell's example in the essay quoted earlier). These are the prototypical examples that populate the literature on incommensurability. In these cases, the incommensurometer gets higher and higher readings, and the cases fall into more easily diagnosed semantic and pragmatic and, at the top end, cosmic incommensurability. The Max Planck Effect kicks in and some scientists go to their graves unreconciled to new pictures, while others go on with their careers unreconciled to their elders—"the old opinion passes away with the old generation: the new theory grows to its full vigour when its congenital disciples grow to be masters; John Bernoulli continues a Cartesian to the last; Daniel, his son, is a Newtonian from the first" (Whewell 1984, 385; see R. A. Harris 1998).

Incommensurability plays a vital role in these processes, and an inevitable one, given the nature of meanings, values, practices, and humans. But it is not always generative, and it only becomes visible with the big dust-ups.

Conclusion

> One familiar kind of conflict is that in which two or more theorists offer rival solutions to the same problem [. . .] often, naturally, the issue is a fairly confused one, in which each of the solutions proffered is in part right, in part wrong and in part just incomplete or nebulous.
>
> —Gilbert Ryle, *Dilemmas*

In this introduction, I have laid out a taxonomy of incommensurability in science, and charted it through the literature, also noting two prominent and related usages. The categories (which are cross-cut with a roughly quantitative notion of incommensurability) are:

- Brick-wall incommensurability
  A form of absolute incommensurability, in which there are no points of contact between the two items held in that relation; it is used in argumentation, falsely, to ascribe craziness to Kuhn or Feyerabend (neither of whom ever advocated it).

- Cosmic incommensurability
  A strong form of incommensurability which rests on linguistic relativity arguments and implicates different perceptual worlds for scientists committed to different (and rival) theories, associated most often with diachronic relations between some theory and its successor; Kuhn argued for a cosmic conception to incommensurability in *Structure,* in a tentative but concerted way—reiterating it frequently, searching for analogies, but also constantly hedging, and undercutting the relativist analogies with realist claims—and then downplayed or ignored it the rest of his career; Feyerabend made only very conservative allusions to cosmic incommensurability in "Explanation" but increasing-

ly embraced it for the remainder of his career. It might be best understood as a radical difference in perspective.

- Semantic incommensurability
  A moderate form of incommensurability which 'only' concerns words, and has spilled the most ink, principally over issues of reference and translation; Kuhn proposed a constrained variant of semantic incommensurability in the 1980s, which he called *local incommensurability,* that identifies the source of the problem in the lexical structure of a few inter-related key terms; Feyerabend, whose "Explanation" was solely about semantic incommensurability, came to be much less concerned with terminological meaning in his subsequent career.

- Pragmatic incommensurability
  A moderate-to-strong form of incommensurability that concerns themes (prominently including values) and practices; it is in a reciprocal relationship with semantic incommensurability, shaping terminological meanings as they shape themes and practices; pragmatic incommensurability is a particularly subtle form, leading to confusion and incomprehension, because terminological meanings can be ferreted out relatively easily, while misaligned themes and practices are largely tacit, forming the background assumptions, but rarely making it to the surface of specific arguments.

In connection with this taxonomy, I noted that all incommensurability is semantic to a degree, but that the more cross-theoretical argumentation approaches a hypothetical brick-wall communicative blockage, the more pragmatically entangled it is.

The two principal usages of *incommensurability* and its related terms outside of science studies are:

- Value incommensurability
  The longest standing non-mathematical usage of the *incommensurability* suite concerns the non-comparability of moral values, which developed into a philosophical leitmotif around the turn of the twentieth century, being picked up by American pragmatists as an element of the case for tolerant pluralism, and flaring again in the 1970s around Isaiah Berlin's conception of liberalism; it stresses the multiplicity of values and value sys-

tems, many of which drive actions (and policies) in opposite directions and cannot be simultaneously maintained, as well as the lack of ranking possibilities for moral, political, and aesthetic values.

- Postmodern incommensurability
  A deployment of the *incommensurability* terminology suite in a way that implicates, but does not technically specify, a range of related notions that intersect with value pluralism, non-trivial cultural divisions, multiple subjectivities, tolerance, alienation, alternate rationalities, and the wholesome corrosion of meta-narratives; while not exactly semantic, because it concerns much larger units than words (or structured lexicons), it is far more closely aligned to language (via language games or phrase regimes) than are usages in the value-incommensurabilist literature.

Overall, I have maintained that incommensurability in science is not the debilitating problem it has often been taken to be (usually by critics who are attempting to refute it). Incommensurability seems to rule out agreement and evaluation in principle. Nothing does. There is always room for creative maneuvering to bring things together or to take them apart, for assimilating them to the same context or for distinguishing them with respect to different contexts. In that light, incommensurability just doesn't obtain. But there are disagreements, sometimes violent ones, which can be traced to semantic and pragmatic misalignments, the best label for whose problem field, because of its history and its ubiquity, is probably *incommensurability.*

I have argued, therefore, that (non-mathematical) incommensurability, so long as we're going to deploy the concept, is best seen as a phenomenon not between things (words, theories, values, frameworks, . . .) but between people who have attitudes towards such things, and creative relationships with them, and who build arguments out of them, about them, and expressing them—between, that is, rhetors and audiences, or between reciprocal rhetors. Any two theories (even two 'notational variants' of the 'same' theory) might be made incommensurable, with enough intransigence by the arguers—even the smallest differences can be built into road blocks.[88] By the same token, any two theories (even, say, corpuscular and undular conceptions of matter) might be brought together with enough ingenuity and goodwill.

I have suggested that there are always in principle remedies to conflicts that arise because of discrepancies among arguers' attitudes, but that such remedies are not always sought because the creative relationships theorists have with their theories, and with congenial theorists, are often enhanced in conflict with rival theories and theorists; frequently in such conflicts, in a general dialectical way, epistemology is the winner, because the theories get stronger and more fully articulated—as Feyerabend maintained for much of his career—and bring us more knowledge. That process, refining positions in the crucible of debate, is deeply rhetorical.

A strongly overlapping project in this introduction, then, sharing many of the concerns and complications of the circles I have been inscribing around the word *incommensurability,* has been my attempt to inscribe circles around the word *rhetoric.* That word escalated into quite wide scholarly circulation over the last quarter of the twentieth century, most notably in science studies, occasioned not coincidentally by the constellation of issues stirred up by Feyerabend, Kuhn, and their monster, incommensurability. While this general-scholarly use of the word goes beyond the most pejorative ordinary-language senses (the ones that collocate with adjectives like *mere* and *empty*), the widespread academic circulation of *rhetoric* is not entirely wholesome. Friends and enemies alike tend to use it as an antonym for *reason.* On the with-friends-like-this-who-needs-Plato? side of the ledger, we could put a substantial proportion of scholars discanting the postmodern idiom, the ones who use *rhetoric* as a nickname for *irrationality,* and mean it as a compliment; on the Plato's-progeny side of the ledger, we have the many late-modernist scholars, mostly philosophers, who use *rhetoric* as a nickname for *irrationality,* and mean it as an insult.

Let's take the latter, an (I'll assume nonmalicious) enemy, Daniel Garber, who argues, *a propos* of our major theme, that there "really" was no incommensurability between Aristotelian and Cartesian views of the body, as Kuhn had suggested (Kuhn 1996, 104–6),[89] going on to propose this discovery should "open the door to a purely rational explanation of the change from the one to the other" (Garber 2001, 417). But his remarkably optimistic locution, "purely rational," reveals him to be heading down the wrong road; *nothing* is purely rational, let alone contesting representations of reality, one essentialist, one mechanist. (The root metaphors of essentialism and mechanism alone are enough to vitiate his adverb, *purely.*) Shocked and dismayed, he does indeed report that his expectation of pure rationality has been frustrated. Garber finds there is, alas, evidence in abundance of "a breakdown of rational argument in the debate" (Garber 2001, 417).

"What this suggests to me," he goes on, "is not that we should give up on reason, and abandon ourselves to rhetoric, non-rational persuasion, and purely [there's that adverb again] sociological analysis of conflict and intellectual change." Instead, we need to build "an account of how arguments are weighed in the context of the complicated situations of real debate and argument" (Garber 2001, 421). I follow Professor Garber on the conflicting relationship between Aristotelian and Cartesian models of the body, though my own diagnosis would be a bit different (low-grade differences that are ripe for assimilation), and I am in full sympathy with his project to explore how specific arguments stand in relation to the specific constraints and allowances, values and practices, meanings and themes, of their specific contexts, rather than in the terms of some abstract notions of proper, philosophically sanctioned logical maneuvering. In fact, so long as he constrained his remarks to "the context of the complicated situations of real debate and argument," I would hire Garber as the publicity agent for this book; that's what bringing rhetoric to incommensurability is about. Here's a lovely piece of copywriting for our project:

> People assent or fail to assent to arguments in part because of their strength or weakness, in part because of the way in which they interact with their own underlying assumptions and beliefs, and in part because of the non-intellectual stakes there might be for accepting or rejecting a belief. (Garber 2001, 419)

But our values and meanings, Garber's and mine, are substantially misaligned. In his framework the real-debate-and-argument project is (1) rational and therefore (2) non-rhetorical. In my framework, and in the general framework of this book, Garber's characterization is a *non sequitur:* rhetoric is concerned with exactly these sorts of context-dependent, real-debate accounts. More than that, rhetoric is rational. How do we commensurate, Garber and I, bringing our contexts into sufficient proximity that we don't have to go out at dawn with pistols for two, coffee for one? We can start with the observations of one of Garber's philosophical brethren, Marcello Pera:

> Scientific rationality is too complex a question to be solved by a few methodological formulas, and rhetoric too subtle and serious an art to be reduced to mere propaganda or indoctrination. To be rational is to put forward good arguments. (Pera 1992, 175)

Garber apparently holds that scientific rationality is complex, since it ought to be pure but isn't, instead giving rise to the complicated situations of real debate. That he wants to investigate how arguments are swapped in those complicated situations suggests that he doesn't think a few general methodological formulas are going to cut it (if they did, we would probably get the pure rationality he venerates). Garber also believes that rational complexities do not mean we should give up on reason, to fall back on non-rational accounts and exclusively sociological analyses (which I presume means issues of power, ideology, status, and the like, all of which I further presume Garber places beyond the reach of reason, involving perhaps mere propaganda or indoctrination). There are people—Feyerabend, for instance, and Foucault—who would not go this far with Garber. But I'm with him—well, most of the way. I don't pine for *pure* rationality, but it's easy to overlook that part of Garber's value system, since he seems prepared to recognize that, whatever we crave, *practical* rationality is the best we are going to get. Nor do I carry the disdain Garber apparently does for sociological analyses, but again the barriers are low enough to step over, since he's prepared to admit sociological factors play a role (it's only the *purely* sociological analyses that worry him).

What's important for both of us (indeed, all three of us, including Pera; indeed-redux, all four of us, since if you have got this far you must have a stake in this business too) is understanding how arguments function—in large part, understanding that they require a quality that Aristotle called *phronesis*, which rhetoricians frequently translate as "practical wisdom," which feeds a major theme in the incommensurability-of-values literature as "practical reason," and which appears in Garber's article as "a question of practical consequences" (Garber 2001, 420).[90] Looking to those consequences means looking to the values of the interlocutors, and how they influence the invention, reception, and appraisal of arguments. What we come down to, then, is a fairly minor semantic incommensurability, Garber and I, chiefly over the word *rhetoric*. Our practices are quite different, since they were delivered to us from different traditions, but if our purposes are the same we can work cooperatively with these instruments, not antagonistically.

If this commensuration is to proceed any further, the next move will have to come from Professor Garber, which minimally calls for him—speaking metonymically—to stop the misguided sneers at the discipline guiding this book. My hope is that the wonderful essays it includes will help some philosophers get over their prejudices.

Proposals and Analyses

Incommensurability is a highly ramified word, which is one of the reasons it has been so immensely generative of scholarship almost from its introduction to science studies; that, and the way it so clearly signals a change in conception of scientific argumentation. The change began long before 1962, but that is the year in which Feyerabend and Kuhn gave it the talismanic term that sponsors this book (and in Kuhn's case, gave it a cluster of such terms—including *paradigm, revolution,* and *normal science*—which goes some way to explaining why he has been so much more widely influential than Feyerabend; he built a good lexicon). But, what, specifically, does rhetoric have to say about incommensurability?

The immediate answer is this book, the analyses and proposals of a group of first-rank rhetoricians (joined by a first-rank philosopher) implicated by the term, *incommensurability.* We've just noted, and repeatedly noted in this introduction, however, the tentacled nature of that term. So, which tentacles are best chewed by whom—in particular, by rhetoricians?

History of science concerns itself, at the core, with past narratives that might implicate incommensurability, and may draw on any number of explicative instruments, as Rudwick (1985), for instance, draws on a jurisprudential model, and Galison (1997) draws on pidgins and creoles. Philosophy of science concerns itself, at the core, with the rationality of encounters that might implicate incommensurability, of the sort we have just seen in my representation of Professor Garber's arguments. Its dominant topics are therefore logic, meaning and reference, and (by way of meaning and reference) translation. Philosophy, too, draws inspiration and instruments from diverse sources, as Thagard (1993), for instance, works out of computational cognitive science, and Biagioli (1993) deploys anthropological considerations. Sociology of science concerns itself, at the core, with cohort allegiances and concomitant behaviors in situations that might implicate incommensurability. There is surely, too, room for such studies as the psychology of science, social or cognitive, in the domains relevant to incommensurability, such as in the approach of H. Margolis (1993) and of Barker (2001a, 2001b).

What remains for rhetoric of science in this problem field? —The investigation of suasion (per- and dis-) and identification/division in arguments that evince or might be expected to evince incommensurability. This focus will, of course, reach into semantic matters, broach perceptual issues, and confront the commonplaces of the problem field—gestalt switches, religious conversion, translation, and the like—but it will naturally gravitate to pragmatic incommensurability and to a degree-based, rather than binary, picture of theme-and/or-practice (mis-)alignments. No doubt this is disciplinary chauvinism on my part, but since the diagnoses of incommensurability are universally in discourses that fail suasively, that do not effect identifications, the rhetoric-of-science focus seems utterly central to the problem field of incommensurability.

A few of the following chapters in this book continue the discussion of incommensurability on the mostly abstract level that I have prosecuted in this introduction, but most of them—pursuing another line that is equally and reciprocally profitable—carry the rhetorical exploration of incommensurability into specific territories, offering case studies of people, attitudes, interests, and arguments in conflict. The book has three major divisions, one of which you are in now. The other two approach the intersection of rhetoric and incommensurability in specific case-driven ways.

The section you are in now includes one other chapter, by **Paul Hoyningen-Huene**, who does us all a service and situates, far more ably than my introduction, the relative contributions of Paul Karl Feyerabend and Thomas Samuel Kuhn to the problem field of incommensurability. His chapter, "Three Biographies: Kuhn, Feyerabend, and Incommensurability," examines the professional trajectories that led both of our principals to converge so significantly in 1962, and diverge thereafter. Hoyningen-Huene outlines the differences in their conceptions of incommensurability, with respect to their concrete content, to their function, and to their implications, and traces Feyerabend's and Kuhn's reactions to one another, and to one another's respective visions of science.

The second section, "Issues" approaches incommensurability from the top down: examining the problem and tracing its ramifications. It has three chapters, by Alan Gross, Thomas Lessl, and Herbert Simons.

Where Hoyningen-Huene offers three biographies, **Alan G. Gross** narrows the focus to "Kuhn's Incommensurability." Gross charts Kuhn's specific refinements to that notion, as he settled more closely on local semantic conflicts and mismatches, offering another important supplement to my general introduction, carefully explicating Kuhn's late notion of a structured lexicon, his incommensurability-driven refinement on *paradigm.* Scientists from the same community may differ in what they know about the concepts in the lexicon, but they all share a professional commitment to the lexicon and its structure. When problems arise with the lexicon—Planck's quanta are the classic example—we have a condition necessary to, though not sufficient for, a scientific revolution. If a revolution comes, its completeness is established when a new structured lexicon is firmly in place so that it is no longer possible to use the central concepts of the old lexicon in the old ways; voila, incommensurability. Gross also has a prescription.

Gross's prescription is not for scientists, but for science-study scholars, to help us clean up our analyses of incommensurability. Drawing on W. B. Gallie's *essentially contested ideas* program—a philosophical therapy that speaks very richly to issues of incommensurability—Gross takes a retrograde stand calculated to infuriate postmodernists. Understanding incommensurability, he insists, depends simply on arguing from historically revealed scientific practice, and then proceeding to engage each other cooperatively. In some respects, this position gives Gross's paper the feel of a dissenting opinion to the strong theme running through this book that incommensurability is something of a chimera which can be dissolved by cooperative argumentation. Gross argues that "Kuhnian incommensurability [. . .] provides a strong conceptual and practical barrier to competing scientists, and applies not only to individual situations, but also to whole fields" (196), illustrating this with the extended example of David Brewster, a nineteenth century physicist whose work sought to splice selection and wave theories of light.

Thomas M. Lessl works from the position that science has operated (roughly speaking) under two distinct ideological canopies since its early-modern origins—a Baconian worldview which came into prominence around the middle of the seventeenth century, and a positivist one that superseded it around the middle of the nineteenth century. The Baconian ideology tied the fate of science explicitly to the millenarian telos of Protestantism. But in the eighteenth century science began to reconstitute itself in terms of secularism, systematized by positivism in the next century. Lessl's chapter, "Incommensurate Boundaries: The Rhetorical Positivism of T.H. Huxley" argues that this ideology fosters the incommensurability of competing scientific conceptions of biological origins. Specifically, he explores the failed alignments of the themes and values saturating the "evolution" and "design" positions in biological speciation. It is usually supposed that design was pushed out of science by the arrival of the Darwinian thesis, but Darwin's advent also coincided with the rise of positivism. Lessl keys on a central actor in the growth of both Darwinian evolution and rhetorical positivism, T.H. Huxley, and finds his blend of the two largely responsible for the exile of design theories. Indeed, Lessl suggests evolutionary and design conceptions might be largely commensurable, outside the ideological hegemony of positivism. Design was "displaced rather than refuted," Lessl argues, a displacement not at the level of paradigms, but "at a superceding level of discourse" concerning the constitutive features of science (205).

Among the conclusions Lessl draws from his exploration of these themes is that there is a third category of science, beyond Kuhn's familiar 'normal' and 'revolutionary' variants. Lessl calls this variant 'ideological science,' and finds it to sponsor "a sort of über-incommensurability that doesn't confound debate so much as preclude it" (234).

Herbert W. Simons, for his part, charts an active course of commensuration, bringing two very unlikely bedfellows together, deconstruction and objectivism, in order to find ways of formulating reasoned judgments and of bringing others to those judgments—a process he calls *rhetorical reconstructionism.* His essay, "The Rhetoric of Philosophical Incommensurability," takes its lead from the familiar observation that communication diagnosed as incommensurable is inevitably circular—but, according to Kuhn (as Simons notes), it is only *partially* so. Since circularity is only partial, opposing paradigms necessarily provide openings for dialogue, and the possibility of common ground. Even without stable foundations or other common measures for adjudicating between philosophical paradigms, Simons argues, there are still possibilities for persuasion across the divide. Reconstructive rhetoric addresses the need for reasoned and reasonable judgments on issues for which there can be no formal or final proof. Where deconstructionists are inclined, for example, to acidly question all reality claims, rhetorical reconstructionists weigh competing reality claims to find which are more credible than others. Simons also provides a very useful appendix, an explicative glossary of pivotal terms.

At the outset of this introduction, I offered two alternatives for why the incommensurability of rival theories has not stymied scientific knowledge making—because it doesn't apply to theories, or because there are ways around it. Lessl's contribution is of the does-not-apply sort, in a meta-incommensurability vein—remove positivism, he claims, and design theories can talk comfortably with evolutionary theories, evolutionary with design (in the way Feyerabend suggests that removing the ideology of realism renders celestial mechanics and relativity commensurable—Feyerabend 1970, 220). Simons falls more into the there-are-remedies camp—specifically for the broad themes of objectivism and deconstruction—taking up the stance of Richards and Bryant and Booth (not to say Cicero and Gallie and Habermas) that our job in arguing is to speak with, not at, each other, to pursue mutual understanding, to overcome misunderstanding. Gross, however, rejects both alternatives, clearly renouncing the does-not-apply suggestion, and arguing strongly against the idea that cases of "profound incommensurability" like that of selection and wave theories of light can be remedied. Indeed, David Brewster's mission to commensurate the incommensurable is what gives his story the poignancy Gross narrates. "Without [the concept of] incommensurability," Gross argues, Brewster's "struggle to accommodate wave theory loses its tragic dimension; it turns the battle to save his science into an example of inept bungling or severe character defect" (197). Gross's analysis takes a somewhat different view of commensuration than that of the other contributors to this volume (see also his overlapping 2004 paper)—seeing it as a matter of *blending* perspectives, rather than of making common ground for *judging* them: choosing when one is more powerful than the other, in global or local ways; choosing which bits and pieces of which perspective are most powerful, in which clusters, for which problems; and/or forging new perspectives from the critical juxtaposition of the conflicting ones.

The remaining six chapters in this volume, in the section, "Cases," all stand more or less in the remedies camp, some more explicitly than others—exploring avenues of consensus and commensurability, bottom-up, in particular scientific disputes. They are rhetorical case studies of specific communicative problems in scientific discourse. All are synchronic (or very-short-range diachronic), and concern mostly pragmatic incommensurability, with a few side orders of semantic incommensurability. They don't always bring such programs to the same table for tea and scones, but they always indicate at least how the places might be set for such reconciliations.

Leah Ceccarelli's chapter, for instance, addresses a debate that has heated many collars over the last two decades, a debate in which participants on both sides seem to be taking Kuhn's revolutionary-war description of scientific controversy not as an account but as a battle plan—one side destined to win only when the last holdouts of the other, incommensurable, paradigm have died out—and envisioning their conflict as a fierce battle to the death. The two sides dispute the proper meaning of key scientific concepts, they engage in resistive readings of each other's discourse, and they impugn each other's values and motives: the usual. Ceccarelli suggests a different plan in "Science and Civil Debate: The Case of Sociobiology," one which rests on a very different vision of scientific controversy. Rather than bloody revolution, she suggests casting scientific controversy in terms of a civil debate, particularly on the model of Cicero's *controversia,* with resolution coming in the form of a negotiated compromise that takes up bits and pieces of each side and thus dissolves the differences between seemingly incompatible contraries. Ceccarelli looks especially at the argumentation of E. O. Wilson, the progenitor of sociobiology, showing that some of his early responses to criticism included admirably Ciceronian themes, but that latterly he has settled into a hostile recalcitrance that forecloses agreement, that promotes incommensurability rather than fruitful cooperative exchange. Intriguingly, she also suggests that like the burden of proof that naturally falls on a rhetor disrupting the status quo, there is (or should be) a *burden of concord* that also falls on such rhetors; the disruptors, in this view, have a larger responsibility for the tone of the debate than the defenders of the established programs (theories, paradigms, ...).

Lawrence J. Prelli's chapter, "Stasis and the Problem of Incommensurate Communication: The Case of Spousal Violence Research," is also prescriptive. Prelli presents and applies a heuristic procedure, known in antiquity as *stasis procedure,* for identifying and resolving communicative blockages confronting scientists operating from different paradigms and orientations. Creatively applied it can resolve the problems routinely diagnosed as incommensurable. Prelli's first move is to redefine the problem of incommensurability from logical into rhetorical terms, downgrading the intransigent adjective *incommensurable* into the more amenable *incommensurate.* Next, Prelli argues that the guided questioning of classical *stasis* doctrine can help frame ambiguities as potentially shareable questions—what he calls *commensurability points*—from which discussants can bridge or realign their perspectives to help achieve commensurate communication about similarities and differences. He then develops a special *stasis* procedure for scientific communication, showing how it can help disputing parties discover how their perspectives converge and diverge on specific situated ambiguities about evidence, interpretation, evaluation, and methodology. Prelli brings these observations and recommendations to bear on the case study that anchors his chapter, an analysis of communicative problems that have emerged among scientists in a significant ongoing controversy over spousal violence research, a heavily value-charged arena.

John Angus Campbell's essay traces Darwin's adaptation of Charles Lyell's anti-evolutionary argumentation. "The 'Anxiety of Influence'—Hermeneutic Rhetoric and the Triumph of Darwin's Invention over Incommensurability," situates Lyell's massively influential *Principles of Geology* as an up-dated Newtonian worldview, in which the implications of weak causes and deep time for recovering the history of the earth are devastating. The history is irrecoverable, in Lyell's project, because the records are being relentlessly destroyed in the earth's furnace by further operation of the very forces that had partially preserved them. Campbell traces how this vision of earth history without history not only failed to daunt Darwin, but (enriched by Malthus's social vision) inspired him. Campbell approaches incommensurability as an artifact of a failed positivist world-project which attempted to control history by thought, opposing it to an open, rhetoriographic understanding of history. With a Feyerabendian twist ("without a constant misuse of language there cannot be any [. . .] progress"—Feyerabend 1978a, 27) on Richards ("rhetoric [. . .] should be a study of misunderstanding and its remedies"—Richards 1936,

3), Campbell argues that evolution is a rhetorical reinvention of Lyell creatively misunderstood.

Jeanne Fahnestock, like Ceccarelli, also investigates a dispute in which the levels of animosity run high. She traces competing perspectives defining the structure and composition of the cell, in her essay, "Cell and Membrane: The Rhetorical Strategies of a Marginalized View," focusing especially on disagreements over the nature of the cell's interface with the world, its membrane or "wall." These controversies have necessarily involved debates over whether the cell is essentially defined by that membrane or by its internal substance—the container or the contained. As the very notion of a "cell" suggests, the container definition has become the orthodoxy we encounter in classrooms and textbooks and TV ads for skin cleansers. But that view was challenged in the fifties and sixties by Gilbert Ling (a chemist holding a suite of heretical views, but someone who has always had supporters), and renewed in the early part of this century by Gerald H. Pollack. The Ling/Pollack view is seen by all concerned as fundamentally incommensurable with the current paradigm of cell structure; as such, it is shunned by the mainstream. But, Fahnestock argues, incommensurability can be a rhetorical investment (as Ceccarelli argues of E. O. Wilson), with arguments arising from an uncompromising hostility in which irreconcilability becomes a self-fulfilling prophecy. Given the productivity of the orthodox view, there is no pressure to listen to, let alone accommodate, the alternate, Ling/Pollack (Associated Induction) understanding of the cell and its membrane. But, as Fahnestock notes, there are parts of this view that are relatively amenable to the mainstream. The differences might be converted rhetorically into less serious differences of emphasis or perspective, and productive talk could ensue. But the Associated Induction side's investment in, and the mainstream's assumption of, incommensurability blocks any usefully reciprocal argumentation.

The following chapter, by **Charles Bazerman** and **René Agustín De los Santos**, takes the notion of incommensurability into an area not traditionally explored in these terms, interdisciplinary convergence. Their chapter, "Measuring Incommensurability: Are Toxicology and Ecotoxicology Blind to What the Other Sees?" shows how the public consciousness of environmental issues that emerged in the 1960s and 1970s led to an increasing demand for information to monitor the changes occurring in our environment. Traditional disciplines in the earth and biological sciences were enlisted and reorganized into new disciplines focusing on environmental concerns. These new applied disciplines were particularly motivated by the regimes of regulation and oversight established by the National Environmental Protection Act (NEPA), with its mandated new genre of the Environmental Impact Statement (EIS). NEPA and the EIS created new markets and forums for information—as well as clients needing their interests advanced and protected within these developing rhetorical regimes. The refigured disciplines encompassed work by diverse scientists—some of them migrants from pre-existing disciplines, some of them maintaining dual loyalties, and some of them newly trained within the emergent disciplines.

With this welter of interests, disciplines, and argument styles competing for results and resources in such a politically charged cauldron, we ought to find incommensurability. Bazerman and De los Santos look. They examine the literatures of these new fields in relation to the prior fields, study the communication networks of current participants, and build upon interviews with people working in the old and new fields, in order to characterize the rhetorical practices at the crossroads of these interests and disciplines. They do not find incommensurability. They find cooperation, negotiation, convergence, a near-textbook case of reciprocal suasion—nothing so grand as philosophic harmony or Habermassian reason, just a case of talking each other into cooperative beliefs and behaviours. Toxicology and ecotoxicology use each other's results and methods, and share practical concerns that foster creativity and flexibility.

Carolyn R. Miller finds the reverse. Like Bazerman and De los Santos, she looks to interdisciplinary developments—or, perhaps *cross-disciplinary* is a more accurate term in her case, since she finds little of the cooperative energy that characterizes interdisciplinarity in general, and that Bazerman and De los Santos find in specific with toxicology and ecotoxicology. Miller complicates Kuhn's picture of science somewhat with rhetorical pressures different from those found in his familiar paradigm-succession story. Her essay, "Novelty and Heresy in the Debate on Nonthermal Effects of Electromagnetic Fields," examines a case that concerns not only change over time but also the problems of inquiry based in multiple scientific fields: the debate over nonthermal effects of nonionizing radiation that has developed over the past fifty years, and gained recent steam from controversies over cell-phone tower sites and research linking brain tumours to mobile phone use. Differences in worldview, Miller argues, derive in this case not only from different levels of commitment to a newly proposed explanatory framework within a single scientific community, but also from different long-standing and deeply engrained commitments in separate disciplines. Cross-disciplinary work raises unique barriers to evidence- and argument-evaluation, because each discipline has a different normal science. "Without a jointly owned orthodoxy," she notes, "the separate disciplines [can] simply contest the other's right to govern the forum."

Research into nonthermal EMF effects involves electrical engineers and physicists as well as biologists and epidemiologists. While Miller finds a situation quite different from the one Bazerman and De los Santos investigate, she argues that what she finds is not incommensurability. Or, rather, since incommensurability brings no explanatory power with it, she argues that it is not a productive way to characterize the differences among these groups. Rather, Miller explores the illuminative virtue of two related tensions she sees operating in this debate, the tension between novelty and tradition and the tension between heresy and orthodoxy. What is at stake in the debate are health and safety standards for exposure of human populations to electromagnetic fields of various strengths and frequencies, both in specific workplaces, for specific consumer products, and in the ambient environment. And, here, these interests pull the debate away from agreement and cooperation, the constituencies concerned only to shore up their own views, denigrate the competing ones.

"What Kuhn called *incommensurability,*" Miller argues in her essay, "can be understood alternatively as the result of [. . .] rhetorical dynamics, which involve a complex combination of both intellectual commitments and socio-political relations" (474)—an admirable epitome of the thrust of this book overall. We are not uniform in our positions with respect to incommensurability. I am pretty sure that in addition to the large swathes of this introduction that most of the contributors agree with, there are claims or perspectives here that make some of them itchy (though only Gross scratches publicly, in the concluding passages of his chapter[91]). But we are largely together that incommensurability, to the extent that it usefully describes a cluster of phenomena that manifest when some arguments clash, is tractable, and while tractability may not be the goal of all the rhetors engaged in such clashes, or any of them, and may not even be always in the best interests of truth and knowledge, what guarantees that tractability is rhetorical pliancy.

Bazerman and De los Santos, on the one side, and Miller on the other, represent two faces of the same coin. When there is effort, a convergence of interests, and goodwill—rhetor-directed charity when listening or reading, audience-directed clarity when speaking or writing—then conflicts in terminology, themes, and values can be managed, and usually are; when there isn't, they can't, or won't be. But the praise for such management, the blame for mismanagement, lies not with theories, but the theorists, not with the values, but the value-holders, not with the arguments, but the arguers. Like incommensurability, rhetoric has often been deplored as an element in scientific inquiry, something that bypasses the rationality needed to certify progress, truth, knowledge. Collectively, however, the contributions to this book offer a picture of argumentation in science that is full-blooded, value-saturated, and plastic enough to do the real, daily work of making knowledge about the natural world; argumentation, that is, which is deeply rhetorical.

Notes

[1] Popper tells this story in *Conjectures and Refutations* in a brief discussion of (mathematical) incommensurability—a curiosity, since it appeared the same year that both Feyerabend and Kuhn proposed incommensurability as a philosophical concept (Popper 1962, 86–87, 149). For discussion of the origins and implications of the story, and the overwhelming likelihood that it is entirely fictitious, see Richard Crew (2000). Another story has Pythagoras committing suicide over incommensurability (Livingston 1986, ix), and various versions have Hippasus killed, sometimes ritually, for boasting the dodecahedron was "his" and not "the master's," or being punished by the gods for one or the other of these offences. As Popper points out, these various rumors probably illustrate nothing more than the deep secrecy of the Pythagoreans. The definitive account of Hippasus as the original discoverer of incommensurability is von Fritz (1945).

[2] The first chronicled use is in the 1350s, in the title of Nicole Oresme's *De commensurabilitate sive incommensurabilitate motuum celi* (the commensurability or incommensurability of celestial motions), which shows that the motions of two or more celestial bodies cannot be precisely related. The *Oxford English Dictionary*'s (second edition) earliest citation in English is 1557: "If thei haue no suche common diuisor, then are thei called incommensu-

rable, as 18 and 25." The first citation of the nominalization, *incommensurability,* is Sir Henry Billingsley's 1570 published translation of Euclid.

[3] See also Kuhn (2000, 36) where he says explicitly that the vehicle of metaphor is "common language" rather than "neutral algorithm," a somewhat different but related notion, which illustrates his increased focus in later work on linguistic accounts of incommensurability and the role that translation played in the incommensurability debates after Kuhn and Feyerabend's initial proposals.

[4] The initial publications are Kuhn (1962, 1970, 1996) and Feyerabend (1962, 1981a), and the initial stink raised about the implications of applying the word to scientific theories (paradigms) is documented best in Lakatos and Musgrave (1970); the next place to stop would be the 1982 proceedings of the Philosophy of Science Association where Kuhn defends (and continues his life-long mission to refine) his use of the term (1983a), where one of his more thorough critics (Kitcher 1983) has a parallel paper, where one of his (qualified) supporters (Hesse 1983) comments on both papers, and where Kuhn (1983c) then responds to Kitcher's and Hesse's observations (all in Asquith and Nickles 1983). Kuhn's contributions to the Lakatos and Musgrave volume and to the Asquith and Nickles volume are reproduced in Kuhn (2000), but seeing them in the original contexts is much more fun. The most systematic treatment of incommensurability as a philosophical issue is Howard Sankey's *Incommensurability Thesis* (1994). Hoyningen-Huene and Sankey (2001) offer a variety of philosophy-of-science perspectives; Chang (1997) offers a variety of ethical-and-aesthetic-values perspectives; and Heidlebaugh (2001) offers a detailed Ciceronian treatment of incommensurability as a general problem of argumentation. All that being said, of course, you're already in the right place. Hoyningen-Huene's chapter in this book chronicles incommensurability in relation to the life and works of Feyerabend and Kuhn, and Gross's contribution looks specifically at Kuhn's refinements of the notion. This introduction offers a more wide-ranging treatment of the topic, in relation to rhetoric. And the other essays all serve the joint project of bringing incommensurability to bear on rhetorical issues and rhetorics to bear on incommensurabilities.

[5] For discussion on the origins and implications of the story, and the overwhelming likelihood that it is wholly fictitious, see Thomas Cole (1992), who even suggests that our two rhetors might have been the same person, Tisias the Crow (nicknamed for a cawing, repetitious oratorical style?).

[6] On Plato's coinage, see Cole (1995) and Schiappa (1999); for a counter-argument, see Pendrick (1998).

[7] As G.B. Kerferd says, "not one barrier but two stand in the way of anyone who seeks to arrive at a proper understanding of the sophistic movement at Athens in the fifth century B.C. No writings survive from any of the

sophists and we have to depend on inconsiderable fragments and often obscure or unreliable summaries of their doctrines. What is worse, for much of our information we are dependent upon Plato's profoundly hostile treatment of them, presented with all the power of his literary genius and driven home with a philosophical impact that is little short of overwhelming" (1981, 1).

[8] Lawrence Prelli, in his essay in this volume, offers a fuller and more technical account of *stasis* than we have room for here, using Cicero's terms—conjectural (factual), definitive (definitional), qualitative (evaluative), and translative (jurisdictional)—and adapting them specifically to scientific argumentation.

[9] For a more succinct account of the use of *rhetoric* as a contemporary disciplinary term than what follows here, see Herbert Simons's definition and commentary in his Appendix of Pivotal Terms (261).

[10] Mara Beller, in this context, argues that Hanson held an incommensurability thesis (1999, 293–300).

[11] Struan Jacobs argues that Polanyi was the source of most of Kuhn and Feyerabend's arguments for incommensurability (2002). Jacobs provides a very compelling case for the influence of Polanyi on both Kuhn and Feyerabend; and for Polanyi's deployment of the themes of Whorfian linguistics and philosophy of science that were prevalent in the fifties. But his implication that Kuhn and Feyerabend "got" incommensurability from Polanyi, and deliberately slighted his influence (Jacobs almost implies intellectual property-theft here) is nonsense. Both of them were pretty frank about their sources, both acknowledged Polanyi's value, and Polanyi was not alone in articulating these issues.

[12] In Feyerabend's case, the ideas that collected under the label *incommensurability* seem to have begun in 1952, while participating in a seminar of Popper's at the London School of Economics (see Hoyningen-Huene's chapter in this volume [153], Feyerabend 1978a, 115; 1995, 192). In Kuhn's case, it's not clear how far back the ideas go, but Steve Fuller (2000, 7n13) traces the incommensurability publication trail to Kuhn's (1959) "Energy Conservation as an Example of Simultaneous Discovery," reprinted in Kuhn (1977a), John Heilbron (1998, 508) suggests that *Copernican Revolution* (Kuhn 1957) shows incommensurability roots, and Alan Gross (this volume, 179) sees incommensurability in Kuhn's first foray into history of science, his 1952 essay on Boyle (the year of Feyerabend's inklings). Aside from a few specific comments here and there, I treat incommensurability as a knot of overlapping conceptions, with little attempt to disentangle Kuhn's threads from Feyerabend's threads. My brief is more to explicate the influence of that knot, and various critical impressions of it. For a careful corrective on my relative indifference, please see Hoyningen-Huene's chapter in this vol-

ume, which meticulously charts Kuhn's and Feyerabend's respective views on incommensurability.

[13] I'm thinking of uses like Richard Rorty's, for instance, in *Philosophy and the Mirror of Nature,* where incommensurability is an important and fairly precise term, partly Kuhn's but with idiosyncrasies of Rorty's project, such that he can casually remark "incommensurability entails irreducibility but not incompatibility" (Rorty 1979, 388). The use is sensible in his context, but trying to plumb such subtleties in a general survey would be all-consuming.

[14] Ian Hacking (1983) lists three: *topic* incommensurability, *meaning* incommensurability, and *dissociation.* Philip Kitcher (1983) concurs in number but not in adjectives or reference (*conceptual, observational, methodological*), as does Lance Simmons (1994) (*global, thin,* and *thick*), and Robert Miner (2001) (just *IT1, IT2,* and *IT3*). Howard Sankey (1999; Sankey and Hoyningen-Huene 2001) has it at two, *semantic* and *methodological,* as does Alexander Bird (2000), *semantic* and *epistemological.* Richard Grandy (1983) does not have snappy labels, but has a very perceptive account of the (for him, six) kinds of incommensurability as a function of the components on which rival paradigms can differ: *symbolic generalizations, metaphysical commitments, models, values, instruments,* and *exemplars.* He also advocates a notion of degree of incommensurability, which correlates with the quantity of these components that are different between the paradigms. I have, of course, poached much of my own account from these various taxonomies. Most of them expand or reduce into each others' categories in fairly obvious ways—Kitcher's *conceptual* is Sankey's and Bird's *semantic,* while his *observational* and his *methodological* correspond jointly to their single categories (respectively), *methodological* and *epistemological*—but each scholar also develops his own specific concerns in different ways, with little overt recognition of others' treatments, so there are lots of small distinctions among them as well. Building a taxonomy beyond my four categories would largely be an exercise in decomposing pragmatic incommensurability into its various component parts—at the least, practical (or methodological), thematic, and (comparative) value incommensurability.

[15] I have in mind here something of a continuum, treating as endpoints the elements in the distinction Thagard and Zhu (2002) make between *weak* incommensurability and *strong* incommensurability); Hacking's (1983) *shallow* and *deep,* Collier's (1984) *weak* (*relative*), *strong* (*absolute*), and *logical* (*universal*), Sankey's (1994) *limited* and *radical,* and Forster's (2000) *weak* and *strong,* all mark similar contrasts. By *continuum,* however, I am not suggesting that there is some commensurability calculus that could give meaningful numbers we could apply to theories (paradigms, worldviews, values, . . .). Thagard and Zhu's definition of the endpoints of such a continuum is very helpful: "Two theories or conceptual schemes are strongly

incommensurable if they are mutually unintelligible, so that someone operating within one conceptual scheme is incapable of comprehending the other. Weak incommensurability, however, does not imply mutual unintelligibility, but only that the two conceptual schemes cannot be translated into each other" (2002, 81). My *total* (or *brick-wall*) *incommensurability* matches their definition of *strong incommensurability* very closely, though my *zero incommensurability* clearly needs to go lower on the scale than their *weak incommensurability* (i.e., low enough to drop the prefix; it is the same as *commensurability).* For those in need of a more technical definition for the endpoints, Hintikka's (1988, 29) can be adapted naturally: "T_1 and T_2 are (totally) incommensurable if and only if there is no question Q_i in a set of relevant questions $\{Q_i\}$, $i \in q$ which is answerable on the basis of both of the two theories." The definition flips nicely for zero incommensurability (or total commensurability): T_1 and T_2 have zero incommensurability (are totally commensurable) if and only if every question Q_i in a set of relevant questions $\{Q_i\}$, $i \in q$ is answerable on the basis of both of the two theories. Hintikka's definition completely formalizes the 'degree' sense of incommensurability, since, in principle, it generates a ratio for any pair of theories representing "the total information of their shared answers to the total information of the answers yielded by the two theories combined"(1998, 25). I, however, would not be first in line to try and calculate such a ratio.

[16] My use of *diachronic* here is a direct import from linguistics, but I am distorting *synchronic* slightly. In particular, synchronic linguistics operates in an idealized frozen slice of time, but what I am calling synchronic incommensurability really occurs over a stretch of time, sometime decades, even (as in the Ptolemaic/Copernican shift) a century or so.

[17] In addition to generating this helpful definition, Hintikka's approach is also very suggestive in its implication that ratios might be generated to yield a measure of relative incommensurability, in accord with our thermometer model.

[18] For the most part, the overlapping-domain criterion is just a background assumption for advocates of incommensurability, but occasionally it is made explicit, as in one of Feyerabend's early formulations: "there exist pairs of theories, *T* and *T'*, which overlap in a domain D' [. . .]" (Feyerabend 1981a, 67).

[19] There is nothing necessarily cosmic about this wholesale shift (it might also attend pragmatic or semantic incommensurability, for instance), but the gestalt-switch analogy (1) is central to Kuhn's thinking about paradigm-change incommensurability and (2) collocates strongly in his work with the different-world, cosmic-incommensurability talk. So, this is a good place to take it up. As Hoyningen-Huene notes in his essay in this volume (152), both the different-world image and the gestalt-shift image were supported, if not suggested, by Kuhn's own Eureka-moment over Aristotle's physics.

[20] Kuhn later blamed much of the confusion over incommensurability on his appeal to gestalt imagery and his exploitation of the conceptual dimensions of the verb, *see* (Kuhn 2000, 34), a point which Nola Heidlebaugh develops, tracing many of the difficulties attending incommensurability to what she describes as collapsing "linguistic understanding into visual" (Heidlebaugh 2001, 12), and arguing (in a way consonant with themes I take up later in this introduction) that visually driven static argumentative practices tend to lead to blockages, while orally driven dynamic argumentative practices tend to keep communicative lines open (see especially Heidlebaugh 2001, 102ff).

[21] In a passage I'm not sure I fully understand but that seems to suggest some scientists at the cusp of incommensurable programs are unwittingly bilingual, Feyerabend implies that scientists *can* flip such switches, unawares: "It is [. . .] possible that being well acquainted with both theories [some scientists] change back and forth between them with such speed that they seem to remain within a single domain of discourse" (1978a, 283). There are some rough analogies to this putative phenomenon with bilinguals and code-switching bidialectals, but the analogies are very difficult to sustain—such subconscious switching only happens in the swing of the moment, almost always when talking with other bilinguals or code-switchers, and the subconsciousness is only temporary. It is very easy to remind such speakers that they have just gone into Spanish for half a sentence or so, or that they have mixed words together from two dialects or registers. Galison justly ridicules Feyerabend's position (Galison 1997, 815).

[22] There are differences, some of them substantive, between the appearance of "Explanation, Reduction, and Empiricism" in the *Minnesota Studies* volume (Feigl and Maxwell 1962), and in Feyerabend's *Philosophical Papers* (Feyerabend 1981a). For convenience (yours and mine; the collected papers are more readily available) I quote from *Philosophical Papers,* for "Explanation" as well as for any others collected there. I have not made a systematic study of the differences between the two versions (though it is clear enough on brief inspection that, unsurprisingly, many of them have to do with erasing his own Popperian allegiances or finding new ways to goad Popper). But, for historiographic reasons, I have cross-checked claims about what Feyerabend said in 1962.

[23] He also uses a conversion-like image in his autobiography, of the time of his first inklings about our major theme. Among his pontifications to Peter Geach and L.L. Hart, in early contact with them at Elizabeth Anscombe's house (c1952), he recalls offering, "Major discoveries [. . .] are not like the discovery of America, where the general nature of the discovered object is already known. Rather, they are like recognizing that one has been dreaming." And adds "Today there is a technical term for such changes—incommensurability" (Feyerabend 1995, 92).

[24] In part, this recurrent example is because of how clear incommensurability appears through diachronic lenses. "The differences become evident when we move to an alien culture of a distant historical period," he says. "The Greek gods were a living presence; 'they were there.' Today they are nowhere to be found" (Feyerabend 1987a, 104). Always aiming for Popper, by the way, Feyerabend was careful to observe that the shift from a Homeric to a classical Greek picture cannot be accounted for "by a method of conjectures and refutations that knows no end" (Feyerabend 1978a, 270).

[25] Hacking's category, actually, is dissociation, but it maps pretty closely (from my perspective anyway) to what I am calling cosmic—high-level communicative blockages with respect to differing views of the same (or overlapping) domain(s).

[26] Hacking's retreat to sixteenth-century proto-medical, alchemical discourse should alert you that he regards these severe conceptual mismatches to be rare; indeed, while they come up for discussion in his chapter on incommensurability, it is not even clear he sees them as a belonging with the same set of problems—for instance, he doesn't even use *incommensurability* as part of his label for the different-worlds situation (calling it *dissociation*) (1983, 69–72), and his overall treatment of incommensurability appears to exclude different-world considerations from science generally and the successive-paradigm cases particularly. He confines dissociation to major historical divides (though he might admit radically different contemporary cultural divides). Certainly he wants to filter off the different-world implications from incommensurability for a closer look at the more easily inspected claims Feyerabend and Kuhn make about communicative barriers and discontinuous theory change.

[27] In yet another taxonomy-of-incommensurability move, Xinli Wang invokes a division linked directly to linguistic relativity (Wang 2002, 482). He doesn't give his types explicit labels, but his descriptions suggest *intra-linguistic* (or *intra-cultural*) and *inter-linguistic* (or *inter-cultural*). The first type is "a radical conceptual shift in a scientific language within the same cultural or intellectual tradition," the second "substantial semantic and/or conceptual disparities between two comprehensive theories embedded in two coexistent, distinct, intellectual/cultural traditions [. . .] (such as Chinese medical theory *versus* Western medical theory)" (Wang 2002, 467n8). I am not aware of anyone else working directly in terms of these categories, though Thagard and Zhu's (2002) study of a putatively 'strong incommensurability' case tacitly makes use of such a distinction (in fact, their study tests part of Chinese medical theory against Western medical theory).

[28] Scientific and technical English, incidentally, is even more acutely thing-oriented than daily English. It is characterized, for instance, by relatively lengthy noun chains (like "recuperation program procedure indication

sheet"), agentless passives (which elide noun phrases responsible for activity), and nominalizational morphology (which turns different lexical categories into nouns, like the–*ity* that turns the adjective *incommensurable* into the noun, *incommensurability*). See, for instance Gross, Harmon, and Reidy (2002), or essays in Halliday and Martin (1993).

[29] It apparently goes back to someone's poor reading of a note in Hanson's *Patterns of Discovery,* which begins with a different (real) quotation of Wittgenstein, but which includes a long (Eskimos-have-N-words!) passage from a psychology book (Crafts *et al.* 1950), whose authors are in fact responsible for the statement. See Hanson (1958,184).

[30] Ian Hacking is a rare exception: "Kuhn's own statements of it [different-world] are very cautious and hedged" (Hacking 1993, 275).

[31] See Hoyningen-Huene (1993, 31–42) for further discussion of Kuhn's locutions in this context.

[32] As Hoyningen-Huene explicates, in his contribution to this volume (and elsewhere), this example does not work under Feyerabend's *non-instantial* construal of incommensurability. I am not as comfortable as Professor Hoyningen-Huene, however, that Feyerabend maintained this construal throughout his career (nor, I hasten to add, am I as intimate with Feyerabend's writings as is Hoyningen-Huene).

[33] Carnap, by the way, is said to have held an incommensurability thesis, dating to at least 1936 (Eaman 1993, 11).

[34] The term *semantic* and its synonyms have attended incommensurability from the beginning, and the majority of commentaries on incommensurability have just tacitly assumed the semantic variant. It's only when one wants to try and isolate meaning-related issues from other aspects (especially the world-changing and value-dependent aspects) of incommensurability that being explicit about the label makes sense. Sankey, in preparation for the 1999 conference that issued in Hoyningen-Huene and Sankey (2001a), drafted the term to discriminate semantic matters (like meaning variance and translation) from methodological issues; see Sankey (1999), Sankey and Hoyningen-Huene (2001b). Bird (2000) also uses *semantic*. Hacking (1983) calls this aspect (or something very nearly this aspect) "meaning incommensurability," Kitcher (1983), "conceptual."

[35] *Radical* here, means, without passing through a known third language: "translation from a remote language on behavioral evidence, unaided by prior dictionaries" (Quine 1969, 45).

[36] Quine, though, really isn't making a point about translation—or rather, he is using these points about translation to make a bigger point, that *all* meaning is indeterminate. I'm staying with translation issues because that's what Kuhn does with Quine.

[37] This is Putnam's formulation: "The incommensurability thesis is the thesis that terms used in another culture, say, the term 'temperature' as used by seventeenth century scientists, cannot be equated in meaning and reference with any terms or expressions *we* possess" (Putnam 1981b, 114), which is certainly at the strong end of the scale.

[38] In fact, Kuhn's position was precisely Popper's on the matter of translation. Translation is possible, Popper argued, but difficult, and the notion of incommensurability "exaggerates a difficulty into an impossibility" (Popper 1970, 56–7). Kuhn always contended that incommensurability made communication difficult, but not impossible, and that the chief route to communicative success was translation (or interpretation). From his perspective, it was his critics who exaggerated the difficulty into an impossibility and then ascribed (like Popper) the impossibility-claim to him.

[39] It's not clear at all to me what Kuhn thinks he is leaving behind to 'translation' when interpretation is removed, strict source-to-target denotative mapping perhaps, a phenomenon that no translator would regard as translation.

[40] The different-world thesis, along with the gestalt and conversion imagery that accompanied its earliest versions, is not just eliminated from his work, it is eliminated in apologetic terms. See, in particular, the speaker's reply postscript to his paper for *Possible Worlds in Humanities, Arts and Sciences* (Allén 1989), reprinted in Kuhn (2000, 86–89), where Kuhn regrets that his own Eureka experience in overcoming the diachronic barriers to Aristotelian physics unduly influenced his conception of the synchronic communicative clashes in science, and his remarks in conversation with Giovanna Borradori, where incommensurability is centralized and exclusively linguicised: "The notion of incommensurability has become increasingly important for me. And my own current work is more closely identifiable as an attempt to explain what incommensurability is and how such a thing could exist, than to explain any other single problem. Now I would put exclusive emphasis upon the linguistic component of it. I talk increasingly of incommensurability as untranslatability" (Borradori 1994, 161).

[41] Howard Sankey terms this variant "taxonomic incommensurability," which is also the title of a paper in which he delineates it very effectively (Sankey 1998).

[42] See Kitcher (1983, 531ff), Kuhn (2000, 40ff).

[43] This position, in one of the more interesting correspondences between Kuhn and Feyerabend, is effectively the one which Feyerabend holds in "Explanation," where he predicates incommensurability directly of terms and concepts as much as he does of theories or frameworks. The following pages are to the 1962 Minnesota Studies version of the paper, to which I had electronic access: the predications to terms and concepts are on pages

31, 59, 75, 78, and 90 (twice); to theories, pages 75, 92, and 93 (3 times); to frameworks, page 83; Feyerabend also says that incommensurability holds between a law (or "conceptual apparatus" which he treats as a synonym for *law* here) and a theory, on pages 58, 59 (twice), and 74 (twice). One difference that I happened to catch between the 1962 and 1981 versions is that he interpolated a few words to reduce the scope of incommensurability with respect to impetus and Newton's framework; where the earlier version was "incommensurable character of the conceptual apparatus" (1962, 58), the later one came out as "incommensurable character of part of the conceptual apparatus" (1981a, 67).

[44] This category bears only accidental links to the way its nominal phrase is deployed in John Collier's (1984) paper, "Pragmatic Incommensurability." The paper is very interesting, and much of it is highly consonant with the position I am arguing here, but Collier does much different things with the notion of "pragmatic incommensurability" than I do. He does not introduce pragmatic concerns for taxonomic reasons (his own categorization follows a three-grade continuous scale). He correlates "strong incommensurability" with "deeper roots" than semantics (Collier 1984, 152)—as do I—but that's not where he sees the workings of pragmatics. He makes incommensurability rest largely in the hermeneutic strategies of theorist collectives—again as do I—but it is not clear how pliable he regards these strategies to be, how open to negotiation, and he does not interest himself in the (inter-paradigmatic) rhetorical space between competing theories, but rather in the (intra-paradigmatic) relation between "interpretive techniques" and the theories they interpret. This concern leads Collier to regard (his) pragmatic incommensurability as falling at the weak end of his scale, whereas I see it (*my* pragmatic incommensurability) as falling at the strong end of my own scale, since the misaligned presuppositions can be deeply subterranean. Collier and I share an important viewpoint, however, despite our very different routes. He sees "weak incommensurability" as fully "sufficient to explain Kuhn's historical evidence," so that we don't need the "strong incommensurability" thesis to account for scientific disputes—more specifically, "assuming that the evidence strong incommensurability is hypothesized to explain can be accounted for by the weaker hypothesis, such mysticism is best avoided" (Collier 1984, 152). I would go somewhat further and say that this weaker thesis can be eliminated as well, that a notion of incommensurability buys us nothing outside of mathematics that non-absolutist notions like "misalignment" or "incommensurateness" don't buy us, at much lower cost.

[45] Under a somewhat more restricted notion of values, localised to theory-appraisal criteria, Newton-Smith also identifies differences in values as a principal source of incommensurability (Newton-Smith 1981, 112). Following his lead, and the discipline-specific work of Douglas Weed (1997) and Robert Veatch with William Stempsey (Veatch and Stempsey 1995),

Lawrence Prelli adopts the term *values incommensurability* in his essay in this volume (294).

[46] See Howard Margolis's revealing account of how the barrier he calls *UCM* (for uniform circular motion) retarded both Kepler's work and then acceptance of Kepler's work. He doesn't use the language of values, however, talking rather of "habits of mind" that become "so deeply entrenched and entangled in expert practice that escaping [them] is difficult" (H. Margolis 1993, 32ff). As he develops it throughout the book (*Paradigms and Barriers*), this sort of entrenchment is very clearly a cognitive correlate of incommensurability.

[47] Holton defines themata as "fundamental presuppositions, notions, terms, methodological judgements and decisions" (Holton 1988, 41); anti-themata are precisely the same, except that they can stand in opposition to other themata—order and disorder, for instance, or atomism and the continuism. Margolis, too, investigates such mental patterns. His term, *habits of mind,* stresses their cognitive dimensions more directly than Holton's *themata*, as does his (idiolectal) observation that they "are not available to introspection and [. . .] are never exactly the same for two individuals" (H. Margolis 1993, 150). But he is considerably more interested in social psychology.

[48] H. Margolis (1993, 134) calls this phenomenon a "breakthrough effect."

[49] As does the use of its antonym, *commensurable,* and the related terms, *commensurate* and *incommensurate.* I am just relying on the Oxford English Dictionary here (OED2), not any philology of my own—which even records *incommensuration* as a synonym with *incommensurable,* but there is only one (seventeenth-century) citation for that one; and much more widely *commensuration,* which has a small range of senses all related to measurement and proportion (Finnis 1997 deploys this term in the incommensurability-of-values literature). The nonmathematical use of *incommensurability* was apparently far more restricted, though I have found at least one early, consciously analogic case (Beaumont 1665, 157), before the twentieth century, when it starts to populate values discourse, still quite sparsely (e.g., Urban 1916, Stuart 1939, Grant 1956). The OED2 editors and compilers do not list a 'general' use for *incommensurability.*

[50] Coke is establishing the autonomy of the Understanding from the Senses (using the technical vocabulary of faculty psychology), a claim which might "at first blush seem a strange Paradox [. . . but which] is as true as any Proposition in Geometry: For the outward Senses apprehend only the corporeity of substance of things represented unto them, but the Understanding only the incorporeity of things so seen, &c, and discerns and judges whether such things so apprehended by the Senses, be pleasant, profitable, just or unjust, reasonable or unreasonable, commensurable or incommensurable" (Coke 1660, 12).

[51] There is a distinctly different notion of absolute incommensurability in the values literature, from Rashdall on, than there is in the science-studies literature. There is no brick-wall, animal-making-noises, zero-common-points-of-contact sense on the part of any of the value incommensurabilists or their critics. Ruth Chang makes this explicit, with her notion of 'covering value.' "Take, for instance, the comparability predicate 'comparable with respect to aural beauty,'" she says. "The pair <fried eggs, the number nine> does not belong within the domain of the comparability predicate, because fried eggs and the number nine do not belong within the domain of 'aurally beautiful.' Similarly, the pair falls outside the domain of the incomparability predicate. We shall say that the value of 'aural beauty' does not 'cover' fried eggs" (Chang 1997, 28).

[52] Actually dating this slogan, however, is another matter. The earliest use I have found is Urban (1916, 678), and it seems to have had at least a moderate currency, which hit obsolescence by the mid-twentieth century. C. K. Grant, in the 1950s, for instance, gives passing mention to "a problem which used to be called 'the incommensurability of values'" (Grant 1956, 407). Lamprecht does not appear to use the slogan, though my search has not been exhaustive.

[53] I am not, please be sure, suggesting any influence of Lamprecht on Kuhn or Feyerabend, direct or indirect, just noting parallels related to their key uses of *incommensurable.* One scholar, incidentally, sees Feyerabend's later views as, effectively, constituting an incommensurability-of-values position. In a review of *Science in a Free Society* (Feyerabend 1978b), Noretta Koertge comments, by way of repudiating such a position, that "in my experience fruitful dialogue with sexists, racists, homophobes, fundamentalists, sociobiologists and punk rock fans is sometimes possible. Of course, it is logically possible that people in opposing traditions share no basic values. But given the shared problem of survival we all face, I think a total incommensurability of values is unlikely on evolutionary grounds. Surely, the best strategy is to assume there is some minimal uniformity in people's morals and cognitive systems. (Of course, unlike Feyerabend, I do not live in California.)" (Koertge 1980, 388-389).

[54] See Dewey (1934, 124): "belief [. . .] always involves valuation, preferential attachment to special types of objects and courses of action [. . .]. The chief role of philosophy is to bring to consciousness, in an intellectualized form, or in the form of problems, the most important shocks and inherent troubles of complex and changing societies, since these have to do with conflicts of value."

[55] Henry Waldgrave Stuart (1870–1951; Ph.D. Chicago, 1900) was a president of the American Philosophical Association, and there is a chair in philosophy at Stanford (where he taught) named after him. Sterling Power

Lamprecht (1890-1973; Ph.D. Columbia) has a graduate fellowship named in his honour at Amherst, where he taught and chaired (1928–1956). Both Stuart and Lamprecht gave Howison lectures at Berkeley, two years apart (Stuart in 1936, Lamprecht in 1938).

[56] Someone who does try to figure out who else might have thought of the value-incommensurability/pluralism blend that Berlin advocates is his literary executor, Henry Hardy, who catalogues a wide number of scholars he identifies as sources of "pluralism before/independently of Isaiah Berlin," including Lamprecht, Stuart, Dewey, and James. Most entries in the list are without comment, and there is no general discussion of American pluralism, but Hardy phrases the relations between Lamprecht's and Berlin's positions in these terms: "[Lamprecht's formulation] is a remarkably precise anticipation of Berlin's ideas, unless of course Berlin drew on it (unawares?) himself" (Hardy 2004).

[57] The essays themselves were written in the period 1948–1959, but the extensive and influential introduction, which centralizes the relationship between incommensurable values and ethical pluralism, was written for the 1969 collection.

[58] In a brief passage which marks the closest one gets in the value-incommensurability literature to an acknowledgement of the influence Kuhn and Feyerabend had on the fortunes of the field's key term, Fred D'Agostino says "although Isaiah Berlin had already in 1958 used the word 'incommensurable' in a way relevant to our concern, recent discussion of incommensurability became intensive and focused, of course, in the 1970s, as the work of Thomas Kuhn and Paul Feyerabend, in the 1960s, was assimilated and dissected" (2003, 27).

[59] Chang's move, though, is not a simple substitution. She defines *incommensurability* as a somewhat more precise variant of *incomparability.* Here is what she says: "Let us henceforth reserve the term 'incommensurable' for items that cannot be precisely measured by some common scale of units of value and the term 'incomparable' for items that cannot be compared [. . .] I am going to set aside the first idea—incommensurability—and focus on the second—incomparability" (1997, 2). In fact, she then spends most of the introduction trying to dispose of incomparability as well. Her introduction is a fine essay on the importance of practical reason for making choices that might seem unmakeable, but it is peculiar as an introduction to this particular collection of essays. The use of *incommensurability* in the title looks a bit like a typical bait-and-switch move, since she has so little interest in it, but all the other essays in the book retain *incommensurability.* Moreover, most of them use that word pretty much the way Chang uses *incomparability,* and many of them use the two terms synonymously.

[60] Paul Hoyningen-Huene's essay in this book takes up this issue more carefully, along with many other technicalities about Kuhn and Feyerabend's intentions with respect to *incommensurability.* In the meantime, I'll just note here that what Feyerabend and Kuhn explicitly say on the topic of comparison is (1) that the empirical consequences of rival theories cannot be compared, so that the criterion of 'factual adequacy' can't adjudicate between theories, but (2) that the lack of consequence comparison does not mean the theories themselves cannot be compared on any number of metrics (particularly values such as simplicity and scope). Kuhn especially devotes much ink to showing how they can be compared, and indeed, the sorts of comparative predicates that concern value incommensurabilists are exactly the sort that Kuhn says new programs are marketed with: "the new theory is said to be 'neater,' 'more suitable,' or 'simpler' than the old" (Kuhn 1996, 155).

[61] Alasdair MacIntyre, though, traces the source of intramural value clash in the divergent genealogy of values, back to incompatible moral traditions. "It is not surprising that there is no rational way of deciding which type of claim is to be given priority," among (for instance) rights, utility, and justice, he says, "or how one is to be weighed against another, [because m]oral incommensurability is [. . .] the product of a particular historical conjunction" (MacIntyre 1981, 68).

[62] The most full-blooded case of an eliminative incommensurability is Joseph Raz's "constitutive incommensurability," which describes certain non-fungible value relationships, where even to contemplate a comparison (or, under a slightly softer interpretation, to contemplate acting on such a comparison) negates the superior value. "It is impoverishing to compare the value of a marriage with an increase in salary," for instance, says Raz, or "it diminishes one's potentiality as a human being to put a value on one's friendship in terms of improved living conditions" (Raz 1986, 22). The key feature is that the act of comparing in such cases constitutes the incommensurability. "We run into 'constitutive incommensurability' of values whenever treating values as commensurable subverts one or both of the values to be entered into the trade-off calculus," Tetlock *et al.* observe, "To compare is to destroy" (Tetlock *et al.* 1996, 37).

[63] A highly notable exception to this strain of attitude-not-methodology in the values literature is Fred D'Agostino, who develops a sophisticated "*technology* of COMMENSURATION" (2003, 20).

[64] Griffin is an exception here, saying "the monism-pluralism issue is not especially central to the issue of incommensurability" (Griffin 1986, 90).

[65] It is perhaps here that the two uses are almost identical "'Incommensurability' was introduced" David Wiggins says "in order to suggest the

heterogeneity of the psychic sources of desire satisfaction and evaluation," which have "a certain liability to fragmentation" (Wiggins 1981, 262).

[66] Elsewhere, he discusses the "crisis of reason" in the sciences, brought to attention by Kuhn and Feyerabend (Lyotard, van Reijen, and Dick Veerman 1988, 280).

[67] That's not to say one cannot do serious work with the suite under the postmodern umbrella. Aristotle characterizes rhetoric and dialectic as having no subject matter of their own, only a manner with which to approach subject matter, and that the closer one approached a specific subject with either of these instruments, the nearer one came to being a practitioner rather than an analyst of that discourse (*Rhetoric* 1358[a]). One might approach mathematics from a dialectical direction, but the nearer one gets to the claims and the equations, the less one is doing the philosophy of mathematics, the more one is doing mathematics. That's the way–*isms* tend to work as well. So, for instance, when legal theorist Pierre Schlag puts the *incommensurability* suite to serious use in *The Enchantment of Reason,* he specifically develops its perceptual and translational aspects, draws directly on Feyerabend and value-pluralists, and uses these resources to advance a postmodernist critique of the grip modernist reason has over jurisprudence (Schlag 1998).

[68] Gross argues in this volume (193) that meta-incommensurability is a misleading notion, because it is not sufficiently Kuhnian (in particular, that the programs Hoyningen-Huene and his colleagues describe as meta-incommensurable do not sufficiently manifest the characteristics of a Kuhnian lexicon). Rather, he argues, the relevant phenomena are better characterized by W. B. Gallie's term, *essentially contested* (e.g., Gallie 1968). I agree with Gross all of the way along, except that I don't want to reserve the term *incommensurability* for situations of the strictly Kuhnian lexical kind. What I am calling *pragmatic incommensurability,* in particular, has much in common with essentially contested ideas, and Gallie's notions of cooperative engagement are very congenial.

[69] See Prelli's essay in this volume for some discussion of this substitution move (298f); Lessl and Campbell make it without comment. Fred D'Agostino expresses a concern related to the one that leads Prelli to abandon *incommensurable,* in his introduction to *Incommensurability and Commensuration*—saying that *incommensurable* doesn't hold "if the modal operator 'able' is given a strong reading" (2003, x), but that suffix consistently has a strong reading (X-able pretty much always means 'able to do X'). Another troublemaker occurs at the other end of the word, another 'strong' marker, the prefix *in-*, which negates the suffix, giving us, effectively, 'unable to do X'—in the particular case, 'unable to measure to a common standard.' Prelli's suffix, *-ate,* on the other hand, has the virtue of being much more neutral; it just makes adjectives without contributing any modal content at all, so that

incommensurate really means something like 'not measured to a common standard,' without any negation of the ability to do so.

[70] Feyerabend quotes this passage in *Against Method* (Feyerabend 1978a, 286–7), though he draws somewhat different conclusions from it.

[71] Attributed variously to Max Weinreich, his son, Uriel Weinreich, and Yoshua Fischman. The earliest published occurrence seems to be Weinreich senior's Yiddish formulation: "A shprakh iz a diyalekt mit an armey un a flot" (M. Weinreich 1945, 13).

[72] The impressive exception to my "aren't the sorts of analogies that have been traditionally pursued" is Peter Galison's astute application of pidgin and creole analogies in his *Image and Logic* (Galison 1997, esp. 49–51), which we will take up somewhat later. His analogy to pidgins is especially illuminating, but it doesn't go through fully.

[73] This incommensurability-in-the-eye-of-the-arguer position is inevitable from the vantage of a sophistic rhetoric. Nola Heidlebaugh pursues it most fully, in her *Judgment, Rhetoric, and the Problem of Incommensurability* (Heidlebaugh 2001).

[74] Nor does it mean, as some early readers of this argument have worried, that incommensurability is therefore *only* psychological. (It could be, of course, depending on how big a basket one labels "psychological"—but so could, by the same token, music, education, the culinary arts, chess, you name it.) Rather, from a standard rhetorical perspective, the territory of psychology in this question is in establishing the rhetorical preconditions of incommensurate misunderstandings—the attitudes that must be addressed and managed and brought into synch, rhetorically, through a discursive meeting ground. This linkage between psychological preconditions and effects on the one hand, and rhetorical management on the other hand, is of very long standing. Gorgias compared rhetoric (more precisely, suasive *logos*) to a drug that induces feelings and behaviours (*Encomium of Helen,* §14). Antiphon of Rhamnus apparently set up the first psychotherapeutic practice, in the agora, and wrote a treatise entitled the *Art of Avoiding Distress* (Kerferd 1981, 51), And Aristotle's extensive treatment of moods and emotions in *The Rhetoric* has been called "the earliest systematic discussion of human psychology" (George Kennedy, in his introduction to his translation of Book 2, Chapters 2–11 of *The Rhetoric*—Kennedy 1991, 122). What is rhetorically relevant about incommensurability is that it is manifest in failures of discourse, and that discourse (under conditions of goodwill, charity, and reasonableness) is required to dissolve it.

[75] The evidence is clear on this point. First, Feyerabend delighted in galling philosophers, and most everyone else. He frequently chooses the more inflammatory *propaganda* over *rhetoric,* to raise the level and the scope of the irritation. Along with his index entry for *rhetoric,* he also has ones for

wicked remarks, bastard subject (under *philosophy of science*), *the unique scientific method as an ideological fairy-tale of scientific chauvinism* (under *scientific method*) and this one: *top dogs,* [. . .]; *see also big shots; Nobel prize winners.* But, more pointedly, he seems not to be very familiar with the tools and schools of rhetoric, and he quite regularly uses *rhetoric* in its widespread, philosophically standard, pejorative sense—as in this peculiarly phrased defense of *Against Method,* "I would say that my book contains 85% exposition and argument, 10% conjecture, and 5% rhetorics" (1988, 405). The term he earlier equated with *logic* now stands in opposition to *argument,* and shows up only as a mild confession of impropriety (mild not because of the designation, but because of the percentage of its use). This usage reveals much about Feyerabend, not the least being his lack of devotion to consistency, and his indifference to rhetoric as an intellectual field. Still, as Tom Cook and Ron Seamon point out (1988), there are many aspects of his work which place it within the bounds of rhetoric (I reject their peculiar conclusion, however, that Feyerabend is primarily useful only for examining what they call "external" rhetoric of science—the arguments that scientists offer to the public for their claims, as opposed to the arguments scientists offer one another about their claims).

[76] Philosophers who appeal to rhetoric include Steve Fuller, Bruno Latour, and Marcello Pera; historians include Mara Beller, Jan Golinski, and Peter Dear; Sociologists include Michael Overington, Trevor Pinch, and Steve Woolgar. Relevant texts for each of these scholars are in the references.

[77] Foundational texts include Prelli (1989a), Gross (1996), Bazerman (1988), and Myers (1990), as well as the papers collected in R. A. Harris (1997); other important texts include Fahnestock (1999), Condit (1999), Bazerman (1999), Ceccarelli (2001), Gross, Harmon, and Reidy (2002).

[78] Maurice Charland (2003) offers an interesting argument to the effect that rhetoric of science is incompatible with Kuhn's view of science *except* during moments of revolutionary transition and incommensurability.

[79] See Perelman's observation that "a curious use of the argument from authority occurs when the qualified authority cannot understand an assertion," which leads people "to conclude that the assertion is incomprehensible, that no one can understand it" (Perelman 1982, 95).

[80] As we saw in Hacking's account of Paracelsus above, these are the criteria for 'getting' historically dissociated perspectives as well. Kuhn, of course, counseled exactly this attitude, calling for imagination, patience, and effort from hermeneuts of earlier paradigms and other cultures (2000, 61, 220). And here is historian Jed Buchwald expressly requesting his reader to adopt such an attitude:

> [The historian's job is] to lead the reader past modern preconceptions, but without entirely abandoning a modern

> perspective, to an understanding of deep foundations—here those of optics just before and just after the development of wave theory. And again I must ask for [your] patience and indulgence. (Buchwald 1989, xxii)

See Alan Gross's essay in this volume for an account of how this kind of sympathetic historical hermeneutics led Kuhn to philosophy and the development of his notion, incommensurability. "The daily hermeneutic struggle of historians of science is philosophically significant [for Kuhn]," Gross writes, "it tells us [. . .] something deep and important about development and change in the sciences" (181).

[81] This is my account of what Galison is doing, not his own account. I develop the analogy in terms of metalinguistic resources and formal pidgin characteristics further than he does; Galison simply uses *trade language, interlanguage, pidgin,* and sometimes *creole* (though he makes a formal distinction between the last two) pretty much as synonymous labels for the medium of inter-cultural encounters and of inter-party encounters.

[82] Richards, in fact, devoted a good deal of energy to an interlanguage developed by his friend, C. K. (Charles) Ogden—Basic English, "an International Auxiliary Language, i.e., a second language (in science, commerce, and travel) for all who do not already speak English" made up of 850 English words (Ogden 1930). See, for instance, Richards (1933, 1943).

[83] To avoid clogging up the text unnecessarily, here are the citations (to Butterfield 1957): "healthy friction" (21), "variations" (29), "rival explanations" and "rival systems" (83), Priestly (218), Boyle (115), Mersenne (83); "truth-freak" is my own Feyerabendian characterization of Descartes.

[84] Gladwell is explicating Randall Collins' book, *The Sociology of Philosophies* (Collins 1998), which explores this kind of synergistic collaboration in the history of philosophy.

[85] More fully, Grandy's very perceptive rendering of Kuhn's vision, defines 'scientific community' as "those scientists whose similarity judgements are more alike than they are like members of other groups. And the 'group similarity relation' is a useful fiction that provides a brief way of speaking about the diverse though rather similar relations belonging to the individual members of the group. Kuhn's point is that the group need not be defined by perfect agreement in their judgements, but only in terms of their relative ease of agreement in contrast with the difficulties of communicating with members of other groups" (Grandy 1983, 18).

[86] Close and detailed case studies—the opposite of vignettes—are the appropriate remedy for this artificiality, but they require time, great effort, and Dickensian amounts of space, to treat properly. Rudwick's densely packed *Devonian Controversy* clocks in at nearly 500 pages, David Hull's *Science as Process* (zeroing in on evolutionary biology and systematics) closer to

600, Galison's *Image and Logic* over 900. Rudwick's book is especially helpful in this regard, at least for my sensibilities, because of the visuals he offers of the relative conceptual "locations" of the arguments and arguers: they look like electrical circuit diagrams.

[87] Robert Olby's (2003) article, "Quiet Debut for the Double Helix," would seem to counter my observation here on the rapid acceptance of Watson and Crick's proposal. The Olby piece is a nice little chronicle of the reception of the 1953 paper that couches it in terms like "muted" and charts the additional "information [. . .] needed to convince the scientific community" (Olby 2003, 402). The acceptance was not, in other words, instantaneous. It was certainly, however, rapid: nine years later Watson and Crick and Maurice Wilkins were on the podium in Stockolm.

[88] Noam Chomsky and some of his followers in a type of transformational grammar that began in the late 1960s, for instance, insisted that the main rival to their program was merely a notational variant of their own work; however, it was also disastrously wrong, Chomsky maintained, while his own views were correct. See, in particular, R. A. Harris (1994).

[89] Garber's case is actually very consonant with my line in this introduction. He argues incommensurability away by shifting the contexts in which the Cartesian and Aristotelian programs are compared, effectively assimilating the Cartesian mechanist picture with Aristotelian mechanics, rather than (as Kuhn had done) distinguishing it from Aristotelian natural philosophy (specifically, its physiological implications).

[90] Another person who joins us on this question—and, in fact, who uses the word *phronesis* as a key term in the commensuration of discrepant standpoints—is Nola Heidlebaugh, in *Judgment, Rhetoric, and the Problem of Incommensurability* (e.g., Heidlebaugh 2001, 80).

[91] One of the privileges of being editor, however, is getting the last scratch in. Gross argues that the unsavory stew Brewster cooks up shows that there 'really is' incommensurability. From my perspective, the case illustrates the reverse. Brewster demonstrates commensurability by bringing the two views into close alignment. It is not, however, an alignment that could be successfully marketed. At most, the Brewster study shows that there are significant incompatibilities in rival frameworks, something I would never dispute, which cannot be brought together satisfactorily for the relevant communities. The difference is that Gross places the lack of reconciliation in the formal products—the frameworks, or, more particularly, their discrepant lexicons. I would place it rather in the community with an interest in those formal products, Brewster's audience, which didn't see the point of contributing to the development of his model when it had another one it already found more productive.

2

Three Biographies: Kuhn, Feyerabend, and Incommensurability

Paul Hoyningen-Huene

Introduction

1962 is the official birth year of *incommensurability*, in its contemporary, extra-mathematical sense; the year it first appears in the science-studies literature.* More properly, I should say "contemporary extra-mathematical senses." *Incommensurability* had two fathers, unusual even for philosophical terms,[1] and the joint paternity of Thomas S. Kuhn and Paul K. Feyerabend has contributed to much subsequent confusion. While the signifier is the same for both of them, each father engendered a different signified, taking the mathematical metaphor in largely overlapping, but subtly dissimilar directions.

In this essay, I focus on the history of this strange coincidence, both in biographical and in theoretical respects. What were the personal intellectual developments of Kuhn and of Feyerabend that led to the development of their respective concepts of incommensurability? What is the relationship between those concepts? How did the two

* I wish to thank the editor, Randy Harris, for suggesting various amendments and for massively improving my style. What else would you expect from a rhetorician?

fathers interact as they made their discoveries? And finally, how did they direct their patrimonies; how did they respond afterwards to each other, and to each other's *incommensurability?*

Thomas S. Kuhn

Thomas Samuel Kuhn was born on July 18, 1922 in Cincinnati, Ohio, U.S.A., and received his undergraduate education in physics at Harvard University from 1940 to 1943.[2] Continuing on to graduate work in theoretical solid-state physics under John H. van Vleck (who became a Nobel laureate many years later, in 1977), he was introduced to the history of science by James Bryant Conant, then president of Harvard. In 1947, pondering the work of Aristotle, Kuhn made a realization that set the agenda for much of his further work. When he read the *Physics* with contemporary physical concepts in mind, much of it seemed either awfully wrong or simply meaningless. But by altering the meaning of some of the key concepts (in some cases only slightly and in others more deeply), it started to make sense. A whole new world of physics appeared before Kuhn's eyes—the cogent, coherent world of Aristotelian physics that had dominated Western thought for over 1,500 years. After completing his Ph.D. in 1949, Kuhn stayed on at Harvard, but turned entirely to the history of science, where he began applying the insights he gained through the semantic realization he made working with Aristotle. From 1948 to 1951, he was a Junior Fellow of the Harvard Society of Fellows, and from 1951 to 1956 Assistant Professor for General Education and History of Science, a program designed by Conant.

Kuhn's career also took him elsewhere, however. From 1956 to 1964, he was a member of the faculty of the University of California at Berkeley, mainly in the History Department. In 1964, he joined Princeton University as a Professor in History of Science. He was also a member of the (Princeton-based) Institute for Advanced Study from 1972 to 1979. Finally, in 1979, he was appointed Professor of Philosophy and History of Science at the Massachusetts Institute of Technology, a position he held until retirement in 1992. He died on June 17, 1996, in Cambridge, Massachusetts, at the age of 73.

Kuhn's path to incommensurability started, according to his own testimony, through his semantically mediated encounter with Aristotelian physics.[3] When he read passages in Aristotle that made no sense to him, he could not believe someone of such extraordinary talents could have written them. But, one day, he began to see patterns in these disconcerting passages. All at once the text made sense to him—a gestalt switch—when he altered the meanings of some of the words. He saw this process of semantic alteration as a method of recovery. He realized that in his earlier encounters he had projected meanings onto the text, and needed now to peel them away. For instance, when he encountered the word *motion* in Aristotle (the standard translation of the Greek *kinesis*), he thought in terms of the change of position of objects in space. But if he expanded that meaning to get more closely at Aristotle's original usage, the word had a much broader range. It covered various sorts of change, of which motion in space is just a special case. Kuhn realized that these sorts of conceptual differences were indicators for breaks between different modes of thought, and he suspected that those breaks must have significance both for the nature of knowledge, and for the sense in which knowledge can be said to progress. Having made this discovery, Kuhn changed his career plans, leaving theoretical physics for history of science in order to pursue the strange phenomenon. It took another fifteen years—during which, among other developments, he wrote and published *Copernican Revolution* (Kuhn 1957)—until the term *incommensurability* appeared, in his classic *The Structure of Scientific Revolutions* (Kuhn 1962, 1970a, 1996). The main target he attacked in *Structure* was the idea of a cumulative development of science, and incommensurability was his weapon of choice.

Incommensurability, for Kuhn in 1962, characterizes the relation holding between two succeeding scientific traditions separated by a scientific revolution, and implicates the following three, different but kindred, aspects of that relation.[4]

1. There is a change in the problems that need to be addressed by any theory in the same domain. Problems whose solution was vitally important to the older tradition may disappear, becoming obsolete or even unscientific; problems that did not exist, or whose solution was considered trivial, may gain extraordinary significance in the new tradition; and the standards alter which are applied to admissible solutions for various problems.

2. There is a change in methods and concepts. After a revolution, many (though not all) of the older methods and concepts are still used, but in a slightly modified way. From the perspective of mutual understanding, this change of concepts is especially

important. This shift of meaning can result in some old elements of the extension of a concept ceasing to be subsumed under it, and new elements can similarly enter into the extension.

3. "The proponents of different paradigms," Kuhn argues, "practice their trades in different worlds" (Kuhn 1996, 150), though he admits that it is not entirely clear how literally "different worlds" can be taken.

These three aspects of incommensurability jointly constrain the interpretation of scientific advancement as cumulative.

Before explicating incommensurability any further, however, we need to look at its other father, and at his path to the concept.

Paul K. Feyerabend

Paul Karl Feyerabend was born on January 13, 1924, in Vienna, Austria, where he also grew up.[5] After his high school exam in 1942, he was first drafted into the working service (*Arbeitsdienst*) and then into the German *Wehrmacht.* In 1945, he was hit by three bullets, one of which struck his spinal chord, leaving him paralyzed from the waist down. For the rest of his life, he had to use a walking stick.

In 1945, Feyerabend studied various performing arts, including play production, the history of drama, and vocals at the Weimar Institute for the Methodological Reform of the *Deutsche Theater.* Starting in 1946, he studied history, physics, and astronomy at the University of Vienna. Feyerabend graduated in astronomy, and went on to graduate work in philosophy under Victor Kraft. Working on observation sentences, he became convinced that their meaning depends on the nature of the objects they describe, an insight that ran counter to prevailing logical-empiricist philosophy, and which would later lead him to incommensurability. He received his Ph.D. in 1951. After brief postgraduate studies in philosophy of science in Copenhagen, Stockholm, and Oslo, he left for England in 1952 to study with Karl Popper at the London School of Economics (LSE). His original plan was to study with Wittgenstein, but he had died in the meantime. His main areas of study were Wittgenstein's philosophy and quantum mechanics.

In 1955, after spending some time back in Vienna, Feyerabend accepted a lectureship in philosophy in Bristol, England. An invitation to spend a year at the University of California, Berkeley, reached him in 1958. He accepted, and the year stretched into more than thirty;

in 1959, he was appointed with tenure. With some interruptions, he taught there until his retirement in 1990. Over his career, Feyerabend held additional positions in Minneapolis, Auckland, Berlin, London, Yale, Brighton, and Kassel. From 1980 to 1990, he held simultaneous appointments at the Swiss Federal Institute of Technology, in Zurich, and at Berkeley—lecturing in Zurich during the Spring term, Berkeley in the Fall. He died on February 11, 1994, in Genolier, Switzerland, at the age of 70.

Feyerabend's path to incommensurability started, according to his own testimony, with his dissertation under Kraft, and his insight that observation sentences depend on the nature of the objects they describe.[6] A change of assumptions about the nature of objects—something that occurs with a change of fundamental theory—triggers in Feyerabend's view a change in the meaning of the respective observation statements, too. Thus, in opposition to logical-empiricist thought, the meaning of observation statements is not given independently of theory.

Feyerabend gave the first oral presentations of this idea in 1952, in Popper's seminar at the LSE, and in Elizabeth Anscombe's house in Oxford. Later, Feyerabend was a frequent guest at the Minnesota Center for Philosophy of Science, which was directed by Herbert Feigl. In response to Feyerabend's new ideas, Feigl produced an obvious objection. If Feyerabend was right about observation statements, this would make crucial experiments impossible. Crucial experiments essentially depend on the existence of a set of observations that at the same time refute one theory and confirm a rival theory. But to fulfill this function the observations require some order of neutrality with respect to the competing theories. Feyerabend, for whom this observation-statement neutrality was impossible, therefore saw them as incapable of fulfilling this adjudicating role. A full treatment of Feigl's challenge occurred in 1962, when Feyerabend published the classic paper, "Explanation, Reduction, and Empiricism," in which he introduced the term *incommensurability* (Feyerabend 1962).[7] The main target he attacked in "Explanation" was the possibility of a formal account of reduction and explanation for general theories, and incommensurability was his weapon of choice.

Incommensurability, for Feyerabend in 1962, characterizes the relation holding between two succeeding general scientific theories.[8] Part and parcel of formal accounts of reduction and explanation were the ideas that (1) the key concepts remain stable when a transition between two fundamental theories occurs, and (2) the earlier theory is incorporated into the later one. But for Feyerabend, something much more radical than incorporation occurs: a replacement of the ontology of the earlier theory by the ontology of the later theory.[9] Consequently, there is a change in the meaning of the concepts of the older theory. But this change is not restricted to concepts of the older theory, it also affects at least some of the observational terms that occur in sentences derived from the theory for the purpose of theory testing, or for explanation.

Thus, introducing a new fundamental theory "involves changes of outlook both with respect to the observable and with respect to the unobservable features of the world, and corresponding changes in the meanings of even the most 'fundamental' terms of the language employed" (Feyerabend 1962, 45; similarly, 1975, 275). More explicitly: "Two theories will be called incommensurable when the meanings of their main descriptive terms depend on mutually inconsistent principles" (Feyerabend 1965b, 227n19; see also Feyerabend 1962, 78; Feyerabend 1975, 269–270, 276; Feyerabend 1981b, 16n38). As a result of this dependence, the main concepts of the older theory "can neither be defined on the basis of the primitive descriptive terms of the [later theory], nor related to them via a correct empirical statement" (Feyerabend 1962, 76).[10] This lack of common elemental descriptive terms and the non-relation via empirical statements, in turn, leads to the consequence that there are no logical relations between the two theories (Feyerabend 1958b, 83; 1975, 271; 1978b, 67; 1981b, 16).

There is no particular problem in any of these implications for science, Feyerabend noted, since scientists have developed ready and productive means for deciding between incommensurable theories. Rather, the problem is for philosophy, since contemporaneous philosophers held rather simple-minded models of theory change and overly abstract views on semantics (Feyerabend 1981b, 16; 1981e, xi; 1981c, 238; 1987, 272).

Kuhn's Incommensurability and Feyerabend's Incommensurability

It must have been around 1960 when Feyerabend and Kuhn realized that they were after something so similar that they baptized it with the same term. Feyerabend described this episode in 1969 in the following way:

> I do not know who of us was the first to use the term 'incommensurable' in the sense that is at issue here. It occurs in Kuhn's '*Structure of Scientific Revolutions*' and in my essay 'Explanation, Reduction, and Empiricism' both of which appeared in 1962. I still remember marveling at the pre-established harmony that made us not only defend similar ideas but use exactly the same word for expressing them. The coincidence is of course far from mysterious. I had read earlier drafts of Kuhn's book and had discussed their content with Kuhn. (Feyerabend 1970, 219)

Kuhn's recollections in 1982 were as follows: "I believe that Feyerabend's and my resort to 'incommensurability' was independent, and I have an uncertain memory of Paul's finding it in a draft manuscript of mine and telling me he too had been using it" (Kuhn 2000, 32n1). In October 1995, Kuhn's story went like this, but he admitted that his recollections were not as precise as he would have liked them to be:

> I think I remember a talk with Feyerabend. He was sitting behind his desk and I was standing at the door of his office, which was very close to mine. Now, I'm not sure this is right, I mean this is the sort of thing I could easily have constructed. I said something to him about my views including the word *incommensurability,* and he said, "Oh, you are using that word too." And he showed me some of the things he was doing, and *Structure* came out at the same year as his big article in *Minnesota Studies.* We were talking about something which was in some sense the same thing. (Kuhn 2000, 297–298)

"In some sense the same thing," of course, implies that there are other senses in which it was not the same thing, but these differences often tend to be overlooked in the literature.[11] It is thus worthwhile to have a closer look at them, against the background of what the two authors share regarding incommensurability. (I will not however attempt a critical assessment of the incommensurability issue here because that would be far beyond the scope of this essay.)

There are at least eight measures by which it is profitable to compare Kuhn's and Feyerabend's respective versions of incommensurability:

- the general *phenomenon* each designated by that term
- the intellectual *route* by which each of them reached their accounts of that phenomenon
- the *clarity* with which each articulated his concept
- the *breadth of the domain* of their concepts
- the *range of theories* to which their concepts apply
- the *pervasiveness* of incommensurability for them between given theories or programs
- the relevance of incommensurability to them for the crucial notion of *theory comparison*
- the relation for them of incommensurability to *truth*

I will now compare Kuhn's and Feyerabend's notions on each of these scores in turn.

The General Phenomenon

Certainly Kuhn and Feyerabend both address roughly the same phenomenon; namely, a particular form of theory change with strange properties ("strange" against the background of expectations nourished by the Anglo-American philosophical tradition in the 1950s and 1960s). For Kuhn, the main example is the genesis of modern physics and the concurrent abandonment of the Aristotelian tradition. On the basis of the incremental image of theory change current at that time, the relation of the Aristotelian tradition to modern physics is peculiar. Modern physics is not the predicted addition to Aristotelian physics, nor a generalization of it. Rather, its core involves a conceptual transformation such that one cannot even render the old intelligible in the terms of the new—hence the difficulties for the historian in truly comprehending the older world-view. There are no direct points of contact, such that an empirically testable statement, A, can be deduced from one theory, and not-A from the other, allowing observation of either A or not-A to decide the case. Instead, large domains of phenomena are conceptualized very differently, such that implying anything about a particular, "same" A is highly problematic.

Feyerabend addresses something very similar. He assumes that observation sentences are strongly theory-dependent and, therefore, that the empirical comparison of two sufficiently different theories poses special problems. Each theory brings its own set of observation sentences to the debate, precluding experience from playing the role of an unbiased arbiter between the two. Again, comparison seems fatally compromised. How can theories ever be compared with regard to their empirical achievements if they cannot be related to a common set of observation sentences? How could one ever say that one theory is better than the other?

Kuhn's and Feyerabend's Intellectual Routes to Incommensurability

The two thinkers arrive at these very similar characterizations, however, by very different routes.[12] Feyerabend's starting point is Vienna Circle philosophy; more precisely, Kraft circle philosophy, in the late 1940s and early 1950s. His discussion is thus abstract and philosophical. Kuhn's starting point is a particular experience that only a historian can encounter.

In the 1950s and the early 1960s, though, there is a marked convergence between Kuhn and Feyerabend, even prior to their interaction at Berkeley, which led to their adoption of the same key term. (I will turn to their interaction in the next section.) Feyerabend turned to history for a concrete historical example to illustrate his abstract thesis, the transition from medieval impetus theory to modern mechanics.[13] When Feyerabend argues his case in the classic 1962 paper, he is a philosopher who, by the standards of the time, is strongly influenced by history of science.

Kuhn didn't know very much history of science, and even less philosophy of science in 1947, but he realized that he had to learn a lot of both in order to explicate his incommensurability experience (which he did not then describe with this term). By 1962, he had made a name for himself as a historian of science, but more importantly in our context, he had also accumulated some philosophical knowledge, which manifests itself in various parts of *Structure*.[14] Thus, in comparison to the beginning of their respective careers, Feyerabend had become more historical and Kuhn had become more philosophical.[15]

The Central Differences

But, in spite of the convergence, there are very clear residual traces of the initial differences in their concepts. In four key dimensions, Feyerabend's concept of incommensurability is much more focused than Kuhn's: the overall clarity, the domain of application, the range of theories involved, and the pervasiveness of the concept.

Clarity

With respect to clarity, Feyerabend has a firm grasp of the nature and origins of incommensurability (in spite of the fact that his assumptions are controversial). In Feyerabend's view, because the nature of objects depends on the most advanced theories about them, and because the meaning of observation statements depends on the nature of those objects, the interpretation of an observation language is determined by the theories we use to explain what we observe. Kuhn, on the other hand, is much less sure about the exact meaning of *incommensurability.* Especially with respect to the world-change aspect that he sees as the "most fundamental aspect of incommensurability," Kuhn frankly confesses to be at a loss: "In *a sense that I am unable to explicate further,* the proponents of competing paradigms practice their trades in different worlds," (Kuhn 1996, 150; my italics) though he is confident enough to add that despite this lack of explication "we must learn to make sense of statements that at least resemble these"(Kuhn 1996, 121).[16]

The Domain of Incommensurability

With respect to the domain of incommensurability, Feyerabend's concept is much more restricted than Kuhn's.[17] For Kuhn, incommensurability had three *prima facie* heterogeneous domains: a change of problem-fields; a change of procedures and concepts; and a change of world-view, including changes in perception.[18] Feyerabend's focus, on the other hand, is exclusively on concepts occurring in fundamental theories, together with their ontological implications. Ironically, however, in developments after 1962, both authors move in opposite directions. Kuhn gradually eliminates everything from his notion of incommensurability that does not concern scientific concepts.[19] Feyerabend, at least occasionally, later includes aspects of perception (Feyerabend

1975, 225–229, 273–274; 1978b, 68n118; 1988, esp. 172–176), and also changes to the set of legitimate problems (Feyerabend 1975, 274–275).

The Range of Theories Subject to Incommensurability

With respect to the range of theories that are subject to incommensurability, Feyerabend's concept is again much more restricted than Kuhn's. For Feyerabend, only *fundamental* (or *non-instantial* or *comprehensive* or *universal* or *high-level*) theories that are *interpreted in particular ways,* mostly via realism, can be incommensurable with respect to each other (Feyerabend 1962, 44, 44n1; 1965b, 216; 1975, 114, 271, 284; 1987, 272). The examples Feyerabend had in mind were the transitions from the Greek archaic (Homeric) world to the world of classical antiquity, from the impetus theory to classical mechanics, from geometrical optics to wave optics, from phenomenological thermodynamics to statistical mechanics, from classical mechanics to the special theory of relativity, from Newton's theory of gravitation to general relativity theory, and from classical mechanics to quantum theory (Feyerabend 1962, 62–68, 79–80, 81–82; 1965b, 227n19; 1965a, 100; 1975, 224–225, 271, 276–277, 279; 1978b, 68n118). Consequently, for Feyerabend, incommensurability occurs rarely (1987, 272). This restriction to fundamental theories follows, of course, from his specific notion of incommensurability: only those theories, and only if they are interpreted in a specific way (mostly via realism), influence the nature of their respective objects. If these theories are interpreted instrumentally, they are commensurable (Feyerabend 1975, 279; 1988, 221).

Kuhn, conversely, included a wider range of theories as candidates for incommensurability (although his very first example of incommensurable theory change concerned two fundamental theories). For him, even smaller episodes, like unexpected discoveries, might be incommensurable to the earlier tradition (Hoyningen-Huene 1993, 197–201). The driving criterion for him was that the new development could not be understood as an addition to the older theory. Rather, there are meaning shifts such that it becomes impossible to articulate the new concepts by means of the old (and the old concepts by means of the new). It took Kuhn some time, however, to arrive at this account. He only achieved clarity in these matters in his work after *Structure* (Hoyningen-Huene 1993, 212–218).

This difference in the range of incommensurability between Kuhn's concept and Feyerabend's finds its most striking expression in the way they regard the transition from the Ptolemaic to the Copernican theory of the planetary system. For Kuhn, the differences between these two theories comprise a cornerstone example of incommensurability. For him, the clear indicator of the incommensurability of the two theories is, for instance, the meaning change of the concept of "planet" as it is manifested in the shift of its extension. Some objects that are planets for Ptolemy, like the sun and the moon, are no longer planets for Copernicus, and an object like the earth that is not a planet for Ptolemy becomes a planet for Copernicus. For Feyerabend, however, because planetary theory lacks the quality of universality there is no incommensurability (Feyerabend 1975, 114); a mere conceptual difference between theories does not suffice for incommensurability. Rather, "[t]he situation must be rigged in such a way that the conditions of concept formation in one theory forbid the formation of the basic concepts of the other" (Feyerabend 1978b, 68n118; similarly in 1975, 269; 1981a, 154n54).

The Pervasiveness of Incommensurability

With respect to pervasiveness, Feyerabend's concept of incommensurability is more extensive than Kuhn's. This claim might superficially appear to conflict with my previous point about the respective ranges of Kuhn's and Feyerabend's concepts of incommensurability, but it is consonant with that point. The claim depends on a different sense of scope—breadth in Kuhn's case (more theories involved), depth in Feyerabend's (fewer theories, but far greater comprehensiveness with respect to those theories). Kuhn insisted in his later writings that the version of incommensurability he championed had always been "local incommensurability," a notion that restricts the meaning shift of concepts to a few, typically interconnected concepts (Hoyningen-Huene 1993, 213, 219).[20] Thus, there may be (empirical) consequences of incommensurable theory pairs that can be immediately compared. For instance, the geocentric and heliocentric planetary theories are incommensurable in Kuhn's sense (as indicated, for instance, by the meaning shift of the term *planet*). But predictions of both theories of planetary positions in the sky are fully commensurable and can be immediately compared regarding their empirical accuracy. By contrast, Feyerabend always thought of his concept in a way that we might term contrastively "global incommensurability," as possibly affecting *all* statements derivable from the two (fundamental) theories. "[I]ncommensurable theories may not possess any comparable consequences," he said, "observational or otherwise" (Feyerabend 1962, 93; similarly 1965c, 117; 1965b, 216; 1975, 275–276; 1981e, xi). Although Kuhn is by far the more cited of the two (on incommensurability as well as on notions like paradigm and scientific revolution), it is Feyerabend's more radical form of pervasive incommensurability that seems to have captured the imagination of the many subsequent users of the word, much more than the modest, local form championed by Kuhn.[21]

Incommensurability and Theory Comparison

There may be no matter that is more confused with respect to incommensurability than comparability of theories, so it is worth explicating Kuhn's and Feyerabend's respective views on this question, which have important convergences and significant differences. Neither of them thought that incommensurability excluded theory comparison *tout court.*[22] In their views, theory comparison is just more complicated and more delicate than imagined by philosophers of science at the time. In particular, it is a far cry from an algorithmic procedure.[23] And there may be—at least for a while in periods of transition—different aspects of evaluation pulling in different directions.[24] Many people read both Kuhn and Feyerabend, it is true, as advancing the view that incommensurability entailed the impossibility of *any* theory comparison, and both authors did deny the possibility of a *specific form* of theory comparison of incommensurable theories. Kuhn denies what he calls *point-by-point* comparison, and Feyerabend denies the (direct) comparison of (or by) content.[25] And basically, they have the same thing in mind; namely, vitiating the comparison of (mainly) empirical consequences of the two relevant theories with regard to empirical data and logical relations (most importantly identity, contradiction, and inclusion). But both thinkers do have positive views about theory evaluation; albeit, very different positive views.

For Kuhn, even if an exhaustive point-by-point comparison is impossible, and even if theory comparison does not reach the strength of a mathematical proof, an evaluation of incommensurable theories with respect to their merits is still possible.[26] *Some* predictions of the two theories can always be directly compared: namely predictions in which the incommensurable concepts play no significant role. The point for Kuhn is that *not all* predictions of the two theories can be compared in this way, but these other predictions must somehow also be part of the evaluation, even if they cannot be matched one-by-one with predictions from the other theory. The sort of comparative theory-evaluation Kuhn has in mind, moreover, can be called "rational," in a means/ends type of rationality. It is rational to choose better achieving theories because they serve the ends of science better.[27] This property of theory choice makes the overall process of science rational and progressive.

For Feyerabend, the case is different. It is neither possible for him to base theory comparison on content, nor on logical relations (nor on verisimilitude; see below); due to the global character of his concept, *all* consequences of both theories are infected by incommensurability. But according to his own testimony, Feyerabend searched for other possibilities of theory comparison from the moment when he stumbled on the phenomenon of incommensurability (Feyerabend 1978b, 68).[28] There is, for one, the possibility of internal (or formal) examinations of the two theories. For instance, the length of derivations leading from first principles to observation statements might be a basis of comparison; or the kind and number of approximations; the occurrence of internal contradictions in one theory or the other might help decide between them; or the inner coherence of each theory might be compared, or their predictive powers; or the theories might be compared on the basis of whether one or the other is linear or not (Feyerabend 1965b, 217; 1970, 228; 1975, 284; 1978, 68n119; 1981b, 16n39). Yet, these criteria must be called "'subjective,' in the sense that it is very difficult to find wish-independent arguments for their acceptability" (Feyerabend 1978b, 68).[29] Furthermore, these factors may pull in different directions, and no objective criteria exist for the necessary weighing of them. In addition, subjective factors in the ordinary sense—such as aesthetic judgments, judgments of taste, metaphysical prejudices and religious desires—may play a role in theory choice (Feyerabend 1970, 228–229; 1975, 285). Thus, theory choice necessarily involves subjective factors of various sorts, and it therefore does not make sense to call it rational. Consequently, neither can the overall process of science be called rational. But in spite of the missing rationality of scientific development, it remains progressive for Feyerabend. Scientific progress consists in the increase of the set of incommensurable theories and viewpoints, each of them forcing the others into greater articulation (Feyerabend 1965c, 107; 1975, 30).

Incommensurability and Truth

Finally, there is one very important substantive point of agreement between the two authors. In spite of the many differences in their concepts of incommensurability, and the consequences following from those differences, both see incommensurability as precluding the possibility of interpreting scientific development as an approximation to truth (or an "increase of verisimilitude").[30] This detachment from truth is due to the changes in ontology that accompany incommensurable theory change.[31] These changes are not just refinements of, or additions to, the older ontology, such that these developments could be seen as cumulative. Rather, one ontology replaces another (in Feyerabend's case, this replacement is necessarily radical, whereas in Kuhn's case, it may be rather localized).

Effectively, this means that—although, in Kuhn's case, there have always been pronouncements to the contrary (Hoyningen-Huene 1989; 1993, 74–77, 267–271)—neither one of them can be read as a scientific realist. Their positions are more adequately described as "globally antirealist."[32] In the current debate about *scientific* realism, the issue is whether theoretical statements may be realistically interpreted or not. In this debate, it is usually more or less explicitly presupposed that a realistic interpretation of the observational statements is legitimate (if it weren't, the debate about the legitimacy of the realist interpretation of theoretical statements would be unnecessary, given a basically antirealist position). In contrast to *scientific* antirealism, *global* antirealism also denies the legitimacy of a realist interpretation of observational statements.

Kuhn and Feyerabend's Exchange on Incommensurability

Not only did different intellectual paths lead Kuhn and Feyerabend into the same problem field, but different career paths also led them into the same geographical area, to the same university, and to appointments (in Kuhn's case partial) to the same department. This concurrence happened shortly before the official birth of incommensurability, in the late 1950s. Kuhn had taught as Assistant Professor for

General Education at Harvard University from 1952 on, but when, to his great disappointment, he was denied tenure (Kuhn 2000, 289), he accepted an offer from the University of California at Berkeley. The Berkeley offer was attractive for him because it was primarily situated in the philosophy department, which was seeking a historian of science (Kuhn 2000, 294). In addition, he also became a member of the history department. Feyerabend had held a position at the University of Bristol since 1955, when he was invited to come to Berkeley as a visiting professor in philosophy. He came in 1958, earning tenure the following year. Thus, from 1958 until 1964, when Kuhn left for Princeton, the two thinkers worked at the same university.

According to Feyerabend, he first read Kuhn's work in 1959 (Feyerabend 1976, 391[33])—although by that time Kuhn had not written much.[34] At the latest, they were well acquainted by 1960, and their most intensive interaction was in 1960 and 1961 (Feyerabend 1970, 197). When Kuhn had finished the first draft of *Structure* sometime in the fall of 1960, he sent it to several people, Feyerabend among them (Feyerabend 1978b, 117n49). Feyerabend reacted to this draft (Kuhn 1960), which I will call *Proto-Structure*,[35] in two letters, comprising thirty-one single-spaced pages of text.[36]

Before discussing in detail the intellectual interaction between Kuhn and Feyerabend, I want to note three characteristics of it that concern its written record. First, there is a striking quantitative asymmetry: Feyerabend refers to Kuhn much more often than *vice versa*. Second, there is no trace of Kuhn having considered the differences between their respective concepts of incommensurability. In fact, there is no trace of Kuhn having considered Feyerabend's overall position at all. Rather, the main thrust of Kuhn's contribution to the debate is his attempt to defend himself against Feyerabend's objections. Third, the bone of contention of their heated debate was normal science, which Feyerabend saw as a sign of Kuhn's conservatism that he deplored.[37] Incommensurability was only a side issue.

Incommensurability comes up in his second letter to Kuhn on *Proto-Structure*, written between Spring and Fall of 1961 (Hoyningen-Huene 1995, 354). Altogether, Feyerabend writes roughly one page on this issue, addressing three aspects of Kuhn's incommensurability in particular. First, he takes up Kuhn's assertion that the proponents of competing paradigms "will inevitably talk through each other," that they will "always [be] slightly at cross-purposes" (*Proto-Structure*, 109,

111, 138; Kuhn 1970a/1996, 109, 112, 148). "This is not true!" Feyerabend exclaims, and "this seems to be a little exaggerated," and "this may happen as regards this or that problem, but not as regards *every* problem" (Hoyningen-Huene 1995, 381, 385, 387). Thus, Feyerabend reads Kuhn as asserting the phenomenon of miscommunication as applying to every single topic of a discussion between the proponents of competing paradigms (i.e. as proposing *global* incommensurability). Second, Feyerabend objects that the alleged form of (global) miscommunication is inconsistent with Kuhn's claim that incommensurable paradigms are in conflict. And those paradigms are indeed in conflict, as the existence of crucial experiments demonstrates. Third, Feyerabend criticizes Kuhn's claim that the transition between incommensurable paradigms "is a conversion experience that cannot be forced" (*Proto-Structure,* 157; similarly 156; Kuhn 1996, 151, 150). Feyerabend objects that since the two paradigms are in conflict, "there must be at least two statements, one from the first paradigm, the other from the second, which are inconsistent." If only one of these statements is empirically adequate, then a decision between them can indeed be forced without any conversion being involved (Hoyningen-Huene 1995, 385, 387).

Feyerabend articulates here what will become the standard objection against the general idea of incommensurability. However, Kuhn would not accept it. First, as I said earlier, Kuhn supports only local incommensurability, not global incommensurability, even if that is not at all clear in *Proto-Structure.* Thus, all objections that derive from the specifics of global incommensurability do not apply, including the purported incomparability.[38] Second, while Kuhn would agree with Feyerabend that there are indeed contradictory statements derived from the two incommensurable theories in question, and that it may be empirically possible to decide between them, this fact (contrary to Feyerabend) would not force the choice between these *theories.* The reason is mainly that there are also empirical statements derived from one theory that find no counterpart in the competing theory. So, an element of judgment has to enter theory choice because different factors that speak for and against a particular theory have to be weighed, resulting in choices that are not absolutely cogent.

Feyerabend's mode of criticism is striking. It seems to be minimally influenced by his own ideas about incommensurability. It rather resembles the sort of criticism that was a little later put forward by people who where much more traditional in their thinking about science, like mainstream logical empiricists or critical rationalists. Feyerabend realized this later and commented on it in his autobiography: "[. . .] my contrariness extended even to ideas that resembled my own. For example, I criticized the manuscript of Kuhn's *Structure of Scientific Revolutions* [i.e., *Proto-Structure*], which I read around 1960, in a rather old-fashioned way" (Feyerabend 1995, 141).

I do not know whether Kuhn answered Feyerabend's objections in any form. It is clear, however, by testimony from both of them that their discussions weren't terribly successful, and that they did not touch upon incommensurability at any deeper level. Kuhn describes his encounters with Feyerabend as follows:

> With Feyerabend I had strange experiences. He was at Berkeley, I gave him this draft manuscript of the book that I'd sent out to Chicago [*Proto-Structure*]. I think he liked it in one sense, but he was terribly upset by this whole business of dogma, rigidity, which of course is exactly counter to what he believed himself. And I couldn't get him to talk about anything except that. And I tried, and I tried: we would have lunch together, or something—he'd always come back to it. I got more and more frustrated and I finally just stopped trying. (Kuhn 2000, 310)

Feyerabend's description of their encounters is not altogether different, except that he also expresses his admiration for Kuhn:

> In the years 1960 and 1961 when Kuhn was a member of the philosophy department at the University of California in Berkeley I had the good fortune of being able to discuss with him various aspects of science. I have profited enormously from these discussions and I have looked at science in new way ever since. Yet [. . .] I was [. . .] less prepared to accept the general *ideology* which I thought formed the background of his thinking. This ideology, so it seemed to me, could only give comfort to the most narrow-minded and the most conceited kind of specialism. [. . . I]t is bound to increase the anti-humanitarian tendencies which are such a disquieting feature of much

> of post-Newtonian science. On all these points my discussions with Kuhn remained inconclusive. More than once he interrupted a lengthy sermon of mine, pointing out that I had misunderstood him, or that our views are closer than I had made them appear. (Feyerabend 1970, 197–198; notes in the citation omitted)

There was another, more public, occasion, five years later, when Feyerabend and Kuhn discussed their differences, at a 1965 conference in London.[39] Again, it was normal science that formed the main topic of this discussion, incommensurability only being side fare. Feyerabend took up the issue of incommensurability at two places. First, in the context of his criticism of Kuhn's notion of normal science, he mounts an argument that is meant to be an immanent criticism of Kuhn.[40] The details of this argument do not matter here, only his ascription of a particular concept of incommensurability to Kuhn:

> Revolutions bring about a *change* of paradigm. But following Kuhn's account of this change, or 'gestalt-switch' as he calls it, it is impossible to say that they have led to something *better*. It is impossible to say this because pre- and post-revolutionary paradigms are frequently incommensurable. (Feyerabend 1970, 202)

Obviously, Feyerabend assumes here that Kuhn's notion of incommensurability implies that there is no progress in science across revolutions. This, however, is definitively not Kuhn's opinion on the matter. In his view, incommensurability does not at all preclude progress across revolutions; it just precludes some *specific forms* of progress (especially an "approach to truth").[41] Thus, Feyerabend gets this essential feature of Kuhn's notion of incommensurability completely wrong.

In the second place where Feyerabend takes up the issue of incommensurability and Kuhn, he is entirely positive about it:

> With the discussion of incommensurability, I come to a point of Kuhn's philosophy which I wholeheartedly accept. I am referring to his assertion that succeeding paradigms can be evaluated only with difficulty and that they may be altogether incomparable, at least as

> far as the more familiar standards of comparison are concerned (they may be readily comparable in other respects). (Feyerabend 1970, 219)

With the expression "they may be altogether incomparable," Feyerabend is again moving in the direction of global incommensurability with which Kuhn would not agree. After invoking his discussions with Kuhn at Berkeley (as quoted earlier), Feyerabend develops the intuitive background of incommensurability:

> In these discussions we both agreed that new theories, while often better and more detailed than their predecessors were not always rich enough to deal with *all* the problems to which the predecessors had given a definite and precise answer. The growth of knowledge or, more specifically, the replacement of one comprehensive theory by another involves losses as well as gains. [. . .] We also saw that it might be extremely difficult to compare successive theories in the usual manner, that is, by an examination of consequence classes. (Feyerabend 1970, 219–220)

Two things are remarkable in this quote. First, Feyerabend contradicts his earlier statement in the same article, quoted above, by saying that the later theory is often better than its predecessor. Perhaps this inconsistency is due to a later date of origin of the latter passage. Second, Feyerabend tends to assimilate Kuhn's notion of incommensurability to his own by referring to comprehensive theories, which are part of his concept but not of Kuhn's.

In summary, Feyerabend's reading of Kuhn regarding incommensurability (and other topics) was rather superficial. Only with respect to the intuitive background that both authors share, does he get Kuhn right. With respect to aspects where the two authors differ, Feyerabend misreads Kuhn or substitutes his own views for Kuhn's.

Kuhn's reaction to Feyerabend in the same volume is not very extensive, and only a few remarks concern incommensurability. In fact, it is the only place where any reaction of Kuhn to Feyerabend and incommensurability can be found. First, Kuhn deplores quite generally that Popper, Feyerabend, Lakatos, Toulmin, and Watkins have badly misunderstood him. He even develops a fantasy that there might be another Thomas Kuhn who has also written a book entitled *The Struc-*

ture of Scientific Revolutions that despite its title, bears only superficial similarity with his own book, especially with regard to its central concerns (Kuhn 1970b, 231).

Second, Kuhn returns the favor of Feyerabend's misreading, stressing that he, Kuhn, does not believe the communication breakdown between incommensurable points of view "is ever total or beyond recourse" (Kuhn 1970a, 232), though he (and not only he) sees Feyerabend as advocating this position. But as far as I know Feyerabend never said that communication or understanding is impossible between incommensurable points of view. It is a misunderstanding to infer from his insistence on the global character of incommensurability between comprehensive theories (as above) that any understanding between them is impossible. Quite the contrary. "I say that scientists from different paradigms can understand one another very well," Feyerabend states explicitly (1976, 391).[42] One reason for this position is certainly that historically, *total* communication breakdown (what Harris calls *brick-wall* incommensurability in the introduction to this volume) has never been reported for any episode in the history of science. Another reason is that Feyerabend was highly aware of the possibilities of indirect and metaphorical speech that may make communication possible where direct and literal communication may completely fail.

Third, Kuhn expresses his agreement with Feyerabend about the non-existence of a neutral observation vocabulary that could serve to articulate the observational consequences of both theories (Kuhn 1970b, 235, 266–267, 268, 269n3). This is, of course, one of the essential aspects of incommensurability that both authors share from the very beginning.

At this point, that is in 1969, the published dialogue of Feyerabend and Kuhn on incommensurability terminates. However, both authors occasionally looked back at the differences of their initial introduction of incommensurability. In a symposium of the Philosophy of Science Association in 1982, Kuhn began his talk about "Commensurability, Comparability, Communicability" by comparing Feyerabend's introduction of the concept with his own (Kuhn 1983a, 669). He correctly located their main contrasts in his own broader use of the term and Feyerabend's more sweeping claims. But their "overlap at that time was substantial" which is true, too. Both were concerned to show that meanings of scientific concepts "often changed with the theory in which they were deployed." And both claimed "that when such changes occurred, it was impossible to define all the terms of one theory in the vocabulary of the other." But whereas Feyerabend restricted incommensurability to language, Kuhn had also spoken of differences in methods, problem-field, and standards of solution as if the latter changes were language independent. Finally, Feyerabend had made use of a notion of primitive terms whereas Kuhn "restricted incommensurability to a few specific terms" without referring to primitive terms. (Kuhn 2000, 34n2) Feyerabend looks back on some occasions in the 1970s and early 1980s and describes what he sees as the main differences between his concept of incommensurability and Kuhn's.[43] First, he notes that Kuhn leans much more towards psychology than he does. This is undoubtedly correct. Second, he believes that for Kuhn every change of concepts leads to incommensurability. For him, only those classes of concepts are incommensurable where the use of concepts from the one class makes the application of those of the other class impossible. This is a consequence of the difference in the range of application of the incommensurability concept: for Feyerabend it is only "comprehensive" theories, for Kuhn it comprises a wider range of theories. Third, he believes that Kuhn asserts total incomprehension between paradigms, whereas Feyerabend denies that. This last point represents in the most ironic way the lack of intellectual contact between these two giants of twentieth-century philosophy of science regarding their shared patrimony, incommensurability. Each of them accuses the other one of believing that incommensurability precludes communication and at the same time, each of them (justly) claims to never have asserted that.

Notes

[1] I am not aware of any other instance where a concept destined to capture the attention of the science studies community for several decades was introduced by two authors simultaneously but not collaboratively (as we will see, they had a loose, collegial sort of collaboration, in the sense of having strongly intersecting scholarly interests which they discussed regularly, but they did not, for instance, jointly author any work).

[2] In this section, I am using material from Hoyningen-Huene (1997a) and Hoyningen-Huene (1997b); especially the latter paper contains much more biographical information. Alan Gross's essay in this volume, gives a slightly different reading to Kuhn's intellectual biography than the one I

offer here, streamlined for the purposes of his exposition, but I see both readings as fully compatible.

[3] See for the following Kuhn (1977a, xi-xiii; 1981, 8–12; 1999, 33–34; 2000, 298).

[4] In this paragraph, I am using material from Hoyningen-Huene (1990); see also Kuhn (1996, 148–150).

[5] In this section, I am using material from Hoyningen-Huene (1994) and Hoyningen-Huene (2000b); especially the latter paper contains much more biographical information.

[6] See for the following, Feyerabend (1958a, 31n21; 1965c, 116n31; 1976, Anhang 4, 389–391; 1978b, 67–69; 1995, 92; 1988, appendix 2, 228–230, 274, 280–281; 1993, appendix 2, 210–213—the appendixes to the different editions of *Against Method* differ considerably from each other).

[7] Feyerabend's first English paper on the problem was published in 1958, but it did not yet use *incommensurability* (Feyerabend 1958a); the note to the title of this paper reads: "This is a very much abbreviated version of my thesis, *Zur Theorie der Basissätze* (Vienna, 1951)."

[8] On the characteristics of Feyerabend's concept of incommensurability I am largely in agreement with Preston (1997, chpt. 4, 99–104). However, there are also areas of serious disagreement with Preston's view of Feyerabend; see Oberheim and Hoyningen-Huene (1999, 2000).

[9] See also Feyerabend (1981a, xi; 1981b, ix): "a theory is incommensurable with another theory if its ontological consequences are incompatible with the ontological consequences of the latter."

[10] Later in the same article, Feyerabend refers to this statement as a definition of *incommensurability* (Feyerabend 1962, 90).

[11] Detailed comparisons of Kuhn and Feyerabend are scarce in the literature. With regard to incommensurability, something can be found in Burian (1984) and Sankey (1994, 6–30, esp. 17). However, I disagree with Sankey's overall analysis with regard to a number of fundamental issues: see Hoyningen-Huene, Oberheim, and Andersen (1996). See also Harris's introduction to the present volume for some scattered comparisons between Kuhn's and Feyerabend's *incommensurability.*

[12] Feyerabend was aware of this difference. See Feyerabend (1978b, 66–68; 1981e, x; 1988, 229–230; 1993, 212).

[13] He develops the example in Feyerabend (1962).

[14] However, for the philosopher of science of the time, it was quite obvious that *Structure's* author was not a professional philosopher. This is, for example, indicated by the tone of many reviews of *Structure,* and by the refusal of Berkeley's philosophy department to appoint Kuhn, jointly with the history department, to full professor (see Kuhn 2000, 7, 301–302), as

well as by some of the criticism Feyerabend mounted against Kuhn's first draft of *Structure* of 1960 (I take up the criticism briefly below; see also Hoyningen-Huene 1995). In retrospect, Kuhn thought of himself as having been philosophically fairly naïve in 1962 (personal communication in September 1995).

[15] For a description of the current state of the incommensurability debate in philosophy, see Sankey and Hoyningen-Huene (2001), and the other essays in Hoyningen-Huene and Sankey (2001).

[16] For an analysis of *Structure's* pertinent passages like those cited, see Hoyningen-Huene (1993, 31–42, sect. 2.1), and Harris, in this volume (29).

[17] Again, Feyerabend noted this difference very clearly (Feyerabend 1978b, 66–67).

[18] For a fuller treatment, see Hoyningen-Huene (1990) and Hoyningen-Huene (1993, section 6.3.a 208–212). For views of other authors, see especially Sankey (1993) (reprinted in Sankey 1997b, chpt. 2) and Bird (2000, chpt. 5).

[19] For this development, see Hoyningen-Huene (1990, 487–488; 1993, 212–218); Sankey (1993); Sankey (1994, 16–30; 1997); Chen (1997).

[20] It must be admitted that on the basis of the textual evidence this had not been clear: there are also passages in *Structure* that fit the idea of global incommensurability well. But for several reasons, above all overall consistency, I see no reason to disbelieve Kuhn that already in *Structure* he meant local incommensurability. See Gross's contribution to this volume for a thorough look at local incommensurability with respect to Kuhn's later concept, the structured lexicon, and Harris's introduction for his take (50-54)

[21] To be more precise, probably the most popular notion of incommensurability is a notion that amalgamates Feyerabend's pervasiveness and Kuhn's different-worlds (thanks go to Randy Harris for reminding me of this fact).

[22] For Kuhn, see Hoyningen-Huene (1993, 218–222), for Feyerabend, Feyerabend (1978b, 68).

[23] For Kuhn, see Hoyningen-Huene (1993, 147–154); for Feyerabend, Feyerabend (1975, 114; 1981a, 238).

[24] For Kuhn, see Hoyningen-Huene (1992, 492–496; 1993, 150–154); for Feyerabend, Feyerabend (1981b, 16n39; 1981c, 238n17).

[25] For Kuhn, see Hoyningen-Huene (1993, 220–221); for Feyerabend, Feyerabend (1970, 222, 227; 1975, 270, 284; 1978b, 68, 170n38).

[26] For Kuhn's view of theory comparison see Hoyningen-Huene (1993, 236–258); for an insightful article on the possibilities of the comparison of incommensurable theories see Carrier (2001).

[27] This sort of evaluation might even be called *objective* (although Kuhn would deplore the term in this context); because it is not based on subjective factors, it is impartial and almost everybody could be persuaded of its result on the basis of good reasons.

[28] To my knowledge, the fullest treatment of this issue can be found in Preston (1997, 115–123).

[29] However, at other places Feyerabend explains his use of *subjective* very differently (Feyerabend 1981c, 238n17), and at still other places he simply seems to use the common sense notion of *subjective* (Feyerabend 1975, 285).

[30] For Kuhn, see Hoyningen-Huene (1993, 262–264); for Feyerabend, Feyerabend (1965c, 107; 1970, 220, 222, 227–228; 1975, 30, 284; 1978b, 68).

[31] Feyerabend explicitly acknowledges that Kuhn's and his (ontological) ideas are "very similar, [. . .] almost identical" (Feyerabend 1989, 26). Moreover, the similarity does not only extend to "Kuhn's as yet unpublished, later philosophy" as Feyerabend adds, but concerns Kuhn's philosophy since the sixties.

[32] This term can be found in Chalmers (1999, 227–232). Unfortunately, Chalmers' treatment of the issue in his otherwise excellent book is, in my opinion, fairly inappropriate and dismissive.

[33] These recollections are repeated identically in Feyerabend (1983), but not in the English editions of *Against Method.*

[34] See the fullest list of his publications in Kuhn (2000, 325–335).

[35] *Proto-structure* has not been published; it is available in the Kuhn archives of MIT library. Many of the differences to *Structure* are minor and merely concern wording.

[36] Published in Hoyningen-Huene (1995).

[37] I have analyzed this debate in Hoyningen-Huene (2000a); the German version of this article contains all the references (Hoyningen-Huene 2002).

[38] See Hoyningen-Huene (1993, 218–220).

[39] Kuhn reports (2000, 299–300) about this meeting but doesn't refer to Feyerabend. The published versions of their papers were only finished in 1969: see the preface to Lakatos and Musgrave (1970). Both Feyerabend and Kuhn substantially reworked their papers after the conference see Feyerabend (1970, 197n1) and Kuhn (1977, xxn8).

[40] For an extended critical analysis of Feyerabend's arguments in this paper see Hoyningen-Huene (2000a; 2002).

[41] For a detailed discussion, see Hoyningen-Huene (1993, 258–264).

[42] My translation. I am not aware that this passage exists in any of Feyerabend's English writings; see also Feyerabend (1965a, 103).

[43] Feyerabend (1976, 389–391; again in 1983, 373–375, a passage that does not seem to exist in the English editions of *Against Method*; 1978b, 66–69; 1981e, x; 1981d, 154n54).

II

Issues

[T]he most serious problem with incommensurability in particular—and, not incidentally, with Kuhn's philosophy of science in general—lies not in conceptual fuzziness per se, but elsewhere: Kuhn never succeeds in making history his philosophical ally in giving his sense of the term *incommensurability* a firm conceptual shape.

—Alan G. Gross, "Kuhn's Incommensurability"

What might be seen as incommensurability in the relationship between the neo-Darwinian paradigm and the proposals of those who would like to see design reinstated in the life sciences may actually be something else, a sort of über-incommensurability that doesn't confound debate so much as preclude it.

—Thomas M. Lessl, "Incommensurate Boundaries"

How does the concept of Kuhnian incommensurability, suitably defined and elaborated, help us to understand continued intractability *and* convergence of belief, including variable patterns of each? Absent foundations or other common measures for adjudicating between philosophical paradigms, what are the possibilities for persuasion across the divide and perhaps even for some degree of reconciliation?

—Herbert W. Simons, "The Rhetoric of Philosophical Incommensurability"

3

Kuhn's Incommensurability

Alan G. Gross

There are at least four Thomas Kuhns: Kuhn the physicist, Kuhn the teacher of history of science, Kuhn the historian, and, finally and most problematically, Kuhn the philosopher.* We certainly can tell a story of development from one Thomas Kuhn to another.[1] In 1949, he completes his doctorate in solid state physics. But he does not pursue a career in physics; instead, two years before the completion of his doctorate, James Conant, the president of Harvard, asks Kuhn to assist him in developing and teaching a course on the history of mechanics. As a consequence, his interests shift; he abandons physics as a profession. Three years after his doctorate—it is now 1952—his teaching concerns have borne scholarly fruit. He publishes an essay on Robert Boyle, the first in a quarter-century-long series that establishes him as a historian of science. In 1959, a final shift occurs: in an essay in which he mentions "paradigms" for the first time, Kuhn launches his career as a philosopher of science, a career that led to fame in the long wake of his book, *The Structure of Scientific Revolutions,* and culminated professionally in 1988 with his election to the presidency of the Philosophy of Science Association. This development from physicist

* I would like to thank Paul Hoyningen-Huene for his generosity in trying to help me understand Kuhn, and Randy Allen Harris for his careful reading of an earlier draft and his many astute comments. They are, of course, not responsible for any failures in comprehension.

to philosopher is consonant with Kuhn's retrospective view in 1995, a year before his death: "my ambitions were always philosophical. And I thought of *Structure,* when I got to it finally, as being a book for philosophers" (Kuhn 2000, 276).[2]

This reconstruction, moreover, makes sense of the development of Kuhn's concept of incommensurability. According to Kuhn, it was motivated initially by the puzzle of Aristotle's view of motion, a puzzle generated by the need to prepare a course on the history of mechanics; it was reinforced by the solution to the puzzle of Boyle's view of the elements. The essay on Boyle was an application of Kuhn's central insight, garnered from the study of Aristotle's mechanics, that the historian's work is fundamentally analogous to that of the anthropologist or archaeologist: examining the existing traces, the historian reconstructs the past only according to the self-understanding of its inhabitants. Although Robert Boyle used our term *element,* he did not have our concept "element." For Boyle, the source of chemical qualities was rather "matter and motion" (Kuhn 1952, 18). A particular state of matter in motion made gold gold, and lead lead. Gold and lead for Boyle were not, therefore, and could not be, elements in our primary sense—the sense in which they form, for chemistry, the basic building blocks of matter. In short, our conceptual sense of the term *element* and Boyle's were distinct—were, in Kuhn's terms, incommensurable.

This claim, of course, puts me somewhat at odds with the editor of this volume, and with many of its other contributors. I hold (1) that there really is incommensurability and (2) that it is an important differentia for science. I don't disagree with Harris, Campbell, and Miller, for instance, that many, if not most disagreements in science, including ones that kick up lots of dust, are best treated as rhetorical and are open to the kinds of analysis and therapy the contributors recommend. I just disagree that we have exhausted the matter with that observation, that we have dissolved the notion of incommensurability.

Incommensurability does not mean that we cannot understand Boyle's signification; it means rather that we cannot understand that signification without a strenuous effort to reconstruct the world of concepts that animated Boyle's program of research. To understand Boyle we must learn to speak his language; that is, we must link terms to concepts in exactly the way he does. The task of historians of science, then, is to provide commentary sufficient for the reader to join them in the task of reconstructing an alien conceptual past. In the words of historian of science, Jed Buchwald (concerning the history of optics), the historian's role is

> to lead the reader past modern preconceptions, but without entirely abandoning a modern perspective, to an understanding of deep foundations—here those of optics just before and just after the development of wave theory. And again I must ask for patience and indulgence [on the part of the reader]. In order to penetrate the core of the issues, we must examine a number of exemplary problems and discuss at some length the arguments of the time that directly illuminate these issues. (Buchwald 1989, xxii)

After von Ranke's invention of modern history and Bultmann's of modern Biblical hermeneutics, this seems hardly an aperçu worth parading before the scholarly world; indeed, Buchwald mentions it not to claim first possession, but simply to motivate and justify some difficulties his readers are bound to face in their encounter with an alien scientific culture. For Kuhn, of course, this aperçu was a necessary personal triumph, given his training as a physicist. Physics is not a hermeneutic discipline devoted centrally to the recovery of meaning in human action; it is devoted instead to interacting with and observing the material world with a view of penetrating the regularities inherent in it. Without this insight concerning the distinction between these two fundamentally different intellectual aims, Kuhn could not have become an historian. Still, this is a matter less of intellectual moment than of intellectual biography. The insight whose articulation involved Kuhn in four decades of strenuous effort—insight worthy of display—was this: the daily hermeneutic struggle of historians of science is philosophically significant; it tells us, Kuhn insists, something deep and important about development and change in the sciences. The major consequence of his forty-year effort is, in Kuhn's judgment, the concept of incommensurability, according to him, "the central innovation" in *Structure* (Kuhn 1993, 315).

But Kuhn's incommensurability has had a troubled career, parallel to his troubled career as a philosopher of science. The philosophy department at Berkeley refused to make Kuhn a full professor; that is, they refused to acknowledge that he was a philosopher of national standing (Hoyningen-Huene 1997, 238–39; Kuhn 2000, 301–302).

Kuhn's fortunes changed appreciably in reputation and influence, but philosophers as a class still tend to deny him any direct value. "To be a card-carrying philosopher of science it is almost obligatory to reject Kuhn's point of view," one writes (Forster 2000, 232). We even have the strange phenomenon of book-length treatments by philosophers that suggest Kuhn's work doesn't really merit a book-length treatment, at least on philosophical grounds (Bird 2000, Fuller 2000). Philosophers in the early years of the twenty-first century still distance themselves from Kuhn's ideas, especially by minimizing the importance he attached to incommensurability.[3]

Part of the problem is certainly Kuhn's. He does not think or write like a philosopher. He persists in thinking in partial metaphors, a bewildering array of conceptual comparisons never analyzed to the point at which their explication could satisfy philosophical scrutiny. These metaphors derive from a variety of conceptual sources: revolutions from politics, paradigms from grammar, disciplinary matrices from physics, exclusive perceptions from gestalt psychology, lexicons from linguistics, evolution from biology, living in different worlds from Whorfian linguistics and possible-world semantics. Incommensurability, of course, derives from mathematics: three and the square root of two are incommensurable because there is no third number that divides them evenly.

It is typical of Kuhn's practice that it is not immediately apparent what mathematical incommensurability has to do with development and change in the sciences. Clearly, it is his job to tell us, to give incommensurability a firm conceptual shape. When challenged on the basis of one of these serial conceptual comparisons, however, Kuhn tends not to advance to analysis, but to retreat to yet another conceptual comparison: to move from paradigms to disciplinary matrices to lexicons. Moreover, when others have provided conceptual clarification, as does Margaret Masterman in her analysis of paradigms, a clarification of which Kuhn approves (Kuhn 2000, 299–300), this approval seems only to underline the irredeemable sloppiness of his philosophical practice.

But in my view the most serious problem with incommensurability in particular—and, not incidentally, with Kuhn's philosophy of science in general—lies not in conceptual fuzziness *per se*, but elsewhere: Kuhn never succeeds in making history his philosophical ally in giving his sense of the term *incommensurability* a firm conceptual shape.

He was an active historian from 1952 to 1978 and an active philosopher for nearly four decades from 1959 until his death in 1996. But although these two careers were pursued simultaneously from 1959 to 1978, they never seriously interacted, a separation endorsed by Kuhn on principle—"no one can practice them both at the same time," he says (Kuhn 1977a, 5). This (gestaltian) division is so definite and so definitive of his habits of work that it is mirrored faithfully in the analyses of philosophers Paul Hoyningen-Huene and Alexander Bird. Bird deals with Kuhn's philosophy of science as if his historical studies were entirely a separate enterprise. Hoyningen-Huene recognizes the centrality of history to Kuhn's project: "Kuhn's theory," he says, "is methodologically dependent on the historiography of science in that it must allow the latter to furnish it with its particulars" (1993, 12). But this centrality, while presupposed, is not a component of Hoyningen-Huene's hermeneutic task.

Kuhn's demarcation is highly principled. He understands the danger of disciplinary amalgamation: history can be seriously misused. The archival record is, he feels, too malleable to resist the well-intended manipulations of philosophers in pursuit of evidence for a favored thesis: "you cannot do history *trying* to document, or to explore, or to apply a point of view that is schematic" (Kuhn 2000, 313–314). "History done for the sake of philosophy," he contends, "is often scarcely history at all" (1980, 183).

Despite these self-imposed strictures, in this essay I will contend that a continuing dialectic between philosophy and history—that is, between incommensurability as a partially specified philosophical kind and incommensurability as a daily consequence of the work of science—is central to Kuhn's project. Without the specification this dialectic supplies, the viability, scope and significance of his concept of incommensurability cannot be fully understood. As I will also point out, I am not sure that Kuhn would disagree, despite his strictures.

Defining Kuhnian Incommensurability

What exactly is Kuhnian incommensurability? In Kuhn's early *Copernican Revolution,* a book that predates *Structure* by five years, incommensurability, though not mentioned, is exemplified at length. Because the revolution is not merely in astronomy, but in the way humankind sees itself in relation to the cosmos, Copernican incommensurability is broad in scope:

> As one critic of Dante has put it, in the *Divine Comedy* the "vastest of all themes, the theme of human sin and salvation, is adjusted to the great plan of the universe." Once this adjustment had been achieved, any change in the plan of the universe would inevitably affect the Christian drama of life and death. To move the earth was to break the continuous chain of created being. (Kuhn 1957, 113)

In terms of impact, some analogous scientific revolutions might be the Darwinian where humankind's relation to other living creatures is at stake, and the Freudian where humankind's relationship to itself is at issue. In both cases, the argument might go, the alteration in the perspective of ordinary people was profound: they found themselves living in a different phenomenal world. After Copernicus, humankind was no longer at the center of creation; after Darwin, there was no longer an unbridgeable gulf between humankind and other living things; after Freud humankind found itself at war with itself.

Kuhn did not sustain an interest in this broad view. By the early eighties, Kuhn had narrowed his concept of incommensurability to matters scientific and later, even in that arena, somewhat minimized its significance. He insisted that "the claim that two theories are incommensurable is more modest than many of its critics have supposed" (Kuhn 2000, 36). He averred that were he to rewrite *Structure* he "would emphasize language change more and the normal/revolutionary distinction less" (Kuhn 2000, 57). He began to conceive of incommensurability in terms only of changes in what he called the *Lexicon*.

The Lexicon, of course, is another Kuhnian metaphor. In speaking about a Lexicon, Kuhn is talking, not about language, but about what language represents, its underlying concepts. The Lexicon "is a mental module that permits us to *learn to recognize* [. . .] *kinds* of physical object (e.g., elements, fields, and forces)" (1993, 315; my italics). Recognition of the scientific kinds that constitute a particular Lexicon is inculcated by means of long periods of apprenticeship; training is by means of standard illustrations "revealed in its textbooks, lectures, and laboratory exercises. By studying them and by practicing with them, the members of the corresponding community learn their trade" (Kuhn 1996, 43).[4] Having learned their trade, practitioners at the research front maintain the commensurability of their Lexicon largely by means of "the concrete scientific achievement [. . .] prior to the various concepts, laws, theories, and points of view that may be abstracted from it" (1996, 11). In *Black Body Theory,* Kuhn's example of such a concrete achievement is Bolzmann's probabilistic derivation of the entropy of a gas: "the derivation was not reduced to rules but instead served as a model to be applied by means of analogy. As a result, when its application was transferred from gases to radiation, Planck and Lorentz could invent different analogies with which to effect the change" (1978, 363; see also 1996, 43–51; 2000, 296).

Scientists learn to identify and re-identify two sorts of conceptual kinds central to the Lexicon of their discipline: those like element, compound, and mixture, they learn to recognize by contrast; those like force, mass, and acceleration, they learn to recognize by kinship. However they are learned, the terms that designate the conceptual kinds of a developing science have five essential characteristics. First, their meaning and their relationships to other terms are not fully articulated. Second, despite this partly unarticulated conceptual core, these terms rigidly designate their objects; that is, a planet or an element can be identified and re-identified as such by every member of the relevant scientific community. Third, these terms are projectible; that is, a new element or a new planet will be identified by virtue of being of the same *kind* as other elements or planets. Fourth, the borders between these terms and the kinds they designate are rigidly enforced: there can be no overlaps, except those in which a species falls under a genus. In pre-Einsteinian physics, for example, there can be no overlap between the conceptual kinds mass and energy; in post-Einsteinian physics, mass and energy are two manifestations of the same conceptual kind. Fifth and finally, the Lexicon itself has a structure, a taxonomy of relationships among kinds. Current chemistry, for example, is unimaginable without the taxonomy of the periodic table; equally, Newtonian mechanics is unimaginable without the relationships set out in the second law (Kuhn 2000, 186). It is this Lexical structure, these fundamental relationships among its conceptual kinds, that is shared by those practicing a science, and that constitutes for them the core of their professional identity.

For those who share a Lexicon of conceptual kinds, the relationships it embodies function as necessary truths (Kuhn 2000, 71). They are not, of course, necessary truths simpliciter; in a different Lexicon, in fact, they may be falsehoods. This radical change in status occurs, for example, in the shift from the Ptolemaic to the Copernican universe; in the latter, it is no longer a necessary truth that the earth is a stationary body at the center of the universe. A revolution, then, is a shift from one Lexicon to another, highly incompatible Lexicon of conceptual kinds. After a revolution, the alterations in the scientific Lexicon have been so great that it has become impossible to do science in the old way. After the wave theory of light is firmly in place, no one can do useful scientific work using selection theory; after relativity theory is firmly in place, classical physics ceases to be current science. Selection and wave theory, classical and relativity physics, are incommensurable.

Incommensurability's Scope and Significance

There are at least two ways to assess the scope and significance of incommensurability as a factor in scientific change. The first consists in applying the concept diachronically, to concrete examples in the history of science. In *Structure,* in speaking about revolutions as changes of worldview, changes that are incommensurable, Kuhn employs this first method:

> Shifts of this sort are not restricted to astronomy and electricity. We have already remarked some of the similar transformations of vision that can be drawn from the history of chemistry. Lavoisier, we said, saw oxygen where Priestley had seen dephlogisticated air and where others had seen nothing at all. In learning to see oxygen, however, Lavoisier also had to change his view of many more familiar substances. He had, for example, seen a compound ore where Priestley and his contemporaries had seen an elementary earth, and there were other such changes besides. At the very least, Lavoisier saw nature differently. And in the absence of some recourse to that hypothetical fixed nature that he "saw differently," the principle of economy will urge us to say that after discovering oxygen Lavoisier worked in a different world. (Kuhn 1996, 118)

There is a serious, perhaps fatal problem with this way of arriving at a philosophical conclusion from the historical facts. In later work, Kuhn warns us that, "if prior expectations are not merely unreasonable, the historian committed to them can usually make a case without consciously forcing the data. Accordingly, the historian is usually well-advised to set expectations aside before beginning research" (Kuhn 1980, 183). While in these animadversions Kuhn does not refer to his own practice, the example I have cited seems *a propos*. The significance of the passage is moderated considerably when we substitute a literal "interprets" for every metaphorical "sees," and when we refuse the leap from the examples to the generalization that "Lavoisier saw nature differently," insisting that "the principle of economy" urges us only to say that in this case the evidence presented massively underdetermines the conclusion Kuhn wants to draw.

The other route to understanding—conceptual clarification—is undertaken by Paul Hoyningen-Huene and Alexander Bird. Hoyningen-Huene is successful at his chosen task, which is to elucidate Kuhn's philosophy of science as it stands. But the self-imposed limitations of this task mean that central concepts, like incommensurability, cannot be weighed in terms of general scope and significance. This latter task is taken on by Alexander Bird. As an example of his argumentative strategy, I cite Bird's argument that the no-overlap principle, central to incommensurability, is less general in application than Kuhn and his followers assert, implying that its philosophical usefulness is far more seriously limited than they assume (I apologize for the length of the quotation needed to convey the flavor of Bird's argument):

> Kuhn seems to think that one reason for respecting the no-overlap principle, and thus for not mixing taxonomies, is that doing so will permit conflicting projections. [. . .] Let us say that "water" has undergone a taxonomic shift. We avoid ambiguity by employing the terms "$water_1$" and "$water_2$." If these violate the no-overlap principle then something might be both $water_1$ and $water_2$. But, Kuhn says, the two terms induce different expectations. Employing both of them is an unstable state of affairs to be resolved only by recourse to matters of fact: "And if the matters of fact are taken seriously, then in the long run only one of the two terms can survive within any single language community."

> This complaint does not exclude overlaps in general. The cases cited of actually overlapping taxonomies show this. Although the differing classifications of the same stuff will, given what is believed, permit differing expectations, those expectations need not conflict. The same stuff can have many properties according to which it is classified, and each of these may involve it in some law. But a thing may be subject to more than one law, like an electron which is subject to gravity, electrostatics, and the laws of motion. So projectibility is no reason to exclude every case of overlapping taxonomy. (Bird 2000, 201)

There is nothing wrong with this argument. It is not a matter of arguing with Bird in his own terms, but of claiming that, because he asks the wrong question, he gets an answer that, whatever its truth, is irrelevant to Kuhn's central concern. In my view, the real question is not whether two differing classifications necessarily conflict, but whether in crucial historical instances, they did actually conflict; it is not a question of principle *per se*, but a question of principle as it is realized in the context of a particular scientific practice. My claim is that Bird's strategy, conceptual clarification at a very high level of abstraction, cannot by its very nature capture the scope and significance of Kuhnian incommensurability because it is too remote from scientific practice to do so. In that sense, this strategy is no better at dealing with incommensurability than Kuhn's own in *Structure,* one that leaves him open to the accusation that he is tailoring history to fit the philosophy he prefers.

I will now present a model of what I take to be the right approach to testing the scope and significance of Kuhn's claims, that of historian Jed Buchwald. Buchwald has analyzed in detail a case in which the conflict is not between the hypothetical $water_1$ and $water_2$ (which Bird adapts from Putnam 1975), but between actual contending conceptual Lexicons deeply imbedded in scientific practice: those of the selection theory and wave theory of light in the second quarter of the nineteenth-century. According to selection theory, light consists of bundles of individual rays; according to wave theory, it consists of transverse waves. In wave theory, therefore, individual rays cannot exist: the concept is meaningless. Since the ontology of light is exactly at the center of both selection theory and wave theory, and since this ontology both constrains and enables the explanation of such key phenomena as reflection, refraction, diffraction, and polarization and their mathematics, selection theory and wave theory are profoundly incommensurable, that is, incommensurable at the level of the conceptual structure of the Lexicon.

The significance of such profound incommensurability is illustrated in the case of David Brewster, an English physicist of considerable talent who never abandoned selection theory, but attempted to integrate the "legitimate" innovations of wave theory into its taxonomy. In this, as Buchwald shows, he largely succeeded. Of Brewster's selectionist account in his explanation of polarization, Buchwald concludes: "it attempts to subvert the vocabulary of the wave theory by appropriating it to selectionist ends, and it does so rather well" (Buchwald 1992, 74). The problem, as Buchwald also points out, is that the price of this accommodation is prohibitively high. It leads to a research program that, while coherent in its own terms, cannot progress because, at each accommodative juncture, a modification has to be made which, however ingenious, is thoroughly ad hoc and, from the point of view of further research, thoroughly useless.

To simplify my exposition, I will focus on Brewster's account of phase. (Points of light having the same displacement from their equilibrium position and traveling in the same direction are said to be in phase.) Selectionists in general admitted phase into their taxonomy—they could hardly do otherwise since it had been demonstrated instrumentally—but "they did not treat the property as primary (in the way that the individual existence of a ray of light was considered primary). Their S-phase was a problematic affection of light rays that can be manifested only under special conditions" (Buchwald 1992, 51n). In his own work, Brewster incorporated from wave theory the usual Fresnel law for the rotation of the plane of polarization, and with it the term *phase.* In so doing, however, "he had expropriated a formula [and a term] from the W[ave]-scheme in a way that strips [them] completely of [their] original significance." From the point of view of wave theory, Brewster's use of the term *phase* is, strictly speaking, meaningless, and his formula works only under "degenerate conditions" (Buchwald 1992, 71–72). His concept of phase is not projectible.

Experimental practice shows that Brewster's attempted accommodations to wave theory lack conceptual robustness. As he moves from instrument to instrument, the categories of his taxonomy of polarized light vary unstably:

> The beam-splitter [. . .] places partial [polarization] *between* elliptic and circular, whereas the special reflection process distinguishes elliptic and circular under a single category *from* partial. This is intensely perplexing because complete and partial are equal siblings [under Brewster's scheme]. Incestuous relationships become possible here. Contrast this with the linear-elliptic-circular of W[ave]-polarized [light]. Here beam-splitters place linear before elliptical and circular, while special reflections group the latter two together. Consequently here the beam-splitter enforces a distinction that is also preserved by another device. This is not true in Brewster's taxonomy, where the beam-splitter's distinctions tend to violate the distinctions enforced by special reflections. Brewster's taxonomy is accordingly unstable. (Buchwald 1992, 54)

Brewster's failure is especially important because he was a talented physicist who tried to integrate the discoveries of wave into selection theory in a most ingenious matter. He failed precisely because the two taxonomies are conceptually incommensurable at the level of structure, just as Kuhn predicts. You have to choose; you can practice optics *either* according to selection theory *or* according to wave theory.

Buchwald's project was one of which Kuhn heartily approved: "Jed has now systematically analyzed numerous aspects of the transition from [the selectionist] to the wave theory of light as a result of changes in kinds. His paper is likely, I think, to introduce a new stage in the historical analysis of episodes involving conceptual change" (1993, 323).[5] Kuhn approves of Buchwald's approach because, while he remains concerned that history and philosophy of science not be amalgamated to their mutual detriment, he does not exclude the possibility of equal partnership if "data can, and [is] permitted to, react back on expectations, make trouble for them, play a role in their transformation" (Kuhn 1980, 182).

Such dialectical fruitfulness is possible in the case of Buchwald because he completed his research of the optics *before* he undertook the application of Kuhn's categories to his data, a methodological precept Kuhn regards as essential if impartiality is to be preserved. In the retrospective introduction to *The Rise of Wave Theory,* still innocent of Kuhn's categories, Buchwald says of Brewster:

> About 1830 he wrote several articles on what he termed elliptical polarization and on the structure of partially polarized and natural light. Brewster claims not to need any hypothesis about the nature of light. He tries to make use of formulas and parameters that are derived from the wave theory in a way that avoids the theory proper. The result in this case is almost complete nonsense, because he creates a hodgepodge that, despite his claim to have avoided theory, mixes the tacit assumption that rays can be treated as individuals with concepts that have little meaning outside the wave theory proper. (Buchwald 1989, xix)

In his later analysis of the same data, Buchwald takes full dialectical advantage of the explanatory potential of Kuhn's categories. Not only does he validate them in an important instance; he modifies his own historical account as a consequence, learning to respect David Brewster as a good scientist, the victim of an incommensurability he never understood. In this later analysis, Buchwald gives Kuhn's concept of incommensurability a firm shape by embedding it deeply in the context of a particular scientific practice.

There can be little doubt of the significance of incommensurability in this case. Moreover, it seems plausible to generalize over the set of cases of which this is a member, those that depend "quite explicitly on the critical role of experimental apparatus" to sort nature into kinds (Buchwald 1992, 57). It is also possible that the scope of Kuhnian incommensurability may extend beyond the relatively narrow band of the experimental sciences after 1800. If we assume, Buchwald suggests, "a somewhat broader notion of apparatus," if we include, for example, "the rules and mathematical methods that [Kepler] was prepared to deploy in accommodating Tycho's observations" (Buchwald 1992, 58), the scope of Kuhnian incommensurability might be legitimately broadened to sciences, like astronomy, that are strictly observational. In this and analogous cases, "theory itself does the sorting work" (Buchwald 1992, 59). Buchwald's case exemplifies a method of which Kuhn fully and rightly approves: the scope and significance of the concept of incommensurability can be best explored through a dialectical interchange between the philosophy and the history of science.

Buchwald's is an inductive argument in favor of a methodological preference, moving from a salient case to a generalization over cases construed as alike in the relevant respects. To justify this preference deductively, I will argue from analogy that Kuhn's analysis of scientific kinds is itself a matter of kinds, philosophical kinds like incommensurability that are not fully articulated, are projectible, rigidly designate, obey the no-overlap principle, and are components of larger, more encompassing philosophical structures. In my view, this means that the difficulties in making precise the central concepts of the sciences are analogous to the difficulties in making precise the central concepts of Kuhn's theory of the structure of scientific revolutions, including that most central of concepts, incommensurability. Just as scientific concepts have to be realized and extended by means of an extended dialogue with nature, so Kuhn's concept of incommensurability must be realized and extended by means of an extended dialogue between it and the history of science, a history whose resistance to philosophical speculation is guaranteed because it was created independently of that speculation. Just as experiment is designed to keep the scientist in close touch with the fine structure of the material world, a strict dialogue with historical reality, reconstructed independently of any philosophical thesis, is designed to keep the philosopher in close touch with the fine structure of past scientific practice. Such disciplined proximity is the best guarantee that the history will not be abused, a possibility that Kuhn feared, and that the philosophy will be sound, as it will be erected on the firm base of the practice it is designed to explain.

Is Incommensurability Limited to Science?

The analogy between Kuhn's philosophical kinds and scientific kinds must not be pushed too far: philosophy is not science. Kuhn's taxonomy of scientific kinds is not a Kuhnian Lexicon. No matter how well established it becomes, Kuhn's theory of scientific change will never dominate to the exclusion of other explanatory schemes, other schools of thought. Although there will always be a normal science, there will never be a normal philosophy. From this I think that it follows that Kuhnian incommensurability is limited to the sciences. In two papers,

Hoyningen-Huene disagrees; in so doing, he introduces a new concept, meta-incommensurability, offering it as an explanation of the difficulty in resolving the philosophical dispute between the defenders of realism and of non-realism.

Though the analogy with Kuhn's incommensurability is deliberate, meta-incommensurability is established independently of its Kuhnian counterpart; in no way is it dependent on the truth or falsity of the latter. Nonetheless, meta-incommensurability shares with its Kuhnian counterpart an essential characteristic: each rival philosophical camp has a different set of ontological commitments, leading to profound differences in the central concepts used, concepts such as "reality," "world," and "fact." These differences have as their most important result that arguments in favor of realism or non-realism are question-begging and circular, that is, they surreptitiously assume what they set out to prove. Hoyningen-Huene offers this as an explanation of our intuition that realists and non-realists talk past each other.

I want to accept the conclusions that Hoyningen-Huene draws, while rejecting meta-incommensurability as the best description for the phenomenon he identifies. I object to his description first because it is misleading: it permits the inference that each philosophical camp is committed to concepts that form a Kuhnian Lexicon incommensurable with the Lexicon of its rival. But it is not the case that each camp shares a Lexicon in Kuhn's sense. Indeed, neither camp is committed to concepts that meet Kuhn's necessary conditions, ones that are projectible, obey the no-overlap principle, and are uniquely organized into structural relationships that, although not precisely articulated, are firmly constrained by particular concrete disciplinary achievements. Nor is it the case in philosophy as it is in science that one "paradigm" replaces another; rather, both philosophical camps continue to coexist and to defend their positions.

Analogous arguments have been made against the existence of incommensurability in linguistics, in the social sciences generally, in geography, and in literary studies (Percival 1976, 1979; Young 1979; Jauss 1969). Jauss, for example, claims that literary studies have been guided by three successive *Weltanschauungen.* In the first, antiquity is the norm; in the second, the romantic view prevails that literature reveals the spirit of the times; in the third, the immanent characteristics of the work itself are stressed. The issue is not whether the characterizations of these *Weltanschauungen* are correct, but rather whether they count as paradigms in Kuhn's sense. They do not, as Jauss himself strongly affirms: "Stress must now be laid on the ways in which revolutions in what counts as knowledge in literature and the humanities differ from corresponding events in the natural sciences. It can't be that a crisis leading to a paradigm change is simply the consequence of an anomaly as in the sciences because literature has no counterpart to empirical verifiability" (1969, 54) The trigger for major change in literary studies is rather "the capacity of works of art through ever new interpretations to translate their essence in a new present that makes their significance available once again" (1969, 55). To this difference must be added a characteristic impossible in the natural sciences: "In the humanities [. . .] a paradigm that is historically over with can have a rebirth" (1969, 47).

The coexistence of rival camps is exactly what we would expect if realism was not meta-incommensurable with non-realism, but rather if the issue was, in W. B. Gallie's phrase, *essentially contested.* Like other essentially contested concepts mentioned by Gallie, realism (and its non-realist counterpart) share five characteristics. First, they are appraisive in the sense that their ultimate establishment would be regarded as a valued achievement on the part of their proponents. Second, they are internally complex in any of their philosophical forms. Third, the object of analysis of both realism and non-realism, the relationship between words and the world, can plausibly be seen, from its beginning, as variously describable. Fourth, realism and its rival are both "open" positions; each is evolving, both within itself, and in tandem with the other. Finally, each party in the dispute between realism and non-realism recognizes that its own way of defining the relationship between words and the world is and remains contested.

For Gallie, these five characteristics are necessary for a concept to be essentially contested. But they are not sufficient; they do not differentiate between these and concepts that are simply confused. When realists and non-realists use such concepts as "reality," "world," and "fact," are they employing two versions of the *same* concept or, rather, two *different* concepts with the *same* name? The former, Gallie says, if they both can trace the concept back to the same exemplar and view their current position as its optimal development. While it is not possible to trace the positions of each philosophical camp to a specific exemplar, in the sense that Christ is the exemplar for all versions of Christianity (to use one of Gallie's examples), it is clear that both realists and non-realists see their positions as anchored in traditions of argument that go back to the quarrel between the pre-Socratics and their immediate successors. Finally, and this is Gallie's last requirement, each proponent sees his position as the best alternative currently available.

When it comes to philosophical dispute, there is a therapeutic advantage to Gallie's alternative to incommensurability. As he himself states:

> a certain piece of evidence or argument, put forward by one side in an apparently endless dispute, can be recognized to have a definite logical force, even by those whom it fails to win over or convert to the side in question; and [. . .] when this is the case, the conversion of a hitherto wavering opponent of the side in question can be seen to be *justifiable*—not simply expectable in the light of known relevant psychological or sociological laws. (Gallie 1968, 185)

This realization that your opponent is sane and sensible, Gallie asserts, might easily raise the quality of philosophical arguments on both sides by curbing the natural human tendency to denigrate, not only the opponent's views, but the opponent himself, and by leading, through cooperative engagement, to the discovery of "where the *real* issue between them lies" (1968, 188).

Conclusion

In this essay, I show that the continuing misunderstanding and apparent lack of intellectual progress in the debate over incommensurability has at least two causes rooted in inappropriate argumentative strategies. In the first place, there is Kuhn's own strategy of conceiving his central concept of incommensurability metaphorically rather than literally. In the second place, there is the strategy of Kuhn's critics who, I contend, debate unproductively by refusing to argue from scientific practice as revealed in its history or by misusing that history, tailoring it as does Bird to a particular philosophical position. To progress in this philosophical debate is to use history of science properly, as Kuhn does in the afterward to his book on black body radiation or as Jed Buchwald does in an essay in the collection, *World Changes.* In both cases, the history of science is reconstructed independent of any Kuhnian philosophical presuppositions, and then tested against Kuhn's final position on incommensurability, a position free at last of Kuhn's metaphorical baggage.

In his obituary to his friend, Paul Hoyningen-Huene remarks that "when the discussion became long-drawn-out, Kuhn could become impatient, especially if he had the impression that the criticism, even if it were correct, was not productive for the development of his theory. (His American colleagues would recount that Kuhn would respond to their criticism with: 'I need your help, not your criticism!')" (1997, 253; translation mine). In doing what Kuhn could not bring himself to do in a systematic manner, Jed Buchwald has helped us by giving us a way of assessing the scope and significance of Kuhn's central concept of incommensurability. He has shown that it is a vital element in conceptual change in the sciences, at least those that depend on experimental apparatus to sort nature into kinds.

This conclusion is, of course, at odds with the view of incommensurability in Harris's introduction. Harris asserts that "the most useful application of *incommensurability,* in the end, is not a name for a condition of theories (things) but of theorists (people)" (92). On the contrary, I argue that the most useful application is to theories. Neither can I agree with the characterization of late Kuhnian incommensurability in Harris's taxonomy: weak, semantic, local. Where Kuhnian incommensurability applies, in my view, it is neither weak, nor semantic, nor local. Rather, it provides a strong conceptual and practical barrier to competing scientists, and applies not only to individual situations, but also to whole fields. David Brewster's failure resulted, not from lack of knowledge or good will, but from the incommensurability of selection and wave optics. Similarly, I am forced to disagree with the presupposition of Bazerman and De Los Santos that incommensurability is a matter of misconstrual: "scientists who inhabit one theory-based perspective are unable to recognize, understand, or accept entities revealed through observations made from an alternative theoretical perspective" (424). David Brewster had no difficulty in understanding wave theory in any ordinary sense of "understanding." Equally, I must dissent from Miller's conclusion that incommensurability is supererogatory: "in the study of scientific change and controversy, incommensurability is an idea we can probably just do without" (502). No other explanation of David Brewster's plight makes sense to me; absent incommensurability his struggle to accommodate wave theory loses its tragic dimension; it turns the battle to save his science into an example of inept bungling or severe character defect. Moreover, to argue from the cases presented in this volume, cases in which Kuhnian incommensurability allegedly does not exist, to the non-existence of Kuhnian incommensurability is simply impermissible.

But the work of these interesting and valuable essays is not philosophy; it is rhetorical analysis. My disagreements, even if sound, in no way vitiate these analyses of the strategies of scientific agreement and disagreement. As such, these essays provide considerable insight into the nature of scientific disputes and the means for resolving them. To these analyses I would only add the observation that the stubbornness of scientists maintaining and strengthening their positions on either side of a theoretical divide—whether incommensurable or not—is not necessarily counter-productive. It can, as Feyerabend argues, and Harris echoes in the introduction, force both sides to refine their arguments and so serve the long-term interests of science, though not of course of amity.

Notes

[1] My quick intellectual-biography overview is only meant to serve the interests of the argument that follows. See Paul Hoyningen-Huene's essay in this volume for a fuller treatment of Kuhn's intellectual development, especially with relation to incommensurability, or his (1997) obituary of Kuhn. See also Heilbron's (1998) obituary, Bird (2000), Fuller (2000), or Sharrock and Read (2002) for various full-length biographies.

[2] But only for philosophers who were initially sympathetic since "this essay is not calculated to convince" those who represented rival positions (Kuhn 2000, x).

[3] A notable exception to this tendency is Rupert Read, who writes "Kuhn was the greatest philosopher of science we have yet known, and one who was, if anyone has yet been, something like a Wittgenstein of the sciences" (Read 2003). See also his very appreciative book with sociologist, Wes Sharrock, *Kuhn: Philosopher of Scientific Revolution* (Sharrock and Read 2002).

[4] Although in this early incarnation, Kuhn speaks, not of kinds, but of theories, and not a lexicon, but a paradigm, I think the transposition is justified. Recall that Kuhn is offering the Lexicon as a way of refining his earlier conceptions, not of discarding them.

[5] This approval is consistent with Kuhn's attempt at a parallel analysis in the appendix to the second edition of *Black Body Theory,* an analysis carried out reluctantly as a consequence of criticism that his history made his philosophy beside the point.

4

Incommensurate Boundaries: The Rhetorical Positivism of Thomas Huxley

Thomas M. Lessl

> The question arises: Is the whole evolutionist current of positivism, the reduction of knowledge to a biological instrument of adaptation, touched off by the Darwinian revolution but already rooted in Hume's critique, merely one variant of positivist thought—a modification, an aberration, a deviation, perhaps an accident? Or could it be that the constitutive, the essential core of positivism contains something that leads inevitably to such biological relativization, for all that one and another variety of positivism fail to draw this dangerous consequence?
>
> —Leszek Kolakowski, *The Alienation of Reason*

The question raised by Kolakowski in this passage might be constructed as a kind of thought experiment—a social as much as an intellectual thought experiment. Most readers will have heard at some time Theodosius Dobzhansky's much quoted claim to the effect that nothing in biology makes sense except in light of evolution. I am most familiar with this statement as an utterance invoked in defense of giving evolutionary theory a central position in science education. It suggests that students deprived of an understanding of this paradigm would be unable to make sense of everything else that is taught under the rubric of biological science. But would this adage still ring true if the principle of natural selection had never been discovered by Wallace, Darwin, or any subsequent theorist? Is it possible, in other words, to imagine a biological science that was not predicated upon evolutionary preconceptions? Kolakowski seems to think that it is not, that the evolutionary perspective was more or less the inevitable direction that science was taking in the nineteenth century—even before Darwin's *Origin*. His metaphor for what Darwin did when he published his theory is "touched off." The volatile fumes of evolutionism had already saturated the intellectual air of Victorian science by virtue of its increased devotion to the kind of anti-metaphysical philosophy promoted by David Hume. To the extent that this was the case in the nineteenth century and still is so today, evolution might be meaningfully described as the application of positivism to biology. We live in a world in which non-evolutionary alternatives to Darwinism, such as those now being advanced by design theorists, are ruled out, not merely because they are incommensurate with the regnant neo-Darwinian paradigm but also because they are incommensurate with the broader positivist stance that has dominated scientific culture since before Darwin's time.

My intention here is to suggest that the apparent incommensurability of these scientific perspectives, those representing naturalistic (neo-Darwinian) and design views of biological origins, is rooted in a distinctive problem that lies outside the purview of incommensurability as it was originally explained by Kuhn and Feyerabend. This is to argue that the evolutionary paradigm that now excludes design arguments from scientific discussion did not triumph through a scientific revolution resulting from the publication of the *Origin of Species*. I will instead argue that Darwinism's triumph represents the success of a broader positivist worldview and that this is the source of its incommensurability with design. In this regard, this essay operates out of an enlarged notion of the conditions that make scientific positions irreconcilable, an understanding that is concerned with the broadest constitutive features of science.

In the wake of the triumph of the neo-Darwinian paradigm its advocates have repudiated alternative design arguments largely in three ways: (1) they assign religious motives to the design position, presumably under the assumption that such motives prejudice whatever status its advocates may have as credentialed scientists; (2) they may also dismiss such claims for failing to take account of a preponderance of evidence favoring naturalistic evolution; and, (3) they may contend that design is, by definition, outside the scope of scientific inquiry.

All three of these objections have a degree of plausibility. But I wish to introduce a fourth one, the unstated objection that design is not incommensurate with science's assumptions and methods but rather with its mode of being. This is to say that the being of science is positivist; its way of understanding its place in the world excludes non-naturalistic considerations of origins. In an effort to understand this broader notion of incommensurability—a species of Harris calls *pragmatic incommensurability* in the introduction to this volume, but one with an especially high reading on his incommmensurometer—I will argue that the issue of ideology, in its original Mannheimian sense, is helpful. Thus the thesis that will be defended here is this: debate about whether scientific inquiry about origins should exclude questions of design goes to the heart of what science is, not merely in a philosophical or operational sense but also in a social and institutional sense. The argumentative impasse that excludes design from scientific thinking about origins is not entirely evidentiary, methodological, nor even paradigmatic; rather it is constitutive, an impasse having to do with the social and institutional situation of modern science. This situation makes meaningful exchange almost impossible, creating a situation not centered on the materials of the debate, but on the jurisdiction of the debaters. The naturalist Darwinian programme holds a massive presumption, such that arguments issuing from a perspective that countenances design are simply stipulated out of the realm of science.

Darwinism, Design, and Ideology

Those who reject design out of hand by appealing to the personal religious motives of the design theorist or the evidentiary *topoi* of modern science make plausible arguments, but their discourses are also filled with rhetorical markers that may be characterized as ideological. In saying this, I hope to show that in this extra-paradigmatic realm of scientific thought the social and the intellectual are fairly indistinguishable. This is to imagine, if my readers will pardon a fanciful comparison, that looking at scientific debate at this level is analogous to moving from a classical conceptualization of the universe in which time and space are distinguishable things to a relativistic one in which they are not.

What I am relativizing here are the social and intellectual concerns that manifest in the rhetoric of those in the positivist movement of the nineteenth century who succeeded in pushing design out of science. Thus I am being deliberate in choosing the term "positivist movement" rather than simply "positivism." I wish to develop a historical picture of positivism here in order to explore the interplay of specific societal and philosophical concerns that were brought together during this particular formulary period in scientific history. As a factor contributing to incommensurability, ideology can best be detected by taking such a historical view. It was the situation of science at this particularly pivotal moment of its history that has made the Darwinian view both scientific and ideological.

The notion of ideology is not employed here in its traditional Marxian sense as false consciousness rooted in class divisions but rather in the broader sense given to it by Karl Mannheim, as an aspect of his sociology of knowledge—"the outlook inevitably associated with a given historical and social situation, and the *Weltanschauung* [or *worldview*] and style of thought bound up with it" (Mannheim 1936, 125). This notion of ideology has fallen into disrepute or at least neglect among postmodern and poststructuralist theorists who object to the concept's assumption of a truth standard against which false consciousness may be measured. But despite their insistence that the pervasively ephemeral and discontinuous characteristics of society and history make a Mannheimian ideological critique impossible, they continue to appropriate the concept tacitly. There does not seem to be an absolute division between the postmodern denial of truth standards and the assumption of an ideological critique that insists that there is always some interrelationship between our interest-motives and our meaning-motives in social action.[1]

Neither does the supposition that the evolutionary position is ideological undermine the integrity of what scientists do. Such a stance was also hinted at by Kuhn when he commented on Max Planck's famous observation that "a new scientific truth does not triumph by convincing its opponents and making them see the light, but rather because its opponents eventually die, and a new generation grows up that is familiar with it" (Kuhn 1996, 151–152). In interpreting this phenomenon, Kuhn goes on to suggest that the stubbornness that would make old timers cling to an outdated paradigm may be rooted in the norms that sustain the work of normal science. Resistance to new paradigms, in other words, is socially conditioned and thus ideological. But this does not mean that it necessarily inhibits scientific discovery. He explains that the scientific culture, in order to keep its members on task in their work of puzzle-solving, must sustain each practitioner's faith in the integrity of its established paradigms, even if this sense of loyalty may ultimately deter revolutionary change. This somewhat localized ideological feature, the Max Planck Effect, upholds scientific investigation; it does not pollute it. In fact, it can be defended by appeal to rational criteria: the past success of the established paradigm in solving puzzles provides a reasonable basis for the hope that it will do so in the future, provided it is not abandoned (R. A. Harris 1998).[2]

In Kuhn's view then, aspects of thought that social theorists would traditionally call ideological may directly serve the interests of scientific inquiry, in spite of their unscientific character. Their qualities as false consciousness only become apparent and only become inhibiting when alternative scientific paradigms are introduced. In this regard my position is in line with Kuhn's, but I am going to contend that, at least where biological origins are concerned, the relevant ideological aspects of scientific culture are distinctive, that they influence its judgment in a qualitatively different way. What makes the symbolic forces that exclude design from biology qualitatively different from those that merely promote resistance to alternative paradigms is the fact that certain human motives tied up with evolutionary science seem to lie at the boundaries that divide science from metaphysical and theological inquiry (see Taylor 1994 for a rhetorical treatment of demarcation issues in science, especially 135–174, which looks at issues involving creation science). Ideology in this case does not serve to keep scientists on task, though it very well might in the cases Kuhn discusses. Rather, it works to keep competing externalized influences at bay.

In defense of this position I will argue that while the regnant Neo-Darwinian paradigm operates as a scientific theory of origins, it is also necessarily tied up with dominant notions about what distinguishes science from other categories of inquiry. Perhaps a more venturesome way to put this would be to say that the Neo-Darwinian paradigm is concerned with *origins* in two distinct senses. In its usual theoretical sense Darwinism proposes a variety of natural mechanisms (primarily natural selection, sexual selection and genetic mutation) which may account for the development of all varieties of life over long stretches of time; but in a second and more unusual sense it is a theory tied up with the most *originative* conceptions of what science is—the boundary conditions of science so to speak. This dual understanding of evolutionary science makes sense of an important fact of history, namely that the triumph of Darwinism as a naturalistic explanation of origins occurred virtually at the same time as the triumph of positivism as a social utopian and philosophical movement seeking the abolition of all non-naturalistic perspectives on knowledge.

Although it would be accurate to say that Kuhn's thinking about scientific revolutions involves something like the notion of ideology, what I am proposing in suggesting that the origins debate is tied up with originative conceptions of science is that there is a broader symbolic concern that must be accounted for here. Even if Kuhnian paradigms denote fixed patterns of representation that determine the course of thinking for those who subscribe to them, these occur within larger frameworks of symbolism that are beyond the scope of Kuhn's theory. Darwinism may be treated as the kind of paradigm described in Kuhn's theory, but it also has a part in this broader orbit of institutional thought.

In his survey of scientific revolutions Kuhn stuck for the most part to his own field of physics and did not examine the development of biological theories. Even so, he seems to have presumed that this model should describe every instance of revolutionary change. However, the Darwinian historian John Greene argues strongly against the explanatory merits of Kuhn's model as applied to the triumph of Darwinism (Greene 1981, 30–59). He contends that Kuhn's thesis cannot account for the establishment of Darwin's theory of evolution in the late nineteenth century and its rebirth after a period of decline in the middle of the next century, because the triumph of evolutionism did not come on the heels of a crisis in some competing paradigm. Greene finds no evidence of crisis within the Linnaean-Cuvierian paradigm that had dominated before the introduction of Darwin's hypothesis. In fact the Darwinian paradigm could not create such a crisis within the Linnaean-Cuvierian camp simply because the evolutionary model presupposes goals for natural history that simply had not been a part of the static model that it replaced. Those who worked within this pre-Darwinian school assumed that the goal of natural history was to name, to classify and describe, and Darwin's model did not challenge its ability to do so. The demand that one finds in Darwinism that the development of biological organisms was the necessary consequence of natural causation could only come from without.

Anomalies in the Linnaean-Cuvierian paradigm were insufficient to bring about the sort of crisis that Kuhn identifies as the necessary impetus for a scientific revolution. This is shown by the fact that when the problem of mass extinctions was encountered in the late eighteenth century, scientists responded merely by adopting the geological catastrophism of Jean Deluc to preserve their static view of creation. The discovery of mass extinctions did not threaten natural history precisely because the Linnaean-Cuvierian paradigm was not concerned with evolutionary change. Neither was its static worldview necessarily threatened by Charles Lyell's influential insistence upon the "principle of uniformity." Despite the popular (and probably Whiggish) view that this was a primary influence upon the Darwinian revolution, uniformity remained, after all, a steady-state concept that could have just as easily been turned against the notion of development with its suppositions of significant change.[3]

The Linnaean-Cuvierian paradigm did not raise the kinds of questions that Darwinism proposes to answer, and so it could not experience internal conflicts that would be sufficient to invite consideration of an alternative evolutionary view. Greene concludes that, in England as well as in France, evolutionary paradigms emerged by way of a variety of intellectual invasions from without rather than by means of a revolution from within. Thus the Count de Buffon's evolutionary paradigm, which was ancestor to the better-known speculations of Lamarck, was in Greene's view a "conscious attempt to introduce into natural history concepts derived from [Cartesian] natural philosophy, from the seventeenth-century revolution in physics and cosmology" (Greene 1981, 36). This pattern was repeated in England as well. The evolutionism of Erasmus Darwin, according to Greene, arose not as an "internal development of systematic botany or zoology," but rather as an external development out of the "eighteenth-century tradition of speculative philosophy of nature" (Greene 1981, 47).

Greene emphasizes that the older natural historians were tied to the assumption that the creation was static, and they were therefore willing to compensate for features of the fossil record that were inconsistent with this view by simply assuming a succession of creations. We might suppose, were we to assume the general applicability of Kuhn's description of scientific revolutions, that the evolutionary arguments that were beginning to make themselves heard in the nineteenth century succeeded because they accounted for more evidence than the established paradigm of natural history. Popular memory has frequently depicted the triumph of Darwinism as such, as a massive assault of fact upon a moribund scientific orthodoxy that was too much under the spell of religion, but, in fact, design seems to be a perspective that was displaced rather than refuted. However, this displacement did not occur in the manner in which Kuhn's theory predicts, which is to say at the level of paradigms. It occurred instead at a superceding level of discourse in which the constitutive features of science were being debated.

In light of Greene's reading of these developments it would be inaccurate simply to say that a naturalistic evolutionary model of origins replaced a static creationist one. For those who worked in the Linnaean-Cuvierian paradigm, design was a metaphysical assumption not a subject of inquiry. The operations of creation, by definition, lay beyond the borders of what scientific investigators were obliged to describe or to theorize about. It was the Darwinian revolution that created the expectation that origins must be explained, but this could only become an operating premise of the scientific enterprise if its boundaries were first redefined in a revolutionary way.

Classical Positivism and the Victorian Professionalization of Science

Up to this point I have argued that a social or ideological view of incommensurability may add a helpful complication to our understanding of the perceived incommensurability of design and naturalistic views of origins. This perspective adds an additional layer of meaning—historical meaning especially—to the more conventional view of incommensurability as a product of paradigm disputes. In taking exception to Kuhn's model, Greene attributes the success of the Darwinian revolution to two factors: the importation of constructs such as uniformitarianism from Lyell and the importation of a naturalistic standard for science. I am now going to claim a third source, the positivist movement. My argument will be that the positivist movement promoted the exclusion of design by creating a social atmosphere conducive to the establishment of naturalistic limits on science.

Before going any further it will be necessary to clarify how I am using the term *positivism.* Typically the term refers either to the classical positivism which reached its highest level of influence around the middle of the nineteenth century, or to the revival of some of its concepts in the Vienna Circle after World War I. I will add a third sense of positivism to these two distinctively historical movements, which were largely a concern of philosophers and social scientists. I associate this third sense, not with any particular group or movement, but with an emerging professional ideology that began to take hold of the natural sciences in the Victorian era. I will use the term *positivism$_1$* to designate what is now typically called *classical positivism,* the body of ideas most closely associated with Auguste Comte's influential *Cours de philosophie positive* and various of his other writings. Although positivism$_1$ is now usually associated with radical empiricism, a philosophical movement in which it was deeply rooted, the main goals of Comte's program were utopian as well as scientific. In all his writings Comte insisted that positivism is both a "Philosophy and a Polity," and that these two aspects of the system "can never be dissevered; the former being the basis, and the latter the end of one comprehensive system, in which our intellectual faculties and our social sympathies are brought into close correlation with each other" (Comte 1953, 1). The philosophy in this case is Comte's version of the radical empiricism or sensationalism that had already been handed down by David Hume and Étienne Condillac, among others; the polity that Comte refers to here denotes two general ideas. The first of these is the movement's expectation that the growth of science was destined to enable it to progressively encompass increasingly complex fields of inquiry, so that a positive science of society, more specifically the field of sociology, would eventually emerge to provide the intellectual basis for a new social order. The second idea, which after Popper I will call *historicism,* is positivism's adoption of a philosophy of history in support of these utopian expectations (Popper 1957; 1962, 1: 1–8, 2: 125, 193, 322, 393, 395). Like his primary intellectual predecessors, Henri Comte de Saint-Simon and Marquis de Condorcet, Comte subscribed to the view that history is progressive and that the course of its evolution can be described in terms of natural law. Although various other renditions of this grand narrative circulated in the nineteenth century, Marxism being the most familiar, all shared the common faith that the operation of reason had been the engine of progress throughout history and that with the advent of modern science, the principles that governed intellectual and social evolution had been laid bare. Science accordingly, as the purest expression of this principle, should enjoy a special place of honor in this emerging order that would be commensurate with its demonstrated powers of cognition.

The second meaning of positivism, $positivism_2$, is the more modern and narrow one associated with its rehabilitation in the Vienna Circle during the early decades of the twentieth century. Under a variety of headings—*logical positivism*, *logical empiricism*, *pragmatism*, *falsificationism*, *foundationalism*, and *logical atomism*—those affiliated with this movement all embraced the same philosophical naturalism as the classical positivists, but they were consistently more emphatic in depicting it as a byproduct of the limits of knowledge or language rather than as a system or polity. Schools associated with the Vienna Circle also placed greater stress on the requirement that the steps of scientific verification be specifiable and that all forms of predication which are incapable of meeting this requirement be dismissed as nonsense, as unworthy of serious discussion. More important for our consideration is the fact that these writers have also downplayed the visionary historical aspects of classical positivism. Even while adopting its views on the limits of knowledge, they have had less truck with the philosophy of history that was openly avowed in $positivism_1$. Karl Popper is a notable example of this. While he clearly aligned himself with the Vienna Circle, despite proposing a methodological focus instead of its emphasis on empirically meaningful statements, he also wrote *The Open Society and its Enemies,* which attacks the kind of historicism most characteristic of the older positivists. Notably, however, it is only in passing in his two-volume treatment of this subject that Popper ever acknowledges Comte's contribution to this historicist pattern. The entire first volume is devoted, as its title indicates, to the "The Spell of Plato," and when Popper turns to the modern renaissance of this historicism in volume two his subjects are Hegel and Marx, while Comte does not even appear in the index![4] In this regard one senses that Popper is acting here as much as an apologist for $positivism_2$ as he is a philosopher. In strongly suggesting that historicism was only some Platonic baggage unwittingly inherited by modern western culture, Popper was also suggesting that it had not even been intrinsic to $positivism_1$ in the first place. Therefore the totalitarian ideologies that had been spawned by the age of Comte and Marx would only seem to have ties to positivist philosophy. In this manner the embarrassing speculative excesses of Comte's totalizing vision could be played down. Hegel and Marx, if not more antiquated philosophers such as Plato and Aristotle, were the true culprits responsible for the aberrant historicism of fascist and communist totalitarianism.

The third category, positivism$_3$, differs from the other two perspectives only in its manner of presentation. It is essentially an informal or rhetorical positivism, one that emerges spontaneously wherever ingredients of the previous two philosophical systems are found. Positivism$_3$ is able to bring aspects of both positivism$_1$ and positivism$_2$ into play, a sense of the high standards of scientific reasoning associated with the Vienna Circle and the historicist vision of classical positivism, precisely because it is informal and because it is a positivism not of philosophy books and monographs but of public discourses.

Positivism$_3$ describes the consciousness of individuals who, though unlikely to ever call themselves positivists, still value what Gertrude Lenzer calls the "increasing consolidation and prevalence of the scientific spirit," either in its "theoretical configurations or in its popular or commonsense forms." According to Lenzer, the substance of this positivist faith endures, even though it is no longer linked to such figures as Comte, J. S. Mill, or even A. J. Ayer and Rudolph Carnap, because it has simply come to be identified with science and no longer has any need of tracing its roots to philosophical movements (Lenzer 1975, xvii).

The focus of my exploration of ideological factors that enforce a split between naturalistic and design perspectives on origins will be on positivism$_3$, positivism in its sense as an informal worldview that manifests not in the discussions of philosophers but rather in the public discourses of science. Henceforth I will simply call this *rhetorical positivism*. I have made mention of the classical positivist movement and the Vienna Circle because the intellectual morphology of these ancestral forms is reflected in the more diffuse attitude spawned by the contemporary rhetoric of positivism. Those who are indwelt by the spirit of this rhetorical positivism may have only a faint conception of what classical and Vienna School positivists taught, but they are likely to recapitulate similar patterns of reasoning nonetheless. This is to say that those who embrace its attitude that knowledge is the exclusive province of science are also destined to appropriate something like the original positivist view of history—even though social theorizing on a grand scale ceased with the decline of the French phase of the movement. Although rhetorical positivism in this sense may attest to accepting only the scientific trunk of the positivist tradition, its distinctly public concerns always incline it to sprout ideological branches more in line with the historicist aspects of positivism$_1$. Thus there is a logical or perhaps psychological connection between the historicist doctrines of classical positivism and the limits-of-knowledge emphasis of positivism$_2$. The movement of the Vienna Circle, if we may regard it as an effort to strip positivism down to its bare epistemological essentials, could purify it of its historicist baggage only if positivists were willing to accept that the triumph of empirical science did not extend out into the social realm. But the scientific culture has never been able to prevent this. The result has been a retreat into rhetorical positivism, a discursive move that involves the disavowal of the world-building aspirations of classical positivism at a formal level, but continues to indulge such world building informally in other circles of discourse.

The distinction here is analogous to one that is often made between Marxism as social theory and Marxism as a political program. Even though communist and socialist political movements have experienced a gradual decline leading to near extinction in the U.S. over the course of the past fifty years, academic Marxism enjoys a more abiding presence. But this does not mean that the elements of thought that characterize Marxist political programs do not enter into the political behavior or thought of merely academic Marxists. Those who imbibe their social theory from Louis Althusser, Antonio Gramsci, and Theodor Adorno may no longer strive to implement political programs designed to fulfill the prophetic expectations enunciated in the *Communist Manifesto,* but this does not mean that their political judgment does not reflect these theoretical influences. Academic Marxists are likely to be embarrassed by the historicist aspects of classical Marxism, no less than the logical positivists were embarrassed by Comte's *Catechism of Positive Religion,* but this does not mean that they restrain themselves from thinking about history and from supporting social policies that best accord with these critical speculations. In all phases of these two parallel (and actually closely related) movements, the thinking of each group has been shaped by a common and abiding spirit. The formal theorizing of Marxist scholars may no longer have a programmatic political counterpart, but this does not mean that this side of Marxism has no presence in the thought of merely academic adherents.

Agnosticism as Rhetorical Positivism: The Case of Thomas Huxley

In order to illustrate these rhetorical dynamics and, more specifically, to consider how these discourses work to enforce within the scientific culture the posture of incommensurability which served to bring about the scientific exile of design in the nineteenth century, I turn now to the case of Thomas Henry Huxley. I focus on Huxley in developing this argument because the rhetorical problems with which he was occupied throughout his public career were especially likely to invite a casual mixing of the positions I have just described. By the middle of the nineteenth century Huxley had emerged as the central actor in a movement that was destined to reconstitute the professional identity of the scientist. This was a social and institutional rather than an intellectual revolution in science, one that was in the process of transforming what had formerly been an esoteric gentleman's hobby closely allied with mainstream religious and political interests, into a muscular leviathan of the new secular and capitalist social order, a profession powerful, independent and institutionalized that was emerging as the prophetic voice of all things modern.

A movement of this kind would need to be sustained by an appropriate worldview, and Huxley's new science found this in an ideology different from positivism only in name. Because it was tied up with certain professional demands, such as the need to ensure that scientific authority would minimize its debts to external influences, this ideology was also strongly disposed to deny itself—lest it should itself appear to be subject to external ideological influences. One might object to this by saying that all ideologies are self-denying, and in some sense this is true. But I am simply arguing that in the case of science this tendency is especially pronounced. Since the seventeenth century, if not since before that time, science has adopted a priestly posture by insisting that its authority lies in sources outside human control or volition. Scientists are priests in the sense of claiming to be only the mediators of information that originates outside of themselves in indubitable fact or in the certainty of logical demonstrations (Lessl 1989). Such claims to authority demand the suppression of symbolic resources that might give any hint of subjective influence, and thus a dominant feature of this priestly rhetoric has been the naturalization of scientific authority. Science gains its distinctive sense of authority by being able to claim that its discourses have a uniquely extra-human origin. It must represent itself as the voice of nature speaking through certain anointed representatives who have been specially prepared to serve as its revelatory vehicles.

In the opinion of Huxley and even of many other English intellectuals who were more supportive of classical positivism, Auguste Comte had betrayed this priestly calling. As a social philosopher, Comte's overriding concern was to find in science the basis upon which a new European civilization might be built to replace the authority structures of feudalism and Christendom that had experienced a cataclysmic collapse in the French Revolution. Such grand aspirations for science were characteristic of positivism$_1$, but because these utopian concerns had led Comte into what English sympathizers regarded as wanton speculation, science had to be divorced from it. Despite Comte's insistence that his philosophy applied to social questions the same principles of rigorous verification that had already made physics and chemistry into positive sciences, it was evident to many of those sympathetic to the positive view of knowledge that he had overstepped its bounds in his zeal to discover a science of society. One of these sympathizers, John Stuart Mill, had complained that Comte, by insisting that untestable claims be regarded as true merely because they fit a desired sociological blueprint, was guilty of "a complete dereliction of the essential principles which form the Positive conception of science." He thus alleged that Comte's later works "contained the germ of the perversion of his own philosophy" (Mill 1887, 58). But notably, it was not the declaration of a scientific priesthood that Mill found offensive. Mill agreed with Comte "that the moral and intellectual ascendancy, once exercised by priests, must in time pass into the hands of philosophers and will naturally do so when they become sufficiently unanimous, and in other respects worthy to possess it." The problem was that philosophers were not yet worthy of such powers. Comte had impatiently advanced a sociological agenda that outreached positive science, and this made positivism$_1$ an embarrassment for those like Mill who hoped to construct analogous systems. In this regard Mill's reaction anticipates by several decades the general position I have dubbed positivism$_2$. He shared Comte's hope that positive epistemology could lead to the scientific reorganization of society, but he also wanted to shift the authorizing basis of this hope back upon the conservative epistemology of positivism—something that could not be achieved were the premature historicist speculations of Auguste Come assumed to have the same priestly authority (Robson 1968, 95–105).

Huxley's embarrassment in being associated with positivism inclined him, not to divide positivism into good and bad representations as Mill and others had done, but rather to vigorously denounce it and subsequently to reinvent it under an altogether different heading—that of agnosticism. Thus the speculative $positivism_1$ of Comte, and the conservative reactionary $positivism_2$ of Mill were in some sense synthesized in the hands of public scientists like Huxley into the rhetorical $positivism_3$ that has subsequently served as the unofficial ideological posture of the scientific community.

Two main features of this new scientific ideology are reflected in the terms that most frequently come to mind when Huxley's name is mentioned, his identity as "Darwin's bulldog," which denotes his prominent role as a public advocate of the emerging evolutionary paradigm, and his coinage of the term *agnostic* to denote the radical limits to knowledge he associated with the triumph of modern science. Huxley's advocacy of Darwinism served to establish the place of $positivism_3$ by amplifying the sense that science's focus is entirely rooted in nature rather than in human constructs. Darwinism makes the physicalist claims of science all encompassing. Once established as the inflexible position of the scientific culture the impression is created that all of its other ideas fit within this compass. Thus by forcibly establishing such a position Huxley could then suggestively reintroduce historicist ideas similar to those of Comte into his rhetoric—but without the same effect. Whereas Comte's exaggerated religion of humanity seemed absurd because it pressed so far ahead of scientific knowledge, the more cautious introduction of similar ideas by Huxley did not have this consequence. Having once given presence to physicalist notions by authorizing them by appeal to evolutionary science, the ideological character of such claims would be occulted. The historicist aspects of $positivism_1$, in other words, seem to have been sublimated into a Darwinian narrative history. This move is complemented in Huxley's rhetoric by his reconstruction of the empiricist ethic of science as agnosticism. The notion of agnosticism is arguably merely a negative expression of the core epistemic principle shared by $positivism_1$ and $positivism_2$. If positivism asserts that the glass of knowledge is half full, filled with the pure water of positive data, agnosticism merely chooses to make reference to the half that is empty. The rhetorical payoff of this inversion is a stronger sense of the epistemic modesty of this intellectual posture. Rather than merely calling this skepticism, it becomes a mark of honesty and humility. But it also is a concept that clearly complements the physicalist arguments that are suggested by his evolutionary stance. If it is the ethical thing to do to reject all claims that lie outside the scope of physical explanation, then this guarantees, more or less, the inevitability of arriving at materialistic explanations of everything.

It bears mentioning, since this may help to illuminate my argument, that the popular impression of Huxley as "Darwin's bulldog" is somewhat inaccurate. There is a notable ambivalence in Huxley's attitude toward Darwinian evolution. His early rush to Darwin's side had more to do with his concern that the *Origin of Species* gain a fair hearing than with his embrace of its main argument. This is borne out by the fact that throughout the remainder of his life Huxley says very little about the theory in his scientific writings and much less than one might expect in his popular lectures and essays. Even when Huxley did defend Darwin in public, he was also given to expressing serious reservations about the explanatory merits of natural selection.[5] The fact that Huxley was more likely to discuss Darwinism in his public lectures than in his scientific ones is symptomatic of the social or ideological interests of historicism that are also detected in his public advocacy. In other words I wish to suggest that Huxley's greater willingness to emphasize the evolutionary paradigm in public settings, despite his rather noncommittal attitude toward the theory as a scientist, is indicative of the constitutive social function of Darwinism. This pattern suggests that perhaps Huxley was embedding a positivist stance within his evolutionary rhetoric, thus fulfilling his role as priest by being more positivist than the positivists. The all-encompassing naturalistic scope of evolutionary theory enabled it to provide the same comprehensive ideological justification for science that was provided by the positivist epistemology. If Darwinism offered an answer to the mystery of all mysteries, then it also elevated the new professional science that Huxley was fighting to establish in England as an authority that was both completely autonomous and completely authoritative.

The idea that Huxley's exploitation of Darwinism provided a scientific cover under which he could advance his positivist worldview is also supported by an examination of the circumstances that led to his coining of the term *agnostic*. Via its passage down to us from the nineteenth century the term has come to denote a specifically noncommittal posture toward religious belief. Certainly this is consistent both with what Huxley meant when he coined the term and with his own attitude toward religious belief. But distancing himself from religious belief was not a concern for Huxley at this time; distancing himself from the positivists was. His participation in the Metaphysical Society, a London symposium that brought together representatives of each of the leading English "-isms" of mid-century, created a bit of an identity crisis for Huxley. Without an "-ism" of his own he found that he was simply thrown in with the Comtists by virtue of the fact that he shared their general orientation to questions of knowledge. It was pressure of this situation, a situation forcing him to name himself or to be named by others, that let to his creation of the term *agnostic* in 1869. His negation of the Greek word *gnosis* was meant to express his more general posture of skepticism regarding all knowledge claims that lay beyond the reach of scientific validation.[6] By thus focusing on what he rejected, namely unverifiable knowledge claims, rather than the materialistic claims he advanced, Huxley was able to affirm an epistemological posture virtually identical with that of the positivists even while actively divorcing himself from their speculative social theories. By calling his position "agnostic," Huxley was simply offering up a negative expression of the positivist creed. Whereas the positivists had asserted that only that which has the affirmation of "positive" evidence should count as knowledge, under the agnostic heading Huxley was asserting the same thing *via negativa.* He was declaring that all that is not based in positive knowledge is simply unknown; science can neither affirm nor deny it but must remain silent.

This close similarity between positivism and agnosticism raises an important question about the consequences of Huxley's terminological differentiation: if Huxley divorced his position from Comte's only by calling it by another name, was he also rejecting in principle the positivist supposition that science ought to be applied to social issues? Clearly Huxley did not like the vision for a scientific society proposed by Comte and his English disciples, but this did not entail the rejection in principle of positivism's assumption that science was destined to found a new civilization. In fact, Huxley could ill afford to abandon this notion, since it was such a powerful motive for the general advancement of science's place in the world. He thus continued to embrace the same general historicism of classical positivism.

I would argue, in fact, that historicist notions were able to enjoy a continuous presence in his discourses precisely by force of Huxley's vocal disavowals of this feature in Comtism. By consistently denouncing the excessive historicism of Comtism throughout his career he was able to create a rhetorical space for advocating similar ideas. This tactic, though unquestionably effective, is so familiar to us in political discourse as hardly to invite critical exploration here. For Huxley it worked, seemingly, by drawing attention away from questions about the logical relationship between the realm of ideas and the realm of human action that was at the core of the positivist movement. Huxley's approach induces his audience to make their judgment at this quite different plane of reasoning. They are discouraged from asking about whether Huxley's position is really different from Comte's at bottom, and instead are asked to consider how the worldview of someone so obviously opposed to positivism could possibly be regarded as advocating a similar social or intellectual agenda. Huxley, in other words, creates a compelling impression of difference, even where there seems to have been little. By shifting the focus so thoroughly from the full half to the empty half of our epistemic water glass—the unknown agnostic half—Huxely effectively creates the impression that it is a different glass altogether. By calling himself an agnostic and by denouncing some particulars of positivism, Huxley could generate the appearance of a more basic separation between his views and those of Comte's English followers. This sense of separation created a cover under which he could advance a version of positivism more suited to the particulars of his own agenda of constructing a professionalized science. The historicist features of Huxley's positivism may seem to bear little resemblance to the grand narrative of the Comtists, simply because his is a more local narrative, designed with only one immediate social purpose in mind, the advancement of science. But once drawn out into the open through critical analysis, its Comtean features become more apparent. The attempted separation of the epistemic aspects of $positivism_2$ from the historicist aspects of $positivism_1$ turns out to be an illusion. The visionary notions of Comte and Saint-Simon may seem far removed from the worldview of Tom Huxley, but this is merely because he has dressed them down in the drab Puritanical attire of agnosticism.

Rhetorical Positivism in the "Physical Basis of Life"

The continued presence of this positivist historicism in the agnostic viewpoint could be illustrated by examining a number of Huxley's public discourses. Here I would like to look at a few key messages that show the extent to which the notions of $positivism_1$ and $positivism_2$ lay along side one another in Huxley's mind and were brought together in his public discourses even as he worked to dissociate himself and the emerging scientific culture he spoke for from the very similar worldview being advanced by Comte's English disciples. A useful starting point for this examination is his landmark speech on "The Physical Basis of Life," in which he advances the naturalistic thesis of "a general uniformity in the character of protoplasm"—Huxley's word for the physical basis of life—and at the same time works to publicly distance himself from the positivist movement (T. H. Huxley 1893, 1:144). It is the juxtaposition of these two concerns that draws my attention to this speech. Huxley's argument for a strictly phenomenalist approach to biology is clearly indicative of his own positivist epistemology, in spite of the fact that he is so completely bent on renouncing that same view within the very pages in which he presents this philosophy of knowledge. He strives to create the impression of a decisive break with French positivism, albeit merely by an apparent rhetorical sleight of hand, simply by arguing that the epistemic framework he himself was advocating had an earlier origin than positivism, in the writings of David Hume. These denials are augmented by his even more successful effort in amplifying, via creative ridicule, the historicist aspects of the movement he was attacking, even as he worked to suppress these aspects of his own worldview. After examining these two features of this speech I would like to consider a few other examples of his public discourse that show how this vital historicist aspect of $positivism_1$ re-enters $positivism_3$ by being sublimated into natural history.

As its title suggests, the manifest thesis of this lecture, given in Edinburgh on November 8, 1868, and published early the following year in the *Fortnightly Review,* was that all living things are reducible to their material bases. In a general sense this is an argument for an exclusively naturalistic approach to biology, though it takes this stand in contradistinction to vitalism rather than design—and by implication at least for evolution as against creationism, in spite of the fact that Huxley, in characteristic fashion, makes no reference to Darwin's theory.

The published version of this speech, which put the *Fortnightly* into seven printings, begins by narrating the epic history of biological science, recounting the process by which competing views have progressively collapsed under the advancing march of materialistic science. Huxley does this more by poetry than by science, a feature of the argument that would make it difficult or impossible for any listener to discern where the science leaves off and imaginative speculation begins. With characteristic literary aplomb he pulls back the veil of appearances to reveal a world hidden from ordinary eyes, the inner complexity and diverse beauty of living things. His readers are invited to imagine an underlying unity known to science that joins together the "giant pine of California, towering to the dimensions of a cathedral spire," with the "invisible animalcules" that "dance upon the point of a needle with the same ease as the angels of the Schoolmen could, in imagination." Physicalism is made to seem without limits in explanatory power. Nothing would seem to be outside the reach of a science that is capable of understanding the "hidden bond" that connects "the flower which a girl wears in her hair and the blood which courses through her youthful veins" (T. H. Huxley 1893, 1:132). Even human speech and "every other form of human action are, in the long run, resolvable into muscular contraction, and muscular contraction is but a transitory change in the relative positions of the parts of a muscle" (T. H. Huxley 1893, 1:132, 134).

Although nowadays this essay is held up as a paradigmatic expression of Victorian materialism, one of its most striking features is the manner in which it both affirms and denies this very stance. In this regard the essay is a bit suggestive of the famous Magritte painting of a pipe accompanied by the inscription "Ceci n'est pas une pipe." In the main it is a solemn declaration of the complete triumph of what positivism$_1$ claims about all arenas of inquiry, that they are destined to be encompassed within the ever-expanding borders of science's emerging empire. Everything in this first section of the lecture seems to proclaim this victorious materialist march upon the field of biology, but then Huxley suddenly turns tables and denies that this is so.

> Past experience leads me to be tolerably certain that, when the propositions I have just placed before you are accessible to public comment and criticism, they will be condemned by many zealous persons, and perhaps by some few of the wise and thoughtful. I

> should not wonder if "gross and brutal materialism" were the mildest phrase applied to them in certain quarters. And, most undoubtedly, the terms of the propositions are distinctly materialistic. Nevertheless two things are certain; the one, that I hold the statements to be substantially true: the other, that I, individually, am no materialist, but, on the contrary, believe materialism to involve grave philosophical error. (T. H. Huxley 1893, 1:154–155)

It is easy to imagine why someone like Huxley would want to take back with his left hand what he gives with his right. On the one hand, while materialism might be attractive to those wanting to advance a scientistic agenda, its association with radicalism would necessarily discourage much needed patronage. It may have been easy for a militant like Frederick Engels to call agnosticism "'shamefaced' materialism" (Engels 1975, 13), and for Vladimir Lenin to ridicule it as prudish, since they made no pretense of maintaining ties to the extant social order (Lenin 1970, 195). But Huxley's aim was only to take hold of existing resources of power not to overturn them, and an open declaration of a position so extreme would weaken his ability to do so.

Materialism was also recognizably a metaphysical stance that, once taken, would concede too much to an external philosophical authority that might impose regulatory claims upon science. The phenomenalist position that Huxley seems to express here thus provides a solution to this danger—a measure of chaste moderation in keeping with the Puritanical culture he represented. By claiming that we know only material phenomena, that we can not say that we know material as material, he avoids the risk of seeming to take a meta-stance that could concede authority to anybody outside of science.

But even as he successfully jumped out of this metaphysical frying pan Huxley still ran the risk of landing in the fire of positivist historicism, not because the epistemic modesty of his anti-realism leads logically to such a position, but because he was taking a position that was virtually identical to that of the positivists, a group notorious for its philosophy of history. Agreement with this first acknowledged hallmark of Comtism could easily suggest that Huxley had some share in this other hallmark of their movement, its world-building aspirations.

Indeed, Huxley's essay stirs up a certain amount of background noise suggestive of a very similar historicist view. It gives an impression of the current state of the biological sciences, as a stage in a process of natural development. Science is imagined to be ascending the rungs of a ladder, first from the presumption of vitalism, now to the presumption of a comprehensive physicalism, and then in the future to an era in which the explanatory potential of this new physicalist paradigm is fully realized. He paints only a few details on the panoramic canvas envisioned in Comte's sweeping outline of the positivist future to come; nevertheless they would fit nicely into any rendition of the classical positivist vision of the three stages of history. This aspect of Huxley's thought may have only a subtle presence in the following paragraph, but it is discernable nonetheless.

> If scientific language is to possess a definite and constant signification wherever it is employed, it seems to me that we are logically bound to apply to the protoplasm, or physical basis of life, the same conceptions as those which are held to be legitimate elsewhere. If the phenomena exhibited by water are its properties, so are those presented by protoplasm, living or dead, its properties.
>
> If the properties of water may be properly said to result from the nature and disposition of its component molecules, I can find no intelligible ground for refusing to say that the properties of protoplasm result from the nature and disposition of its molecules. (T. H. Huxley 1893, 1:153)

The argument here is not predicated upon existent science at all but rather upon a theory of the development of scientific knowledge that is virtually identical with the one advanced by the positivists. Huxley assumes the same pattern of progressive development for the various sciences that was the cornerstone of Comte's historical vision, his notion that the more basic sciences provide the platform upon which more complex sciences are built. Thus, the above passage prognosticates that just as chemical knowledge had been constructed in the eighteenth century upon the scientific platform built during the revolution of classical physics, so also would a physical understanding of biological phenomena emerge upon the foundations of chemistry.

This pattern of appealing to past successes, not so much to instruct as to provide a basis for making promises about the future of biological inquiry, is a thread that runs through much of Huxley's discourse. What makes it different from the discourses of positivism$_1$ is mainly a matter of style. Comtian rhetoric is openly meta-scientific; Huxleyan rhetoric, by contrast, occults its meta-science by seeming to collapse it back into scientific arguments. Because of this a casual reading of the following, especially from our vantage point, in an era of advanced biochemistry, might give the impression that Huxley is making a claim on the authority of established scientific verities; however, a closer look reveals that its scientific sounding language merely draws the reader's attention away from what is actually a purely speculative statement about the destiny of biological science.

> Carbon, hydrogen, oxygen, and nitrogen are all lifeless bodies. Of these, carbon and oxygen unite in certain proportions and under certain conditions, to give rise to carbonic acid; hydrogen and oxygen produce water; nitrogen and other elements give rise to nitrogenous salts. These new compounds, like the elementary bodies of which they are composed, are lifeless. But when they are brought together, under certain conditions, they give rise to the still more complex body, protoplasm, and this protoplasm exhibits the phenomena of life.
>
> I see no break in this series of steps in molecular complication, and I am unable to understand why the language which is applicable to any one term of the series may not be used to any of the others. (T. H. Huxley 1893, 1: 150–151)

The historical bridge from chemical knowledge to biological knowledge that Huxley crosses with verbal ease in this passage—a paradigm case of the figural logic embodied by incrementum (Fahnestock 1999, 86-121)—is one that biochemistry, in fact, has not managed to cross in the thirteen decades since this utterance was first made, even though its assumption that biology may be reduced to chemistry has become normative. I point this out not to claim that Huxley is some kind of false prophet but merely to draw attention to the fact that his argument operates upon both the empiricist and historicist axes of positivism$_1$. The historicism of his position may easily go unnoticed, since it is embedded within what seems to be scientific language, in phrases such as "I see no break in this series of steps in molecular complication," and in Huxley's claim to see no reason why "the language which is applicable to any one term of the series may not be used to any of the others," but it is discernable nonetheless. Because Huxley here plays upon the vagueness of the word *see,* these have the appearance of being empirical or logical claims, but it would be more accurate to regard them as expressions of faith in the destiny of science to progress in its comprehension from the domains of knowledge upon which it currently stands up ever higher into the realms of biological complexity.

Comte calls his a "general view of the progressive course of the human mind, regarded as a whole" and openly acknowledges that the historical vision he is constructing charts the course that humanity is destined to follow (Comte 1893, 1: 1). Huxley risks much less. He plots the future development of knowledge only by appeal to specific particulars of scientific thought as his examples, and he never pulls back—at least not in this essay—to reveal the more general historical narrative that is implied in all of this. Such may befit the temperament of Huxley, who was accustomed to working in the details of scientific research, but the end result is much the same as what Comte proposed, a vision of knowledge's ascent up several successive ladder rungs and leading up into the heavens of global scientific comprehension.

This judgment is supported by the fact that Huxley was also willing to acknowledge the macro-historicism that Comte openly proclaimed, albeit only privately, in the following words penned to his wife in 1873:

> We are in the midst of a gigantic movement greater than that which proceeded and produced the Reformation, and really only the continuation of that movement. But there is nothing new in the ideas which lie at the bottom of the movement, nor is any reconcilement possible between free thought and traditional authority. One or other will have to succumb after a struggle of unknown duration, which will have as side issues vast political and social troubles. I have no more doubt that free thought will win in the long run than I have that I sit here writing to you, or that this free thought will organize itself into a coherent system, embracing human life and the world as one harmonious whole. But this organisation will be the

> work of generations of men, and those who further it most will be those who teach men to rest in no lie, and to rest in no verbal delusions. I may be able to help a little in this direction—perhaps I may have helped already. (L. Huxley 1900, 2: 111)[7]

Here are all the hallmarks of positivism$_1$: (1) its progressive view of history in which science emerges from and succeeds religion; (2) the abandonment of traditional authority for rational authority; (3) the inevitability (naturalness) of this evolutionary process; (4) the destiny of this movement to encompass all arenas of human life—including governance—in one vast system; and, finally, (5) the hallmark notion from which positivism takes its name, that language should only reference empirical data.

I will assume that it was the intensity of Huxley's zeal for the professionalization of science that made such acknowledgments of his positivist frame of mind a matter strictly for private disclosure. Still, there was much to be gained by advancing similar ideas, if only he could do so while still keeping them closely tethered to the labors of science. One way to sustain the impression of this at least, returning again to the 1868 lecture, would be by accompanying the advancement of such ideas with a forceful repudiation of Comte. This counteracted any danger that might arise from the fact that his own position so strongly suggested a similar epistemology and historical vision.

Accordingly, even as Huxley envisions the unyielding progress of science in his Edinburgh speech he also goes out of his way, in its best remembered passage, to repudiate Comte's philosophy. This departure from his main theme was warranted, as he explains it, by the fact that a lecture on "the Limits of Philosophical Inquiry," just given by the Archbishop of York and reported in an Edinburgh newspaper on the eve of Huxley's speech, had attacked the very same understanding of science that Huxley was set to advance in this lecture—but credited positivism as its source. Since this struck so close to home, a defensive or even anti-clerical response might have been expected, given that the Archbishop was attacking what Huxley acknowledged to be his own view on the limits of knowledge. But such a response would have conceded the clergyman's claim that Huxley's views aligned with those of Comte. In the face of this dilemma Huxley, with characteristic political deftness, opted to join the Archbishop's attack—even as he also challenged what it alleged.

> Now, so far as I am concerned, the most reverend prelate might dialectically hew M. Comte in pieces, as a modern Agag, and I should not attempt to stay his hand. In so far as my study of what specially characterises the Positive Philosophy has led me, I find therein little or nothing of any scientific value, and a great deal which is as thoroughly antagonistic to the very essence of science as anything in ultramontane Catholicism. In fact, M. Comte's philosophy, in practice, might be compendiously described as Catholicism *minus* Christianity. (T. H. Huxley 1893, 1: 1 56)

This passage has the effect of pushing into the background, though certainly not out of the picture, the fact that Huxley's crucial notion of the limits of knowledge was virtually identical to Comte's. Rather than addressing this suggestion head on, which he could not really do without also aligning himself with other aspects of positivism, Huxley instead offers an attack on Comte's scientific credentials. In doing so he shifts the argument from the issue of means, from the conditions of scientific knowledge which had been the basis for the Archbishop's alignment of positivism with the materialistic science emerging in England, to the issue of effects, to a concern with the scientific understanding that might be presumed to follow from the employment of such means. By suggesting that Comte's scientific understanding was deficient he could then also play down the significance of the many ways in which his own worldview overlapped with that of the French philosopher.

Huxley then adds some emotional weight to his argument by appealing to the anti-Catholic feeling of his Scottish Presbyterian audience.[8] The idea that Huxley is referencing in calling positivism "Catholicism *minus* Christianity," was positivism's cardinal premise that since the scientific mode of consciousness was innate to human thought, it had always been at work in human societies—as was made evident by the steady progress of knowledge throughout civilized history. The organizational genius of the Catholic Church, Comte believed, exemplified this. The positive spirit had been at work even in the Middle Ages, he argued, in spite of that period's domination by the false consciousness of theology and metaphysics, and had produced an understanding of social order that had some scientific merit. Consequently, it made sense to expect that the new and fully positive social order that was beginning to emerge in Europe would be Catholic-like, a new civilization carrying forth the sociological discoveries of the former age.[9]

Neither here nor in any of his subsequent and more extensive commentary on positivism does Huxley address his specific reasons for rejecting this aspect of Comte's philosophy. That would largely be unnecessary, given his audience. There is no doubt that Huxley, a notorious hater of all things Catholic, speaking to an audience of Protestants in what he in a subsequent letter to Ernst Haeckel called "holy" Edinburgh, meant to exploit religious prejudice in this effort to demarcate his own worldview from that of the Comtists. In making this appeal Huxley seems to satisfy himself with the idea that by depicting Comte as a scientific illiterate whose reasoning was polluted by Popish infatuations, he had satisfactorily refuted the Archbishop's alleged linkage between positivism_1 and his own views on science.

Although Huxley's appeal to religious prejudice could easily be dismissed as a mere sophism, both the presence that this *mot* achieved in the subsequent history of this debate and the ultimate fading of English interest in Comte's views in the later decades of the nineteenth century suggest that it may be worthy of some attention, especially since it specifically attacked the historicist aspect of positivism_1 that is so clearly sublimated in positivism_2. Comte's expectation that the new civilization prophesied in positivism would reflect the sociological genius of the Catholic Church was grounded in his more rudimentary philosophy of history with its doctrine that social evolution coincided with the ever-widening scope of positive knowledge. The open acceptance of such a doctrine would give the Comtists an authority superseding that of the scientific profession Huxley was building in England. It would be tantamount to accepting that movement's role as the rightful architect of scientific history, thus giving it a deliberative authority above that claimed by Huxley's scientists.

Since acceptance of this view would concede to social science an authority higher than that given to the class of professional scientists Huxley sought to raise up, he needed to divorce his worldview from that of the Comtists, despite the fact that in most respects they were the most like-minded of his peers. In constructing such a demarcation Huxley was not at liberty to renounce the epistemic aspect of positivism, nor did he really desire to abandon its evolutionary view of history. As a representative instance of this rhetoric of demarcation, his "Physical Basis of Life" address preserves this sense of historical authority by merely shifting it into the background. It continues to have presence even as it remains largely unseen. Huxley's attack on Comte's scientific credentials and on the Catholic aspects of his historical vision, accordingly, works in concert with his own effort to construct a similar philosophy of history.

Symbolic Camouflage

As is evident in the disclosure to his wife we visited a moment ago, in private Huxley expressed a view of history more like that found in positivism$_1$ than he was ever willing to acknowledge in public pronouncements on the subject. The chief difference is that he regarded the "gigantic movement" in which he was participating as a "continuation" of the Reformation rather than of Catholic history. In this regard, Huxley's charge against Comte could very easily be turned back on him. We could very easily allege that his worldview is "Protestantism *minus* Christianity." This in fact is something that is reflected in the shift of emphasis to the epistemic side of positivism that manifests so consistently in Huxley's rhetoric. Notably, for instance, in an earlier and once again private disclosure, this time to his friend the liberal clergyman Charles Kingsley, Huxley openly avows a linkage between this positivist notion of the limits of knowledge and the Bible-centered view of Protestantism:

> Science seems to me to teach in the highest and strongest manner the great truth which is embodied in the Christian conception of entire surrender to the will of God. Sit down before fact as a little child, be prepared to give up every preconceived notion, follow humbly wherever and to whatever abysses nature leads, or you shall learn nothing (L. Huxley 1901, 1: 235).

The "received view" of positivism blends quite nicely here with the "received view" of Protestantism. Just as Luther rejected the Catholic notion that spiritual knowledge is passed down via the teaching authority of the Church and substituted for this traditional notion of authority the claim that it is simply rooted in the Word of God, so also in appealing to this ideal of high empiricism Huxley suggests that scientific knowledge springs directly from the book of nature and can thus be grasped independent of any mediating human agency.

As we know from the history of Protestantism as much as from the history of positivism, such shifts do not undermine so much as they reconfigure human authority. The outcome of the Reformation was not the destruction of Church-based authority so much as its decentralization and displacement into more localized institutions. But even as this occurred in endless sectarian proliferations, the Protestant churches have held fast the ideal of *sola scriptura,* the denial that their teachings depend upon any authority outside the Word of God. The doctrine of *sola scriptura* in this regard operates as an argumentative resource that shifts attention from the merely human voices that continue to hold sway in religious institutions—the interpretive authority of learned exegetes, of dominant historical figures and of established theological tenets in determining what meanings may and may not be taken from the Bible. Because these meanings can always be linked to the Bible in some fashion, the invocation of Scripture is likewise always ready at hand wherever there is need to dismantle the allegation that a merely human authority has come into play.

That Huxley should recognize the availability of an analogous pattern of authority-building with which to support the exclusivity of empiricism is not surprising, given the Protestant atmosphere in which he worked. In fact this is a *topos* that had been prevalent in English science since the beginnings of the modern scientific movement, in the "two books" argument of Francis Bacon. In his *Advancement of Learning* and scattered throughout his other discourses promoting the scientific cause, Bacon forged a deliberate alignment between the modern scientific movement and the back-to-the-Bible ethic of the Reformers, even to the extent of situating the commandment to undertake the empirical study of nature in the words of Christ (Matthew 22:29). Science, Bacon declares,

> is a singular help and a preservative against unbelief and error; for, saith our Saviour, *Ye err, not knowing the Scriptures nor the power of God;* laying before us two books or volumes to study if we will be secured from error; first the Scriptures revealing the will of God, and then the creatures expressing his power; for that latter book will certify us that nothing which the first teacheth shall be thought impossible. (Bacon 1968, 3: 221)

Although the "two books" metaphor dates to the patristic period, it was revitalized by Bacon in the seventeenth-century in order to construct a more specific alliance between science and the Reformation. By linking empiricism to the spirit of reform, Bacon elevated the professional standing of those who pursued science by making them priestly exegetes of nature—a theme dramatized in his posthumous novel the *New Atlantis.* But even more vital than this, at least in the long run, was the effect that this argument had in simultaneously grounding science in Protestantism even as it established science's intellectual independence from theology. To the extent that the Baconian argument appealed to the authority of the Bible but did so in order to bring that authority to bear on the argument that science was an independent source of revelation, it also declared itself immune in some degree from the corrective influence of those whose authority lay in the interpretation of sacred Scripture.

Although the explicitly Protestant origins of this Baconian argument might have fallen out of favor by the time Huxley began his rhetorical career, this pattern of reasoning retained much of its original potency, if not its religious resonance. John Angus Campbell, recognizing what I take to be a similar pattern in Darwin's writing, has called this science's "Baconian grammar," a theistic pattern of argument that persists even into an era of purportedly secular or naturalistic science (Campbell 1986). What is most notable about this rhetorical grammar in Huxley's discourses is that it seems to operate to suggest meaning at one level of consciousness that it denies at another. In this regard the meanings set up in accordance with this grammar work as a kind of symbolic camouflage, blending into a symbolic background elements of nonscientific belief that the scientific culture prefers not to expose, so that their extra-scientific character will be invisible. These blended ideas retain their presence but not their visibility. This is how camouflage works. It keeps foreground elements in the foreground because they have some function there, while also making them indistinguishable from their background. The background in this instance is the Protestant culture of Huxley's English constituents, and the elements that are blended into this background are the historicist ideas that Huxley wanted to retain because they created desirable meanings that would advance the scientific cause. These in fact are notions very similar to those that Huxley wished to attack in French positivism, so as to undermine its claims of scientific authority. But for the same reason that Catholic aspects of positivism$_1$ were easy targets for his attack on Comte, the Protestant aspects of positivism$_2$ worked in Huxley's favor. For Huxley's culturally Protestant audience the Catholic aspects of positivism$_1$ could not be blended into the cultural background of religion—camouflage being only camouflage in its appropriate context.

Clearly Huxley was at least aware, as the religious language of some of the previous excerpts would suggest, that his empiricist arguments harmonized with certain elements of Protestant thought. But even without such markers, utterances such as the following from one of his public lectures a few years later show the same historicist tendencies as we saw in the letters to his wife and to Charles Kingsley. Here Huxley turns aside momentarily from a lesson on scientific method in zoology to prognosticate on the larger implications of the fact that science is now "teaching the world," that the,

> ultimate court of appeal is observation and experiment and not authority; she is teaching it to estimate the value of evidence; she is creating a firm and living faith in the existence of immutable moral and physical laws, perfect obedience to which is the highest possible aim of any intelligent being. (T. H. Huxley 1873, 144)

Unlike the private utterances examined earlier, this passage does not use any distinctively religious language—except perhaps in the phrase "living faith." What it does exhibit, however, are patterns of valuation that bring scientific empiricism into tacit association with the religious thinking of Protestants. The assertion that observation and experiment rather than human authority represent the "ultimate court of appeal," might be viewed as constituting a stance quite distinct from Protestant notions of epistemic authority, but we can still trace out the Protestant form of the argument. When Huxley contends that the special virtue of science lies in the fact that it rejects all human authority, and in the process of doing so receives truths which are ultimate and inviolable because they are written into nature, he is expressing an epistemological ethic analogous to the Protestant hermeneutic. He is voicing a version of the Baconian two books argument but leaving the theological side of the analogy to the imagination.

This is augmented by the noticeably prophetic tone of this utterance, its declaration that science is in possession of some "ultimate" saving truth. Huxley seems to throw up as the background against which the auditor is asked to measure his words, a vista of hope and progress stretching off into a distant horizon—a new scientific millennium which is the reward for the "perfect obedience" he describes as the price of scientific discipleship. Set against this millenarian backdrop the scientific acts of discipleship that Huxley depicts take on an extraordinary value as actions destined to form a universal guide.

Conclusion: Über-Iincommensurability and Ideological Science

I have argued in this essay that, no less than the French positivists and their English disciples, Huxley was actively engaged in a world-building project that coincided with his scientific advocacy of a naturalist view of origins. Thus his evolutionary perspective is always already both an ideological and a scientific stance. The question for us to consider in conclusion is: Fom which of these two stances does the apparent incommensurability of design and evolution proceed? A Kuhnian viewpoint would suggest that it proceeds the scientific stance on the origins question—that it was a crisis of the creationist biological paradigm that brought forth the new and incommensurable evolutionary paradigm. As I noted at the outset, this interpretation has already been challenged by the historian John Greene. What I have tried to do in building upon Greene's position is to suggest the relevant operation of a different sort of incommensurability, one having to do with the respective standing of science in two competing worldviews.

In supposing that it was a concern with the need to build a new institutional culture for science rather than merely a concern with biological questions that led to the abolition of design, I have offered a critical historical examination focusing on Thomas Henry Huxley's role in this transformation of science's cultural identity. Huxley's rhetoric supports my thesis on three counts. First, Huxley was undoubtedly committed to a positivist position that excluded design on philosophical grounds even before he heard of Darwin's theory. Second, contrary to popular opinion, he was only a lukewarm supporter of Darwin's theory and was not willing to risk much personally by linking himself to its scientific particulars. Third, Huxley was the central actor in what might fairly be regarded as one of the pivotal movements in scientific history, the great enlargement of professionalized science in the Victorian era, and the complementary restructuring of higher education that was needed to sustain this.

Although I do not have the space to examine them closely here, the institutional obstacles faced by science during Huxley's lifetime support these contentions. The Victorian era marked the beginning of the end of the clerical domination of higher education that had existed in the western world since the Middle Ages. By the nineteenth century the secular forces of capitalism, industrialism, and science had gained sufficient strength in alliance to undo the traditional ecclesiastical hold on education. The natural sciences were best suited to lead the charge in this battle. Scientists already had a foothold in higher education and a powerful priestly *ethos* that could compete with that of the established church.

In England at least, Huxley emerged as the rhetorical figure most capable of carrying this battle forward. For this reason it is not surprising that he should have been closely associated with the emerging evolutionary paradigm. The triumph of materialistic science symbolized by Darwinism would potentially undermine the traditional patronage claims of ecclesiastical bodies and would pave the way for turning over those responsibilities to industry, to the emerging secular state, and to scientists themselves. Thus to the extent that the triumph of Darwinism could symbolize the ascent of the new educational order, it in some sense had to be embraced. If science could seem to undermine the core doctrine of the religious powers that traditionally controlled higher education, it would have a critical advantage in its campaign to restructure English universities.

The fact that Huxley would come forward in 1880 to commemorate the twentieth anniversary of the publication of *The Origin of Species* illustrates this symbolic role. As with his lecture on the "Physical Basis of Life," this speech completely omits mention of Darwin's theory of natural selection. Huxley celebrates Darwin's "great work" as something from which "a host of young and ardent investigators seek for and find inspiration and guidance," but he makes no mention of the very mechanism upon which Darwin staked his claim to account for origins in naturalistic terms. A similar ambivalence is suggested by Huxley's warning that though Darwinism began as "heresy," it might easily end up as "superstition," so that, "as matters now stand,"

> it is hardly rash to anticipate that, in another twenty years, the new generation, educated under the influ-

> ences of the present day, will be in danger of accepting the main doctrines of the "Origin of Species," with as little reflection, and it may be with as little justification, as so many of our contemporaries, twenty years ago, rejected them. (T. H. Huxley 1897, 229)

We might presume that this is merely a cautionary note lest scientists become so enamored with Darwin's theory that they lose sight of science's skeptical and inquisitive spirit, and its aversion to dogma. But why would someone so eager to caution against the danger of a Darwinian heresy also be equally eager to call himself "Darwin's bulldog"? How could Darwin's bulldog also be, in the words of the Darwin biographer Peter Bowler, a "pseudo-Darwinian" (Bowler 1988, 72–77)? Sherrie Lyons has recently provided a book-length answer to the scientific part of this question, by taking us more deeply into Huxley's own scientific explorations (Lyons 1999). But this does not account for the fact that Huxley has, by his own design, come down to us as the nineteenth century's foremost advocate of Darwinism.

A clue to solving this puzzle may be found in another brief passage from this same commemorative essay in which Huxley proclaims that Darwin's work has provided "a firm base of operations," from which science "may conduct its conquest of the whole realm of Nature" (T. H. Huxley 1897, 229). Although this might easily be interpreted merely as a claim about the heuristic merits of Darwin's work, in light of Huxley's reservations about this hypothesis, it would make more sense to interpret it as an statement concerning the "revolution" about which Huxley privately wrote his wife. The embrace of Darwin's theory symbolizes positivism's promised conquest of all realms of knowledge.

Ultimately this becomes a question about which came first. Is naturalism the outcome of the triumph of evolutionary science, or is evolutionary science the only avenue of research tolerated in an age dominated by naturalism? I have argued here that the positivist framework, which clearly was in ascendance long before Darwin's theory appeared—which, as John Angus Campbell shows in his contribution to this volume, Darwin also capitalized on brilliantly—was the more likely cause of design's exclusion. But even if the embrace of a positivist worldview had been a chief factor leading to the exclusion of the design hypothesis, we can also predict that the scientific culture would be reluctant to acknowledge this. Positivism, by definition, is a philosophy that denies itself. Its claims therefore are always already scientific claims.

If this is an accurate view of the conditions that gave rise to the hegemony of naturalism, it is not one that necessarily diminishes the importance of Darwin's contributions to science. What it does suggest is that as we scrutinize the scientific culture that was created by Huxley's generation and which we have inherited, we will find a preponderant tendency to overgeneralize from what Darwin taught us and in some instances a tendency to distort the strength of the evidence that supports his theory. This is precisely what modern critics of the Darwinian paradigm allege.[10] While they uniformly accept that evolution occurs—in the sense that natural processes are known to bring about at least minor changes in species over long periods of time—they just as consistently find that any such merely conditional or partial acceptance of the neo-Darwinian paradigm is loudly condemned as unacceptable.[11]

Thus when Professor Dean Kenyon's superiors removed him from the basic biology course at San Francisco State University in 1992, it was not because he rejected Darwin's theory (Johnson 1995, 29–30). Neither was it because Kenyon was not qualified to address problems with the standard paradigm. His scientific credentials were impeccable and his understanding of evolutionary biology thorough. After receiving his Ph.D. from Stanford University in 1965, Kenyon had gone on to do post-doctoral work at Berkeley and Oxford, and to co-author *Biochemical Predestination* (Kenyon and Steinman 1969), an extensive critique of chemical abiogenesis. Kenyon's offense was that he questioned the comprehensiveness of the evolutionary paradigm by openly confessing to his students his doubts that a naturalistic explanation of the origins of life would ever be possible.

It was the contention of those who removed him from the classroom, both the chair of his department and the dean of his college, that Kenyon should teach only the evolutionary paradigm. But from the standpoint of science this was precisely what he had done. He did not offer an alternative to Darwinism; he merely pointed out some of the ways in which its claims are not borne out by the evidence. Kenyon was doing what virtually all science textbooks advocate—at least in the abstract. He was exposing gaps in the evidence. It just so happens that those gaps undermine the neo-Darwinian paradigm.

Such behavior as Kenyon's would hardly raise an eyebrow if his subject were quantum mechanics or Einstein's general theory of relativity. The comprehensiveness and empirical consistency of these, like most general theories, is frequently challenged, but university professors are never chastised for acknowledging this. One would think that open discussion of the limits of Darwinism would be celebrated, but Darwinism is rarely criticized in public from within the orthodoxy. Teaching the weaknesses and limitations of scientific ideas is, as John Angus Campbell has recently argued, the very heart and soul of good science, something we would want teachers to model for their students (Campbell 2003). But evolution is different. Once aligned with positivism, any challenge to evolution threatens not just the naturalistic scope of science but also the positivist ideology of history that assures the ascendancy of science.

The inflexibility of this stance sometimes puts the scientific community in the position of having to defend standpoints that are at best simply tautological and at worst incoherent. A few years ago the most prestigious scientific association in the U.S., the National Academy of Sciences (NAS), decided to do its part in defending the univocality of naturalism by publishing a handbook advising teachers on how to deal with the creationist threat. One problem faced by such teachers is the demand of school boards and parents that weaknesses of the evolutionary model be acknowledged in science classrooms. The NAS manual answered this simply by reassuring teachers that "there is no debate within the scientific community over whether evolution occurred, and there is no evidence that evolution has not occurred" (National Academy of Science 1998, 4). What they meant by this, of course, is simply that evolutionary theory has absolute presumption, so that even when facts are inconsistent with what neo-Darwinism predicts, that could never constitute anything approaching "debate." This becomes evident as the writers go on to say that while some of the "details of how evolution occurs are still being investigated" debate about these details only concerns "the particular mechanisms that result in evolution, not the overall accuracy of evolution as the explanation of life's history" (National Academy of Science 1998, 4). But how can the accuracy of evolution be known if the mechanisms of evolution are in doubt? Does not the truth value of Darwinism stand or fall by its ability to demonstrate that its postulated mechanisms are capable of creating every feature of every living thing? One does not need an advanced degree in microbiology to see that these writers merely beg the question. The authors of the booklet may speak the truth when they state that a majority of scientists argue from the presumption of evolution, but this is not the same thing as showing that it is a position established beyond doubt.

If the neo-Darwinian paradigm was not established in the fashion that Kuhn describes in his influential theory of scientific revolutions, it is unlikely that it could ever be undone through either the processes of normal or revolutionary science outlined in *Structures of Scientific Revolution*. Even revolutionary science presumes open debate, and deliberation about alternatives to evolution is really not even allowed in the scientific world. Those who do challenge it do so outside normal scientific channels, and even if they have a mainstream scientific education they are silenced as *personae non gratae.*

In this regard what might be seen as incommensurability in the relationship between the neo-Darwinian paradigm and the proposals of those who would like to see design reinstated in the life sciences may actually be something else, a sort of über-incommensurability that doesn't confound debate so much as preclude it. What we see is not a consequence of different understandings of the formal features of theory, or of the relations among various theories, but an incommensurability that is centered in the *stasis* of jurisdiction (what Lawrence Prelli treats in his contribution to this volume under the heading of "translative" [305]; see also Gross 2004, for an argument on the importance of jurisdiction to incommensurability). The failure to recognize this creates much confusion, because of the appeals made to questions of fact, definition and quality in the debates that rage between design theorists and neo-Darwinians. But even though advocates of design challenge evolutionary claims about matters such as common descent, the fossil record and the constructive probabilities of genetic mutation, these are not the discursive behaviors that result in their being banished from the courtrooms of science. Neo-Darwinians also debate these questions. But when the challenges on these matters come from design theorists or even from perceived design theorists, they are seen to threaten more than just the current paradigm; they are seen to point up the inadequacy of science's naturalistic assumptions.

Disagreements concerning jurisdiction in this case have to do with law, as they would in forensic matters—but not to questions of law that can be settled in a courtroom. They have to do with the most primary questions of law as encoded in the Constitution or perhaps even more deeply in collective memory or tradition, and thus their association with ideology. Similarly, the challenge of design is still a scientific challenge, but it is perhaps about a third order of science, one distinct from those of normal and even revolutionary science outlined by Kuhn. In addition to the knowledge that evolves through the ordinary processes of normal science, and occasionally through creative theoretical insights that bring about scientific revolutions, there is also a science that is so deeply tied to the identity of the scientific culture as to be almost sacrosanct: thus, far more intractable than the scientific constructs that create the familiar roadblocks for revolutionary science that we see in the textbook cases of incommensurability. Issuing such challenges is tantamount to rejecting the ideological grounds of debate. From the naturalist perspective, therefore, the challenges have no jurisdiction. They cannot be entertained. They are ruled out of court with stipulations like "there is no debate."

The idea of evolution developed into an ideological construct of this kind, not through the influence of Darwin so much as through the influence of positivism that he channeled (in large part through his use of the *vera causa* logic that Campbell documents in his essay later in this volume) and that Huxley championed. The linkage of Darwinism to this ideology was something of an afterthought, though certainly a logical and inevitable one. Once the evolutionary worldview was tied to this scientific hypothesis it attained an authorization that lay within the control of scientists themselves.

Notes

[1] I am here paraphrasing Albert Salomon (1946). Issues of postmodernism and ideology, as for almost any x in the expression 'postmodernism and x,' are subtle, ramified, and, therefore, not particularly easy to chart; my position here aligns with Jorge Larrain, who offers a useful summary of the postmodern view of ideology (Larrain 1994). I recognize of course that the word *ideology,* with a subtle, ramified, and therefore difficult to chart range of senses, remains very important in postmodernist argumentation. My "disrepute or at least neglect" here refers to the original Mannheimian use of the term.

[2] One philosopher denies the existence of the Max Planck Effect; See David L. Hull (1988, 379ff).

[3] This is the position taken by Peter Bowler (1989, 109–10). Bowler states that "We are thus led to the conclusion that the emphasis on Lyell's debate with the catastrophists represents an interpretation of nineteenth-century geology that has been shaped by our modern desire to see this episode as a prelude to the Darwinian revolution" (Bowler 1989, 110).

[4] Popper traces historicism back to Plato, but he also recognizes that it was reincarnated in Comte's social theory. Popper even goes so far as to declare that Plato was a "sociologist." See Popper (1962, 1: 35).

[5] See especially, Michael Bartholomew (1975), Peter J. Bowler (1989). Thus in his 1887 essay on the "Reception of the *Origin of Species,*" Huxley professes doubt concerning whether Darwin's theory will stand the test of time: "Whether the particular shape which the doctrine of evolution, as applied to the organic world, took in Darwin's hands, would prove to be final or not, was, to me, a matter of indifference. In my earliest criticisms of the 'Origin' I ventured to point out that its logical foundation was insecure so long as experiments in selective breeding had not produced varieties which were more or less infertile; and that insecurity remains up to the present time" (in Francis Darwin 1904, 1:551); Huxley's speech of seven years previous, in celebration of the twentieth anniversary of the *Origin*'s publication, celebrates the general influence of Darwin's theory but is completely silent about the content of that theory, making no mention of its central doctrine of natural selection ("The Coming of Age of 'The Origin of Species'" in Huxley 1897). This was a fact that was not lost on Darwin. Huxley followed up with an apologetic letter but one nevertheless forthright about his doubts: "I hope you do not imagine because I had nothing to say about 'Natural Selection,' that I am at all weak of faith on that article. On the contrary, I live in hope that as paleontologists work more and more [. . .] we shall arrive at a crushing accumulation of evidence in that direction also. But the first thing seems to me to be to drive the fact of evolution home into people's heads; when that is once safe, the rest will come easy" (Huxley to Charles Darwin, in L. Huxley 1900, 2: 13).

[6] The origins of this concept are summarized by Bernard Lightman (1987, 12–13). This explains the interesting historical fact that, while Huxley was doing battle with religious interests throughout most of his career, at the particular time that he proclaimed himself an agnostic, his most pressing concern was to distance the kind of science he was advocating from that of the positivists. The full context of this crisis is presented by Adrian Desmond (1994, 372–5). See also Sydney Eisen (1964, 337–58).

[7] Thomas Huxley to Henrietta Huxley, August 8, 1873.

[8] This at least was the accusation leveled against him by the positivist lawyer, Vernon Lushington, who wrote to his friend Charles Darwin asking him to urge Huxley to reconsider the harshness of his treatment. The fact

that Huxley was giving this speech to a religious audience certainly supports that suspicion. Even though Huxley denied that he was playing the religion card, in a letter to the German materialist Ernst Haeckel the following January 20, 1869, Huxley boasted that he had shocked "holy Edinburgh" with this speech (L. Huxley 1900, 1: 431). See also Sydney Eisen (1964, 343).

[9] Certainly the most colorful exposition of this idea will be found in Comte's *The Catechism of Positive Religion,* (Comte 1973).

[10] A special issue of the journal *Rhetoric and Public Affairs* 1 (1998) was devoted to the topic of design in 1998. Representative contributions of the work of leading design theorists, such as William Dembski and Michael Behe, were included in that volume. See also the book, *Darwinism, Design, and Public Education* (Campbell and Meyer 2003), which evolved from that special issue.

[11] A rhetorical history of the design movement which has arisen since in the 1990s is chronicled by Thomas Woodward (2003).

5

The Rhetoric of Philosophical Incommensurability

Herbert W. Simons

> [F]or rhetorical man the distinctions invoked by serious man are nothing more than the scaffolding of the theater of seriousness, are themselves instances of what they oppose. And on the other side, if serious man were to hear that argument, he would regard it as one more example of rhetorical manipulation and sleight of hand. [. . .] And so it has gone; the history of Western thought could be written as the history of this quarrel.
>
> —Stanley Fish, *Doing What Comes Naturally*

> [O]nce [. . .] the omnipresence of the rhetorical dimensions of language [are] recognized, philosophical discourse can no longer be misconceived as logical rather than literary, literal rather than figurative, argumentative rather than rhetorical—in short, it can no longer be conceived of as philosophical in any emphatic sense of that term.
>
> —Thomas McCarthy, *The Rights of Publicity and Privacy*

A good deal of quarrelsome rhetoric about rhetoric circulates through contemporary philosophical discourse.* On one side are those for

* My thanks to the following for their thoughtful suggestions based on careful readings of an earlier draft of this paper: David Depew, Gregory Fahy, Randy Harris, Nola Heidlebaugh, Edward Schiappa, Barbara Herrnstein Smith, Alan Soffin, and Michael Truscello.

whom rhetoric continues, as in the Socratic dialogues, to mean deception, manipulation, exploitation, and to stand in stark opposition to knowledge, truth, logic, coherence, reality, authenticity, "the good," and even philosophy itself. These historic divisions find their way into the writings of contemporary rationalists, analytic philosophers, logical empiricists, and others committed to one or another universalist, objectivist or foundationalist orthodoxy (Harré and Krausz 1996). The modern-day sophists whom they oppose are far more inclined to embrace deconstructive rhetorical theory's anti-foundationalism (Fish 1989), or at least to find common cause with rhetoricians in suggesting that there is a realm of judgment in a domain between the alethic certitudes of True and False (Burke 1969; J. Margolis 1992) where "hard" fact and "cold" logic are insufficient.[1] Rhetoric for these neo-sophists involves persuasion rather than demonstration, a process of coming to judgment, and of bringing others to that judgment. They add that their relativism, or perspectivism, or skepticism, or constructivism is bolstered by recent theory and research, suggesting, for example, that what gets called "knowledge," even in the hallowed realms of science, is far less rule-based or data-derived and far more a consequence of interpretation, social construction and rhetorical choice than had previously been imagined (J. Margolis 1990; 1992; Simons 1990).

This essay offers a rationale for a reconstructive rhetoric, one that builds on, but moves beyond rhetorical deconstruction's critique of traditional philosophy in ways that answer to the need for reasoned and reasonable judgments on issues for which there can be no formal or final proof. As preparation for this concluding argument, I make a case for neo-sophistic as the convergence of several contemporary strands of philosophical thought which, in their challenges to traditional philosophy, suggest consanguinity with ancient Greek sophistic. I then pit neo-sophistic against traditional philosophy (itself a congeries of positions), finding evidence not just of incommensurability between paradigms, but of incommensurability about the very idea of incommensurability—this consistent with neo-sophistic's anti-foundationalist stance.[2] The absence of common measures for adjudicating between sophistic and traditionalist paradigms helps to explain the longevity of the debate. Recalling that incommensurability derives from but partial circularity, the essay explores the possibilities for convergence between the paradigms and for convergence in clashes between other seemingly incommensurable positions. This leads to an examination of factors accounting for both impasse and convergence in the face of apparent incommensurability, including variations in the nature and extent of partial circularity. The essay ends on a note of "theory-hope," suggesting why it is that the antagonists in this ancient struggle are not fated forever to repeat their litanies of reciprocal accusation and resistance.*

Neo-Sophists and Traditionalists

I use the term "modern-day sophists" (or "neo-sophistists") to suggest continuities between contemporary relativisms, including perspectivism, postmodern skepticism, and social constructionism (see Appendix), and ancient Greek sophistic. Nelson (1987) has argued that while the "new sophistic" is for the most part a twentieth century phenomenon, it bears the imprint of earlier critics of modernism, particularly Vico and Nietzsche. As against the fixed essences and eternal verities which Plato and his disciples sought to discover by way of philosophical inquiry, the ancient Greek sophists seemed to rejoice in the flux and uncertainty of daily life, with its possibilities for constructing equally appealing arguments for opposing positions (Schiappa 1991). Exhibiting apparent indifference to universal Truth, they maintained that words differed from the things they named, responded skeptically to arguments about the existence of the Gods, insisted on the inherent variability of human tastes and judgments, employed flexible category schemes, and utilized situation-dependent logics of justification (Leff 1985; Poulakos 2001).

So, too, are today's neo-sophists inclined to question universalist, objectivist and foundationalist orthodoxies (see Appendix), including such fundamental modernist presuppositions as the correspondence theory of truth, the mind as a glassy essence, the potential for language to serve as a mirror of reality or clear window-pane upon the world, and rules of rationality as indubitable and stable across contexts (Bernstein 1983; Megill 1994; McCloskey 1985; Potter 1996; Rorty 1979,

* It should already be apparent that comparisons between sophistic and traditionalist terminologies lead into a thicket of troublesome terminology. So as not to interrupt the flow of the text, I offer definitions and commentary on modernism and postmodernism, relativisms and anti-relativisms, and a few other key concepts in the Appendix of pivotal terms at the end of this essay.

1989). Out of these criticisms has come renewed interest in rhetorical *technē*, both in its productive aspects as a set of procedures by which groups of individuals might be induced to arrive at wise and prudential communal judgments (Billig 1987; Heidlebaugh 2001), and as a set of critical tools by which, for example, to discern the rhetorical devices that maintain illusions of objective representation (Billig 1987; McCloskey 1985; Nelson, Megill and McCloskey 1987; Potter 1996; Simons 1989; 1990). What I am calling deconstructive rhetorical theory refers to postmodern questioning of every kind, and not simply to the movement known as deconstructionism led by Jacques Derrida (e.g., 1976) and Paul de Man (e.g., 1978). As applied to modernism's foundationalist framework—its assumption that observation, or reason, or formal logic, or analytic philosophy provides an Archimedean court of last resort—deconstructive rhetorical theory is skeptical, relativistic, in some cases nihilistic. At its most nihilistic, it reduces reality to appearance, substance to form, theory to ideology, fact to opinion, truth to trope, sincerity to impression-management, sacred to profane. Nothing is obviously, unconstructedly, objectively true; everything is subject to disputation. Relativism's positive role in this stance of negativity is to prompt further exploration of assumptions that have too quickly achieved the status of truisms.

Relativism's "negatively positive" role is well illustrated in the essay, "Death and Furniture," in which Derek Edwards, Malcolm Ashmore and Jonathan Potter (1995) debunk "bottom-line arguments against relativism" by refusing to take their taken-for-granted assumptions for granted. High on their list of anti-relativist rhetorics is the dramatic thump on a table, offered as proof positive of an unmediated "out-there": a reality that is not to be denied. But, suggest Edwards, Ashmore, and Potter, the table-thumping is better understood as an argument without words: an act of semiosis. "The very act of producing a non-represented, unconstructed, external world is inevitably representational, threatening, as soon as it is produced, to turn around upon and counter the very position it is meant to demonstrate" (1995, 27).

Thus, the table-thumping, far from being above rhetoric, or an end to rhetoric, is a rhetorical device, akin to the persuader who purports to "state the obvious." Moreover, "furniture" arguments—and here we can also include other staples of realist arguments, such as rocks, natural disasters, and death—are realism working on its chosen safe ground. Furniture is emblematic of "the reality that *cannot* be denied" whereas Death signifies "the reality that *should* not be denied" (26). But, in focusing on these cases, realists appear to be setting aside, conceding to relativism even, a vast array of more contentious cases, such as language, social order, cognition and science. And on careful, admittedly nitpicky analysis, it is not clear what the table-thumping signifies—the presence of a table, solid and substantial? But only a bit of it is being hit, and physicists might well question its solidity.

> What makes it a bit of a *table?* And for whom? How does the rest of the table get included as solid and real? And how exactly is this demonstration, here and now, supposed to stand for the table's continuing existence, then and later, and for all other tables, walls, rocks, ad infinitum, universally and generally? A lot is being taken on trust here, however "reasonably" (29).

As with Furniture, the alleged undeniability of Death—and of death marches, holocausts, and the like—appeals to a transcendent truth beyond deconstruction. Moreover, references to its brute reality and to the brutal means by which it is too often achieved, add moral weight to anti-relativism and implicit condemnation for relativism's alleged amorality and refusal to "face facts." As a practical matter, Edwards, Ashmore, and Potter concede realism's utility for most purposes. But what emerges from the "Death and Furniture" essay is a semiotic realism, a symbolic realism, and ultimately a rhetorical realism—a realism that calls into question the naive assumptions by anti-relativists that some things simply are as they are, unmediated by language, by inference, and by the possible need for persuasion.

Beyond deconstruction lie the possibilities for rhetorical reconstruction; that is, for proposing constructive alternatives to traditionalism, not just debunking its most sacred assumptions. Whereas deconstructionists are inclined, for example, to call all reality claims into question, rhetorical reconstructionists are more likely to weigh competing reality claims on the assumption that some will be found more credible than others.[3] And rhetorical reconstruction is also concerned with what John Lyne (1990) has called the "art of the sayable"; that is, with inventional resources for arguing and thinking. But rhetorical reconstruction begins with deconstruction, and in its critique of modernism's attempted chokehold on intellectual discourse—modernism's insistence, for example, on a divorce between logic and rhetoric—neo-sophistic is equally uncompromising.

Readers may object that this account paints contemporary philosophy with too broad a brush. And, yes, variations and exceptions need to be given their due. Yet the idea of two rival camps within philosophy, clearly opposed in their attitudes toward rhetoric, is given strong support by Edwards, Ashmore, and Potter—as well as, among others, by scholars as diverse as Kenneth Burke, Stanley Fish, Richard Lanham, Thomas McCarthy, Allan Megill, John S. Nelson, Richard Rorty, and Barbara Herrnstein Smith. By its pervasive intractability across issues and over time, the debate seems to exhibit what Smith calls the "microdynamics of incommensurability" (B. H. Smith 1997, xiv, 125ff). Operating from what they presume to be the superior logics of their own positions, each side speaks past the other while finding good reasons for dismissing what the other has to say. Differences in starting points also find their way into the nooks and crannies of the unfolding and seemingly unending arguments, creating the impression of an unbridgeable divide. As I hope to show, these differences have a recursive or reflexive quality, extending even to incommensurability about incommensurability. For example, neo-sophists such as Burke, Fish, and Lanham are apt to see in the entrenched argumentation of both sides the workings of "rhetorical partisanship," and might even suggest that partisan-angled seeing and saying are part of the human condition. However, for those whom Smith calls the traditionalists, this seemingly detached observation about the human condition, however ironic, might be construed both as self-refuting and as proof-positive of the possibilities for objectivity, even in one so unserious as *homo rhetoricus*.

Thus does the debate between traditionalists and neo-sophists display "symptoms" of the problems that argumentative exchange is ideally designed to resolve. And indeed, there is no relief from the symptoms, not even in my own pretensions to evenhandedness in this essay. For to speak about philosophical rhetoric about rhetoric is already to suggest that something more than, or other than, objective rules of rationality are at play in the debate (Geller 1997); specifically, that the philosophers in question are engaged in persuasion rather than demonstration. And to take seriously the very idea of incommensurability, let alone to assert that traditionalist and neo-sophistic paradigms are incommensurable, is already to take sides with the rhetoricians against those among the traditionalists who either claim that incommensurability is an untenable notion, or who argue from within their paradigm that there are in fact common measures for adjudicating between competing claims. Again and again we will have to confront not just first-order incommensurability in the debate between traditionalists and neo-sophists, but incommensurability about incommensurability, reflexively understood. Still, as Smith argues, "the reduplicative—echoing, mirroring—structure of this situation" is instructive. It raises the general question "of how to understand the cognitive intractability (or, as it may seem, the blindness, stubbornness, and folly) of those who disagree with us" (B. H. Smith 1997, 126).

As Smith notes, this question has ethical and political resonances as well as extensive theoretical implications. It takes us beyond metaphysics and epistemology to rhetorical theory's traditional preoccupation with matters of direct practical concern, as in conflicts between advocates for and against gun control or abortion rights (MacIntyre 1981; Simmons 1994). And it raises as well the question of whether there can be something on the order of progress in the face of incommensurability between traditionalists and neo-sophists.

Kuhnian Incommensurability Revisited

As regards both these questions we can profit from revisiting the literature on Kuhnian incommensurability (Kuhn 1970b, 1977a, 1983a, 1983c, 1993, 1996, 2000; Sankey 1992; Gross, this volume; Harris, this volume; Hoyningen-Huene, this volume). Kuhn remains central to what has come to be called the "rhetorical turn" in both science studies and in philosophy (Barnes 1982; Rorty 1989; Simons 1990), and his concept of incommensurability in the face of paradigm clash has played a pivotal role. Never mind for the moment that the concept was rendered ambiguously, or that it was not easily distinguished from incompatibility or even simple misunderstanding, or that one of its versions, that of untranslatability, seemed to render it incoherent or perhaps contradictory (Davidson 1984; Harré and Krausz 1996; Putnam 1981b; Sankey 1992). These problems notwithstanding, Kuhnian incommensurability seemed to assert the centrality of rhetoric to the scientific process and in ways that helped account for both intractability *and* eventual progress. Untrained in rhetorical theory, Kuhn nevertheless hit the right notes for rhetoricians in asserting that communication across the revolutionary divide was a matter of persuasion, involving "argument and counterargument in a situation in which there could be no proof" (Kuhn 1996, 152). Absent were algorithms for theory choice. In fact, the process of inquiry—the search for facts—was highly theory-laden. And the scientist presented with a new paradigm had to weigh multiple values or criteria, thus entering the rhetorical realm of judgment rather than the modernist realm of fact. As against the traditional distinctions between rhetoric and logic, knowledge and belief, truth and fiction, Kuhn, armed with a stunning assemblage of evidence, had characterized paradigm clash as rhetorical through and through (Simons 1980).

Kuhn also raised the intriguing question: by what manner of rhetorical means could proponents of a new paradigm win converts from among adherents to the old? As part of his effort to set revolutionary science clearly apart from normal science, Kuhn's own answer to this question tended in places toward the melodramatic. The adherents to the old would not so much be persuaded as "die off" and be replaced by more receptive newcomers. Or those persuaded would have experienced something akin to a gestalt switch or religious conversion.

Arguably, these melodramatic images aptly described some major perceptual and cognitive realignments in the history of science. Kuhn left little doubt but that radical rethinkings were sometimes necessary and that conversion entailed confrontation with multiple and interlocking components of the new paradigm, not just point by point comparison.

But, as Michael Malone has argued, Kuhn's idea of conversion in the face of incommensurability did not require the notion of a gestalt switch (Malone 1993). Nor, on a reading of "religious conversion" as entailing a leap of faith, was that analogy entirely appropriate. What was required for the very idea of incommensurability to have meaning was Kuhn's insistence that the new and old paradigms were in some ways rivals and at the same time not fully comparable. But if the paradigms were not fully comparable, this did not mean that they completely failed of comparison, as Kuhn frequently attempted to make clear (e.g., Kuhn 1983a). Drawing on Kuhn's writings in different contexts, Gerald Doppelt (1982) distinguished four ways in which paradigms might be incommensurable:

1. they do not speak the same scientific language,
2. they do not address, acknowledge, or perceive the same observational data,

3. they are not concerned to answer the same questions, or resolve the same problems, and/or

4. they do not construe what counts as an adequate, or even legitimate explanation in the same way. (Doppelt 1982, 114)

A given paradigm clash might involve all four of these components, but still be somewhat comparable in the sense that intelligible and reasonable arguments could be made in favor of one or the other. This is a key point that I want to return to in contrasting traditional and neo-sophistic readings of Kuhnian incommensurability, but suffice it to suggest here that Kuhn need only have committed himself to a view of incommensurability as derived from partial circularity with respect to the privileging of language, data, problems, or explanations, as identified by Doppelt. This in fact is what Kuhn claimed—some of the time. I take the following as representative of Kuhn on incommensurability as based on partial circularity:

> When paradigms change, there are usually significant shifts in the criteria determining the legitimacy both of problems and proposed solution. [. . .] To the extent [. . .] that two scientific schools disagree about what is a problem and what is a solution, they will inevitably talk through each other when debating the relative merits of their respective paradigms. In the partially circular arguments that regularly result, each paradigm will be shown to satisfy more or less the criteria that it dictates for itself and to fall short of those dictated by its opponent. (Kuhn 1996, 109–110)

This conception of incommensurability is viable in the sense of helping to explain both impasse and convergence on matters scientific and philosophical, including the talk between neo-sophists and traditionalists over incommensurability. But the notion of partial circularity is extremely vague and needs pinning down. What kinds and degrees of circularity over what issues? Questions of this sort bid fair to help us understand local variations in the life histories of scientific and philosophical controversies. Nothing requires that we see all scientific and philosophical controversies as ending in either checkmate or stalemate. In what follows I provide evidence of incommensurability beginning with the debate between traditionalists and neo-sophists over

Protagorean relativism and then in their takes on Kuhnian incommensurability itself. I then return to the issues just raised. How does the concept of Kuhnian incommensurability, suitably defined and elaborated, help us to understand continued intractability *and* convergence of belief, including variable patterns of each? Absent foundations or other common measures for adjudicating between philosophical paradigms, what are the possibilities for persuasion across the divide and perhaps even for some degree of reconciliation? What in general are the benefits of a rhetorical, neo-sophistic perspective on the comparison-making process, as opposed to a traditionalist perspective?

Incommensurability in the Debate about Protagorean Relativism

Contemporary anti-rhetoric has its antecedents in the Socratic dialogues. For those whom Smith labels the "traditionalists," the ur-text is clearly Plato's *Theaetetus,* its representative anecdote is Socrates's undoing of Protagorean relativism, and its master trope is *peritrope,* a turning of the tables (B. H. Smith 1997). As was his wont, Socrates invites Protagoras's student, a mathematician named Theaetetus, to put forward a proposition. Theaetetus obliges with an affirmation of Protagoras's "man is the measure of all things." Through further examination, problems with the doctrine are explored. In denying the possibility of objective, absolute truth, the Protagorean cannot justifiably assert that his own doctrine is true or the product of objective knowledge. And if man is the measure, so too must opinions to the contrary also be counted as true for those holding them. Theaetetus is forced to concede that the Protagorean doctrine is self-refuting, hence incoherent, hence without philosophical value. And in so far as Protagoras was esteemed among the Greek sophists as a critic of established orthodoxies, the undoing of his pupil becomes metonymic of sophistic irrationality more generally (Bostock 1988; Polansky 1992). The proof of relativism's incoherence in the form of self-refutation has a dramatic symmetry to it: "a threat averted, an outlaw brought to book, order restored, orthodoxy vindicated" (B. H. Smith 1997, 81).

But, as Smith maintains, the traditionalists are not alone in having reason to claim the last and decisive word. A modern-day Protagorean might have asked of Socrates whose interests were being served by his dubious paraphrase of Protagoras (surely Plato's, we might surmise),[4] and whether his tricksterism was conscious or unconscious.

Less aggressively, he might have pleaded for greater understanding: "What Protagoras really intended to say, dear Socrates, was that man is the measure of all things, *in his opinion.*" Or, "That the Gods (about whom Protagoras had been discoursing) were *not* the measure of all things" Hence, that Protagoras's comments had been taken out of context. Or "That all knowledge claims, Socrates' included, are built on one or another epistemic framework, paradigm, or form of life" (Bernstein 1983; Harré and Krausz 1996).

Taking specific aim at the epistemic framework undergirding Socratic rhetoric, Smith charges the traditionalists generally with question-begging. Often left unstated, and even unexamined, are assumptions about what can be "taken as," or "reduced to," or in other ways seen as "logically implicit." This extends to assumptions about the very nature of assertion itself. One traditionalist who does examine them declares that all assertions, however formulated, are implicit claims of absolute truth. This could not be otherwise, it is asserted, for why else would the assertion be listened to and possibly believed? Another asserts as self-evident that "Relative rightness is not rightness at all." Says Smith, question-begging assumptions such as these—about assertion, about truth, about rightness—form an incontestable web. For the traditionalist, they cannot be otherwise.

So too has it been assumed, she says, that to accede to relativism is to succumb to the conclusion that "every opinion, every scientific theory, every artwork, every social practice," etc., is as good as any other. But, says Smith, what follows from relativism is not necessarily the view that all theories, social practices, and so on, are equally valid, but that none is valid "*in the classic sense*" (B. H. Smith 1997, 77). The error in what she calls the "Egalitarian Fallacy" is the assumption that differences of "better" or "worse" must either be objective, or they cannot be measured. To the contrary, argues the relativist, they can be evaluated in other, non-objective ways, based on criteria of applicability, coherence, connectability, and so forth. Says Smith, "These measures are not objective in the classic sense since they depend on matters of perspective, interpretation, and judgment, and will vary under different conditions. Nevertheless, they appear to figure routinely, and operate well enough, in scientific, judicial, and critical practice" (B. H. Smith 1997, 78).

A variant of the "Egalitarian Fallacy," says Smith, is the "Anything Goes Fallacy."

> This is the mistaken idea that a theory that does not ultimately affirm the absolute force of certain relevant constraints (for example, determinate meanings, an objective morality, or an objective reality) implies that in the relevant domain, anything—any utterance or interpretation, any social practice, any belief, and so on is acceptable. The fallacious assumption here is that there can be no *other* explanation for why we do not all talk nonsense, or run amok, or believe ridiculous things. [Both] the Egalitarian Fallacy and the Anything Goes Fallacy take for granted as unquestionable or irreplaceable the orthodox concepts or explanations at issue. (B. H. Smith 1997, 78)

In general, suggests Smith, traditionalists "experience their own beliefs as intuitively right and then pass them on as prior, necessary, and properly universal" (B. H. Smith 1997, 83). Rejected out of hand is the possibility that the assumptions they pass on were passed on to them through socialization in a certain epistemic tradition.

To be sure, the neo-sophists also assume what needs to be proven, also regard the Other as misguided or worse, also misread or miscast the Other's pronouncements—or so it seems to the traditionalist. What Smith dismisses as a web of incontestable interlocking assumptions, traditionalists might praise as evidence of coherence—of a system of thought that hangs together. When Smith invites speculation on the possibility that traditionalists' assumptions are a product of socialization, she in effect avoids, they would say, the real issues by focusing on questions of upbringing, a sure sign that her central argument is unsustainable. But, paradoxically, these criticisms of the neo-sophist can be seen as further support for Smith's microdynamics of incommensurability, in which, as Jeffrey Geller (1997, 376) puts it, "every phrase is in dispute and no rule, whether of logic, forensic etiquette, or fair play, can be taken for granted."

Incommensurability in the Debate about Kuhnian Incommensurability

There appear to be striking similarities in the forms of argumentation used by Socrates in his assault on Protagorean relativism and the critiques leveled against Kuhnian incommensurability. Meanwhile, neo-sophistic defenders of Kuhnian incommensurability, or some version of same, respond to his critics in a manner not unlike Smith on Socratic foundationalism. As seen by the traditionalists, incommensurability is an untenable thesis, at once incoherent and self-refuting. As seen by the neo-sophists, incommensurability is ubiquitous in paradigm clash but not necessarily disabling. Centrally at issue for both are conceptions of what Israel Scheffler calls "rhetorical persuasion": whether inherently subjective and irrational (Scheffler 1970), or capable of yielding wise and prudential judgments under conditions of uncertainty (Booth 1979; Heidlebaugh 2001; McCloskey 1985). In the battle over meanings, including interpretations of what the Other has to say, each side employs the tools of persuasion. For example, James Harris reads Kuhn in a manner conducive to his case against incommensurability. Kuhn, he asserts, "seems to make of science a matter of nonrational, intuitive flashes of insight, or political clout, or mass psychology" (J. F. Harris 1992, 79). Similarly, Imre Lakatos identifies conversion to a new paradigm with "mob psychology" (Lakatos 1970, 178), as opposed to a process of individual reflection that is far more open to evidence. Counters Kuhn, "To say that, in matters of theory-choice, the force of logic and observation cannot in principle be compelling is neither to discard logic and observation nor to suggest that there are not good reasons for favoring one theory over another" (Kuhn1970b, 234).

The arguments for and against Kuhnian incommensurability are embedded within the epistemic frameworks of each camp. For Joseph Margolis, the case for incommensurabilism gains credence from what he takes to be new "truths" in support of Protagorean relativism (J. Margolis 1992). We now "admit," says Margolis, that "the most comprehensive schemes we can imagine are all partial, fragmented, contingent, historically bounded, ultimately blind, and radically ungrounded" (J. Margolis 1990, 310). For James Harris, Kuhnian incommensurability, and indeed relativisms of every kind, cannot withstand the test of self-contradiction. Like Protagoras's "man is the measure," Margolis's "admission" implicates his own comprehensive scheme of schemes, rendering it partial, fragmented, ungrounded, and the like, and hence no more credible than its antithesis. Similarly, Harris hoists Kuhn's "meta-paradigm"—his paradigm of paradigms—on its own petard, proposing the same tests of intelligibility, comparability, and decidability as Kuhn offers for scientific paradigms, and concluding that, by Kuhn's own logic, his meta-theory must either be rejected or accepted on blind faith (J. F. Harris 1992, 82–83).

But this, again, is not the end of the story. Kuhnian incommensurability (and incommensurability about incommensurability) draws sustenance from the larger intellectual movement against traditionalist orthodoxies that I have called neo-sophistic. Absent foundations of knowledge, for example, it is difficult to see how rival paradigms can be objectively compared. So too does incommensurability become more credible if the very ideas of truth and falsity are context-dependent. and if one need not be forced to choose between bivalent, "true-false" alternatives, such as those proposed by Harris. These possibilities form a part of Margolis's incommensurabilism.

Suppose, suggests Margolis, that in place of the traditionalist's stable, knowable reality, we entertain a sophistic theory of flux, one that calls into question as well the assumption that the traditional, bivalent sense of such words as "true" and "false" needs to be maintained through all transformations of culture, language, place and time (J. Margolis 1992). Margolis's "alethic relativism" offers the possibility that the very ideas of "T" and "F" may be context-dependent, different at the Italian Renaissance section of the Metropolitan Museum than in arguments before the Supreme Court, or in articles published in *Scientific American;* different too when viewing Leonardo da Vinci's *Last Supper* as a Christian icon, or as portraiture, or as an example of horizontal design (Harré and Krausz 1996). Thus, one might endorse pairs of incongruent or incommensurable or incompatible beliefs, each in its own context, without brooking contradiction. Challenged by Margolis is the ubiquitous assumption of bivalence, that there are only truth and falsity, and that they necessarily exclude each other. Other alternatives—all within the realm of rhetorical judgment—include aptness, plausibility, reasonableness, and appropriateness, which "do not stand in polar oppositions to their negations" (Harré and Krausz 1996, 143). As though to anticipate traditionalist objections of the sort offered by Harris, Margolis hastens to add that bivalent (T-F) and many-valued logics should not be used conjointly over the same range of application. It is bivalently true, for example, that Leonardo painted the *Last Supper*, but a multi-valued matter of judgment as to whether it succeeds or fails as a Christian icon (Harré and Krausz 1996). Margolis shares with traditionalists here an aversion to contradiction and a commitment to a certain kind of realism (e.g., Leonardo painted the

Last Supper) but in other ways he stands outside the circle of traditionalist assumptions, rendering his own incommensurabilism incommensurable from theirs.

Explaining Intractability and Convergence of Belief

Why do proponents of competing theories or paradigms so often talk past each other, rendering convergence of belief and even reconciliation all the more difficult to achieve? Why, on the other hand, do adversaries in such debates sometimes achieve mutual understanding and move toward consensus? It is time to return to these issues, and, in doing so, to help clarify the notion of partial circularity.

Continuing impasse is perhaps easier to explain than convergence of belief. Recall that I endorsed a version of Kuhnian incommensurability based on partial circularity in support of rival paradigms, a version, I alleged, which helped account for both continuing impasse and convergence. I added that Kuhn's notion of partial circularity (as quoted), while useful, was also quite vague and hence in need of elaboration.

By *circular*, I mean arguments (or premises in arguments) that are presupposed and then privileged as the basis for subsequent argumentation. As suggested by Smith, the resultant question-begging may not be conscious, or if conscious, not thought to be problematic. "Everyone knows that our senses tell us true." Hence, the unquestioned premise may be treated enthymematically and left unarticulated along with other premises that form a web of belief. It may also be reinforced by social support from those whom one values (and values in part because of their agreement). Then, as opposing arguments are joined, the taken-for-granted premises may insinuate themselves into the entire debates, exhibiting Smith's microdynamics of incommensurability.

This is not to say that belief change is precluded by incommensurability, or that reconciliation is impossible. For example, debates can bring circular reasoning into light, and this can be a source of convergence, but question-beggars are often well defended and sometimes legitimately so. It would be question-begging on my part to assume that circular reasoning is necessarily bad, or even that it is possible to avoid it.

The dynamics of resistance and change in the face of incommensurability need not be the same for all controversies. Still, as with Kuhn on scientific incommensurability, it seems reasonable to assume that adherents to rival philosophical paradigms will seek ways to resist wholesale changes in their positions even as they are forced by their opponents to acknowledge problems or anomalies and as they see erosions of support from within their own camp. Humans, suggests Smith, are cognitive conservatives; that is, they tend to seek confirmation of their existing beliefs, expectations, and the like, even to the point of dogmatism and inflexibility. As with individual human beings, so too with scholarly communities; reigning beliefs tend to be privileged over challenges to them. Thus, even as adherents to a paradigm are confronted with anomalies or problems in their ways of thinking, they are likely to respond with the "yes-but" of partial agreement, partial resistance. Presumption falls with the status quo, burden-of-proof with the challenge (Whately 1963, 112–115).

As we entertain the possibilities for convergence of belief in the face of philosophical paradigm clash, it is well to remember that the process of persuasion is dialogic rather than monologic, involving, as Protagoras would have put it, not just *logoi*, but the *dissoi-logoi* of counterargument. The results of such debate may not be apparent immediately, or even over the course of a generation. Rather than yielding gestalt-like conversion, the dialogue may create incremental self-doubt, confusion, then ambivalence—all this in Margolis's multivalent region of judgment, between the alethic certainties of "T" and "F." Says Smith, it is impossible to engage in protracted debate with an opponent and not be influenced by what he or she has to say. This applies, she suggests, even to debate over incommensurables. "Nothing said here implies a permanent structure of deadlock. On the contrary, what has been said explicitly and implied throughout is that no orthodoxy—or skepticism—can be totally stable, no theoretical closure complete, no incommensurability absolute" (B. H. Smith 1997, 87).

By the same token, she adds, "one cannot interact with a theoretical closure and remain totally 'outside' of it, even if the interaction is skeptical or adversarial. Thus one disputes 'logic' with logic (or logic with 'logic'), neither identical but each, over time, shaped by the other" (87). Smith is eloquent in characterizing the process of interaction as taking place both in public theaters (classrooms, conferences, journals, and the like) and in the private theater of the mind, "where the 'self' takes all the roles—truth-deliverer and truth-denier, master and disciple, chorus or mixed voices and motley audience—and every self-refutation is, simultaneously, the self's triumph and transformation" (87).

I suggested earlier that Kuhn's notion of partial circularity as the basis for incommensurability was quite vague, and that the processes by which impasse was maintained or convergence achieved were probably quite variable from issue to issue and community to community. Just how the dialectic of change versus resistance will play out will vary with such factors as the nature and extent of circularity as well as the availability of some degree of common ground. We can see this best, perhaps, with respect to familiar social issues. I have chosen three.

1. *Reincarnation.* Depending on what kinds of lives we have led, says the Hindu, we may return to life after death in a superior or inferior state. This is supported by Hindu scripture and reinforced by evidence from dreams and divinations. For those reared in the Western, Judeo-Christian tradition, this assumptive framework makes little sense. We in the West, says Richard Shweder (1992), not only privilege our own preferred mind states (e.g., systematic observation over dreaming) and reality-posits ("life does not beget new life after death"), but then shift to the role of the existential *ubermensch,* presuming to be able to explain from that same Western perspective, for example, Hindu India's "real" reasons for clinging to belief in reincarnation, and what interests the Hindu's belief in it "really" serves. Westerners and tradition-oriented Hindus may have much in common, but as regards the issue of reincarnation, the premises that they privilege are likely to lead to continued impasse.

2. *Abortion Rights.* In another, somewhat more hopeful case, some "pro-life" religious conservatives and "pro-choice" liberals were led through a process of facilitated dialogue to discover common ground over a preference for adoption over abortion, and were thus persuaded to join together in a campaign for more adoptions of unwanted children. The agreement to cooperate required abandonment by each group of long-held misperceptions about the other. The religious conservatives now saw that the "pro-choice" liberals favored abortion *rights,* not abortion *per se.* For their part the "pro-choice" advocates now acknowledged that their "pro-life" counterparts cared

about life after birth, and not just life (if there be such) between conception and birth. Still, this is convergence at the edges of the debate; as regards the issue of abortion rights, the two positions have for the most part remained incommensurable (Chasin *et al.* 1996).

3. *Capital Punishment.* As regards the ongoing debate in the United States over capital punishment, the introduction of DNA technology seems to be making a more fundamental difference. Like the debates over reincarnation and abortion rights, the capital punishment debate exhibited incommensurability in the sense that each side's moral or normative conceptions provided "different, noncomparable, and incompatible units of measurement for use in judgment" (Heidlebaugh 2001, 18). But that incompatibility has proven partial; both sides endorsed the common and very central principle that people should not be executed by the state for capital crimes that they did not commit. With DNA and other evidence establishing widespread violations of that principle, opposition to capital punishment has increased (Turow 2002).

In the foregoing cases we saw one instance in which acceptance of another's beliefs was virtually precluded by a totalistic ideology that privileged one's own mind-states and reality-posits while explaining away the other's way of thinking. These are signs of extreme intractability, perhaps symptoms of cosmic incommensurability. In another case we saw convergence at the margins, continuing impasse at the core, not unlike Lakatos's (1970, 133–134) characterization of scientific research programs as having a "hard core" which is insulated against change by a "protective belt."[5] In the third case we saw more opportunity for convergence because of well evidenced threats to an important shared principle for proponents and opponents of capital punishment alike. To be sure, some true believers have remained unshaken by the evidence, convinced that the judicial system can do no wrong. It matters, then, how open or closed a partially circular system of belief happens to be, how central and how peripheral are the privileged, taken-for-granted premises and, similarly, where the common ground is that can provide a basis for convergence—whether it is central or peripheral.

Impasse and Convergence in the Debate between Traditionalists and Neo-Sophists

It seems reasonable to assume that incommensurability between traditionalists and neo-sophists, as well as incommensurability about incommensurability, will play itself out in the manner just described. That is, it will involve persuasion, not formal proof or demonstration, in which the dialectical processes of resistance and change, continuing impasse and convergence of belief, will be situationally variable: dependent on such factors as the nature and extent of circularity, the availability of common ground, the degree of social support from valued others, the skill with which arguers and counterarguers deploy the available means of persuasion, and the good will of the interactants. Belief modification can be piecemeal and gradual, unlike the conversion experiences and gestalt shifts associated with Kuhn. As cognitive conservatives, adherents to the opposing paradigms will resist wholesale change, pairing concessions to opponents with elaborate defenses.

The "yes-but" of concession/resistance was manifest, for instance, in our look at some of the debates over Protagorean relativism and Kuhnian incommensurability. Implicitly or explicitly, Smith concedes that self-refutation is not a good thing, just as Margolis grants that it is illogical to contradict oneself. Yet neither owns up to self-contradiction and each finds ways in which the Protagorean or the Kuhnian, suitably understood, is not logically inconsistent. Likewise, traditionalists such as James Harris are unlikely to champion question-begging, but rather to deny at considerable length that they are guilty of it. Or, as regards the argument that their views are socially conditioned, a product of their times, traditionalists may concede the point but deny its significance.

Similarly, as regards ontological questions, anti-relativists may force from Protagoreans the concession that the "man" of "man is the measure" is an implicitly acknowledged existent, not some social construction (Harré and Krausz 1996). But the Protagorean may call attention to the epistemic framework by which the anti-relativist unquestioningly equates words and things. Social constructionists such as Steven Woolgar may likewise grant that reality is "really" socially constructed, while at the same time distinguishing between "reality" in quotation marks and reality "unmanned" and essentialized (Woolgar 1983).

Conclusion: Where Does the Debate over Philosophical Incommensurability Leave Us?

What are the possibilities for progress in the debate between traditionalists and neo-sophists? We have observed that their paradigms are partially circular and that they appear to be diametrically opposed, extending to incommensurability about incommensurability. Under the circumstances, the very idea of "progress" will be seen through different lenses. For traditionalists, progress in paradigm clashes rests on belief in the possibility of objective comparison and evaluation. Without it, communication across the divide is meaningless.

Says Scheffler, on the traditionalist side, "to put forth *any* claim with seriousness is to presuppose commitment to the view that evaluation is possible, and that it favors acceptance; it is to indicate one's readiness to support the claim in fair argument, as being correct or true or proper" (Scheffler 1967, 73). Moreover, fair-minded comparison of paradigms requires a "stepping back" for Scheffler. Such an evaluation, he asserts, "is itself not formulated within, nor bound by, the paradigms which constitute its objects. It belongs rather to a second-order reflective and critical level of discourse" (Scheffler 1967, 74).

But the neo-sophist responds that Scheffler's "second-order level" is itself paradigm-dependent. Further, no system of logic is ever self-validating; it depends on some other system for its support, and that system must be validated by another, that system by yet another, and so on, in an infinite series. If, as neo-sophists claim, the criteria for evaluation in the debate between traditionalists and neo-sophists are paradigm-dependent, and if second-order reflections are themselves infirm, then there exists no outside standpoint from which to draw up a common measure by which the paradigms can be compared. Clearly, the ideas of progress by the rival camps must themselves be incommensurable.

We could leave things hanging at this point, but I do not think we should. For deconstructionists, interminable impasse is not a bad thing. It is far superior to self-refutation—what Geller (1997) cleverly calls the death of internal causes. Deconstruction's object, moreover, is a stalemate of traditionalism, not a checkmate. But for neo-sophists committed to a reconstructive rhetoric, non-objective comparison of incommensurables can yield reasoned and reasonable judgments on terms that may even be acceptable to some traditionalists. Let us attend further to the distinction between deconstructive and reconstructive rhetorics.

Rhetorical Deconstruction

Deconstructionism as a methodology is committed to a naysaying or skeptical relativism, one that calls into question what it alleges is question-begging by traditionalism of the assumptions underlying its own epistemic self-privileging. Deconstruction therefore makes a space for rhetorical reconstruction in suggesting that there is no escape from rhetoric, not even in table-thumping demonstrations of a supposedly unproblematic "out there."[6] Deconstruction is sophistic in suggesting that reason is rhetorical, that "reality" is appearance, that knowledge and power are joined at the hip, and that every representation is in a certain sense a misrepresentation. This sophistic influence is at work, for example, in Burke's playful conjecture that language does our thinking for us (Burke 1966), in his more literal-minded insistence that every terministic reflection is also a deflection (1966), in Derrida's (1976) "no escape from the text," and in Jameson's (1972) "prison house of language." From provocations such as these, we are led back to Rosenmeyer's (1955, 232) construal of Gorgias as suggesting that "speech and reality are not commensurable."

Thus does rhetorical deconstruction offer a counter to traditionalism's logic, all the while as it seeks to evade the charge of self-contradiction. It attempts the latter in a variety of ways, for example, by recasting apparent contradictions as paradoxes, or as aporias or antinomies (Harré and Krausz 1996), or by bracketing issues of truth or falsity in the name of "programmatic relativism" or "methodological quizzicality" (J. Margolis 1992; Simons 1990). Thus, for example, do Woolgar and Dorothy Pawluch (1985) identify rhetorical techniques by which language is used to constitute reality, not simply to describe it. And thus does Woolgar (1983) problematize his own accounts, as well as accounts of those accounts, leaving nothing settled, except perhaps his own claims to the advantages of leaving nothing settled.

But thus does rhetorical deconstruction take itself out of humankind's ongoing conversations about matters practical and philosophical that require movement beyond skepticism. How in our "prison house of language" can we justifiably reclaim a sense of human agency? How, in the face of limits on our ability to know, and to know that we know, can we adjudicate between competing reality claims, and competing ways of rendering them? How, if the human sciences are not simply an extension of the physical sciences, can we best reconfigure social thought? How can we choose between competing values (e.g., freedom vs. equality, national security vs. civil liberties) in the face of changing exigencies? How can we structure these and other conversations so that they are most likely to lead to wise and prudential judgments? And having arrived at these judgments, how can we convince others—traditionalists included—to join with us? Questions of this sort are the province of rhetorical reconstruction.

Rhetorical Reconstruction

Drawing on the insights of deconstructionists, rhetorical reconstructionists nevertheless join with traditionalists in attempting to formulate reasoned judgments and to bring others to those judgments on a wide range of issues. They do so, moreover, in a manner not altogether unlike that of the traditionalist. If scientific methods seem to work best for purposes of testing the efficacy of a new drug, then why not use them? If the methods of the law tribunal seem most suitable for purposes of determining what constitutes a missing person, or child abuse, or date rape, then why not use them?

Moreover, as regards incommensurables, reconstructionists could join with Scheffler when he asserts, along with Kuhn and Feyerabend, that "lack of commensurability does not imply lack of comparability" (Scheffler 1967, 125). Even works of art, he says, "may be reasonably discussed, criticized, compared, and evaluated from various points of view." (See Harris's introduction and Hoyningen-Huene's essay in this volume, both of which take up Kuhn's and Feyerabend's views on the comparability of incommensurable theories or paradigms.)

But Scheffler then adds that the incommensurability of these works does not establish that paradigm debates "must consist of empty rhetoric alone . .. and is no bar to their reasonableness or objectivity" (Scheffler 1967, 125). The rhetorical reconstructionist would take issue with Scheffler's conflations of reasonableness with objectivity, and rhetoric with emptiness of thought or argument.

What Scheffler might mean by "empty rhetoric" is the sophistic style of argumentation and neo-sophistic's refusal to endorse the epistemic framework of traditionalism's anti-relativism. Rhetorical reconstruction is relativistic, but it is not anti-empirical or unreasonable. Rather it rejects pretensions to objectivity on issues of judgment where, for example, competing interests may be involved or definitions legitimately in dispute.

Consider, in this regard, the many kinds of cases in which the quest for an objectively correct or true or proper answer is likely to end in failure—Determinism as opposed to free will? I cannot claim to have chosen it without appearing to have contradicted myself; moreover, my commitment to human agency may itself be determined—A best estimate of the extent of child abuse in America, or date rape, or persons reported missing? But the estimates available, even from putative experts, turn out to be wildly discrepant, and to reflect differing definitions and conflicting interests (Miller and Holstein 1993)—A best way to report the results of a clinical drug trial wherein forty of one thousand subjects in the placebo treatment condition died over the course of the study and just twenty of one thousand subjects died during that period in the active treatment condition? By one reading the drug treatment was twice as effective at preventing morbidity. By another reading, one that compared proportions of deaths to sample sizes in each group, the drug treatment was only two percent more effective. Wherein does objectivity lie (Simons 1993)? Finally, are there objectively correct metaphors for characterizing "the way we think about the way we think" (Geertz 1980, 21)? Clifford Geertz identifies a number of such metaphors drawn from the humanities, comprising what he calls "the refiguration of social thought," but he would be hard-pressed to demonstrate objectively that they are right or correct.

On some issues, and for some purposes, the reconstructionist may propose criteria for evaluation, but they may not be the same as those of the objectivist. Smith makes the important point that

> People who are (formally or *de facto*) non-objectivist in their conceptual commitments can and do argue, credibly and persuasively, that their judgments and proposals (political critiques, judicial rulings, legislative programs, and so on) reflect not merely their individual or partisan preferences but the interests and values of larger relevant groups, including, sometimes, the entire relevant community. Moreover, such people can and do argue the superiority of their judgments on similarly compelling, though not classically objective, grounds: the fact, for example, that those judgments reflect more extensive, more current, or

> more pertinent information than rival or prevailing judgments, or that they take into account a wider range of significant factors or subtler and longer-range implications. (B. H. Smith 1997, 4)

Thus might the rhetorical reconstructionist argue reasonably but not objectively. But, as I argued in my introduction to *The Rhetorical Turn* (Simons 1990), such a rhetoric will of necessity be unstable, self-questioning, reflexive—always in the process of reconstituting itself in light of new historical saliencies and new habits of conviction. Its "truths," if there be any, will be situated, contextual, contingent—true for particular purposes; true under a given set of circumstances; true assuming the validity of taken-for-granted premises. And it will continually be engaged in a politics of competing pluralisms, a parliament of voices about which voices to privilege, and about how to construct, array, compare, and assess the objects of its scrutiny, including the multiple and competing rationalities about rationality with which it must contend.

Can rhetorical reconstruction move us beyond deconstruction in ways that help bridge the divide between traditionalism and neo-sophistic? I think it can. The key is in recognizing that the opposing paradigms are only partially circular, and thus provide openings for dialogue based on appeals to common ground.

Appendix of Pivotal Terms (Further clarification of key concepts in the text)

Modernism (vs. postmodernism, poststructuralism)

By contemporary philosophical discourse, I refer broadly to twentieth century philosophy, a period sometimes referred to as "late modernity." Centrally at issue in contemporary philosophy has been the future of "modernism," understood as a post-Enlightenment project, animated by progress in science, and the example set by Kantian and Hegelian rationalism, to undergird all disciplinary inquiry with grand theories, built on firm, objectively derived foundations, and offering stable, universal truths. Fissures in the post-Enlightenment project appeared from its very beginnings (Habermas, 1987), but it was not until the late twentieth century (a period sometimes referred to as "post-modernity") that "post-modern" challenges to modernism came into full flower. At issue for such influential thinkers as Jean Baudrillard, Stanley Fish, Michel Foucault, Jean-François Lyotard, and Richard Rorty were questions, said Pauline Rosenau (1992, ix),

> that pertain to the deepest dimensions of our being and humanity: how we know what we know, how we should think about individual endeavor and collective aspirations, whether progress is meaningful and how it should be sought. Post-modernism questions causality, determinism, egalitarianism, humanism, liberal democracy, necessity, objectivity, rationality, responsibility, and truth.

In fundamentally challenging modernist presuppositions, postmodernism, and here I include (controversially) post-structuralism (e.g., Barthes, 1957, 1975; Derrida 1976), prompted reconsideration within contemporary philosophy of "pre-modern" rhetorical theories and perspectives, which modernism had thoroughly repudiated. As Allan Megill (1989, 143) observed, based on a thoroughgoing review of postmodernist literature, the postmodern turn appears as a 'weakening' of philosophy into rhetoric—as a turn from, say, the Aristotle of the *Posterior Analytics* to the Aristotle of the *Ethics* and the *Rhetoric* (143). Said Megill, "where 'opinion' is at issue and where appeal is made to community, the rhetorical tradition comes clearly into play" (143). Poststructuralists of a "deconstructive" bent (e.g., de Man 1979; Derrida 1976) have suggested an even more radical connection: that between postmodern skepticism and those among the ancient Greek sophists who questioned existence, as well as the human capacity to know what is, and to reason logically about it, and to provide accounts of it in purely literal terms (Culler 1982; Norris 1982; on *sophists*, see more below).

Rhetoric (rhetorical perspective; sophistic and neo-sophistic thought)

As used here, *rhetoric* includes the practice of persuasion (this essay included); theory about that practice (encompassing, for instance, rhetorical invention, argumentation, audience adaptation, stylistic figures and tropes); and the variety of perspectives on rhetoric (some celebrative, some highly critical, some "global" in scope, some focused on the civic arena) which figure in current controversies about matters practical and philosophical (Simons 1990; Simons 1998). My own take on the rhetoric of philosophy and the philosophy of rhetoric has been deeply influenced by Kenneth Burke (e.g., Burke 1966,1969a), whose ideas about rhetoric's relationship to the debate between traditional philosophers and their sophistically oriented critics has been ably summarized and defended by Timothy Crusius (1999), among others. With Crusius I hold that Burke came earlier than most to the problems with philosophical modernism and responded with more judicious correctives. With Burke I hold that there is no escape from rhetoric for us symbol-using, "symbol-misusing" animals. This holds true, I maintain, for philosophers of a traditionalist bent who purport to demonstrate (or "prove") rather than persuade. Therefore, when I claim that quarrelsome rhetoric about rhetoric circulates through contemporary philosophical discourse, I do not mean that the discourse on either side of the divide is intentionally deceptive or dishonest; rather that it uses such tools and techniques of persuasion as selective naming and framing, highlighting and downplaying, positioning and privileging of premises, and that it does so in highly contentious ways. As should be made clear in context, I sometimes explicitly identify in this paper with the modern-day sophists in opposition to traditional philosophy and sometimes position myself as evenhanded rhetorical analyst of the discourse of both sides in the debate. (On *rhetoric*, see also the introduction to this volume, which sketches out a rough history of the rhetorical tradition, singling out key words and characteristic obsessions. Several essays in Part III of this volume also develop rhetoric in certain ways—such as Ceccarelli on Cicero and Prelli on *stasis* theory—and all of them exemplify it in analysis of specific cases.)

Sophistic and Neo-Sophistic; Deconstructive vs. Reconstructive Rhetorics

For purposes of "contemporary appropriation" of sophistic thought, as opposed to "historical reconstruction" (Schiappa and Swartz 1994), I look past differences among the Greek sophists for what appear to be their common denominators. (My apologies in advance to classical Greek scholars for whom these distinctions assume greater significance.) Of particular interest for me are the sophists' philosophical musings, which were probably of less importance to them than their practical concerns. Moreover, I *selectively* appropriate sophistic thought, rejecting, for example, Gorgias's apparent indifference to matters of ethics in rhetoric (White 1984, chpt. 4), while appreciating their concern for the local, the contingent, the situated, and their corresponding suspicions of Platonic idealism. Stanley Fish captured the spirit of what I am after when, in a wide-ranging interview with Gary Olson that is reproduced in *There's No Such Thing as Free Speech* (Fish 1994), he allowed—even applauded—a label that was pinned on him by a critic: "contemporary sophist." In his defining essay on "Rhetoric," Fish (1989) characterized contemporary critiques of foundationalism and objectivism as part of a counter-tradition, begun with the Greek sophists, that had helped move rhetoric from periphery to center.

Something of that counter-tradition also runs through Burke, as Crusius (1999) has noted, and through Burke's (1969) appreciative reading of rhetorical theory's "underground" history, as Gaonkar (1990) has noted. As the premier contemporary American pragmatist in the tradition of John Dewey, William James, and Charles Sander Pierce, Richard Rorty's (1979) early critiques of objectivism and foundationalism from within the language-analytic philosophical tradition gave modern-day sophistic an enormous boost, and his subsequent championing of theorists like Sigmund Freud (Rorty 1989) who, in the manner of the Greek sophists had disarmed traditionalist opponents with novel language (Heidlebaugh 2001), provided grounding for what I am here calling "reconstructive" rhetoric. Smith (B. H. Smith 1997) is less sanguine than I am about the possibilities for convergence between traditional philosophers and their sophistically oriented critics, but she has been my principal guide nonetheless, both to the debate between them, and to the case for sophistic rhetoric.

Relativisms and Anti-Relativisms

Relativisms range from those which hold (or are seen by their critics to hold) that one view is as good as any other, to those such as Smith's which explicitly reject that formulation, preferring a view of relativism as emphasizing the possibilities, in a world of flux and uncertainty, of still arriving at reasoned and reasonable contingent (i.e., conditional, probabilistic) judgments. Neo-sophistic of the kind I endorse is inherently relativistic in Smith's sense of the term, although its deconstructive side paradoxically questions such notions as "reasoned" and "reasonable" even as it uses reason to make its case (Rosenau 1992).

Relativisms—cultural, aesthetic, ontic, epistemological, etc.—are best understood in relation to the absolutisms they oppose. For an introduction to relativism and its absolutist opponents, I have found Rom

Harré and Michael Krausz (1996) to be particularly useful. In the latter category, they distinguish belief in the existence of trans-cultural, trans-historical universals (universalism), belief in their knowability by way of impersonal controls over human error and prejudice (objectivism) and belief in the possibility of adjudicating philosophical disputes by appeals to rock-bottom truths or principles (foundationalism). These together constitute what J. Margolis (1992), a relativist, calls the "archic canon," and they are the bedrocks of what B. H. Smith (1997) labels *traditionalism*.

From a traditionalist perspective, Protagorean relativism is the view that "man is the measure," or, more formally, that "a Proposition P is true for the person propounding it, under the unique circumstances in which he or she propound[s] it" (Harré and Krausz 1996, 72). Truth, thus relativized, is taken to be self-refuting. Relativists respond, says J. Margolis, in one of two ways. "Relational" relativists accept the terms of the traditionalist's assumptions about truth as bivalent, but maintain that "circumstances" include perspectives, habits of mind, epistemic frameworks, and the like, which cannot be eliminated from any theory of knowledge. This, I take it, is B. H. Smith's (1997) line of argument. Margolis grants that relativism is self-contradictory on the alethic assumption that truth is necessarily bivalent, but not when bivalence is replaced "in a formal way with a logically weaker set of many-valued truth-values or truth-like values" (J. Margolis 1992, 8). This alethic alternative undergirds Margolis's "incommensurabilism," a non-relationalist Protagorean relativism, he says, that simultaneously rejects the necessity of bivalence while acknowledging that alternative conceptual schemes are "moderately incommensurable" (J. Margolis 1992, 18).

Contemporary relativisms include *perspectivism* (or *perspectivalism*), *constructivism* (or *social constructionism*), and *postmodern skepticism*. They come in many varieties and blend into one another.

Perspectivism (or perspectivalism)

Perspectivism (*perspectivalism*) denotes for me a kind of relativism which holds that "truths" are inevitably partial, limited, a function (at least in part) of one's epistemic framework or way of seeing (Booth 1979; Burke 1966, 1969a; Goodman 1978; Kuhn 1996). The perspectivist need not be committed to any one perspective on a matter and may indeed hold that insight is gained by viewing it from multiple perspectives. Likewise, the perspectivist may shift perspectives depending on what's at issue and may concede that some things are indisputably real or true or logical, and hence not a matter of perspective. As with relativisms, then, perspectivisms are not all of one kind. By way of defense against the charge of self-refutation, the perspectivist may assert that any given perspective is subject to critique (Booth 1979; Burke 1969a), but, to qualify as a perspectivist, must insist that every correction, and every correction of that correction (etc.) is itself but a perspective and is thus ultimately ungrounded.

Skepticism

Skepticism has its roots in ancient Greece and possibly earlier. Rosenau (1992) characterizes postmodern skepticism as dark and gloomy, but I read its questioning and demystifying as ironic, perhaps playful, and not necessarily pessimistic, except for those who fully believed in the post-Enlightenment project. There is, to be sure, a cynical side to "po-mo," reflected in such notions as "the demise of the subject, the end of the author, the impossibility of truth, and the abrogation of the Order of Representation" (Rosenau 1992, 15). But these might also be seen as provocations. Patrick Morag, for instance, finds a more positive strain in the questioning mode of deconstruction than many others, a search for ethical and political significance in the discourses it interrogates, a "productive [. . .] transformative" enterprise, not the simple dissolution of significance that all of deconstruction's opponents and most of its friends find (Morag 1997, 4).

Constructivism (including social constructionism)

Constructivism (here I also include *social constructivism*) stands in opposition to traditionalist belief in what J. Margolis (1992) characterized as the archic canon (i.e., the accurate, veridical representation of a fully autonomous reality). Against that thesis, the constructivist (and the social constructionist) views the "real" as at best a product of "reciprocal coordination" (Latour 1993) between observer and observed, never free from the influence of discursive and institutional processes. On some matters, constructivism isn't terribly controversial; traditionalists agree, for example, that five dollar bills gain currency as "institutional fact" through a constructionist process (Searle 1995). But what of the natural world, and of efforts by scientists to describe and explain it? Here, too, B. H. Smith views constructivism in a positive light, believing that its accounts of the scientific process hold together coherently and reliably, providing a "rhetorical/pragmatic" alternative to traditional philosophy of science.

But as with the other -isms discussed in this appendix, constructivism (and social constructionism) can be deconstructive to the point of calling into question all that we thought we knew. Who "we" earthlings are as a people and as persons, what we take to be normal vs. deviant, real vs. imagined, may now be seen as having been *constituted* by rhetoric (Charland 2001; McGee 1975). The term "child abuse," for example, is of relatively recent coinage. As with other putative social problems, moral crusaders, health-professionals and other claims-makers participated in a persuasive process of disquotation by which "child abuse" was named, defined, categorized, etc., ultimately becoming child abuse (Gusfield 1989; Pfohl 1977). Charland is correct, then, in suggesting that this has important implications for rhetorical theory, but its implications are wider still. What do we know for sure? "Strict" constructionists might propose that all is rhetoric (Woolgar and Pawluch 1984). Is everything we might reasonably claim to know a matter of inference and interpretation, by ourselves and by those who exert influence upon us? This brings us closer to the views of "contextual" constructionists (e.g., Best 1990). To what extent is our thinking hostage to the "terminologies" available to us (Burke 1984b)? To what extent are what Burke (1966) calls our "terministic screens" socially constructed (Crusius 1999)?

Notes

[1] Alethic questions are those bearing on the meaning of "true" (Harré and Krausz 1996).

[2] [Editor's note: Simons means " incommensurability about incommensurability," a recurrent phrase in his argument, somewhat differently from what Paul Hoyningen-Huene, Eric Oberheim, and Hanne Andersen (1996) mean by that phrase (which they use as a gloss for *meta-incommensurability*). In particular, Simons's usage is more general. For Hoyningen-Huene and his colleagues (following Feyerabend 1970, 220) theoretical frames can differ with respect to whether incommensurability holds in a specific case—A and B may be incommensurable if examined under a realist philosophy of science, but not an instrumentalist one (see also the introduction to this volume for some discussion—80). Simons is more interested in the differences between opposing philosophical frameworks which have antithetical answers

to the question, "is there such a thing as incommensurability (outside mathematics)?" The concept is also taken up in my introduction (80, 145), Alan Gross's chapter, "Kuhn's Incommensurability" (193-194), where he disputes the usefulness of the concept (*RAH*)]

[3] My distinction between rhetorical deconstructionists and reconstructions parallels that between "strict" and "contextual" constructionists of social problems. For essays on both sides of this divide, see Miller and Holstein (1993).

[4] One version of the original proclaims that "Of all things, the human being is the measure, of things that are, that they are, and of things that are not, that they are not" (Johnstone 2001, 595). Note the parallels to his expression of religious agnosticism: "Concerning the gods, I cannot know either that they exist or that they do not exist" (Johnstone 2001, 595).

[5] Says Lakatos, a "negative heuristic" forbids direct attack on the core of the program. Instead, a "protective belt" of auxiliary hypotheses is "adjusted and re-adjusted, or even completely replaced, to defend the thus-hardened core" (Lakatos 1970, 132).

[6] While it is hardly a recurrent theme of his work Derrida, has actually argued for this space-making point about deconstruction himself. In responding to the "post-structuralist" framing that deconstruction has received in Anglo-American theory, for instance, and to the uniformly corrosive connotations it has in that milieu—being associated only with eating away at communicative action, breaking it down—Derrida remarked that

> the undoing, decomposing, and desedimenting of structures, in a certain sense more historical than the structuralist movement it called into question, was not a negative operation. Rather than destroying, it was also necessary to understand how an "ensemble" was constituted and to reconstruct it to this end. However, the negative appearance was and remains much more difficult to efface than is suggested by the grammar of the word (de-), even though it can designate a genealogical restoration [*remonter*] rather than a demolition. (Derrida 1988, 2)

As always with Derrida, we now need to do some hermeneutical work. In particular, it is not at all clear how deconstruction, as it is constituted, and largely as it is practiced, is anything other than a "negative operation." But this passage seems to suggest that while such an operation necessarily forces the breakdown of signification, it is, in Derrida's scheme, preliminary to some building up. See Morag (1997) for a highly charitable reading of Derrida (with equally charitable nods to de Man) in this regard. Morag argues that a Derridean reading "aims to *produce* a certain signifying structure" (Morag 1997, 19). To the extent that he finds these productive moments, Morag sees in Derrida impulses toward the sort of rhetorical reconstruction that I am exploring with respect to incommensurability; in my own readings of Derrida, however, and of deconstruction generally, those impulses are incredibly few and far between.

III

Cases

Kuhn's description of persuasion as an inexplicable conversion experience seems highly *un*rhetorical.

—Leah Ceccarelli, "Science and Civil Debate"

Realignments can occur at points of commensurability that take the form of shared, situated questions.

—Lawrence J. Prelli, "Stasis and the Problem of Incommensurate Communication"

[T]he problem of incommensurability is [. . .] an artifact of Cartesian anxiety—a prohibition imposed by formal logic on inventional ingenuity.

—John Angus Campbell, "The 'Anxiety of Influence'"

Incommensurability can [. . .] be an active rhetorical investment.

—Jeanne Fahnestock, "Cell and Membrane"

Incommensurability gives philosophic warrant to a kind of intellectual stubbornness within a knowledge space wholly occupied by rational considerations.

—Charles Bazerman and René Agustín De los Santos, "Measuring Incommensurability"

In the study of scientific change and controversy, incommensurability is an idea we can probably just do without.

—Carolyn R. Miller, "Novelty and Heresy"

6

Science and Civil Debate: The Case of E. O. Wilson's Sociobiology

Leah Ceccarelli

When I began studying the rhetoric of science, I was advised to read Thomas Kuhn's *The Structure of Scientific Revolutions,* as one of the first books to open up the possibility for such a line of research.* Excited to discover the entryway to this field, I followed that advice, becoming most interested by what I found in the chapter alliteratively titled "The Resolution of Revolutions." While the metaphor of revolution suggested a somewhat violent shift from one scientific paradigm to another, Kuhn's use of the word "resolution" hinted at the possibility of a mutually satisfying settlement that was negotiated carefully through rhetorical means by the participants. Asking how a shift between paradigms was induced and resisted, Kuhn wrote,

> What sort of answer to that question may we expect? Just because it is asked about techniques of persuasion, or about argument and counterargument in a situation in which there can be no proof, our question is a new one, demanding a sort of study that has not been previously undertaken (Kuhn 1996, 152)

* I would like to thank Randy Allen Harris, Mike Leff, and Michael Truscello for their comments on an earlier version of this essay.

To my eager eyes, this looked like a call for scholars like myself to examine scientific controversies through the tradition of rhetorical inquiry, a field that is centered on techniques of persuasion and devoted to analyzing argument and counterargument in situations in which there can be no proof.

But gentle prodding by one of my rhetoric professors got me to look more closely at what Kuhn was saying and the language he was using to say it. Kuhn describes the competition between paradigms as a "battle" between proponents who "must fail to make complete contact with each other's viewpoints" (Kuhn 1996, 148). This failure to make contact is, of course, due to the incommensurability of the two sides, the lack of a common measure to take stock of each paradigm in any mutually referential way; or, in the rhetorical terms Randy Harris uses to frame the problem in the introduction to this book, the lack is of a common ground on which the two sides can identify with each other and begin the process of resolving their differences through rhetorical negotiation. Their standards and definitions differ, they misunderstand each other and thus "communication across the revolutionary divide is inevitably partial," and they might even be said to be "practicing in different worlds" (Kuhn 1996, 148–50). Because of the incommensurability of the two sides, the resolution of a scientific revolution cannot occur through a negotiated settlement but only from the conquest of one side by the other.

Kuhn agreed with the general sentiment, if not the specific terms, of Max Planck's statement that "a new scientific truth does not triumph by convincing its opponents and making them see the light, but rather because its opponents eventually die" (Planck 1950, 33; Kuhn 1996, 151; see R. A. Harris 1998 for some discussion). The young may be open to conversion, but the old frequently resist. Kuhn described the acceptance of a new scientific truth not as a rhetorical matter of negotiated consent that carefully reshapes a challenger paradigm until it persuasively answers the objections of its opponents, but as the total triumph of that challenger. An individual scientist accepts the new paradigm in a "gestalt switch" moment of truth (Kuhn 1996, 150), which Kuhn equated to a "conversion experience" (Kuhn 1996, 151), talking about the "faith" sparked by "personal and inarticulate aesthetic considerations" (Kuhn 1996, 158) that made shifts of allegiance in science similar to religious epiphanies. Here was where Kuhn located the 'resolution' of revolutions: in the grace of inexplicable and immediate ideological transformation, combined with the inevitable passing of those opponents who are unable to see the new truth. "Conversions will occur a few at a time until, after the last holdouts have died, the whole profession will again be practicing under a single, but now a different, paradigm" (Kuhn 1996, 152).

For one who understands rhetoric in its popular sense, as mere talk divorced from reason, the challenge Kuhn leveled against the logical empiricist view of science seems a radical move toward rhetoric. But for one who understands rhetoric as an academic tradition devoted to the study of how arguments are made reasonable to audiences when absolute truth is unavailable, Kuhn's description of persuasion as an inexplicable conversion experience seems highly *un*rhetorical. The Kuhnian vision of scientific debate as a battle between incommensurables conceives of conflict as a zero-sum game and considers reasoned argument to be hopeless. According to this view, the winners in a scientific debate are the ones who motivate the most young people to make a leap of faith and join their side of the crusade, not the ones who are most successful at using the available means of persuasion to make their theory seem reasonable to their peers.

Kuhn's view of debate in *Structure* is a cynical one, but it is not an unusual view in our contemporary culture. Argument is often envisioned through the metaphor of war, in which controversialists attack and defend positions, attempt to shoot down their opponents, and make points that either hit or miss their target (Lakoff and Johnson 1980, 4–5). Rejected as aggressive, argument is considered a corrosive component of our culture (Tannen 1998). Our experience in the public sphere shows us what happens when groups get stuck in futile argument; they resort to escalating, reciprocated diatribe or the use of violence because debate fails to persuade speakers from incommensurate positions to alter their beliefs (Freeman, Littlejohn, and Pearce 1992; Pearce, Littlejohn, and Alexander 1987; Pearce and Littlejohn 1997).

But this cynical view of argumentation as the clash of incommensurables is not the only way of thinking about the phenomenon of contentious debate. For example, George Lakoff and Mark Johnson suggest that the metaphor of argument as dance is an intriguing hypothetical alternative to the metaphor of argument as war (Lakoff and Johnson 1980, 5). In this paper, I explore a similar alternative, but unlike Lakoff and Johnson, I suggest it not as a vision of what a hypothetical culture unlike our own might believe, but as a competing vision of how successful arguers in specific arenas of our culture, like the world of science, think and behave. I contend that the resolution of a scientific revolution often takes the form of a rhetorically negotiated truce between seemingly incommensurable positions, and thus, a view of argument as a cooperative creative process—the Isocratean perspective that Harris calls in the introduction, *reciprocal suasion*—is a more accurate way of describing scientific debate than the negative view of argument as aggressive battle. In fact, those scientists who envision debate as a conflict between two incommensurables, a battle that cannot be resolved except through the death of the old guard and the conversion of the young, are operating from a limited view of rhetoric that may actually hurt their prospects for successfully sponsoring revolutionary change in science.

To develop this alternative image of scientific argument, I will begin with a sketch from the rhetorical tradition of a form of debate in which controversialists engage in contentious argumentative exchange that is both civil and productive. Cicero, the most influential inheritor of the Isocratean tradition, articulated such a model in a series of his mature dialogues, of which *De Oratore* is both the clearest and best known. It presents a model for civil debate that differs from the vision of argumentation as a zero-sum game. After introducing this prototype of civil debate from the rhetorical tradition, I will briefly review some research in the rhetoric of science that suggests points of connection between this model and successful scientific debate. I will then explore the case of E. O. Wilson's attempt to lead a scientific revolution around his concept of sociobiology. Jeanne Fahnestock, in her contribution to this volume, investigates a theory of cell physiology propounded by scientists who regard their work as incommensurable with the prevailing theory, scientists who engage in an aggressive form of argumentation that has contributed to the marginalization of their view. Wilson's view is similar, and the latter sections of my essay illustrate the ways in which Wilson's vision of scientific debate imagines a war between incommensurables, and results in an aggressive, imperialistic rhetoric that looks to be seriously impairing his chances of achieving a positive resolution to the conflict.

Civil Debate in Cicero's *De Oratore*

Like Plato's treatises about rhetoric, Cicero's *De Oratore* is designed as a dialogue between historical characters who were familiar to the author. But the differences are more striking than the similarities. It is usually clear to the reader of Plato's *Gorgias* and *Phaedrus* that Socrates speaks for the author. He is the protagonist whose wisdom leads his inferiors to admit the error of their ways (or if an opponent in debate is intractable, at least allows the reader to see the obstinacy of that opponent's position). In contrast, there is a sense when reading Cicero's dialogue that the author shares the conflicting views of two wise protagonists. Cicero pits oratorical colleagues Crassus and Antonius against each other in a debate over the proper scope of the orator's art, a dispute that offers equally strong arguments on both sides of the case. The reader of *De Oratore* finds his or her allegiances shifting from one side to the other at each turn of the argument. This pro/con argumentation, or *controversia*, eventuates not in one side demonstrating the falsity of the other, which then loses the case, but in the challenger responding to criticism from the other side by incorporating it into its own position.[1] As a philosophical dispute taking place in the tranquility of a private courtyard, it is a debate that is civil and friendly, seeking not the binary judgment of a courtroom, but the creative birth of a new perspective that encompasses aspects of both original positions.[2] As Michael Mendelson puts it in his analysis of the method of debate modeled in *De Oratore,*

> The aim of antilogic/controversia is not to impose one position on another; rather, the goal is to continue the dialectical process, to accommodate alternative logoi, to modify (instead of abandon) an initial position, and to (re)formulate new arguments by considering opposing alternatives as potential components in a more comprehensive claim. (Mendelson 1997, 41)

Nola Heidlebaugh describes this method as a type of active artistic judgment that responds to the apparent incommensurability of a controversy with a metaphor of weaving or stitching rather than weighing or measuring (Heidlebaugh 2001, 142). It is the approach of a master rhetor.

To get a better sense of the give and take, revision and defense, accommodation and friendly contradiction, that characterizes the argumentative method of controversia displayed in *De Oratore,* consider the beginning of the dialogue. One of the main characters, Crassus, offers an encomium of oratory, praising it as the most significant and all-encompassing of human enterprises. His position is that the orator must have a broad education, knowing not only the techniques of effective public speaking, but enjoying knowledge of every possible subject matter as well. The debate is engaged when a senior colleague, Scaevola, disputes Crassus's claim, arguing that it is too extreme, giving too much intellectual territory to the orator. After politely identifying a point of agreement with Crassus, he offers a counterargument that defends the status quo in which philosophers are the ones who enjoy knowledge of subject matters and orators are specialists in the skill of public speaking. "I should be at the head of a multitude who would either fight you by injunction, or summon you to make joint seizure by rule of court, for so wantonly making forcible entry upon other people's possessions," he concludes (Cicero 1942, 1.10.41). Crassus, sensitive to the concerns of his opponent, responds by accepting the critique, eventually giving up his claim to two of the three branches of philosophy he would like the orator to possess (the branches that we might call *science* and *logic*), but keeping a tight hold on the third, namely, the broad study of "human life and conduct" (Cicero 1942, 1.15.68). His opponent concedes the debate (which is later taken up by Antonius), but he responds with a smile,

> in this very speech you have made against me, you have by some trick so managed matters as both to grant me what I said did not belong to the orator, and then somehow or another to wrest away these things again and hand them over to the orator as his absolute property. (Cicero 1942, 1.17.74)

This metacommunicative comment is a hint to the reader, although the interlocutors do not call attention to it again, that this strategy of argument-through-incorporation is the pattern that is followed throughout the debate. Each time Crassus offers a claim that is too controversial to be accepted by his opponent, he is confronted with a persuasive counterargument; his response in each case is not to admit defeat nor resort to tactics that seek to destroy his opponent, but instead to accept his opponent's position while subtly assimilating it to his own. The result at the end of the dialogue is an understanding on the part of the readers (and one suspects, the interlocutors as well) that Crassus has won the debate—that the orator must have a broad education, encompassing both style and matter, rhetoric and philosophy. But at the same time, one does not get a sense that Antonius, Crassus's main opponent, has lost; instead, Crassus has been compelled to alter his position to adapt to his critic's legitimate and reasonable counterarguments. Crassus must admit that although the orator should have a broad education, he cannot become a philosopher, nor can he reasonably be expected to know the details of every subject matter.

This friendly tone of academic debate modeled in Cicero's *De Oratore* is different from the more hostile approach toward argumentation that Cicero took when arguing legal cases. There is less room for accommodation of an opponent in a courtroom where the goal is determining guilt or innocence within a limited time frame before a jury of uninformed and disinterested citizens. In his speech defending Marcus Caelius Rufus, Cicero implies that the accuser, Clodia, is a whore, an unstable nymphomaniac, a pornographic woman whose purposes were nothing more than degraded whims (Cicero 1969); the humorous but biting invective of this speech is hardly the tone of civil debate adopted by Crassus and Antonius when they dispute the proper scope of rhetoric in the bucolic setting of a Tusculan villa. But such a difference in tone is expected. The situation determines the form of rhetoric employed, and the two situations are quite different. The outcome of a debate on philosophical matters has less urgency than the outcome of a court case: the audience does not have to vote yes or no on the proper scope of rhetoric at the end of the day and no one will be going to jail depending on what they decide. They have the freedom to adjust their positions and explore ways of negotiating their disagreements without fear of losing both case and client when the rules of the system foreclose further debate and demand a decision in favor of one side or the other. The nature of the audience differs as well: in academic debates like the one between Crassus and Antonius, the judges are peers who are experts in the matter (or at least students who have some knowledge of the matter), and they hold a vested interest in the outcome of the debate; in courts, the jury is a group of impartial observers who are chosen for their lack of previous knowledge of the case. Because decision-makers in the academic arena have specific beliefs, experiences, and interests in the case, as well as the time to reflect and the freedom to make a judgment that is more nuanced than a yes or no decision, persuaders in this arena have the opportunity, indeed, the incentive, to engage in a type of argumentation that resembles a dance more than a war, a friendly game more than a life-or-death struggle. The give and take of offering a position, listening to the opposition's response, and accommodating the legitimate counterarguments of that opposition without abandoning the crux of your initial position is a rhetorical art that is particularly appropriate in such a setting.

In short, a recognition that the form of rhetoric varies with the context suggests that where you have "argument and counterargument in a situation in which there can be no proof," but where a verdict does not immediately need to be handed down and where the audience is knowledgeable and invested in the outcome, the type of pro/con argumentation exemplified in Cicero's *De Oratore* is more appropriate for dispute resolution than a pitched battle between declared incommensurables. In fact, in such contexts, a more aggressive form of argumentation that refuses to give ground would seem inappropriate, an ill-mannered move that causes observers to turn away from the offending party. If scientific argumentation is more similar to the sort of academic debate modeled in Cicero's *De Oratore* than to the contests Cicero encountered in the law courts, then when scientists mistakenly assume that the sort of rhetoric appropriate to one setting can be transferred unchanged to the other, they are misreading the rhetorical situation and are likely to fail in their persuasive efforts.

Controversia in Science

There are many cases in the history of science of serious conflict between a revolutionary theory, and the status quo for which Cicero's vision of civil academic debate seems more appropriate than Kuhn's vision of antagonistic opposition, where revolutionaries and traditionalists pit incommensurable positions against each other in a battle to the death. One of the best examples of civil debate in science is outlined in Martin Rudwick's detailed study of the Great Devonian Controversy, an analysis that shows that the terms used over the course of a geological debate were malleable, as were the concepts they were used to express (Rudwick 1985, 447). Like Crassus, these geologists were willing to alter their arguments in the face of opposition, but not so much that the revolutionary nature of their claims was totally destroyed. As a result, the final resolution of that controversy utilized parts of both sides, "involving something akin to diplomatic negotiation" (Rudwick 1985, 448).

Studies of the Darwinian revolution have made a similar point about how new paradigms achieve success. In subsequent editions of *The Origin of Species,* Darwin responded to critics by emphasizing his more moderate claims and attenuating his more controversial claims. Rather than argue that natural selection was the *sole* directive force in evolution, Darwin reminded readers that he had always allowed for other forces as well (Gould 2000, 101–102). Rather than claim that his theory of evolution proved there was no divine Creator, Darwin suggested that it revealed the supreme power of God (Campbell 1975, 383–385; 1989, 72–73).[3] This kind of response in a debate transforms a refutation into a confirmation by judiciously adapting one's arguments. David Hull's study of the scientific reception of Darwin's theory shows that among those who accepted the idea of evolution in the ten years after Darwin began the debate, "older scientists were as quick to change their minds as younger scientists" (Hull, Tessner, and Diamond 1978, 722). This empirical challenge to the Max Planck Effect (R. A. Harris 1998) may indicate that the observation made by so many scientists (Darwin included—e.g., 1958 [1892], 258) that persuading the old guard is futile because the establishment is incapable of seeing things in a new way is a myth. When scientists need an excuse for why an argument might not be accepted by an audience, an appeal to the *topos* of age may seem a reasonable answer. But in fact, whether there is an incident or two that resembles a conversion experience, the bulk of scientists, of all ages, are brought to accept a new theory not because their hearts are open to being converted to a radical new idea as it is originally set out, but because the ensuing debate includes the transformation of that theory to respond to criticism, and such transformation persuades them that the new perspective is reasonable.

My own studies of the interdisciplinary revolutions in science that led to molecular biology and the evolutionary synthesis have shown that seemingly incommensurable fields of study were brought together by scientific rhetors who used integrative metaphors to allow scientists with radically different values and beliefs to see things from the perspective of their counterparts, and strategic ambiguity to suppress disagreement by making each side believe that acceptance of a new interdisciplinary field would achieve its own most desired goals (Ceccarelli 2001a). This sort of creative rhetoric of negotiation seems to be the most successful way of motivating radical change in science, rather than Kuhn's vision of revolutionaries and the forces of convention facing each other in battle across the great divide of incommensurability. Alan Gross makes the case for the success of rhetorical negotiation in science when he contrasts Newton's first paper on optics with his final *Opticks* (Gross 1996, 111–128; reprinted in R. A. Harris 1997, 19–38). The first paper failed to persuade because it was not adapted to its audience; it relied on confrontation rather than on attempting to find points of connection with the previous tradition. Newton's later paper succeeded because it was rhetorically designed to emphasize continuity with its opposition, redescribing the author's insight as evolutionary rather than revolutionary.

These studies and others like them show that scientific colleagues who are originally opposed to a new theory can be persuaded if the argument is adapted to respond to their concerns. Controversies often end in a negotiated settlement in which both sides win something. And even in those instances where there are clear winners and losers, the winners are likely to adopt some aspect of the losers' case, adapting their original position in response to the challenges of their opponents (R. A. Harris 1990, 29). The notion that conflicting positions in a scientific revolution are incommensurable is a stance taken when one views a conflict with a fatalist attitude. It can often become a self-fulfilling prophecy because it encourages a rhetoric of diatribe or violence (after all, why should one try to accommodate an intractable opponent?). But it is not the only, nor even the best, way of interpreting what happens when a scientific revolution is introduced and resolved in favor of a new paradigm. In fact, this vision of scientific debate may actually contribute to the failure of scientific revolutions led by people who carry out the implications of this cynical perspective. In what follows, I will explore a case where a scientist's inability to recognize the possibilities of civil debate may have contributed to the less than satisfactory outcome of the controversy he initiated.

The Case of E. O. Wilson's Sociobiology

In previous research on E. O. Wilson's case for sociobiology, I have argued that the rhetorical tactics he employed were not well designed to persuade his readers (Ceccarelli 2001a, 2001b). But I have been unable to adequately explain why he would choose to adopt those strategies in the first place (Lyne 2002, 76–77).[4] This time, rather than look primarily at the books that frame his attempt at revolution, I trace the argumentation he offered in the early moments of the controversy, when he was still trying to decide what sort of response he should make to his adversaries. This analysis shows that Wilson toyed with the possibility of adopting a view of scientific change as civil debate, but soon abandoned that perspective for a vision of scientific revolution as a battle between incommensurables. In fully embracing the latter approach, I believe he lost what could have been a promising path toward resolution of the sociobiology revolution.

The State of the Sociobiology Controversy

In 1978, just three years after E. O. Wilson's *Sociobiology* was published, philosopher of science David Hull made a prediction about the future of Wilson's theory. To ground this prediction, he compared Wilson's proposal with two cases from the history of science, Charles Darwin's theory of evolution and Franz Josef Gall's theory of phrenology. When both of these theories were first proposed, they included claims that were later determined to be accurate and claims that were later determined to be inaccurate, and both enjoyed immediate acceptance from some influential scientists and immediate rejection from others. The main difference between these two cases was the way their supporters conducted themselves in the scientific debate that followed; in the face of criticism, phrenologists kept a tight hold on the weakest part of their theory, while evolutionists were willing to loosen their grip on the parts of their theory that were most controversial in order to get that theory accepted. I include a long extract from Hull's article because it establishes so well my point in this essay: that civil debate modeled after Cicero's controversia in *De Oratore*, rather than diatribe in the face of incommensurability, is the most effective rhetorical approach for scientists proposing a paradigm change.

> A strong tendency exists to conclude that Darwin's great achievement was to devise a theory that was basically correct and that Gall's failure was to come up with a set of ideas that were crudely mistaken. [. . .]

> However, if the history of phrenology and evolutionary theory have anything to teach us, it is that the truth of new theories *as they are originally set out* is not all that important. Phrenology in the first half of the nineteenth century was no further from the truth than the theory of evolution, which became widely accepted in the second half. What really determines the success or failure of new scientific theories is how advocates of these views continue to conduct themselves. They must be conceptually flexible, socially cohesive, and terminologically rigid. The role of evidence in science is too obvious to belabor, but evidence never totally constrains the freedom of scientists in formulating their theories. The fudge factor is just as important. Any scientist who is not a 'master wriggler,' to use Darwin's phrase, will see his views refuted almost immediately. [. . .] To be successful, a scientist must be able to recognize clear threats to his or her position and respond appropriately. But the proper response to imminent refutation is not admitting defeat; it is changing one's position while retaining one's original terminology. Successful scientists are those who master the art of judicious finagling. (Hull 1978, 137–38)

Hull went on to describe evolutionists as "masters at transmuting refutations into confirmations"; in contrast, "the chief failing of the phrenologists is that they were not sufficiently adept at finagling" (Hull 1978, 162).

So when asked to predict the future for sociobiology, Hull concluded that it would depend on how Wilson and his supporters continued to conduct themselves; to succeed, "they must be willing to modify their views, no matter how extensively, in the face of various considerations that threaten to undermine their position, while maintaining that these are *modifications* not *abandonments.* The results are new *versions* of their theory, not *replacements*" (Hull 1978, 161).[5]

Today, over twenty-five years after Wilson first proposed his theory of sociobiology, its long-term success remains uncertain. There is no consensus in the scientific community about the adequacy of Wilsonian sociobiology; as Ulica Segerstråle put it, there are still two sides to the sociobiology debate, and both see themselves as defenders of the truth (Segerstråle 2000, 1, 5). While the debate is not over, there is reason to believe that Wilsonian sociobiology is traveling the path that Hull described for phrenology, and may soon share its fate. Sociobiology does not enjoy social cohesion, and in fact, the very term *sociobiology* is being discarded by people doing work in areas that fall within Wilson's original proposals. As Segerstråle, a supporter of Wilson's, is forced to admit, the Human Behavior and Evolution Society has dropped the term *sociobiology* from the title of its journal, and a recent book telling the history of the application of evolutionary biology to human social behavior almost completely omits Wilson from its pages (Segerstråle 2000, 318, 319). While there are some sociobiologists who look to Wilson as the father of their discipline, others working in this area distinguish their field from Wilsonian sociobiology, preferring to identify their work as evolutionary psychology, ethology, or biopolitics (Segerstråle 2000, 320). As Ernst Mayr notes, most active people who work on the problems that Wilson lists under the term *sociobiology* do not use that term to describe their work: "they do not call themselves sociobiologists" (Mayr 1997, 205).

I think the current lack of social cohesion around the term *sociobiology* is due in part to the failure of Wilson and his supporters to maintain conceptual flexibility in the face of criticism of his theory. In Hull's terms, Wilson has not mastered the art of judicious finagling; or, as I would put it, Wilson has not been a particularly good rhetor.[6] In 1978, it was easy to imagine a different future for Wilson's sociobiology, for as I will illustrate, Wilson *was* showing signs of being a judicious rhetor at that point in time. But he was simultaneously struggling with a more cynical view of the adequacy of debate, and by the early 1980s, Wilson had fully adopted a more hostile stance that further polarized the two sides and destroyed the potential for negotiating a civil resolution to the conflict. I believe Wilson settled on such a posture because he accepted the common perspective that sees debate as the antagonistic clash of incommensurables.

The Art of Judicious Finagling in Wilson's First Response to Critics

To suggest that Wilson, a two-time Pulitzer Prize winner, is not a particularly good rhetor may seem odd. But this assessment assumes that to be a good rhetor one must do more than neatly turn a phrase, write a book that is deemed "distinguished" by a board of judges, or make a lot of money for a publisher. As Aristotle would put it, one must possess the faculty of observing the available means of persuasion in a particular case (*Rhetoric* 1355[b]), and then deploy those means effectively. By this standard, Wilson, who ultimately proved himself to be more adept at sparking controversy than persuading his peers, does not qualify as an accomplished rhetor.

Such was not always the case. When the debate over sociobiology was first engaged, there were hints of judicious rhetorical responses from Wilson. For example, one of the biggest critiques of Wilson's theory from the very beginning was its disciplinary reductionism. In *Sociobiology,* Wilson had made the rather startling claim that the new field and its related discipline, neurophysiology, would "cannibalize" those disciplines that take psychological or sociological approaches to the study of human behavior (E. O. Wilson 1975a, 6, 574–575). One passage in particular envisioned a world where "the humanities and social sciences shrink to specialized branches of biology" (E. O. Wilson 1975a, 547). Another defined sociobiology as the "systematic study of the biological basis of *all* social behavior" (E. O. Wilson 1975a, 4).

Critics of Wilson's theory—like Scaevola responding to Crassus in Cicero's *De Oratore*—immediately complained that Wilson was claiming too much territory; they suggested that he would do better to abandon his reductionist stance, and take on the more reasonable and promising claim that sociobiology could study the biological basis of *some,* but not all, social behavior. As John Maynard Smith put it, "sociological theories cannot be expressed in neo-Darwinist terms, any more than evolutionary theories can be expressed in solely biochemical terms" (J. M. Smith 1975, 496). Smith said he was persuaded by Wilson's argument that we can learn a lot about human social behavior from the study of animal societies and our ancestors, "but this is a long way from wishing to incorporate sociology into biology. In a similar way, I think a knowledge of biochemistry is essential to an understanding of evolution, but I do not expect to see evolution theory become a branch of biochemistry" (J. M. Smith 1975, 496). Other critics saw

Wilson's disciplinary reductionism as a consequence of his "biological determinism," a position that they opposed as too extreme; while "not denying that there are genetic components to human behavior," they thought that Wilson went too far in claiming that human social behavior is wholly "determined by evolutionary imperatives operating on inherited predispositions" (Allen et al. 1975, 43; Sociobiology Study Group of Science for the People 1976, 182).

In response to this criticism, Wilson adopted the stance of a master finagler, judiciously altering his stance without abandoning his theory. Engaging in a species of rhetorical criticism, Wilson explicated a passage in *Sociobiology* that his critics had interpreted as evidence of his extreme genetic determinism, arguing that the real meaning of the passage was not that genes are primary in determining human behavior, but merely that genes should be studied because they might have *some* influence (E. O. Wilson 1975b, 60). In fact, he said that his "wholly empirical conclusion" in the book was that the degree of genetic determinism of human behavior is moderate, and therefore his book was "closer to the environmentalist than to the genetic pole" when it came to explaining human social behavior (E. O. Wilson 1976, 183). Backing off from his vision of the humanities and social sciences being reduced to specialized branches of biology, Wilson pointed out that this part of his book was merely a thought experiment, meant to suggest the potential of future extensions of the theory to human behavior "even though our vastly more complex, flexible behavior will make the application technically more difficult," and "the degree of success [of that application] cannot yet be predicted" (E. O. Wilson 1976, 187).

The following year, Wilson described the relationship between sociology and biology as the "tense creative interplay" of discipline to antidiscipline. Here he offered a perspective toward controversy that envisioned opponents not as enemies out to destroy each other, but as creative partners in the production of something new. Again he responded to the criticism of his disciplinary reductionism not by admitting defeat or lashing out, but by shifting his position while retaining the core of his theory. Echoing Smith's critique as if it were his own position, Wilson proclaimed that sociology would benefit from an association with biology but "sociology is not destined to be cannibalized by the antidisciplines, any more than cytology was absorbed by biochemistry" (E. O. Wilson 1977, 137). He admitted "the limits of reductionism," and maintained that the laws of biology would be useful but not "sufficient" for the sociologist who has to deal with "emergent properties" that "cannot be predicted from a knowledge of the genetic programs" of human beings (E. O. Wilson 1977, 137). Not long after his 1975 book said that biology would soon cannibalize sociology, Wilson argued the opposite position as if it has been his stance all along. "Biology is the key to human nature, and social scientists cannot afford to ignore its emerging principles. But the social sciences are potentially far richer in content. Eventually they will absorb the relevant ideas of biology and go on to beggar them by comparison" (E. O. Wilson 1977, 138). In another essay written around the same time he admitted that "sociobiology might at best explain a tiny fraction of human social behavior in a novel manner" and he looked forward to the "true, creative debate" that had just begun over the applicability of the theory (E. O. Wilson 1978b, xiii-xiv) .

In short, when the sociobiology debate was just beginning, Wilson endorsed a notion of argument as reciprocal, generative suasion and he showed his commitment to the give and take of civil debate by demonstrating conceptual flexibility, re-envisioning sociobiology as one approach to the study of social behavior that shows great promise, but is not all encompassing.

Wilson's first response to criticism was the sort of rhetorical negotiation that we see modeled in Cicero's *De Oratore.* Rather than write off his opponents and assume that scientific revolution is a clash of incommensurables that can be resolved only by converting the uninitiated on the basis of personal and inarticulate aesthetic considerations, he operated under the wisdom of Crassus, altering his position to accept the criticism of his opponents while at the same time incorporating their claims into his own. His vision of disciplines and antidisciplines existing in creative tension followed a Ciceronian model of pro/con academic argumentation as a positive, creative force. But Wilson's tune soon changed.

The Clash of Incommensurables in Wilson's Response

Unfortunately, Wilson did not maintain the rhetorically sensitive response for long. In fact, even in the beginning, when Wilson was demonstrating conceptual flexibility and figuring debate as a creative process between friendly colleagues, there were also signs that he envisioned debate as a vicious fight between foes. For example, in the same article where he described his position as a moderate one and consonant with that of his opponents, he characterized critiques of his theory as "attacks" of "vigilantism" (E. O. Wilson 1976, 183). But, he warned, he would not be the victim for long; the battle would be engaged on both sides. He threatened those who ignored his theory with the prospect of human geneticists "coming up on their blind side" (E. O. Wilson 1978a, 7). The violence of these metaphors suggests a perspective toward controversy that perceives a battle between enemies rather than a conversation between colleagues. It is the sort of perspective that envisions theory change in science as a bloody revolution, and the two sides of a controversy as incommensurable.

By the early 1980s, these themes had hardened and expanded in his writings, strongly shaping what has become his consistent subsequent stance; Wilson began re-interpreting his arguments in *Sociobiology,* describing them now in extreme terms, exactly as his critics had when they first read the book, identifying sociobiology as the biological explanation of "all forms of social behavior." Unlike his conciliatory 1976 claim that his book was closer to the environmentalist than to the genetic pole when it came to explaining human behavior, he now said that *Sociobiology* had "suggested that much of human behavior is under the control of genes." "I argued [in my book]," Wilson said, " that the core of human nature, including not only the ability to create language and culture but also our drives and the predisposition to learn certain ideas, is prescribed by genes that were assembled in stepwise fashion over millions of years" (E. O. Wilson, in Berman 1985, 210–11; see also his Forward to Barash 1980, xi). By the 1990s, Wilson was embracing reductionism as the "unchallenged linchpin of the natural sciences" and "precisely my view of how the world works" (E. O. Wilson 1994, 345–346). In fact, he went so far as to argue that "all tangible phenomena, from the birth of stars to the workings of social institutions, are based on material processes that are ultimately reducible, however long and tortuous the sequences, to the laws of physics" (E. O. Wilson 1998, 266). The metaphors of violence that existed in tension with metaphors of creative play in Wilson's early commentary on the controversy in the 1970s became even more pronounced, drowning out the softer metaphors and resulting in the fully imperialistic stance he took toward the social sciences in *Consilience* (E. O. Wilson 1998; see Ceccarelli 2001a, 129–133).

In short, the flexibility of his early response hardened into an extreme reductionism, and unsurprisingly, it has been received with hostility by almost all who have responded to it (Ceccarelli 2001a, 142–45). It is as if Crassus had followed his careful weaving of argument and counterargument with a bold statement that his initial concessions were a mistake or a lie, and that the orator should indeed have possession of all domains of knowledge.

In the long term, Wilson was unable to conduct himself as a judicious rhetor; the conceptual flexibility with which he initially reinterpreted his reductionist posture, to improve the acceptability of sociobiology as a program of research, was abandoned soon after he adopted it. Although he began with the more moderate and rhetorically sensitive response that served Darwin so well, Wilson's subsequent reaction was like Gall's, the entrenchment of his most controversial claims. So why did he abandon the art of persuasion and instead adopt such a rigid and hostile tone? I believe Wilson adopted this approach because he was influenced by the perspective toward scientific debate that envisions theory change as a zero-sum game, a Kuhnian philosophy of revolutionary change stemming from the clash of incommensurable positions, in which compromise is weakness and reasoned arguments are hopeless. If the winners are the ones who inspire the most young people to join their side of the battle or who happen to convert the most people on the basis of personal and inarticulate aesthetic considerations, not the ones who are most successful at persuading their peers that their theory is reasonable, then the sensible thing for someone like Wilson to do is to attack and belittle his opponents while making his own position appear most desirable for the new generation of scientists. The give and take of pro/con argumentation is not an optimal approach when the goal is inspiring the uninformed; instead, the more effective strategy is to emphasize the revolutionary nature of one's claims, while characterizing opponents as fearful and dogmatic. If, on the other hand, the resolution of a scientific revolution is more likely to take the form of a negotiated settlement from a civil academic debate, then Wilson's approach was misguided.

Of course, Wilson was not the only one to adopt a hostile attitude in this scientific debate. Many of his critics used the same tactics, envisioning this not as a civil debate between equals that called for negotiation, incorporation, the weaving of point and counterpoint, but as a conflagration that needed to be swiftly and decisively doused. Perhaps he just ran out of patience with his critics and adopted their methods.[7] But Darwin managed to hold his public cool in the face of intemperate criticism. Moreover, because Wilson was the one who began with the startling claim to disrupt the status quo, initiating the debate, the burden of proof fell on him, not just for the evidence but for the tone. If the debate was to proceed civilly it was up to him to adopt the role of Crassus, moderating his claim in response to the vigorous criticism of his opponents. It is a commonplace in rhetorical theory that presumption and burden of proof reflect inertia. Presumption resides with the status quo; burden of proof falls on the rhetors attempting to disrupt it (Whately 1963, 112–115). The fact that the controversy continues suggests that Wilson has not met that burden to everyone's satisfaction. But perhaps there is a burden of concord that should fall in such debates as well. Once a new claim or theory has broken the ice, as Darwin's *Origin* and Gall's *Anatomie et physiologie* broke the ice, perhaps it falls (or should fall) on the ice-breaker to begin the job of integration and accommodation. After brief indications that he might rise to the challenge, Wilson has failed the burden of concord.

That Wilson only pursued concord transiently, and in combination with a more concerted, hostile approach toward the opposition, helps to explain why Wilsonian sociobiology has not been accepted by the scientific community as a viable new paradigm, and why it continues to lose popularity as time passes. If successful scientific debate is most often characterized by civil argumentation leading to the modification of each side, rather than a bloody battle to the death over incommensurables, then Wilson's first response of adaptation might have been more effective had he applied it consistently and maintained it in the long-term, rather than resorting to a strategy of hostility and the entrenchment of his most controversial views.

Conclusions

In critiquing the attitude that views scientific debate as a battle between incommensurable positions, I am not suggesting that scientists could or should stop quarrelling with each other, abandoning the rough and tumble of debate in order to engage in friendly discussion. Adopting an adversarial perspective is not the problem. The problem arises when one engages in debate with the presupposition that victory cannot be achieved through argumentation but only through an epiphany or the somewhat longer process of waiting for your opponents to die (or speeding up that process by metaphorically killing them off with *ad hominem* attacks). The view of debate that envisions a clash of incommensurables leading to the triumph of a new theory excludes another view of debate; namely, a Ciceronian perspective toward academic debate where contrary positions are effectively integrated in a contentious but civil controversy.

Incommensurability may be for Wilson what Jeanne Fahnestock calls "an active rhetorical investment" (this volume, 290), one which has prevented him from maintaining the conceptual flexibility he needed to achieve social cohesion around his version of sociobiology. At the beginning of a controversy, strategies of dissent and assent must be carefully modulated if they are to lead to a satisfying resolution for the initiator of a conflict. To draw attention to a new area of research, the startling claim is appropriate. Just as Crassus began with the proclamation that all knowledge is the domain of the orator, so too did Wilson begin the sociobiology debate with a broad claim about its scope, a claim calculated to spark controversy. But as Randy Allen Harris puts it, the rhetoric of dissent "is highly unstable and either rapidly evolves into arguments which seek broader adherence, or it just dies out" (R. A. Harris 1990, 34). The prudent approach for Wilson would have been to adopt and maintain a flexible position toward reductionism and the scope of his theory once the debate was enjoined, seeking the adherence of his opponents by scrutinizing and modifying (but not abandoning) his original position. Instead, after a brief foray into the realm of the judicious rhetor, he chose to abandon that attitude, posturing for his allies and the uninitiated by restating and even strengthening his most controversial claims.

As long as both sides of the sociobiology debate continue to think about their positions as incommensurable, they will continue to speak past one another and there will be no consensus around a mutually acceptable position. But this state of affairs is not necessary. Scholars who have closely examined the controversy point out that logically, the two positions are much closer than they make themselves out to be (Segerstråle 2000, 296, 297, 299) and rather than subscribing to incompatible theories, they each hold a half-truth that can and should be combined with the other to create a more complete picture (Midgley 1980, 21). But neither side tries to communicate with the other; instead, they happily continue to emphasize their differences, playing out the debate as a zero-sum game while drawing public authority and moral capital from their critiques of each other (Myers 1990, 243–46; Segerstråle 2000, 301).[8] In short, they keep the battle lines drawn and act as if they are incommensurable, even though they need not see themselves in this way. As a result, most scientists working in proximal or overlapping domains have responded with the equivalent of "a pox on both your houses!" Not wanting to be associated with the bitter rivalry, they continue to do their research while carefully avoiding the name "sociobiology." Just as scientists continued to study the ways in which specific parts of the brain correspond to mental traits without calling it phrenology, so too will scientists continue to study the ways in which social behaviors are influenced by the evolution of genes without calling it sociobiology. But Wilson will have lost his chance to christen the area of research, attach his own name to it as its founding father, or positively influence its future development. And more importantly, both sides of the controversy will have wasted time and resources in ineffectual invective, rather than investing their energies toward a civil debate that could lead to the productive development of their viewpoints and an eventual resolution of a revolution.

Of course, I will be the first to admit that there is a problem with too much conceptual flexibility; it can result in a theory that is so watered down that it is no longer original or worthy of attention. Some who have seen or imagined a more moderate position for Wilson have suggested that such a position is "banal" or fails to remain "noteworthy" (Todorov 1998, 33; Lyne and Howe 1990, 145). This is certainly a danger that a scientific rhetor faces when modifying a claim to respond to the complaint that an original position is too extreme. Positioning a scientific claim as both original and as compatible with standards of reasonableness shared by the rest of the scientific community is a persistent problem that scientific rhetors face (Myers 1990, 67).[9] But it is, in part, their ability to adapt to these conflicting constraints that distinguishes successful scientists like Darwin from unsuccessful scientists like Gall, and, I am arguing, Wilson.

The perilous image of two incommensurable sides competing to win in a battle gives scientists like Wilson little encouragement to treat debate as a creative process that judiciously weaves aspects of both the pro and con positions into a stronger collective cloth. Without the conceptual flexibility that allows a Ciceronian civil debate to end in a broader perspective that combines apparently incommensurable sides, the social cohesion and terminological acceptance that could one day result in scientists agreeing to adopt a Wilsonian sociobiology is unlikely to occur. This, I believe, is at least partially the result of Wilson's acceptance of the perspective toward argument popularized by the incommensurability thesis, a perspective toward argument that does not sit well with scholars who are familiar with the richness of the rhetorical tradition.

Notes

[1] For readings of Cicero's *De Oratore* that explain this tactic of incorporation well, see Thomas O. Sloane (1997, 41–46), and Michael Mendelson (1997).

[2] For more on Cicero's antithetical method in other texts he wrote, see Nola J. Heidlebaugh (2001).

[3] Of course, Darwin was also careful in setting out his original argument, showing restraint by offering a much more moderate position than he had presented to himself in his notebooks. For example, see John Angus Campbell (1987, 77; reprinted in R. A. Harris 1997, 8f). See also Campbell's essay in this volume, for further attention to Darwin's rhetoric.

[4] I have suggested that one explanation for Wilson's rhetorical choices is that he was operating from an implicit theory of persuasion consonant with his sociobiological theory. See Ceccarelli (2001a, 137–139). The argument I am making here is compatible with that earlier suggestion, but tries to make it more satisfying by connecting Wilson's implicit theory of persuasion to a philosophy of science that believes paradigm change is the result of a pitched battle between incommensurables. I do not know if the connection between Wilson and Kuhn is a causal one (for instance, that Wilson read *Structure* and was influenced by it), or if both were influenced by a larger cultural attitude about the futility of debate.

[5] In a later essay, Hull gives another example of how a scientific theory as originally introduced differed greatly from the theory that was ultimately accepted: "within five years after the rediscovery of the principles of Mendelian genetics, all but one of these principles had been abandoned or modified extensively. The exception, the law of segregation, has been modified since. Yet Mendelian genetics remains 'Mendelian genetics' in the face of all this change. Similar observations hold for every other scientific paradigm I have ever studied" (Hull 1980, 86).

[6] Like the term *rhetoric*, *finagle* has a negative popular connotation, implying deceitfulness. But one can attach a more positive connotation to the word as well. To finagle is "to achieve by indirect or involved means," so

"judicious finagling" is to exercise sound judgment in achieving something by indirect or involved means. This is similar to the concept of rhetoric that I favor in this paper.

[7] Wilson's critics demonstrated an extreme measure of incivility when they dumped a pitcher of water on his head at an academic debate. For a thorough critique of the critics, see Segerstråle (2000).

[8] A recent entry into this debate, addressed to allies and the uninitiated, and designed to gain public authority and moral capital for one side is Alcock (2001). Alcock sees criticism coming from outside the field of sociobiology as costly, not beneficial, and looks hopefully to a future in which critics of the adaptationist program are no longer heard and we can truly celebrate a day of "triumph for sociobiology."

[9] Lawrence J. Prelli makes this point when he describes conflicting scientific *topoi*, like "significant anomaly" and "external consistency." See Prelli (1989a, 196–198, 200–201). Michael Leff argues that the tension between individual inventiveness and solidarity with the audience is a perennial one that faces all rhetors and requires a "special, delicate, and contextually sensitive relationship between self and community." See Leff (2003, 145).

7

Stasis and the Problem of Incommensurate Communication: The Case of Spousal Violence Research

Lawrence J. Prelli

Introduction

In the introduction to this volume, Harris argues that incommensurability is not relevant to linguistic disputes, that it is not extensible from its mathematical homeground, and that the sorts of disputes that have been traditionally labeled *incommensurable* in the history, sociology, and philosophy of science are, in principle, remediable.* This essay continues and particularizes those arguments: specifically, that "incommensurable" problems can be remedied when incommensurability is understood as a rhetorical problem of practical communication rather than as a mathematical or logical problem of formal translation. To avoid that confusion, I modify the terms of analysis to address the problem of "incommensurate" communication. Incommensurate

---

*The author appreciates Christopher Hutton's and Lara Varpio's insights on a draft of this chapter and, of course, the able editorial guidance of Randy Allen Harris.

communication is the failure of discussants to address the same situated ambiguities so that they, in effect, argue at cross purposes. Restoring commensurate communication requires that discussants address the same kinds of questions, but without presupposing common standards for adjudicating preferred answers.[1]

I argue that the systematic, guided questioning of classical *stasis* doctrine can help frame situated ambiguities as potentially sharable questions from which discussants can "bridge" or "re-align" or "calibrate" their perspectives and, thereby, achieve commensurate communication about similarities and differences. I articulate a procedure, founded on classical precepts, that is designed to address obstacles to communication about science and whose constituent questions correspond with the very sources of incommensurate communication that commentators on "incommensurability" have identified. The procedure contains sixteen kinds of questions that can function as possible points for realigning communication which, otherwise, would remain at cross purposes.

I then examine problems of incommensurate communication manifested during a controversy concerning survey evidence about women-on-men violence within American couples, applying the *stasis* procedure to map those problems and specific efforts to mitigate them. Dr. Murray Straus, the researcher whose work generated the controversial data, intervened in the rhetoric of the controversy to render communication between the quite radically divided sides commensurate. Straus's efforts are suggestive of opportunities and limitations for transforming a polarized controversy marked by incommensurate communication into commensurate communication about perspectival similarities and differences. In effect, Straus applied what I call *commensurability points* to sort out alternative perspectives as offering different but compatible rather than contradictory answers to situated questions. His efforts culminated with the invention of an argument about the interdependence of the otherwise divided parties on practical problems of mutual concern.

Incommensurability Reconsidered

Incommensurability is rooted in the absence of uniform, shared and neutral standards for making meaningful comparisons and evaluations of competing claims, the lack of a common measure. Feyerabend

(1962, 58–60) and Kuhn (1996, 125–130, 149) argued that lack of neutral vocabulary rendered theories incommensurable since their constituent terms presupposed meanings and references that were not transferable from one theory to another. Both (Kuhn 1996, 103, 109; Feyerabend 1975) also amplified the absence of shared, neutral, rule-governed methods for testing and evaluating claims as a source of incommensurability. Competing theories and paradigms presupposed their own distinct methodological strictures and criteria. Kuhn (1977; 1999, 199–200) also claimed that there are no super-ordinate, uniformly applicable criteria for rendering theory choice neutral or objective. Distinctive value preferences for attributing significance were involved. These often overlapping positions are the well known "semantic" thesis, the "methodological" thesis, and what we might call the "values" thesis. All three theses can be invoked to suggest that the absence of uniform, shared, neutral standards makes different perspectives formally incomparable and, thus, incommensurable.[2]

I contend that incommensurability becomes an insoluble problem when the concept's mathematical meaning is applied to sources of linguistic, methodological, and valuational conflict or confusion. "Incommensurability" originates in mathematics from the strict incomparability of quantities or measures derived from different standards, rules, or algorithms. Commensurability, in contrast, presupposes that comparable quantities are derived from the same procedures. For example, if the same algorithms are used to measure and calculate chlorophyll data collected by a satellite sensor, the resulting quantities are commensurable and thus strictly comparable. We can, for instance, note that concentrations are higher for days in June than for days in May. But suppose algorithms used for calculating concentrations for June and for May combined different light intervals? The resulting quantities would be incommensurable and, thus, incomparable. June's chlorophyll would not then be the *same* as May's chlorophyll, so we could not with confidence use those data to determine which month had higher concentrations since they were not measured with the same standards.

Comparison of different perceptual, linguistic, valuational, or methodological perspectives is not the same as comparing quantities calculated from different algorithms.[3] Practical opportunities are always available for situated comparison of even the most radically different points of view, once we set aside the presumption that formal sameness is required for meaningful communication. As Harris outlines in the introduction to this volume (26), Hilary Putnam (1981b) challenged Feyerabend's strong version of the semantic variability thesis for this and other reasons. According to Putnam, we can compare and understand meanings across *different* perspectives because "interpretive success does not require that the translatees' beliefs come out the *same* as our own, but it does require that they come out intelligible to us" (Putnam 1981b, 109). Intelligibility across perspectives occurs because we still share a wealth of possible "commensurabilities" as human beings facing similarly situated problems (118–119).

Putnam's criticism points to an important difference between algorithmic and perspectival applications of incommensurability. Quantities require the same algorithms for "commensurable" comparisons. But perceptions, terms, values, and methods do not require the same theories, paradigms, frameworks, cultures, or other perspectives for intelligible comparisons. If they did and two cultures truly were incommensurable we could not translate another culture's language in ways that made it intelligible to us. Yet we do this all of the time. Putnam contended that proponents of the incommensurability thesis make such translations themselves and, thereby, refute their own position:

> If this thesis were really true then we could not translate other languages—or even past stages of our own languages—at all. And if we cannot interpret organisms' noises at all, then we have no grounds for regarding them as *thinkers, speakers,* or even *persons.* In short, if Feyerabend (and Kuhn at his most incommensurable) were right, then members of other cultures, including seventeenth-century scientists, would be conceptualizable by us only as animals producing responses to stimuli (including noises that curiously resemble English or Italian). To tell us that Galileo had "incommensurable" notions *and then go on to describe them at length* is totally incoherent. (Putnam 1981b, 114-115)[4]

Incommensurability, in short, doesn't hold outside of mathematics. Further, situations fitting the descriptions of incommensurability are amenable to remedy when understood as rhetorical problems of practical communication rather than as logical problems of formal translation. Barnes and Mackenzie argued for this reframing of the "incommensurability" problem nearly twenty-five years ago:

> Historical study of incommensurability should not address itself to formal problems of translation, of what can properly and "rationally" be rendered as what, of what meanings in two systems are "logically" identical. What should be asked is how far incommensurability presents a *practical* barrier to communication, how far discourse and effective communication is actually established between exponents of different paradigms, and what in fact are the characteristics of such discourse. Given their apparent historical significance, what are the properties of "imperfect communications" and "persuasive argumentation" as phenomena in their own right? (1979, 50)

Kuhn (1996, 198–204) himself tried to tame incommensurability's irrational implications by treating it as a problem of communication rather than of logic. According to Kuhn, participants could mitigate "incommensurability" by acting as members of different language communities who could better understand each other by attending to recurrent sources of ambiguity that divide, confuse, or otherwise frustrate communication. Those mitigating efforts required a shift away from abstract and formal problems and toward situated and substantive obstacles to effective inter-paradigmatic communication. Kuhn (1996, 202) even articulated a rudimentary translation procedure for resolving incommensurability through systematic, guided exploration of ambiguities conducive to communication breakdown. Discussants, he suggested, should probe similarities and differences both in how they perceive facts and in the terms they used to define their meanings.

We can minimize confusion of practical communication problems and problems of formal analysis and translation with some minor terminological revision, dropping *incommensurable* in favor of *incommensurate.*[5] This term has the benefit—morphologically speaking—of licensing concord, of leaving the door open to resolution. *Incommensurate* is used when things are askew: when someone's salary is incommensurate with her responsibilities, when a treatment is incommensurate with an illness, an appeal with its audience. When X and Y are incommensurable, there is no recourse. When they are incommensurate, something can be done. The salary can be raised, the treatment modified, the appeal reworked. In the first case, X and Y are immutably at odds; in the second, there are remedies.

Incommensurate communication does not result from the presence of different perspectives *per se*, but because those who hold them are at cross purposes about situated problems and ambiguities. They find that their communicative efforts are somehow askew. But that problem can be remedied. Discussants need to realign their communication so that they address the same situated ambiguities and problems from the distinctive vantages afforded by their perspectives. Remedies for incommensurate communication, thus, presuppose the presence of perspectival differences rather than presume that those differences constitute problems to be avoided through recourse to some overarching, neutral standard.

Incommensurate communication can concern perceptions, data, conceptions, values, or methods, but commensurate communication always can be restored provided that discussants work to find and address the same kinds of situated questions. Addressing the same kinds of questions does not presuppose common standards for adjudicating preferred answers. The task is to clarify situated ambiguities that are at the source of communicative confusion and conflict. When clarified as types of questions, the *questions* can then serve as stable, neutral, and shareable points of reference from which participants can take "proportionate" measure of perspectival differences and similarities in their answers. In the next section, I show how the systematic, guided questioning of classical *stasis* doctrine can help frame situated ambiguities as shared questions from which discussants can "bridge" or "re-align" their perspectives and, thereby, achieve commensurate communication about similarities and differences.

Stasis Doctrine and the Problem of Incommensurate Communication

Classical *stasis* doctrine provides a method of guided questioning that is well suited to probing similarities and differences among perspectives. *Stasis* doctrine presumes that whenever anything is discussible or debatable, there are predictable kinds of questions that are implicitly or explicitly at stake in ongoing, situated communication. The Greek term *stasis,* the Latin term *status,* and the English term *issue* all refer to that phenomenon (Dieter 1950). As expressed in classical dialectical and rhetorical theory, philosophical discussion was likely to center on one or more of the following general kinds of question:

i. problems of fact ("*Is* it?")

ii. problems of definition ("*What* is it?")

iii. problems of nature or quality ("Of what *sort* is it?")

iv. problems of action ("What *action (if any) is appropriate* in the given case?").

Though differing on details, ancients and moderns generally agree that these or similar questions are at the source of deliberation, discussion, and debate (Prelli 1989a, chapter 4).[6]

The Roman rhetorician, statesman, and advocate, Cicero, believed the general procedure of guided questioning was universally applicable to any subject susceptible to controversy. In *Topica,* he explained that the general *stasis* questions—"Is it?" "What is it?" "Of what sort is it?" and "Is action required?"—constituted methods of invention and of judgment in philosophical and other disputes (Cicero 1949b, 21.81–23.90). As a method of judgment, the four questions fixed attention on the kinds of ambiguities or problems being addressed and decided.[7] As a method of discovery, the four questions helped differentiate and inventory the range of alternative, possible arguments and opinions offered as answers to those kinds of questions. Thus, the general question of *fact* ("Is it?") allows discrimination of alternative opinions about existence and nonexistence. The question of *definition* or *name* ("What is it?") brings into view alternative grammars for assessing the meaning of the purportedly existent and enables discrimination of semantic variation from genuine disagreement. The question of *kind* ("Of what sort is it?") reveals additional predicates and propositions that qualify the significance and value of what exists and is definable. And the question of *action* raises the practical issue of how ideas can guide public and private actions once the reality, its definition, and its kind have been determined (Buckley 1970, 150–154).

Roman rhetoricians adapted the general *stasis* questions to the special circumstances of legal pleading. Cicero stressed that *stases* occurred at the points of clash between conflicting pleas. In *De Inventione,* he explained that all legal controversy involves dispute over questions of fact (conjecture), definition, kind (quality), and procedure. His discussion of *legal stases* points to the general usefulness of guided inquiry into recurrent kinds of situated ambiguity as a method of judgment and of discovery:

> Every subject which contains in itself a controversy to be resolved by speech and debate involves a question about a fact, or about a definition, or about the nature of an act, or about legal processes. [...] The "issue" is the first conflict of pleas which arises from the defence or answer to our accusation, in this way: "You did it"; "I did not do it," or "I was justified in doing it." When the dispute is about a fact, the issue is said to be conjectural (*coniecturalis*), because the plea is supported by conjectures or inferences. When the issue is about definition, it is called the definitional issue, because the force of the term must be defined in words. When, however, the nature of the act is examined, the issue is said to be qualitative, because the controversy concerns the value of the act and its class or quality. (1949, 1.8.10)[8]

In addition to the conjectural, definitional, and qualitative issues "there is a controversy when the question arises as to who ought to bring the action or against whom, or in what manner or before what court or under what law or at what time, and in general when there is some argument about changing or invalidating the form of procedure" (1.11.16). That sort of controversy is designated the *translative* issue of procedure.

For Cicero, then, the four special *stasis* questions of legal pleading constituted a procedure that could effectively guide the processes of discovering arguments and adjudicating points at issue. But a *stasis* procedure's heuristic effectiveness depends on its strengths as a category system. Overlapping categories are seldom clarifying categories. The four *stases* are clarifying categories because they are individually discrete in their respective scopes of application. Moreover, non-exhaustive categories are less useful than exhaustive categories. The four *stases* are exhaustive since they collectively encompass the *types* of questions susceptible to controversy. As Cicero put it, "There will always be one of these issues applicable to every kind of case; for where none applies, there can be no controversy. Therefore, it is not fitting to regard it as a case at all" (1.8.10).[9]

Philosopher and historian of ideas Richard McKeon drew from the categorical strengths of *stasis* procedure when formulating his "new rhetoric" adapted to the pluralistic world of the twentieth century (McKeon 1971). Under conditions of pluralism, communication requires mutual understanding of similarities and differences on questions of practical concern (McKeon 1957). McKeon thought *stasis* doctrine was the kind of category system needed to sort out philosophical, cultural, or ideological perspectives brought into communicative contact and controversy over situated, practical problems (McKeon 1971, 53–58). He followed the Romans in believing that *stasis* doctrine identified individually discrete and collectively exhaustive ***kinds of judgments*** that could conceivably be made about controversial matters (1966, 370). The four general questions, then, could be used to examine similarities and differences across perspectives by sorting out the range of alternative answers those perspectives afforded. Discovery of perspectival similarities and differences is a prelude to possible creation of mutually acceptable grounds for collaborative thought and action regarding situated problems under conditions of pluralism.

McKeon not only took philosophical, cultural, and ideological pluralism as facts of twentieth century life, but he viewed those facts as strikingly preferable to monistic alternatives (Hauser and Cushman 1973, 213–218). His views find parallels in criticisms of monistic doctrines of scientific theory and method (Sankey 1997, 149–164; 2000). Both Kuhn and Feyerabend warranted with historical examples the conclusion that science exhibited theoretical, methodological, and valuational pluralism (Sankey 1997, 2001; Feyerabend 1965b 1975; Kuhn 1977; Sankey and Hoyningen-Huene, xiii-xv, 2001). Feyerabend (1965b, 149; 1975, 35–46; 1981a, 104–109, 139–145), in particular, stressed that pluralism was required for science to progress and, accordingly, urged wide proliferation of distinctive theories over and against purportedly dogmatic demands for uniform theory formation. As he summed the general point, "Unanimity of opinion may be fitting for a church, for the frightened or greedy victims of some (ancient or modern) myth, or for the weak or willing followers of some tyrant. Variety of opinion is a feature necessary for objective knowledge" (1975, 46). These commentators on science, very much like McKeon, saw perspectival diversity as a potentially productive resource rather than a problem that had to be surmounted.

Incommensurate communication, I have said, presupposes the presence of pluralism. The task is to render that communication commensurate. McKeon identified the methodological qualities of *stasis* procedure that make it useful for addressing the problem of incommensurate communication under conditions of pluralism.[10] First, *stasis* questions are universal in scope and thus applicable for analysis of any situated controversy regardless of circumstantial and substantive details. Second, *stasis* questions are neutral since they identify the *kinds of ambiguities* susceptible to controversy without presupposing, distorting, or biasing how controversialists should frame or answer them in particular, situated cases. Third, *stasis* questions can operate as "reference points" where controversialists can enter each other's perspective and, thereby, discern their similarities and differences. And fourth, *stasis* questions furnish potential sites for transforming controversy into collaboration. They can function as *loci* for negotiating new grounds in common for addressing particular, situated problems of mutual concern.

Stasis procedure is a promising method for addressing the problem of incommensurate communication for the added reason that the kinds of ambiguities that the procedure identifies correspond with what commentators have identified as the major sources of incommensurability. Kuhn located one kind of ambiguity in *perceptual* differences, turning to gestalt psychology to show how the same phenomenal data could be structured differently during acts of perception (1996, 85). He and Feyerabend pointed in various ways to *semantics* as a source of incommensurable language use across theories. Kuhn and others also directed attention to *value* differences when assigning significance to different theories and paradigms as another source of incommensurability. Kuhn and Feyerabend, among others, pointed to *methodological* differences across paradigms and theories as a source of incommensurability. These four sources of ambiguity, taken together, constitute a web of possible incommensurate relationships whenever alternative perceptions, meanings, values, and methods are brought into communicative contact under conditions of perspectival diversity.[11] These four ambiguities correspond directly with the kinds of questions identified in *stasis* doctrine:

1. *perceptual* problems or ambiguities about *existence* in a phenomenal domain ("Is it?")
2. *semantic* problems or ambiguities about conceptual *meaning* ("What is it?")
3. *value* problems or ambiguities about *significance* ("Of what sort is it?)
4. *methodological* problems or ambiguities about appropriate *action* ("What shall be done?").

The fact that commentators on incommensurability did not intend this correspondence is a rather striking confirmation of the claim that the constituent categories of *stasis* doctrine are universally applicable and exhaustive of the kinds of question susceptible to controversy.

Communicative obstacles encountered when discharging the grand functions of doing science also correspond with these four major sources of ambiguity: adducing evidence, interpreting constructs and theory, evaluating significance, applying methodologies. I call these "places" of intersection between *stasis* doctrine and the recurrent ambiguities of "doing science" the *superior stases* of science communication: (1) *evidential,* (2) *interpretive,* (3) *evaluative,* and (4) *methodological* (Prelli 1989a, 145–147). *Superior stases* identify the *general* sources of ambiguity during communication about science. And, of course, they apply to identify the general sources of incommensurate communication among scientists who adhere to different theories, paradigms, or other perspectives. The *superior stases* can be used to track whether discussants and debaters frame and respond to different kinds of ambiguities and, thus, communicate at cross purposes. Discussants can, however, frame the *same* general kind of problem differently. Even so, options available for such "framings" are finite and identifiable in the form of one or more of four familiar issues: (i) *conjectural,* (ii) *definitional,* (iii) *qualitative,* and (iv) *translative.* Since these *stases* identify alternative ways of framing each of the four superior kinds of ambiguity, I call them *subordinate stases* (Prelli 1989a, 147–158). The relations among the superior and subordinate *stases* are mapped out in Table 1.

Superior Stases

Subordinate Stases	Evidential	Interpretive	Evaluative	Methodological
Conjectural	Is there scientific evidence for claim x?	Is there a scientifically meaningful construct for interpreting the evidence?	Is claim x scientifically significant?	Is procedure x a viable scientific procedure in this case?
Definitive	What does the evidence mean?	What does construct y mean?	What does value z mean?	What does it mean to apply procedure x correctly?
Qualitative	Which empirical applications of the evidence are more warranted?	Which interpretive applications of construct y are more meaningful?	Which evaluative applications of value z are more significant?	Which methodological applications of procedure x are more appropriate?
Translative	Which from among alternative evidence better addresses ambiguities about existence?	Which from among alternative constructs better addresses ambiguities about meaning?	Which from among alternative values better addresses ambiguities about significance?	Which from among alternative procedures better addresses ambiguities about scientific action?

Table 1: Rhetorical Stasis Procedures of Scientific Discourse

Communicative problems might arise along either dimension, even when there is agreement about the nature of the issue along the other dimension. Communication, for instance, can become incommensurate even if discussants address the *same* general ambiguity if they frame that ambiguity with different subordinate *stases*. The potential framings are as follows (mapping the superior *stases* against the subordinate *stases*).

1. *Evidential* problems can be framed as

   i. *conjectural* issues about the existence of evidence for alleging that some phenomenon exists

   ii. *definitional* issues about what existing evidence means
   iii. *qualitative* issues about significant, situated uses and applications of evidence
   iv. *translative* issues about which from among alternative bodies of data and evidence provides the best criteria for resolving evidential problems

2. *Interpretive* problems can be framed as
   i. *conjectural* issues about the existence of meaningful concepts or theories
   ii. *definitional* issues about the meanings of concepts and theories
   iii. *qualitative* issues about significant, situated applications and uses of theories and concepts
   iv. *translative* issues about which from among alternative theories and concepts provides the best criteria for addressing ambiguities about interpretation

3. *Evaluative* ambiguities can be framed as
   i. *conjectural* issues about the existence of relevant values
   ii. *definitional* issues about what values mean
   iii. *qualitative* issues about significant situated uses and applications of values
   iv. *translative* issues about which from among alternative values provides the best criteria for addressing evaluative ambiguities

4. *Methodological* ambiguities can be framed as
   i. *conjectural* issues about whether there is a viable procedure or methodology
   ii. *definitional* issues about how to operationalize a procedure or method
   iii. *qualitative* issues about significant, situated applications and uses of procedures and methods

iv. *translative* issues about which from among alternative procedures or methods provides the best standards for addressing problems of methodology

Communication can also become incommensurate if discussants use the same subordinate stases to address different kinds of situated ambiguities, but lack agreement about the nature of the general issue. The potentials here, if we map the subordinate stases back against the superior stases, are as follows.

i. *Conjectural* issues locate points for decision in questions about the existence of

1. data or evidence (the general ambiguity is *evidential*)
2. meaningful constructs or conceptions (the general ambiguity is *interpretive*)
3. significant values (the general ambiguity is *evaluative*)
4. viable methods or procedures (the general ambiguity is *methodological*)

ii. *Definitional* issues locate points for decision in questions about the *meaning* of

1. data or evidence (the general ambiguity is *evidential*),
2. constructs or concepts (the general ambiguity is *interpretive*)
3. values (the general ambiguity is *evaluative*)
4. "operationalized" procedures (the general ambiguity is *methodological*)

iii. *Qualitative* issues can focus the points for decision in questions about the *significance* of situated uses or applications of

1. data or evidence (*evidential*)
2. concepts or constructs (*interpretive*)
3. values (*evaluation*)
4. methods or procedures (*methodological*)

iv. *Translative* issues locate points for decision in questions about the relative *appropriateness* of alternative possible criteria, standards or "grounds" for acting upon problems of

1. existence (*evidential*)
2. meaning (*interpretive*)
3. significance (*evaluative*)
4. action (*methodological*)

Finally, communication can become incommensurate if discussants disagree about *both* superior and subordinate *stases*. Altogether, the *stasis* procedures outlined in Table 1 can track 120 possible ways in which two discussants can address different kinds of questions and, thus, find themselves at cross purposes. Without genuine *stasis*—the authentic clash of opposing positions on the same question—there can be no meeting of minds about perspectival similarities and differences. Incommensurate communication can then occur due to unacknowledged perspectival similarities and differences in perceptions or observations, theories or concepts, values, methods, or even comprehensive worldviews. If discussants address different *stases*—whether subordinate, superior, or both—communication becomes partial or "breaks down." Since communication is not "about" the same questions, participants are at cross purposes when discussing, arguing, or otherwise adducing their claims.

The sixteen subordinate *stases* are possible places for realigning and bridging perspectives so that genuine *stasis* is achieved and communication becomes commensurate. Indeed, commensurate communication across perspectives is potentially achievable at any one of those sixteen points. The specific ambiguity—the *locus* identified by the *stasis* category—becomes the common reference point from which to "measure" perspectival similarities and differences. Incommensurate communication, then, can be remedied by identifying the *kinds of questions* giving rise to perspectival conflicts and confusions and by tracking how discussants respond to those questions, but without presupposing shared, neutral standards for arriving at the "right" kinds of answers. I call these commonplaces for enabling commensurable communication *commensurability points* since they designate places where those holding different perspectives can compare similarities and differences from a common vantage[12].

The rhetorical *stasis* procedures of scientific discourse thus serve both diagnostic and prescriptive functions: discussants and debaters can (1) discover how they frame and respond to different situated questions and, thus, how they come to argue at cross purposes; and (2) find "places" for realigning or calibrating their communication so that they frame and respond to the same situated questions. The second, prescriptive function requires discovery of commensurability points for restoring commensurate communication between frameworks under conditions of perspectival diversity. The next section examines a controversy over evidence about women-on-men violence as incommensurate communication among alternatively argued perspectives.

Framing Evidential Issues: the Women-on-Men Violence Anomaly

For nearly twenty-five years, Dr. Murray Straus and his colleagues at the University of New Hampshire's Family Research Laboratory (FRL) have debated critics concerning the facts of violence in the family and the best methods used to acquire those facts. Straus, a pioneer in using quantitative survey methods to explore problems of domestic violence, developed an instrument called the Conflict Tactics Scales (CTS). The scales incorporate the premise from social conflict theory that social groups, such as the family, are beset with conflicts. The CTS measure the extent to which family members—spouse with spouse, parent with child, sibling with sibling—attempt to settle conflicts using tactics of reasoning, verbal aggression, and physical violence (Straus 1979; Straus, Gelles, and Steinmetz 1980, 253–266; Grotevant and Carlson 1989, 293–298).

The physical violence scale operationalizes violence as an "act carried out with the intention, or perceived intention, of causing physical pain or injury to another person" (Straus, Gelles, and Steinmetz 1980, 20). The scale consists of nine such acts, distinguished by the relative degree to which they entail risk of physical injury. "Minor" assaults are throwing objects; pushing, grabbing, and shoving; and slapping. "Severe" or "abusive" assaults, which possess "high potential for injuring the person being [assaulted]" (22), are kicking, biting, or hitting with one's fist; hitting or attempting to hit with an object; beating up the other person; choking; threatening with a gun or knife; and using a gun or knife. Indices for measuring "wife beating" and "husband beating" focus on "severe" violence data and indices for measuring "normal" spousal violence center on "minor" assaults data. The indices are combined to yield incidence rates for all spousal assaults.

Straus, Gelles, and Steinmetz applied the CTS in the first of two National Family Violence surveys that sampled 2,143 couples (married or cohabiting). The three researchers' reportage of the survey's data framed and addressed three kinds of subordinate *stases* about evidential ambiguities. First, they framed and addressed the *conjectural* issue by contending that the survey data were unique since they were based on the first systematic investigation of violence in the American family using a representative probability sample. The newly existent data thus promised to fill important gaps in factual knowledge about the empirical phenomenon of violence in the family:

> Knowledge concerning how much violence exists, what kinds of families are violent, and what causes violence has been difficult to come by since most of the research to date is based on small, unrepresentative samples from limited geographic areas. Consequently, even the question "What proportion of American families are violent?" could not be answered with any degree of certainty. (Straus, Gelles, and Steinmetz 1980, 4–5)

The facts were surprising. The "wife beating" indices yielded the statistic that 3.8 percent of couples experienced one or more acts of wife beating during the twelve months prior to the interview or, as typically rendered, that 1.8 million women are beaten by their husbands annually (Straus, Gelles, and Steinmetz 1980, 40). When data from the "severe" violence indices are combined with those concerning "minor" violence, 12.1 percent of the men surveyed admitted to using at least one violent tactic against their spouses (36).

The "husband beating" indices, however, yielded even more surprising and highly controversial data about women-on-men violence in couples. The data showed that 4.6 percent of couples experienced one or more acts of husband beating during the year of the interview. Put otherwise, more than two million men were beaten annually by their wives (Straus, Gelles, and Steinmetz 1980, 41)! When combined with the "minor" violence indices, 11.6 percent of couples experienced at least one act of woman-on-man violence during the year of the survey (36).

The three researchers' reportage then framed and addressed the *definitional* issue about what these data meant by identifying the symmetrical pattern in men-on-women and women-on-men violence at both "severe" and "normal" levels. Women use violence to settle conflicts with men as much as men do to settle conflicts with women. The marriage license, as the researchers put it, is something of a "hitting license" (Straus, Gelles, and Steinmetz 1980, 31)—and for both parties. (In fact, the act of cohabitation generally seems to grant such a license.) This symmetrical pattern conflicted with psychological and cultural accounts of violence in the family that portrayed women as peaceful victims of violent men; instead, the pattern suggested that the problem was not *violence against women* but the mutuality of *spousal* violence (36).

Straus and associates framed and addressed a third kind of evidential ambiguity in the form of *qualitative* issues. They adduced a range of potentially mitigating variables that should qualify situated applications or uses of the symmetrical violence pattern (Straus, Gelles, and Steinmetz 1980, 43–44, 246n6). Among those variables: (1) men's violence is likely to be more injurious than women's due to their greater average physical strength; (2) women's violence might be used in self-defense more than men's violence; (3) women are more "locked into marriage" than are men, since they are more likely to lack economic and other resources to escape violent situations; and (4) men's reportage might exaggerate women's violence and minimize men's violence. Based on these and other qualifying considerations, Straus and his colleagues urged that public policy should assign "first priority to aiding wives who are the victims of beatings by their husbands" (Straus, Gelles, and Steinmetz 1980, 44).[13] Still, they maintained that the surprising amount of women-on-men violence could not be ignored either when theorizing about the causes of wife abuse or when considering the range of options available for intervention in abusive situations:

> The violence *by* wives uncovered in this study suggests that a fundamental solution to the problem of wife-beating has to go beyond a concern with how to control assaulting husbands. It seems that violence is built into the very structure of society and the family system itself. [. . .] Wife-beating is related to other aspects of violence in the family. It is only one aspect of the general pattern of violence, which includes parent-child violence, child-to-child violence, and wife-

> to-husband violence. To eliminate the particularly brutal form of violence known as wife-beating will require changes in the cultural norms and in the organization of the family and society which underlie the system of violence on which so much of American society is based. (Straus, Gelles, and Steinmetz 1980, 44)

According to this systemic view of the wife abuse problem as part of a cycle of violence within the family, remedies must involve ways of breaking the cycle.[14] Thus, Straus and his colleagues included among their proposals the controversial clinical recommendation that women in shelters be advised against hitting back lest they perpetuate the very cycle that has culminated in their misery.

These qualifications did little to minimize the controversy that erupted over the claim that man-on-woman and woman-on-man violence rates were nearly equal in the American family. Some critics raised evidential ambiguities about the survey data by turning to studies of shelters, hospital and treatment centers, and police and court records. Those studies showed that severe violence was radically asymmetrical along gendered lines in American couples. Specifically, studies showed that women (1) seldom hit their partners while themselves being victims of repeated physical assaults (Dobash and Dobash 1979, 14–20; Pagelow 1981); (2) used violence almost exclusively in self-defense (Saunders 1986, 1988; Schechter 1982); and (3) suffered more injuries than their abusive partners (Berk, Berk, Loseke, and Rauma 1983). Critics used these studies to warrant the claim that the survey data *had* to be wrong.

The survey data about women-on-men violence clashed with received frameworks for interpreting problems of violence in the American family. That clash was magnified when family violence researchers called evidence of severe women's violence "Husband Beating" (Straus, Gelles, and Steinmetz 1980, 40–41). One researcher (Steinmetz 1978a) used severe violence data to warrant the existence of a "battered husband syndrome," a purportedly unreported phenomenon due to cultural presumptions against the very idea that women could batter men with the severity associated with men's battery of women. That expression capitalized on widely circulated claims about the existence of a "battered woman syndrome" (Walker 1983; also see Walker 1988; 2000).

For cultural, clinical, and theoretical reasons, the women-on-men violence data were seen as strikingly anomalous. Common sense, for instance, associates severe battering with women victims of violent men, rather than with male victims of violent women; it is difficult to imagine that the all-too-familiar images of brutalized women appearing in newspapers and news magazines have close but largely hidden parallels with large numbers of brutalized men. Advocates for battered women, health care providers, criminologists, and others who work with abused women, perceived "husband beating" or "battering" and the associations those terms evoked as wildly anomalous given their own experiences. Researchers, advocates, and clinicians who used feminist theory (for examples, Dobash and Dobash 1979; Brienes and Gordon 1983; Schechter 1982) to inform their work rejected the family violence researchers' terms for describing violence data—whether the acts were "severe" or "minor"—because theoretical expectations inclined them to assume that violence in the American family is predominantly "violence against women." These criticisms extended to gender neutral expressions such as "spousal violence," as well as to gender neutral uses of terms such as "beating" or "battering" which, from a feminist perspective, only mask the reality of gendered power imbalances that pervade patriarchal society's social structures and institutions, including the institution of the family (Dobash and Dobash 1979, 11–12; Breines and Gordon 1983, 530; Schechter 1982).

Activists for battered women also feared that the women-on-men data would be put to dangerous political uses. The clinical recommendation that women in shelters should be cautioned against using violence was perceived by some as a form of victim blaming. Others feared the data were especially harmful since they could be used to block efforts to secure public resources on behalf of battered women (Pleck *et al.* 1977; Fields and Kirchner 1978; also see Steinmetz's response to Fields and Kirchner 1978b). It was clear to some that family violence researchers had questionable political values since they attached significance to purportedly spurious and counter-factual data that threatened women's interests.

The data, again, were strikingly anomalous from the vantage of received theoretical, clinical, and cultural perspectives, which led to attacks on their accuracy and even veracity. Some critics accounted for the anomaly by blaming the methods used to generate the data. They contended that the CTS are too narrowly empirical and quantitative to yield insight into the complex social problem of wife abuse. That problem's complexity requires "context-based" methodologies capable of yielding more textured insight into the lived experience of violent families. The CTS, in contrast, rendered violence abstract, divorced from its lived social context. The CTS, some critics concluded, exemplified an antiquated positivism that served only to obscure the social, historical and cultural circumstances from which violent actions take meaning (Dobash and Dobash 1979, 24–30; 1983; Brienes and Gordon 1983, 502–503, 508–510).

The 1985 National Family Violence Resurvey of a representative sample of 6,002 families did little to settle the controversy. The resurvey data not only corroborated the claim that women were as violent as men within the family (Straus and Gelles 1990b, 96), but also weakened initial qualifications of that claim. Women were about as violent as men according to their own reportage; thus, removing the possibility that this pattern was due to men under-reporting their own violence and exaggerating their partners' violence (Stets and Straus 1990, 156–57, 162). The data showed women evidently *initiated* violence as much as did men, weakening the "self-defense" consideration (Stets and Straus 1990, 154–155, 161). And, aside from actual injury data confirming the view that severely assaulted women sustained actual injury at a greater rate than did severely assaulted men (they were seven times more likely to need medical treatment) (157), differences between women and men in suffering physical injury as a consquence of violence overall were "not particularly strong or large" (Stets and Straus 1990, 158). The resurvey, then, strengthened Straus's and his associates' initial conjectural and definitional stands, but weakened circumstances (other than severely assaulted women's greater need for medical treatment) that they had used to qualify how evidence about women-on-men violence should be applied or used.

The controversy thus continued after the resurvey. "Violence against women" researchers continued to turn to other data to challenge the resurvey, arguing, for instance, that "findings [. . .] based on contemporary and historical studies using intensive and extensive methods persistently reveal a pattern of asymmetrical violence between men and women" and that whenever women did use violence they did so "in a context in which they have been repeatedly assaulted by the man and are trying to defend themselves and/or stop his violence" (Dobash and Dobash 1992, 271–272). All credible evidence, critics maintained, demonstrated that pattern; only survey data suggested gender symmetry in domestic violence (Dobash and Dobash 1992, 264–270; Kurz 1993, 93). Critics also continued to attack the CTS as a flawed and antiquated positivist methodology, and the data they yield as counterfactual and counter-intuitive (Dobash and Dobash 1992, 274–281). Critics reiterated concern about possible misuses of resurvey data to undercut public policy supporting battered women (Dobash and Dobash 1992, 262–263; Kurz 1993, 89, 98–99), with some directly admonishing family violence researchers for "victim blaming" (e.g., Kurz 1993, 96, 98). Some continued to draw sharp theoretical divisions between gender neutral and gendered labels and concepts (Dobash and Dobash 1992, 258; Kurz 1993, 96–97). Limned throughout the entire controversy were statements implying that family violence researchers displayed a political *ethos* that lacked feminist sympathies (Dobash and Dobash 1992, 262–263; Kurz 1993).[15]

Criticisms of the claim that woman-on-man and man-on-woman violence had comparable incidence rates in the American family were attempts to bring discussion back to *conjectural* issues about all four kinds of ambiguity in science communication. Those criticisms deny the survey-based approach to family violence research the basic grounds needed for consideration as a legitimate mode of inquiry into the problem. Specifically, the challenges assert that the survey-based research program is founded on untrustworthy evidence, uses distorting and misleading terminology, applies antiquated methods, and pursues questionable values. There is little left as remainder for subsequent discussion. These *conjectural* stands about evidence, interpretation, methodology, and evaluation would, if accepted, place the entire survey-based research approach beyond the pale of legitimate inquiry into social problems involving women and violence. The legitimacy of the research program that generated the anomalous claims was thus put at issue, rather than received theory and practice.

The controversy became polarized whenever either of the two sides was perceived as denying legitimacy to the claims of the other. For instance, Straus defended his research by turning the tables on his critics, arguing that they, rather than he, practiced narrow and monistic methods that yielded predictable results (Prelli 1996, 412–416). He also insinuated that his critics' political commitments were questionable while extolling his own values, since these "alleged" feminists willfully distorted research that could be used to improve women's lives (Straus 1993, 81–84). On the other side, Kurz exemplified this polarizing approach. She contended that "family violence research" and "violence against women research" were "irreconcilable" (Kurz 1993, 98). As one observer put it, each side to the controversy engaged in "inflammatory rhetoric" aimed at "polarizing, de-legitimizing, and discrediting" the other (Mann 2000, 13).[16]

Can this controversy be addressed as a problem of incommensurate communication? As it turns out, Straus himself approached the evidential dimensions of the controversy in that manner. That attempt involved demonstrating that communication between the quite radically divided sides could become, in a word, "commensurate." In the next section I examine Straus's attempt and assess its relative success in fostering commensurate communication.

Addressing the Problem of Incommensurate Communication

Commentators often describe the controversy about the two family violence surveys in terms consonant with what I am framing as incommensurate communication between people who adhere to different perspectives. For example, Ruth Mann wrote that participants "talk past, rather than to, each other" (Mann 2000, 14). Straus even drew on my own work, using terms drawn from rhetorical studies of science to articulate a similar view:

> Much of the controversy over the findings of the National Family Violence Surveys involves confusion of scientific legitimacy with the use of scientific research to legitimize claims making. As Prelli [1989a] shows, standards for judging scientific claims and policy claims vary from one community to another. Social scientists and social movement advocates tend to apply different standards when assessing the reasonableness of claims. Advocates tend to ignore the scientific legitimacy of the information they use, and scientists tend to ignore the political legitimacy of the information they produce. This is one of the reasons

> that the debate over the National Family Violence Surveys persists after more than a decade. Each community thinks it is addressing the same issue, but each is addressing different issues, using different criteria, and often talking past each other. (Straus 1992, 231)

Stasis analysis of the sources of incommensurate communication during this controversy showed how some of Straus's critics framed the points of conflict as *conjectural* issues and, thereby, raised ambiguities about whether family violence researchers possessed basic evidential, terminological, methodological, and valuational resources for maintaining that their survey-based approach is a legitimate mode of social inquiry. Whenever Straus and his associates framed and addressed evidential ambiguities as definitional, qualitative, and translative issues they would find themselves at cross purposes with those critics. On the other hand, Straus raised similar conjectural ambiguities about his critics' preferred orientations toward social inquiry into the problem of violent couples. They, too, cannot proceed beyond conjectural issues without ultimately finding themselves at cross purposes with Straus and his associates.

Efforts to overcome polarization must move beyond conjectural impasses about evidential ambiguities. One way is to waive, at least provisionally, the conjectural points at issue by presuming the legitimacy of data deployed on both sides of the controversy. Though skeptical about the legitimacy of survey data, Mann attempted to direct attention to what I have called the *definitional* point of commensurability between opposed perspectives:

> Interlocutors agree that women are the principal victims of domestic violence, and that interventions need to prioritize women's safety. However, they are not able to address the substantive issue of how to interpret contradictions between *family violence* findings that women also perpetuate violence, and *violence against women* evidence, drawn from police and emergency-room records, homicide statistics, and abuse testimonials, that domestic violence is *violence against women*. Instead, contenders denounce empiricist methodologies, on the one side, and feminist fundamentalism, on the other. (Mann 2000, 13–14)

It is clear, then, that the interlocutors should address the *definitional* issue about evidence to advance deliberation about problems of abuse of women. As she put it, "*Family violence* studies consistently 'find' that women as well as men perpetuate violence. The real issue is what these findings *mean*" (Mann 2000, 267n21). This framing of the issue begrudgingly "waives" the *conjectural* issue about the facticity of those "findings."[17]

Straus himself addressed the very same definitional question—"What do the conflicting data mean?"—but in contrast with Mann's efforts he also attempted to sort out different perspectives about the meaning of evidence. Indeed, his article on the "representative sample fallacy" is a rare example of a participant making an informed rhetorical intervention into actual discursive practices that have culminated in polarizing the controversy (Straus 1990b). In that article, Straus identified definitional, qualitative, and translative points of commensurability between the conflicting bodies of data concerning women-on-men and men-on-women violence. He, like Mann, waived *conjectural* questions about the credibility of either set of discrepant data, but assumed their mutual credibility as a common point of departure for sorting out opposed perspectives on evidential ambiguities.[18]

From the vantage afforded by the *definitional* point of commensurability, both sets of otherwise conflicting data can appear empirically reliable if they are taken to describe different populations of abused women:

> Women in shelters experienced a frequency of assault that is so much greater than that experienced by assaulted women in the general population that it is reasonable to assume a qualitatively different experience for these two groups of women. This difference could explain why studies based on women in shelters show that very few report assaulting their partner, whereas survey findings reveal that women tend to assault their spouses at about the same rate as husbands assault their wives and indeed often hit first. (Straus 1990b, 86)

The controversy, then, is partly due to confusion over what the two bodies of credible data *mean*. That confusion could be resolved by recognizing that the two kinds of data were not contradictory but actually compatible accounts that described the experiences of different populations of abused women.

Both sides had generalized from characteristics found in their preferred samples to the *wrong* populations. Critics of survey data committed the "clinical fallacy," which generalized from a "clinical" sample of people getting assistance for a problem to all people who might exhibit that problem to some degree but not seek help for it. For example, Pagelow committed the clinical fallacy when she claimed that "one of the few things about which almost all researchers agree is that batterings escalate in frequency and intensity over time" (qtd. in Straus 1990b, 86). That pattern fits the experiences of women who seek the safety of shelters, but it cannot be generalized to include the experiences of assaulted women in surveys since most of them never sought help or treatment. Many violent spouses in the survey might actually have *stopped* abusing their partners after one incident, rather than escalating the abuse to levels experienced by women who seek safety from shelter services. Straus also admitted that he and his associates committed what he called the "representative sample fallacy" when they generalized from characteristics of abused women in representative community samples to the characteristics of clinical populations of women who sought help from shelter and other services. Straus admitted he committed that fallacy when using survey data to warrant his controversial clinical recommendation that women in shelters be counseled against using violence themselves (1990b, 86–87; also see 1993, 77).

From the vantage of a *qualitative* point of commensurability about alternative uses and applications of these previously misinterpreted data, a significant obstacle to commensurate communication had been removed. The "representative sample fallacy" had additional but unmentioned implications for dislodging a related source of political controversy swirling around the notion of the "battered husband syndrome." That notion was a major source of confusion and controversy about survey data (Johnson 1995, 285). Survey data suggested nearly equal incidence rates of women-on-men and men-on-women violence in the general population, but use and application of those data now were qualified. One could not generalize from those data to the existence of a largely concealed population of brutalized men whose sufferings were as great in magnitude and frequency as those of women who sought security from shelters. To do that is to commit the representative sample fallacy. That implication, though unmentioned, opens opportunities for commensurate communication about evidential issues.

Straus also indicated that each of the discrepant sets of data established the best evidential grounds within its own practical jurisdiction. Practical purposes, he contended, defined which bodies of data were best for deciding the evidential basis of the problems addressed:

> [T]he appropriateness of the sample depends on the purpose for which the information is used. On the one hand, findings based on a random sample of assaulted women may be misleading if the goal is to uncover relationships that can be the basis for assisting battered women in shelters. [. . .] This requires knowledge based on the experiences of the population being assisted, regardless of whether their experience is representative of the total population. The experience of other populations may or may not be relevant.
>
> On the other hand, findings based on a "treatment sample" do not necessarily apply to the community at large. The experience of women who have sought assistance from battered women's shelters may not be relevant for designing intervention in the larger community to *prevent* marital violence because, unless the program is based on information obtained from the experiences of a representative sample of the community, one cannot know if it fits their life circumstances. (Straus 1990b, 89)

From the vantage of a *translative* point of commensurability, then, the two sets of discrepant data constituted the best evidential grounds for settling ambiguities about the existence of patterns of violence between men and women, but primarily within their own respective, practical domains.

Straus thus sorted out alternative positions and showed that the two sides offered different but nonetheless *compatible* answers to the definitional, qualitative, and translative ambiguities which, otherwise, could divide them. The question now is whether this effort to create commensurate communication can open opportunities for collaboration between the two parties to the controversy. Straus invented an argument structured by what I elsewhere called the "interdependence" *topos* (Prelli 1996) to motivate such collaboration, "bridging" otherwise conflicting perspectives at the *translative* point of commensurability. Applied to the circumstances of this controversy, the interdependence *topos* creates the possibility that community survey researchers and shelter researchers are interdependent insofar as each needs the evidence furnished by the other to achieve practical goals conducive to bringing the wife abuse problem to comprehensive solution. Shelter researchers cannot use clinical data to address the practical goal of furthering general community violence prevention efforts, but Straus and his associates furnish data that can be used toward that end. Straus and his colleagues cannot use survey evidence to contribute to counseling efforts for women in the shelters, but shelter researchers can furnish data that are useful for that objective. Both sides make important contributions that should be recognized as mutually valuable insofar as they are all interdependent in motivation to bring the wife abuse problem, and spouse abuse generally, to comprehensive resolution.

In summary, Straus reframed the evidential controversy as involving complementary rather than contradictory perspectives. He did so by sorting out alternative perspectives and bringing them into relationship at evidential commensurability points. Not only did this create opportunities for commensurate communication, but it also generated a new possibility—the interdependence argument—for coordinating thought and action under conditions of what we might call *empirical diversity* (Prelli 1996). Straus's rhetorical efforts thus correspond with what rhetorical theorists identify as the primary functions of *stasis* analysis under conditions of pluralism (Cushman and Tompkins 1980, 57–58; Prelli 1989a, 57–58, 59–60).

I can only speculate about the effectiveness of Straus's rhetorical efforts to intervene in this still unfolding controversy. He himself believes the controversy will not end soon, since "[. . .] neither side is motivated to understand the other. Rather, each side seeks to impose its perspective because they believe the preferred definition is vital to advancing their moral agenda and professional objectives. In my opinion that will continue" (Straus 1999, 39–40). Straus is not alone in that assessment. Other commentators see an absence of sufficient motive to engage in commensurate communication rather than to maintain the polarizing stance that opposed sides are "incommensurable" and, thus, irreconcilable. Mann expressed this point bluntly when she claimed "family violence" and "violence against women" researchers are caught in a struggle for "control" and "ownership" of domestic violence: "Neither chooses to recognize her or his opponent's position because both have something very real to lose. [. . .] Winning is not just about rhetoric, theory or ideology; that is, it is not just about 'conceptualizations.' It is also about research funds, social service jobs, and, perhaps most importantly, professional prestige" (Mann 2000, 14). Thus (as we turn the other cheek to her slight on rhetoric), we can see that interdependence, as a motive, might not be sufficient to induce controversialists to suspend preferred moral, professional, and personal motives for adhering steadfastly to one perspective or the other within the particular circumstances of this controversy. However, there is some indication that Straus's rhetorical efforts have had some influence in shifting controversy away from continued polarization and toward deliberation, if not collaboration.

Johnson (1995) treats the evidential controversy as though it was settled sufficiently to move on to address interpretive ambiguities by formulating a more rigorous and nuanced conceptual vocabulary for describing the different kinds of data as different kinds of violence in couples. Indeed, his reasoning (Johnson 1995, 288–289) resembles Straus's effort to reconcile conflicting perspectives on survey data and clinical data, but on *interpretive* rather than evidential issues.[19] Johnson, much like Straus, approached the controversy over discrepant data as confusion of a part of the problem of violence in couples for the whole. Both sides depict their evidence as describing *the* problem of violence in couples when the discrepant data actually concern different *kinds* of violence experienced by couples. Thus, in Johnson's view, a new—and purportedly more rigorous—conceptual vocabulary is needed to disentangle two kinds of violence: "common couple violence" and "patriarchal terrorism" (Johnson 1995, 284–285).

From the vantage of a *conjectural* commensurability point about interpretation, this new terminology is offered as preferable to old terms such as "wife beating," "wife battery," and "battered women" for both empirical and clinical reasons. According to Johnson, the old terminology was responsible for confusing particular types of violence with the overall problem itself (Johnson 1995, 285). The new conceptual vocabulary enables researchers to distinguish those different kinds of violence.

From the vantage of the *definitional* point of commensurability, we can see how Johnson sorted out the new terminology's meanings. "Common couple violence" occurs when "conflict occasionally gets 'out of hand,' leading usually to 'minor' forms of violence, and more rarely escalating into serious, sometimes even life threatening, forms of violence" (Johnson 1995, 285). In comparison, "patriarchal terrorism" is "a product of patriarchal traditions of men's right to control 'their' women," and "a form of terroristic control of wives by their husbands that involves the systematic use of not only violence, but economic subordination, threats, isolation, and other control tactics" (Johnson 1995, 284). That conceptual distinction disentangled previously confused meanings.

From the vantage of the commensurability point of *quality* about interpretation, we can see how the new terminology applies to the otherwise discrepant data. "Common couple violence" applies to Straus's survey data. "Patriarchal violence" applies to clinical evidence (Johnson 1995, 286). Neither side to the controversy, then, offers an evidential basis that is sufficient for understanding the entire problem of violence that couples experience in its full complexity. Both perspectives are required:

> Unfortunately, this debate has been structured as an argument about *the* nature of family violence, with both sets of scholars overlooking the possibility that there may be two distinct forms of partner violence, one relatively gender balanced (and tapped by the survey research methodology of the family violence tradition), the other involving men's terroristic attacks on their female partners (and tapped by the research with shelter populations and criminal justice and divorce court data that dominates the work in the feminist tradition). (Johnson 1995, 285)

Johnson's sorting of perspectives for addressing *interpretive* ambiguities along conjectural, definitional, and qualitative lines thus paralleled Straus's use of those same points of commensurability to distinguish alternative perspectives on evidential ambiguities. Johnson's shift from evidential to interpretive problems presupposes the legitimacy of Straus's handling of conflicting perspectives on evidential problems along conjectural, definitional, qualitative, and translative points of commensurability. Perspectival differences, then, are not the problem that must be resolved by proving one perspective right and the other wrong. Perspectival differences are important resources for resolving complex social problems provided that communication about those differences can become, as I have suggested, commensurate. Johnson's interpretive arguments not only mirror Straus's evidential arguments, but also imply that the two sides were interdependent in their mutual need to address interpretive ambiguities about violent couples in their full complexity.[20]

Summary and Conclusions

The problem of incommensurate communication occurs because discussants frame and respond to different questions and, thereby, argue at cross purposes. Moreover, those engaged in incommensurate communication are likely to premise arguments with the presumptions or proscriptions of their own preferred perspectives. As Kuhn noted, partial communication results and that kind of communication inevitably is circular (Kuhn 1996, 94, 148–149; see also Herbert Simons's essay in this volume, for a fuller exploration of partial circularity in light of Kuhn). But the problem of rendering communication commensurate is not to avoid perspectival circularity so much as to find "commonplaces" for entering each perspective's circle of understanding in the right ways. Those commonplaces are located at points of intersection between otherwise different perspectives. Addressing the problem of incommensurate communication requires finding those points of intersection or mutual entry between perspectives.

I have presented and applied a procedure for finding commensurability points that is founded on classical *stasis* doctrine. The kinds of ambiguities that commentators identify as sources of communication breakdown—what I have called *incommensurate communication*—correspond with the kinds of questions identified by classical rhetorical and dialectical *stasis* theory as the bases of all controversy. *Stasis* procedure applies to situated ambiguities not only to clarify how disputants argue at cross purposes when they address different questions, but also to find ways of realigning communication so that disputants can address the same questions. Those realignments can occur at points of commensurability that take the form of shared, situated questions. From the common vantages furnished by those questions, discussants can sort out how their perspectives yield similar or different answers without presupposing shared standards for evaluating which answers are "right." The sixteen kinds of question identified by the *stasis* procedures presented in this essay can help disputants locate potentially shareable commensurability points within specific kinds of situated ambiguities that give rise to controversy, confusion, or conflict about science under conditions of perspectival diversity. Those commensurability points can thus be applied to (1) locate situated questions that discussants need to address in common; (2) sort out perspectival similarities and differences in relation to those situated questions, and when applicable, (3) discover new grounds in common to overcome polarization and foster problem-solving deliberation.

I have examined the controversy about women-on-men and men-on-women violence data as involving incommensurate communication whenever disputants failed (intentionally or not) to address the same situated questions and instead argued at cross purposes. *Stasis* analysis tracked ambiguities at the source of incommensurate communication, and revealed how specific strategies fostered commensurate communication about different but not altogether incompatible answers to evidential questions. Those strategies opened up prospects for participants to suspend, at least temporarily, polarizing tactics and efforts at mutual recrimination so that each side to the controversy could consider pressing situated questions from the other's vantage. In particular, Straus's rhetorical intervention revealed what I have called points of commensurability from which the different sides to the controversy could differentiate their similarities and differences.

That analysis illustrated how polarizing argumentation tended to frame questions as conjectural about evidence, interpretation, evaluation, or methodology. Such framings frustrate efforts to collaborate or to otherwise sort out similarities and differences among the perspectives. Each side portrays the other as lacking legitimate data, terminology, value commitments and methods. In overall effect, the adversary's perspective is denied any legitimacy as a mode of inquiry into social problems. This pattern of conjectural framing, it is worth noting, is exhibited in the attacks of scientific establishments on the legitimacy of purportedly "fringe" or "pseudo" sciences. For example, creationists were denied scientific legitimacy during a court case when their opponents took conjectural stands against their evidential, interpretive, evaluative, and methodological claims (Prelli 1989a, 219–236). In the controversy examined here, each side sought to deny the other's legitimacy when making claims about social inquiry into complex problems of violence in the family.

Once controversy becomes polarized, the disputing sides argue as though the issues dividing them are ultimately resolvable by proving their preferred perspectives conclusively right and the adversary's perspective conclusively wrong. Such argumentation presumes that there are standards—neutral or otherwise—which, once applied, would definitively settle the issues. Those presumed standards usually turn out to be the preferred values and assumptions of one or the other clashing perspective. That sort of argumentation is not without rhetorical advantages, but any advantages are purchased at the cost of foreclosing prospects for *commensurate deliberation* about divisive issues (Prelli 1996, 424). As Mann put it, strategies that "polarize," "de-legitimate," and "discredit" fail to address or "resolve issues" (Mann 2000, 13).

The analytic and methodological response to polarizing tactics should not be to present the controversy as evidence that the clashing perspectives are "incommensurable." That designation, I have argued, carries the corollary implication of formal incompatibility and incomparability. Though that designation might be used to urge a "both/and" rather than an "either/or" framing of the controversy (Mann, 208), it still fails both to furnish means for achieving mutual understanding between different perspectives and to locate motivational grounds for collaboration, should such understanding be achieved. Framing the controversy as involving problems of incommensurate communication between different perspectives does not foreclose those prospects. The way remains open to move beyond polarization toward situated, problem-solving deliberation under conditions of perspectival diversity.

It should be clear, then, that addressing the problem of incommensurate communication is as much a matter of possessing the right attitude as the right method. Perspectival diversity must be acknowledged as fact from the start. We saw this with respect to the controversy about violence in couples. Though periodically making polarizing appeals, Straus also stressed that society needs the moral agendas and professional objectives of both perspectives. Indeed, he expressed the "hope that the controversy will be resolved by recognizing the need for both perspectives, and that this will bring an end to attempts to discredit those whose agenda and professional contribution requires a different approach and a different perspective" (Straus 1999, 40). He is not alone in that attitude. Mann similarly wrote that prospects for communication between the two sides are favorable if only the two sides were willing:

> As a growing number of researchers across those two perspectives recognize, alternate language games develop in contradistinction to each other, not in ignorance of each other. For those who are willing there is little difficulty translating across languages, cultures, perspectives, or paradigms. To accomplish this, however, interlocutors must share a desire to engage in honest or "authentic" communication. They must be willing both to seek and find common ground, and to acknowledge and respect differences. (Mann 2000,14)

Efforts to address problems of incommensurate communication, then, are predicated on the mutual willingness of the parties involved to at least provisionally acknowledge perspectival diversity as a potential constructive resource rather than as an obstacle to be overcome. Should the parties fail to so acknowledge the potential value of perspectival diversity, no method, *stasis* doctrine or other, could afford them viable communicative alternatives to continued polarization and mutual recrimination.

Stasis doctrine can help disputing parties discover how their perspectives converge on and diverge from specific situated ambiguities about evidence, interpretation, evaluation, and methodology. Such discovery might give rise to improved mutual understanding, but that alone surely will fail to resolve the controversy. Mutual understanding is not tantamount to mutual commitment. *Stasis* doctrine *might* help generate new grounds for mutual cooperation based on mutual understanding under conditions of perspectival diversity in particular, situated cases, but any new grounds for motivating cooperation are always in potential tension with the strong pull of preferred, perspectivally driven, motivations (Cushman and Tompkins 1980, 51). Straus's interdependence argument, for instance, gives at least temporary and limited motive for cooperation on practical problems concerning abused women. But, as we have seen, that motive could prove suboptimal when placed alongside professional, political, or moral motivations for adhering to one or the other conflicting perspective. The problem of mutual motivation looms large in communication about practical problems under conditions of perspectival diversity and it cannot always be resolved through comforting appeals to "both/and" thinking alone.

The problem of incommensurate communication will become increasingly salient as inter-disciplinary, inter-paradigmatic, and inter-theoretical perspectives for pursuing scholarship proliferate in the social and the natural sciences. The fact that theoretical, methodological, political, or other kinds of "bridging" or "straddling" *can* be done is not at issue, since it is done all of the time; in this volume, the case documented by Bazerman and De los Santos is a clear example. Sorting out similarities and differences among them can, nevertheless, become an especially keen communication problem. Under conditions of perspectival diversity, *stasis* analysis offers a promising approach both for mapping the features of incommensurate communication and for finding ways to render that communication commensurate.

Notes

[1] An early version of this essay entitled "Stasis and Incommensurability" was presented on the panel, "Competitive Papers in the History of Rhetoric," for the International Society for the History of Rhetoric—American Branch, at the Speech Communication Association (SCA) (now National Communication Association) meeting held at New Orleans, Louisiana, November 3–6, 1988. Also developed and incorporated in this essay are ideas about applying stasis doctrine to Kuhn's understanding of incommensurability that first were formulated in "The Rhetorical Perspective in Thomas S. Kuhn's Analysis of Paradigmatic Science," a paper presented on a Rhetorical and Communication Theory panel at the SCA meeting at Anaheim, California, November 13, 1981. This and subsequently published work on *stasis* procedure for the study of scientific controversies in *A Rhetoric of Science: Inventing Scientific Discourse* (Prelli 1989a), chapter 8, formed the basis for collaboration with Nola Heidlebaugh on a SCA seminar on the problem of incommensurability entitled "Rational Discourse and the Problem of Incommensurability," which occurred in November 1990 at Chicago, Illinois.

[2] These three sources of incommensurability should not be confused with the distinction between so-called "global" and "local" or "partial" incommensurability (Simmons 1994, 120–121; Veatch and Stempsey 1995, 258). One or more of the three sources can be involved regardless of whether incommensurability concerns comprehensively distinct frameworks or differences in specific elements of otherwise compatible perspectives. Sankey

and Hoyningen-Huene (2001, ix-xv) elaborate on "semantic" and "methodological" incommensurability. To those two I add "values incommensurability." "Values," as understood in this essay, are not restricted to Kuhn's (1977) inventory of criteria for theory choice (accuracy, consistency, etc.) but could also include any conceivable conception of the good, technical or otherwise. For instance, Veatch and Stempsey (1995, 260–265) characterized medical researchers', clinicians', and patients' worldviews as permeated with value preferences (both technical and non-technical) that culminate in "partial incommensurability" when assessing the relevance of facts. Similarly, Weed (1997, 116–120) points to often subtle differences in epidemiologists' values (again, both technical and non-technical) as an important source of incommensurability during appraisals of causal inferences, even when those inferences are warranted from the same data.

Newton-Smith (1981, 112–116, 149–150; Veatch and Stempsey 1995, 257) identifies "values variability" as one of three sources of incommensurability formulated in Kuhn's work and, accordingly, restricts it to the values of scientific theory choice. A second source he calls "radical standard variance" (Newton-Smith 1981, 110–111, 150), which concerns questions about criteria for problem significance during paradigm debates. Since they, too, are questions about values (Newton-Smith 1981, 111; see Kuhn 1996, 110), I subsume both under "values incommensurability." (The last source Newton-Smith labels "radical meaning variance," which resembles semantic incommensurability (150–151, 151–156).) What I am calling *values incommensurability* is fully consonant with Harris's *pragmatic incommensurability,* as expressed in the introduction.

[3] Kuhn (1996, 199–200; 1977, 326, 330–331) was at pains to make this point when he defended his earlier position on incommensurability from charges of irrationalism. His defense depicted those charges as founded on the false analogy between comparison of values claims from different perspectives and comparison of quantities derived from different algorithms.

[4] Putnam argued that the two then dominant theses in philosophy of science were self-refuting: incommensurability and the logical positivist criterion of meaning. The critique of logical positivism is on pages 105 to 113 (1981b).

[5] I am pleased to note that Greg Myers has independently adopted *incommensurate* as a matter of course, not *incommensurable,* for programs that manifest competing "views of reality" (1990, 215), though I can't be sure that there is a principled choice behind the usage. I am equally pleased to find that both Thomas Lessl and John Angus Campbell have adopted the usage in this volume, for much the reasons I have. Henry Johnstone's (1959, 32) use of *incommensurate* (for "philosophical statements") is, however, as Finoc-

chiaro notes (1977, 122n6), essentially the same as Kuhn's and Feyerabend's *incommensurable.*

[6] In addition to sources cited in Prelli (1989a, chapter 4), see also Fahnestock and Secor's 1988 application of classical *stasis* doctrine to arguments in the "disciplinary contexts" of science and of literary criticism, and Gross's 2004 case for the importance of the translative stasis for incommensurability, though his understanding of incommensurability is markedly different from mine

[7] Quintilian expressed the judgment function of *stasis* doctrine clearly:

> [E]very question has its basis [*status*], since every question is based on assertion by one party and denial by another. But there are some questions which form an essential part of causes, and it is on these *that we have to express an opinion;* while others are introduced from without and are, strictly speaking, irrelevant, although they may contribute something of a subsidiary nature to the general contention. It is for this reason that there are said to be several questions in one matter of dispute. Of these questions it is often the most trivial which occupies the first place. [. . .] A simple cause, however, although it may be defended in various ways, cannot have more than one point in which a decision has to be given, and consequently the *basis [status] of the cause will be that point which the orator sees to be the most important for him to make and on which the judge sees that he must fix all his attention.* For it is on this that the cause will stand or fall. (1920, 3.6.7–9; emphasis added)

[8] Cicero elaborated on these three questions: "As to the dispute about a fact, this can be assigned to any time. For the question can be 'What has been done?' [. . .] 'What is being done?' [. . .] and 'What is going to occur?' [. . .]" (1.8.11); "The controversy about a definition arises when there is agreement as to the fact and the question is by what word that which has been done is to be described. In this case there must be a dispute about the definition, because there is no agreement about the essential point, not because the fact is not certain, but because the deed appears differently to different people, and for that reason different people describe it in different terms" (1.8.11); and, "There is a controversy about the nature or character of an act when there is both agreement as to what has been done and certainty as to how the act should be defined, but there is a question nevertheless about how important it is or of what kind, or in general about its quality, e.g., was it just or unjust, profitable or unprofitable?" (1.9.12).

[9] The only exceptions were "asystatic" questions not susceptible to *stasis.* On the features of *stasis* procedure as a categorical system see Prelli (1989a, 58–60).

[10] Methodological features discussed here are based on McKeon (1971, 47–48, 54–58; 1966, 368–371) and Prelli (1989a, 57, 60).

[11] Kuhn (1996), arguably, offers the most comprehensive view of sources conducive to incommensurable communication.

[12] I am aware of the terminological inconsistency of moving from *incommensurable* to *incommensurate,* but then adopting *commensurability* points rather than *commensuration* points, or the like. But my purpose here is to propose ways that disputes traditionally labeled *incommensurable* might find their way towards resolution. Part of that purpose is accomplished by exposing the weak undergirding premises for the notion of incommensurability in order to talk about real communicative blockages, rather than the imaginary incompatibilities that result from over-extending the mathematical notion of incommensurability. With that exposure, the adoption of the ordinary-language term, *incommensurate,* helps both to distinguish real communicative issues from imaginary formal issues, and to raise the possibilities of overcoming communication breakdowns. But once the distinction and the possibilities are on the table, there is no reason to avoid the more familiar technical language. Indeed, returning to that language helps further to make my case, since *commensurability* is the natural antonym of *incommensurability.*

[13] Straus followed his own advice when giving testimony before the United States Civil Rights Commission by directing attention to the dramatic statistic about wife abuse as warrant for increased federal funding for battered women's shelters and treatment programs. Women-on-men violence data appeared in the written report (1978).

[14] Straus is representative of family violence researchers who use systems theory to account for violence in the family. Indeed, Straus wrote one of the early efforts to map the dynamics of family violence in systemic terms (Straus 1973, 105–125).

[15] An important ambiguity that permeated this controversy is captured with the question, "What qualities of thought and conduct constitute a 'real' feminist *ethos*?" Efforts to delineate the boundaries of "legitimate" feminism in this case call for a more comprehensive study than can be offered here. One might pursue that study along lines parallel to my work on creating and challenging qualities of thought and conduct that purportedly constitute a legitimate scientific *ethos* during situated conflicts about "demarcation exigencies" (1989a, 221, also see 105–109, 126–127, and 218–236; and 1989b, especially 60–61). Also see Taylor's (1992; 1996a, especially 101–134; and 1996b) work on the rhetoric of demarcation. In this controversy, qualities of

feminism and of science are variously combined, differentiated, and intermingled in the struggle to constitute a suitable *ethos* for inquiry into important social problems such as wife abuse.

[16] For one sharp manifestation of this controversy in the context of family therapy see Avis (1992), Kaufman (1992), Bograd (1992), Meth (1992), and Erickson (1992) in a special section on "Violence—The Dark Side of the Family," for the *Journal of Marital and Family Therapy.*

[17] Even so, Mann's framing suggests that there is something less than real about data pointing to the existence of an unanticipated, anomalous, and, evidently, stubbornly recalcitrant "finding" (note the scare quotes) concerning violence in the American family. Why is the conjectural issue not as real as the definitional one? After all, *stasis* doctrine tells us that the framing of definitional issues is contingent upon how conjectural issues are framed and, ultimately, waived or settled. Regardless of how that question might be answered, from the vantage of *stasis* analysis Mann missed an opportunity to enter into the conflicting perspectives and discover how they each would respond to both conjectural and definitional issues about evidence.

[18] Straus's waiver of the conjectural issue does not mean that his critics now accepted the CTS data as credible. According to Mann, much criticism of the CTS is "a consequence of the findings—especially findings that women and men express and sustain minor and severe acts of violence at equivalent rates" (Mann 2000, 11). The "'act-based' operationalization of abuse" was "highly contentious," it appears, because it yielded politically undesirable data (11). That concern did not stop others from modeling their studies after the act-based approach, provided that alterations were made to reduce "contentiousness." For instance, the 1993 Statistics Canada Violence Against Women Survey used an act-based scale, but dropped questions about women-on-men violence from the survey and, thereby, avoided the "politically contentious findings" that concerned critics (Mann 2000, 266–267n18). This methodological tweaking "fixed" the data but in a way that presupposed that the data and the method that generated them were, in fact, reliable, albeit not always politically desirable.

In a separate article, Straus (1990a) defended his methods and his data by arguing for the CTS's reliability and validity. His responses did address conjectural, as well as other evidential and methodological issues. His arguments in that article indicated that the controversial data about women-on-men violence, though anomalous from several theoretical vantage points, were reproduced in numerous other studies and, thus, constituted a stubbornly recalcitrant empirical pattern.

[19] Johnson (1997, 289) acknowledges Straus's effort to help reconcile the controversy with the representative sample fallacy, but does not believe that he went far enough. For instance, he claimed that Straus failed to ac-

knowledge that the "representative sample" might miss a significant segment of the general population that undergoes severe battering. In fact, Straus (1990b, 85) did acknowledge that victims of severe abuse and those who severely abuse them are not the kinds of people who are likely to stop and fill out a questionnaire or submit to a telephone interview.

[20] Johnson and Ferraro (2000) later revised this new interpretive language by dropping the "patriarchal terrorism" concept in favor of the more gender neutral "intimate terrorism." They also added two new categories: "violent resistance" and "mutual violent control." This terminological development is thought conducive to even more precise readings of available evidence on violence involving couples.

8

The "Anxiety of Influence," Hermeneutic Rhetoric, and the Triumph of Darwin's Invention over Incommensurability

John Angus Campbell

If ever two paradigms were "incommensurable"—built on logically incompatible assumptions—one could scarcely hope for a sharper contrast than that between Charles Lyell's actualist geology, based on non-progressionism, and Charles Darwin's evolutionary biology, based on natural selection.

Lyell's epochal *Principles of Geology* (1990 [1830–1833]) was not only anti-evolution (a large section of the second volume is spent refuting Lamarck[1]) but its doctrine of non-progressionism positions evolution, alongside Scriptural Geology and other *a priori* speculative systems, as beyond the purview of science.[2] As an intellectual project, Lyell's *Principles* offers an updated Newtonian worldview in which the implications of weak causes and deep time are made relentlessly and appallingly clear. Lyell takes Newton's emphasis on lawfulness and offers a view of the earth as a hydraulic perpetual motion machine, the processes of which not only show no sign of a beginning or prospect of an end but, most importantly, are going nowhere (Rudwick, in Lyell 1990, 1: xv-lv). Uplift and subsidence, the two great powers, have operated at their present rates and degrees of intensity over trackless time, and are the only forces of which geological science can have knowledge. The great intellectual triumph of *Principles,* from Lyell's perspective, based on a complex argument about the constancy of the earth's diameter, was that subsidence had a slight edge over uplift (Herbert 1968, 11). What this meant was that the history of the earth, in any chronological sense, was irrecoverable, because its records were being destroyed in the earth's shredder/furnace by further operation of the very forces that had partially preserved them. The surviving geologic column was hopelessly fragmentary.

Wherever history defeats our presumptions about the logic that leads to our present, and the *Principles/Origin* connection is an example, we confront the gulf of incommensurability—the logical disjunction between two historically related but distinct systems of thought. How Lyell's vision of earth history without history could inspire evolutionary theory is, from the perspective of logic, highly unclear. Yet Lyell inspired not only Darwin, but Alfred Russell Wallace as well, Darwin's "co-discoverer" of natural selection (A.R. Wallace 1855, 184–196). How? Everyone in the nineteenth century read Thomas Malthus's *Essay on the Principle of Population* (1992 [1798]), including the sailors on the Beagle and the London literati who attended the dinner parties thrown by Darwin's brother Erasmus: to arrive at evolution by natural selection (twice no less), the key is not just Malthus. The indispensable text—without which the catalyst of insight would not activate—was Lyell's *Principles,* with its principled anti-evolutionism.

If evolution was a lock, an identical key was forged twice—first by Darwin and then by Wallace when, though separated in space by half a world (Darwin in England and Wallace in Borneo), and in time by two decades (respectively 1838 and 1858), each man used his knowledge as a field naturalist to combine Malthus's laws of population with Lyell's law of the succession of type.[3] That key was forged rhetorically in the historical moment.

Viewed logically (that is, absent any sense of context), the gulf between Darwin and Wallace, on one side, and Lyell, on the other, is so great one would think that only a gestalt switch, a sudden irrational insight—a Kuhnian conversion—could cause a person to change from one way of thinking to the other. Viewed rhetoriographically (that is, through "the application of rhetorical methods to the study of historical objects, agents, movements, arguments, [and] ideas"—Campbell 2002), matters make more sense. Incommensurability dissolves. Systems of thought develop over time, and even radical innovators inevitably share much in common with their more conventional colleagues. When one understands the process of paradigm shift to incorporate at its very heart the task of making insight intelligible and persuasive, incommensurability is revealed as an odd quirk of positivist philosophy. Paul Feyerabend, who admired a great deal about Kuhn's picture of scientific change, did not buy the conversion story. "That is not what we find," he said, "when we look at history," following up with a dramatic example:

> The transition from classical physics, with its objective space-time frame, to the quantum theory, with its attendant subjectivities, certainly was one of the most radical transformations in the history of science. Yet every stage of the transformation was discussed. There were clear problems; they worried both the radicals and the conservatives. Many people suggested solutions. These solutions, too, were understood by the contending parties, though not everybody liked them or regarded them as important. The final clash between the new philosophy and its classical predecessor found its most dramatic expression in the debate between Bohr and Einstein. Did Bohr and Einstein talk past each other? No. Einstein raised an objection; Bohr was mortified, thought intensely, found an answer, told Einstein, and Einstein accepted the answer. Einstein raised another objection; Bohr was again mortified, thought intensely—and so on. Looking at such details, we realize that the conversion philosophy simply does not make sense. (Feyerabend 1999, 267)

Feyerabend's example is the most dramatic scientific shift of the twentieth century (see Beller 1999 for a detailed explication of the transformation, very consonant with Feyerabend's observations). Mine will be the most dramatic scientific shift of the nineteenth century (which, admittedly, did not culminate until the twentieth). His example is largely dialectical (Beller's book is entitled *Quantum Dialogue*). Mine will be largely rhetorical, on the level of personal invention.

My essay approaches the incommensurability between Darwin and his immediate mentor—indeed, more generally, between Darwin and the larger tradition in which he, Lyell and Wallace stood—as a special instance of what Harold Bloom (1975) calls "the anxiety of influence," Michael Leff (1997) calls "hermeneutic rhetoric," and Marcello Pera (1994) just calls "scientific argument" (though this term for Pera has special dialectical and rhetorical characteristics). Bloom, Leff, and Pera are at one in seeing historical transitions as profoundly commensurable. Appearances to the contrary in the history of science gave rise under the probing of Kuhn and Feyerabend to the label *incommensurable,* but at least in the particular case of Darwin's rhetorical assimilation of Lyell, the term *incommensurate* is more appropriate; the appearance of incompatibility dispels when we look at the inventional processes. (See Prelli's discussion of *incommensurable/incommensurate,* in his essay in this volume.)

From Lyell, for evolutionary theory, Darwin prominently took actualism (the extrapolation of present forces to historical operation), uniformitarianism (the extrapolation of present intensities to historical operation), and loads of time. (He also learned well Lyell-the-advocate's relentless argumentative manner.) And he adopted the philosophy-of-science themes collected by John Herschel under the label *vera causa.* But once he brought them into contact with Malthus's ideas about the population pressures on the struggle for existence, he applied all of them in ways that neither Lyell nor Herschel would sanction, in ways that they might, if they had read *The Structure of Scientific Revolutions,* have called *incommensurable.* In particular, both of them, along with many others, saw the *vera causa* principle as incompatible with biology.

I will chart Darwin's inventional course, relying substantially on his notebooks, from his geological apprenticeship under Lyell, beginning with his voyage on the HMS Beagle, through his very instrumental paper on the parallel roads of Glen Roy, into his 1842 *Sketch* and his 1844 *Essay,* where he worked out the rhetorical strategies that resulted in *The Origin of Species.* What we need to keep in mind, every step of the way, is that it was Darwin's thorough assimilation of Lyell (and of Herschel) that prepared him for his momentously productive disengagement.

Anxiety, Hermeneutics, and Scientific Argumentation

For Bloom, a literary critic, the anxiety of influence occurs whenever a new "strong writer," coming late in a long tradition, faces the challenge of overcoming his or her predecessors (Bloom 1975, 5–16). The key even to poetic novelty for Bloom is not "inspiration" (in the sense of creation out of nothing) but the rhetorical skill by which a confident new talent reinvents tradition—turns tradition against itself and makes it say something new. For Leff, a rhetorical critic, the problem is to understand how new civic discourse is produced when prior tradition and its proper interpretation are in dispute. For Pera, a philosopher of science, the problem is to understand how arguments growing from, and governed by, factors intrinsic to scientific questions account for the rationality of science—specifically for the development of novelty (the very pride of scientific reason)—better than any formal model of scientific method. Despite the differences in the genres and terminologies that focus the thought of these theorists, Bloom, Leff and Pera are clearly allied in their unequivocal rejection of incommensurability (as, indeed, is Beller). Each scholar sees invention as following an intelligible, if unforeseeable, route shaped and sharpened through the agonistic contest between tradition and innovation.

Bloom offers what might be called the closest to a formalist theory of invention in that he breaks down the process by which a strong writer asserts himself over prior tradition to a six-step process. While all six steps are not central to our concerns, the first two (*climamen* and *tessera*) and the last (*apophrades*) are particularly suggestive as they place in relief his leading idea that poetic innovation rests upon systematic misreading. Following the *clinamen* or swerve Lucretius attributed to the atoms in *De Rerum Natura,* the strong new writer, Bloom argues, follows along the path of the predecessor but then executes a swerve "which implies that the precursor's poem went accurately up to a certain point, but then should have swerved, precisely in the direction that the new poem moves" (Bloom 1975, 14). In Bloom's second step, *tessera,* a simultaneous completion and antithesis, in which "a poet antithetically completes his predecessor by so reading the present poem so as to retain its terms but to mean them in another sense, as though his predecessor had failed to go far enough" (Bloom 1975, 14). The final stage, *apophrades,* or "the return of the dead," places the predecessor in a new, and historically subordinate role. This stage identifies the effect that is so marked in the relation between Darwin and Lyell, at least in the history of biology; after Darwin, Lyell can no longer be understood except as Darwin's predecessor, as the man whose thought, properly understood, is not his own but intelligible only as completed in Darwin's.

In contrast to Bloom, but in consonance with his chief source, the mature Cicero, Leff has no formal theory of invention. Following the classical rhetorical tradition, Leff sets out an account of how the demands of rhetorical situations trigger revisions of prior civic tradition. Beginning with a critique of foundationalist views of rhetorical theory and method (which has strong affinities with Pera's critique of the same views in science), Leff examines how foundationalism has obscured the very heart of rhetorical reason located in invention. Welcoming the recent hermeneutic turn in literary studies with its denial of fixed meanings in texts, Leff notes how this shift "opens space for concepts of probability, social exchange and situated meanings to become matters of serious concern" (Leff 1997, 201). His recuperation of classical rhetorical criticism stresses how the classical understanding of rhetorical criticism parallels the emphasis in contemporary literary criticism on the intimate relation between reading and message production.[4] As Leff explains, the main strand of classically grounded rhetorical education was never to provide mere theory or rules (in the modernist scientistic sense) but "to impart the practical judgment and linguistic resources needed to adapt to changing circumstances" (Leff 1997, 201). Given that "propriety in rhetoric like prudence in ethics could not be reduced to mere theoretical rules" the examples and rich store of strategies characteristic of a rhetorical education "could be realized only as they were actually used in rhetorical production" (Leff 1997, 202). In a rhetorical education, Leff notes, a reader read not just as a consumer of discourse but as a rhetor in training. As reader, one identified strategies. As rhetor, one re-embodied them in new situations.

With Bloom and Leff, Pera opposes both modern foundationalism and postmodern relativism and sees in rhetorical argumentation a viable, if demanding, *via media* (Pera 1994, 11). Incommensurability for him is largely the unexploded legacy of Cartesianism (Pera 1994, 10), which has gained an unnatural new lease on life by having been adopted by a form of postmodern thought that itself is but an uncritical reversal of Cartesian foundationalism (a form of philosophic life I elsewhere call "Punk Cartesianism"—Campbell 1998). Pera sees Cartesianism as consisting of three propositions:

1. that science has a secure method demarcating it from any other endeavor,

2. that the rigorous application of this method guarantees the fulfillment of the aims of science, and

3. that if science had no method it would not be a cognitive endeavor. (Pera 1994, 4)

The generating point of Pera's program is his argument that propositions 1 and 2 have been shown by the contemporary philosophy of science to have failed, and that proposition 3 does not follow. Very much in the spirit of Leff, though drawing parallels more with Aristotle than with Cicero, Pera argues that the interests of science would best be served were science to return to Aristotle's *topoi* and informal logic. By giving up the pretence of grounding its claim to rationality on the basis of a formal method that it clearly and habitually violates whenever fundamental novelty is afoot, Pera contends science can relinquish Cartesianism and positivism without thereby giving way to casuistry (Pera 1994, 4–12, 129–152). Pera's argumentation-based science does not reduce science to popular oratory because, following the insights of Perleman, the arguments of science, however non-formal, are conducted according to factors of scientific and dialectic and are addressed to an audience of experts. Pera shows how fundamental novelty, the essential feature of what traditionally has been called invention, has emerged time and again in scientific debates from the time of Galileo to the present through the supple and even opportunistic play of the factors of scientific rhetoric/dialectic.

The insights of these three scholars are valuable, and their data is helpful, but their shared perspective is not new. One can usefully characterize Bloom's "anxiety of influence," Leff's "rhetorical hermeneutics," and Pera's "factors of scientific rhetoric and dialectic" as a restatement and appropriation of Cicero's mature approach to discourse. For all of them, as for Cicero himself, reading, no less than writing or speaking, is a "*multiplex ratio disputandi*"—a contest between two (or more) voices that inherently begets new discourse (Conley 1985; 1990, 37; see Heidlebaugh 2001 for a specific application of Cicero to the question of incommensurability; see Ceccarelli, in this volume, for a look at the failure of one scientist, E. O. Wilson, to approach scientific argumentation in Ciceronian terms). In the Ciceronian method (if we may use so systematic a term for so supple a procedure), propositions whether scientific or practical are situated in their respective frames of reference, no matter how divergent these may be, and brought into the contest of public debate. Out of this contest (by witnessing it and participating in it, even if vicariously) an audience tests the various likelihoods that emerge from the clash of divergent propositions. For a Ciceronian, "incommensurability," rather than constituting a bar to reason, is the *stasis* that gives rise to the argumentation and practical judgment that—if only in this case and for now—can resolve it. (See Prelli's essay in this volume for a systematic exploration of just this point.)

Four points are particularly worth noting in relation to the perspective common to Bloom, Leff and Pera. First, all three cut the Gordian knot of incommensurability in a single decisive blow by the simple expedient of resituating the logic of controversy in the context of its occurrence. Second, all three locate the center of innovation in the existence of questions, whether poetic, political or scientific, that require interpretation, that occasion contestation, and that suasory discourse mediates. Third, all three exemplify how invention, whether in poetry, politics, or in science, is less the fruit of inspiration than of invention, reinvention, appropriation and misappropriation. Fourth, all three find that received interpretations, through such situated argument, are made to turn in directions no one could have anticipated by knowledge of any deep structure, formal logic or method.

Darwin, Tradition and Invention

Darwin was a strong writer in Bloom's sense of one prepared to assert himself in the face of predecessors whose achievements would have overawed a lesser talent. Darwin certainly felt acute anxiety over the relationship of his project to prior tradition, yet on one point he seemed supremely confident. Ever sensitive to what others would think of them, and of him, if his ideas were announced too soon, Darwin seems to have entertained very little doubt that he was the legitimate heir of a methodological tradition that stretched from Francis Bacon and Isaac Newton to Herschel and William Whewell. He believed that his peers and teachers had misconstrued this tradition on one decisive point, the application of the *vera causa* principle to the organic world (Kavalovski 1974). It is here that Darwin's situationally motivated and creative "misreading" (as Bloom would put it) joins Leff's hermeneutic rhetoric and Pera's rhetorical and dialectical scientific argumentation.

Darwin avoided (or managed) the anxiety of influence by throwing it back on the received scientific tradition and specifically on his teachers,—Lyell, Herschel, and Whewell. Darwin invited his teachers, and the generation of scientists under their sway, to see his evolutionary project as the fulfillment of their common philosophy of science, rather than as the ruin of natural theology, centered in the design argument, which received philosophy supported. In effect, scrupulously evoking what he took to be the central methodological principle of the British scientific tradition, Darwin proposed to use it as a fulcrum from which to move the scientific world (Kavalovski 1974, 40–72). Darwin correctly saw that the future direction of science would be materialistic, and by redefining the "design" argument as tantamount to a defense of miracle, he put his opponents, however unfairly, at a rhetorical disadvantage. In Bloom's terms, Darwin was no "heretic," because he did not aim at merely changing the emphasis of tradition; rather, he was a "revisionist," one who identified a specific doctrine—the *vera causa* principle—as the point at which tradition veered from itself (Bloom's *clinamen*) and which he proposed to set right (his *tessera*). (See Bloom 1975, 28–29.) Viewed in this light, the anxiety of influence, hermeneutic rhetoric and the rhetorical/dialectical factors of scientific argument Darwin used to justify his interpretation of tradition, all unite to dissolve the paradox of incommensurability. I shall argue that Darwin's theory of evolution can be viewed either, as T. H. Huxley held, as a logical extension of Lyell rightly understood, or, with equal plausibility, as a rhetorical reinvention of Lyell creatively misunderstood (Darwin 1887, 190).

Defer and Defy: Portrait of the Revisionist as a Young Man

Lyell's *Principles of Geology* is the first major work of science Darwin read without benefit of a teacher (de Beer 1983, 44). It made a deep impression, confirming and claiming Darwin as a scientist, becoming his *Bible.* He read it in that deeply imitative, contemplative sense in which the *Book of Common Prayer* counsels the faithful toward the scriptures (and Darwin was then a candidate for Holy Orders): to "read, mark, learn, and inwardly digest" (Seabury Press 1979, 184). Beginning on the Beagle with initially light marginal notes, on his return *Principles* became the most extensively annotated of all his books (DiGregorio and Gill 1990, 1: xxxiii).[5] Darwin's reading and rereading of Lyell marks the proximate beginning of that psychological/creative process of the anxiety of influence and that hemeneutic/rhetorical process of invention that together formed Darwin into an "original" thinker. Starting from an imitation so close as to be mere copying, Darwin's Lyellian discipleship, in the end, made his ideas incommensurate with those of his master and of his colleagues.

The first volume of the three-volume *Principles* had been published in 1830, the year before the Beagle set sail, and subsequent volumes were published in 1832 and 1833. John Stevens Henslow, Darwin's Cambridge mentor, had recommended it to him but cautioned Darwin not to believe it. Several things positioned Darwin to look upon Lyell, and the world, with fresh eyes: the literal meta-phora of his physical presence aboard the Beagle, his wrenching sea-sickness and self-doubts, his rallying passion for tropical adventure from his constant rereading Alexander von Humboldt's *Personal Narrative* (which first had inspired him at Cambridge) and his sense of opportunity (Barlow 1987, vol. 1; Darwin 1952, 604). In the early weeks of the voyage and beyond, when Darwin was not reading von Humboldt, he was pouring over Lyell. While neither Henslow nor Adam Sedgwick, with whom Darwin had just completed a short geological trip to Wales, were extreme catastrophists in the sense of William Buckland, let alone in the sense of the contemporary French and continental theorists, they certainly believed that earth processes occurred much more rapidly than did Lyell. It did not take Darwin long to pay his old teachers the supreme compliment of breaking with their thought.

As the ship approached Porto Praya on St. Jago off the west coast of Africa, Darwin saw a bright white line of limestone sixty feet above the sea, running the length of the island. Closer inspection revealed a thick layer of ancient crystalline volcanic rock beneath a twenty-foot-thick layer of limestone capped by a layer of basalt (de Beer 1983, 47). The upper layer of limestone was particularly dramatic in that the basalt had clearly cooked the limestone to a rock hard firmness to the depth of a foot below the line of juncture between the two formations. From its base to its top, the limestone revealed a complete gradation of form from the remnants of shells in the soft lower chalky portions, to a completely transfused upper section in which no trace of shell remained. The limestone band continued around the island at an average height of sixty feet, reaching eighty in other points and occasionally dipping to forty feet. At one place, near an extinct volcano, all three formations disappeared beneath the level of the sea. Using what he had learned from Lyell, Darwin recognized that St. Jago had begun its formation in a shallow sea, that molten basalt had covered and transfused the upper portions of the limestone when the island was yet submerged and that through subsequent periods of uplift and subsidence, the island had risen to its current height. Darwin explained the variations in the height of the limestone by the unevenness of the sea bottom. The place near the volcano where the formations dipped under the sea Darwin attributed to local subsidence (de Beer 1963, 57–59).

In his *Autobiography,* Darwin records that it was on St. Jago that he first resolved to make geology his special subject and to write something on the geology of each of the places that he visited (de Beer 1983, 46–47). St. Jago marks the earliest point at which Darwin's personal ambition to distinguish himself in science becomes explicit. As he was celebrating the vegetation of the island in the language of von Humboldt and describing its geological structure in the language of Lyell, he was making St. Jago his personal intellectual colony. He was the first competent geologist to land on the island and he knew it. Darwin's pattern of using tradition as a repository of inventional resources for advancing innovation had begun.

From Geology to Species: the Lyellian Platform of Darwin's Inventional Program

In becoming a Lyellian geologist, Darwin was making his own perspective a theory-driven framework that advanced a specific agenda of research encompassing far more than geology in any narrow sense. The opening sentence of Lyell's masterwork indicates the range of its subject matter:

> Geology is the science which investigates the successive changes that have taken place in the organic and inorganic kingdoms of nature; in enquires into the causes of these changes, and the influence which they

> have exerted in modifying the surface and external structure of our planet. (Lyell 1990, 1:1)

The central feature of Lyell's legacy to Darwin is not only his breadth of vision but his polemical agenda. When Lyell wrote *Principles of Geology,* he was thirty-two years old, a barrister turned geologist, and was out to make a point. *Principles* is, in fact, a three-volume brief that aims to place geology on a philosophic footing by establishing the first principles of the science. Lyell's over-arching principle is:

actualism

explanation must follow currently known causes.

And governing actualism for Lyell are three other methodological principles:

uniformitarianism[6]

currently known causes have operated in the past uniformly at their currently observed rates of intensity;

deep time

geological change occurs over abysms of time that stagger the imagination; and

non-progressionism

geological processes are non-directional and consist of endless repetitions of the processes we currently see.

Much of Lyell's message reaffirms commonly held interpretive principles, but as Martin Rudwick has shown, *Principles* is governed by its rhetorical aim—to create a new consensus by demarcating geology from Biblicism and pseudo-science (Rudwick 1970, 5–33; Lyell 1990, 1: xvi-xix). On the one hand, the target of the *Principles* is the Biblical flood. Based on the *Principles* a literal worldwide flood is incredible, though a regional catastrophe in the Middle East would still be plausible given what is known of the current world order. On the other hand, Lyell saw early on that progressionist accounts of creation could easily be turned to the support of evolutionary theories, and Lyell (with many Victorians) apparently found the idea of an animal ancestry for humans morally repugnant. By making the world-process cyclical and uniform, and by denying scientific status to any claim to knowledge of a world order prior to our own, Lyell was attempting to cut theories of evolution off before they could capitalize on his geological insights. His affirmation of a world governed by providentially grounded laws made his viewpoint orthodox, at least from the standpoint of the conventional deism, and integral to the received Baconian/Newtonian tradition of English science.

Rudwick has stressed four points as essential to understanding Lyell's work. First, Lyell's detailed examples are illustrations not merely of the processes of earth formation and deformation but of geological reasoning—paradigms, in Kuhn's more restricted sense of the term (1996, 187ff). Second, Lyell emphasizes two defining principles as demarcating science from superstition. The geological causes that we are to credit are not only (1) those presently acting around us but are also (2) those presumed to have operated in times past at the same rate and intensity they currently possess. That is, actualism and uniformitarianism go beyond defining Lyell's operating assumptions; they define science, or at least scientific geology. The key distinction in Lyell's text is thus between kind and degree. Throughout his work, Lyell persistently conflates these two different meanings. Considered abstractly, earth history might, for all anyone knows to the contrary, have been the theater of forces of a magnitude beyond anything presently occurring or having been witnessed in recorded history. But Lyell rules out "catastrophism"—the belief in unprecedented magnitudes of force or of different kinds of force acting in the past and brands it as the mark of superstition and pseudo-science. Third, Lyell's conflated kind/degree pair with its begging of an important question, sets the philosophical and rhetorical agenda for his work. Philosophically, Lyell will treat historically unprecedented degrees of force as unscientific. Rhetorically, he must be prepared to explain away all evidence in their favor.

Giving the unmistakably regulative character that Lyell assigns to currently operating forces in the body of his text, the subtitle of his opus— *An Attempt To Explain The Former Changes of the Earth's Surface, by Reference to Causes now in Operation*—reveals the cunning of the advocate (a cunning Darwin would thoroughly assimilate and turn against his mentor). Lyell's colleagues shared with him the desire to push currently observable forces as far as they could go. The issue in dispute between Lyell and his colleagues was the adequacy of these causes. Rudwick's fourth and final point, that Lyell is advancing a specific program for geological science, reveals the scope and object of Lyell's explanatory ambition. What *Principles* offers is a "system," a view of the geological world as a steady-state ensemble of inter-related processes in which there can be no overall direction (Rudwick 1970, 7–8). In short, given the built-in epistemological constraint which confines knowledge to present forces acting at present intensities, the sense of the world as having had a history, as having proceeded from a known point of beginning to the more progressive or advanced present, vanishes.

As for extinctions and the emergence of new forms of life, Lyell insisted that though particular forms of life have disappeared over time, the general types have persisted and thus no story of progress can be wrung from the imperfect and constantly disappearing records of the earth. Life forms are no more directional, or permanent, than the shifting shapes of sand sculpted by the wind. The one great exception to non-progressionism that Lyell allows is humanity, which he acknowledges as a comparative late arrival (Lyell 1990, 1:155). On the issue of humanity, Lyell is as much a catastrophist or creationist as any of his colleagues. But what is different for Lyell about humanity, and what partially softens the import of his capitulation on the recency and unprecedented advent of humans, is his focus on the moral as opposed to the physical nature of humans. While morally humans represent a clear break without precedent in physical nature, physiologically humans remain continuous with other vertebrates and mammals (Lyell 1990, 1: 155, 156, 157–166). Though Lyell's anti-progressionism was always controversial and never very convincing to his peers, in the 1830s and even beyond, it was a position that could still be maintained (Hooykaas 1966). The last explicit appeal to Lyell's anti-progressionism as a defense of religious orthodoxy appeared in the (originally anonymous) early review of Darwin's *Origin* by the Bishop of Oxford, Samuel Wilberforce, of the Wilberforce/Huxley debate fame (Wilberforce 1860, 138).

Paradoxical as was Lyell's narrative that dramatized earth history while denying that history could be reconstructed empirically, his vision of a balance of continuity/change, destruction/reconstruction, extinction/creation, and of the intimate fit between organism and environment, was an enormously important achievement intellectually and rhetorically. Intellectually, Lyell's *Principles* gave geology an updated Newtonian world picture in which the earth was no longer unchanged and constant since its comparative recent creation, but was a dynamic, self-contained, hydraulic, perpetual motion machine, vastly old yet ever new, the processes of which formed the subject matter of geologic science. Rhetorically, his insistence that geology was not about the past but about the present, gave his thought the cachet of modernity—a kind of geo-presentism well adapted to an age of monumental and heroic industrialization, while his equal insistence on the unfathomable abysms of deep time gave it an almost irresistible romantic appeal to the imagination.

Lyell was a nearly pure scientific rhetorician. He introduced no novel idea or explanatory principle, in the manner of Copernicus, Galileo, Newton, Darwin, or Einstein; rather (almost in the manner of Freud) he drew a powerfully affecting picture of forces long recognized but untheorized. Unconvincing as was Lyell's non-progressionism, Lyell convinced Darwin (even after Darwin became a transmutationist, and even after Darwin had reread Malthus) to see evolution as fluctuating and nondirectional (Herbert 1968, 196–203). In bequeathing to Darwin and Wallace a mode of geological explanation that was imperfectionist, localist, gradualist and modernist, Lyell had formulated the problem of species in a way that—with Malthus, luck, and a little invention—it could be solved. Darwin did not solve the problem at once and probably not on the Beagle. The most important thing Darwin did with Lyell on the Beagle was appropriate his perspective by learning to see the geology of South America through Lyellian eyes.

Patagonia, Western South America, and Darwin's Emergence as a Lyellian Geologist

Darwin's appropriation of Lyell occurred in three phases, of which the most difficult and time-consuming was the first, or South American, phase; the second phase, his theory of coral reefs, and the third, his subsequent post-voyage transition to transmutation, were relatively brief and comparatively painless. The story of Darwin's assimilation of Lyell's geology in South America has been told many times.[7] In briefest compass, one can say that beyond Darwin's all-too-easy victory at St. Jago and his sophisticated, if highly technical analysis of St. Paul's rocks, which immediately followed, his career as a Lyellian geologist begins in earnest on the shores of Patagonia. In September of 1832,

Darwin's trip with Captain Robert Fitzroy and various crew members up the Santa Cruz river in April of 1834 enabled him to confirm his suspicion that South America had risen from the sea in gradual start-and-stop steps. His subsequent geological observations along the coast of Chile in the summer of 1834 confirmed his gradual-rise theory. The climax of his assimilation/reconstruction of Lyell came at Uspallata Pass in April, 1835, where, just one step ahead of winter, at 7,000 feet Darwin found in a "compact greenish sandstone [. . .] a small wood of petrified trees" (Darwin 1986, 1: 442). This discovery confirmed that the Andes had risen by the same start-and-stop processes as had the coast of Patagonia. Darwin had added an entire continent to Lyell's geological manifesto, broadened the reach of Lyell's theories, and fleshed out his dynamics with South American case histories (Browne 1995, 287; Desmond and Moore 1991, 167). Having thoroughly assimilated the teaching of his theory-mentor, Darwin now began the first stage in the process of extending and modifying Lyell's teachings that would ultimately take him to mature theories of his own—and beyond geology altogether. It is in this very process that we can see how incommensurability begins innocently, intelligibly and at home.

Darwin Reinvents His Mentor

By the time he quit the shores of South America on September 7th, 1835, in a process common to intellectual history (Jung vs. Freud, Lenin vs. Karensky, Postal vs. Chomsky) Darwin was both a Lyellian and a Lyellian rival. Darwin was a Lyellian in that his work in the Andes brilliantly added fresh confirmation to the principal features of Lyell's geology. He was a Lyellian rival in that no sooner had he confirmed the fit between Lyell's geology and the South American evidence, than he found in one of Lyell's arguments a critical flaw that provided an opening for an original theory of his own. Lyell had argued that coral atolls were formed from corals growing upward from the rims of sunken volcanoes (Lyell 1990, 2: 290). This theory seemed unlikely to Darwin for a variety of reasons—not least of which was that the size of some atolls was far too large to be the rim of a volcano. Darwin developed a more comprehensive theory by combining Lyell's broad vision of a rising and falling earth with his own close study of corals (Herbert 1968, 34–35). Darwin distinguished three kinds of coral formation: fringing reefs, barrier reefs and atolls. Fringing reefs are coral formations that grow parallel to an island and develop when coral grows outward from the land. Barrier reefs are walls of coral at some distance from the land separated from the shore by a shallow lagoon or channel. Atolls are circular coral islands surrounded by a shallow channel protected at its outer edges by a barrier reef. Darwin reasoned that were the land of an original island to sink, its fringing reef would grow upward, becoming in time a barrier reef (Darwin 1952, 555–557). If the sinking continued, coral would grow upward from the edges of the disappearing central island leaving a remaining circle of coral that would appear as the atoll.

The nub of the theory was elegantly simple. Coral will survive only to a maximum depth of 120 feet. The question then becomes, how is one to account for the platform on which the coral rests? To believe that originally there were subaqueous volcanic mountains of roughly uniform height scattered across similar zones in the Pacific (and the Pacific is predominantly a zone of subsidence)—as Lyell's argument required—is far fetched. Why had not several of these mountains risen to the surface? Again, the coral atolls could not be standing atop eroded sea mounts, for the result would have been a flat disc-shaped fragment of land. Subsiding land with the coral growing upward remained the only logical possibility (Darwin 1952, 560–569). Darwin captured the process through which he developed this theory in his *Autobiography.*

> No other work of mine was begun in so deductive a spirit as this, for the whole theory was thought out on the west coast of South America, before I had seen a true coral reef. I had therefore only to verify and extend my views by a careful examination of living reefs. But it should be observed that I had during the two previous years been incessantly attending to the effects on the land, together with denudation and the deposition of sediment. This necessarily led me to reflect much of the effects of subsidence, and it was easy to replace in imagination the continued deposition of sediment by the upward growth of corals. To do this was to form my theory of the formation of barrier reefs and atolls. (de Beer 1983, 57)

Darwin's development of his coral reef theory illustrates the close interrelation of tradition as the ground of innovation that is the signature of his mind. Darwin's theory was formed, as his own language indicates, both through "deduction" and through "imagination." Darwin had imitated his mentor so closely that now he was prepared to correct him on his own ground (Bloom's *tessera*)—and this on a breathtaking scale.

From Stones to Species

According to Lyell, species became extinct because gradual, incremental changes abolished their environments. Yet, as Darwin had struggled in January 1834 at Port St. Julian to make out the succession of the extinct and extant plant and animal species along the coast of Patagonia, it did not seem to be the case that the mastodon whose bones he discovered had inhabited an environment materially different from the present one. What had been unclear to Darwin in January 1834 in Patagonia gradually became clear to him in February 1835, most likely when he was sailing between Chile and Valdavia on the Pacific (Hodge 1982, 20). While the passage in which he wrote out his views is too long and involved to cite in full, the following excerpt captures the central element of his reasoning.

> The following analogy I am aware is a false one; but when I consider the enormous extension of life of an individual plant seen in grafting of an Apple tree & that all these thousand trees are subject to the duration of life which one bud contained. I cannot see much difficulty in believing a similar duration might be propagated with true generation—if the existence of species is allowed, each according to its kind, we must suppose deaths to follow at different epochs, & then successive births must repeople the globe or the number of its inhabitants has varied exceedingly at different periods. –A supposition in contradiction to the fitness which the Author of Nature has now established. (qtd. in Hodge 1982, 20)

Though some of the reasoning in this passage is involved and draws on a non-Lyellian strand of Darwin's thought, the analogy that grounds it is straightforward. Darwin rejected Lyell's environmentally determined account of extinction and was pursuing a biological analogy between the limit to the duration of the individual life and the limit to the duration of species. In his South American travels, Darwin had seen the remains of whole orchards that had been generated by grafting and had long been productive, but whose trees had all died at the same time. Darwin was claiming in this passage that even as artificial grafting can transmit a limit to the life of the individual plant or animal, and thus to all the individuals generated by graftings from it, so also might a limit to the duration of life be passed on through sexual generation.

With tentative and apologetic language, Darwin drew on these observations to replace the traditional contrast between the artificial process of grafting and the natural process of sexual generation with an analogy between them. In terms of what Jeanne Fahnestock (1999) calls *figural logic,* Darwin has replaced an antithesis with an implicit metaphor—the grafted orchard now becoming a single organism on the order of the zoophytes he was simultaneously examining in his microscopic work. Darwin's move may be framed inventionally as a figural form of Burke's "casuistic stretching," where "one introduces new principles while theoretically remaining faithful to old principles" (Burke 1984a, 229). Lyell himself had called attention to the suggestion of Giovanni Brocchi that there might be a genetically imposed limit to the duration of species, as there was to the life span of individuals. While Lyell unequivocally rejected Brocchi's view as unsupported by evidence, he accorded it status by recognizing it as a potential alternative to his own (Lyell 1990, 2: 128–129). In wanting to bring together Lyell and Brocchi, Darwin holds alternative inventional resources in tension. In breaking with Lyell on species extinctions, Darwin did not deviate from Lyell's broader explanatory paradigm. Moreover, evidenced by his phrase "each according to its kind," Darwin continued to speak the very language of *Genesis.* In his final comment—"Successive births must repeople the globe or the number of its inhabitants has varied exceedingly at different periods. –A supposition in contradiction to the fitness which the Author of Nature has now established"—Darwin reaffirmed with Lyell that species extinctions are followed by special creations to maintain the divinely instituted order of the world.

Galapagos

When the Beagle arrived off Chatham Island on September 15th, 1835, Darwin initially saw the Galapagos, just as he had seen the Andes, as a further extension of the Lyellian geology he had accepted in south-eastern South America. We do not see Darwin on the Galapagos, any more than he was in the Andes, attempting to extend his critique of Lyell's views on extinction to species origins, nor do Darwin's notes show any trace of an attempt to explain why the islands had the species that they had, or why they lacked the ones they lacked. The Galapagos finches, for example, until the work of Sulloway, were advanced as part of the evidence that presumably convinced Darwin of the truth of transmutation. Yet, as Sulloway has shown, Darwin personally collected only a very few specimens and did not bother even to record the island from which they came. (Why collect many if God created but a few original types, and when the local birds so clearly were variants of the South American ones?) Only on his return to England, when specialists helped him to appreciate the value of this evidence, did he borrow the carefully labeled collections of his shipmates from the voyage. Unfortunately for ornithologists, he guessed at the locations for his own birds, mixed them with the locations given by the sailors, and created a taxonomic nightmare at the British Museum.[8]

A similar illustration of how thoroughly Darwin's first view of the Galapagos was filtered through the lenses of creationism is given by the Galapagos tortoises. According to the traditional classification of the tortoises as *Testudio indicus,* the tortoises had been transported to the islands from the West Indies. Though Darwin had been told by the vice governor of the islands, Nicholas O. Lawson, that the tortoises were so distinct that he could tell from which island any particular one came, the fact of mere variability in these supposed colonists did not seem to him scientifically important at the time (Darwin 1952, 337–351). He would labor over the next 10 years to reconstruct from secondary sources proof of species development he had let slip by.

The Lyellian inventional resources embedded in Darwin's initial creationist account of the Galapogos, however, were far richer than his initial reading of the fauna of the islands would suggest. Darwin's first note questioning the fixity of species was written between June and July of 1836—scarcely nine months after his visit to the Galapagos islands—as the Beagle, having rounded the horn of Africa, was heading toward England. Darwin was in the process of putting his Galapagos collections in order and, in light of the imminent end of the voyage and his own growing confidence in himself as a seasoned naturalist, naturally had an eye out for potentially significant generalizations. Reflecting back on the islands especially in light of his collection of mockingbirds, Darwin observed:

> The specimens from Chatham and Albermale [sic] Isd appear to be the same; but the other two are different. in each isld. each kind is exclusively) found: habits of all are indistinguishable. When I recollect, the fact that from the form of the body, shape of scales & general size, the Spaniards can at once pronounce, from which island any Tortoise may have been brought. When I see these islands in sight of each other, & possessed of but a scanty stock of animals, tenanted by these birds, but slightly differing in structure & filling the same place in Nature, I must suspect they are only varieties. The only fact of a similar kind of which I am aware, is the constant asserted difference—between the wolf-like Fox of East & West Falkland lslds.—if there is the slightest foundation for these remarks the zoology of Archipelagos will be well worth examining; for such facts [*would*] undermine the stability of Species. (Barlow 1963, 263)

While the meaning of this note is disputed among Darwin scholars—is he saying he believes in evolution, or only that he is considering it?—the extent of his evolutionary speculations is less important than their presence for our purposes. The note establishes Darwin's awareness of a timely and radical opening in the fabric of convention. Even if we agree with Sulloway that Darwin here is only considering the possibility of transmutation to set it aside for want of evidence (Sulloway 1982, 351), that he would regard transmutation as an attractive explanation at all, let alone in such a calm and matter-of-fact manner, indicates his received explanatory framework was in flux. What stirred Darwin to rethink his commitment to Lyell's explanatory program was a change of venue—Tahiti, New Zealand, Australia—and the surprise of flora and fauna stranger than any he had ever seen. It is worth pointing out that even Lyell's commitment to species permanence was weakened by his visits to oceanic islands (The Canaries) in the mid-1850s (L. G. Wilson 1971, 43–55).

While Darwin had found Tahiti of geological interest, and joined with Fitzroy in writing in support of the work of Anglican missionaries there, he found little in it of botanical or zoological interest. New Zealand, the Beagle's next port of call, was a different matter. Here was a very large island with varied and contrasting "stations" capable of supporting abundant life (Darwin 1952, 511). Yet, in the thick woods Darwin discovered few birds, and only one indigenous animal, a native rat; and it was in the process of being exterminated in competition with the European rat, introduced but two years before! A similar fate was overtaking the native plants (Darwin 1952, 511). Lyell had accounted for New Zealand and Australia by offering them as illustrations of the expected lack of mammals during his secondary age of northern archipelagos (Lyell 1990, 1:129). Perhaps it was the sheer size of the land involved, but "adaptation" to conditions no longer seemed to Darwin a credible explanation for what was missing, particularly in a land that was not geologically young.

In the wake of his growing awareness of the failure of Lyell's account of species origins, it is highly probable, as circumstantial evidence strongly suggests, that Darwin followed the same course of invention that he had earlier when he broke with Lyell's account of extinctions: he picked up the alternative theory Lyell had rejected—evolution (Hodge 1982, 60–66). When one adds the negative evidence of Darwin's abandonment of the Lyellian explanations he had used earlier in south-eastern South America, to the positive evidence of Darwin's use of Lyell's categories in novel ways it seems clear that something in Darwin's thinking is changing. When one notices in addition the continuity between Darwin's final speculations recorded while still on the voyage in his *Ornithological Notebooks* and his first unequivocal evolutionary speculations, the inference that he became an evolutionist on the voyage, while not proven, becomes a distinct possibility. Even if Darwin only converted to evolution in London, after the experts had rendered their verdicts on his collections, Hodge makes an excellent point in observing that Darwin first became a transmutationist when, as a Lyellian geologist, he recognized that transmutation had to be added to Lyell's other criteria for his biogeography to make sense (Hodge 1982, 64). The geology Darwin had learned from Lyell was beginning to lead him to speculations potentially incommensurate with the perspective of his mentor.

Darwin's Debut

Darwin's official debut in the world of high science was 4 January, 1837, when he gave his paper on the relentless, irregular rise of the coast of Chile out of the sea (Desmond and Moore 1991, 207–208). The same day Darwin read his debut paper before the Geological society, following Richard Owen's suggestion, and the advice of others, he deposited a substantial cache of specimens—80 mammals and 450 birds—with the professionals at the Zoological Society at Leister Square (Desmond and Moore 1991, 208). The Zoological Society did very well by Darwin. His fossils and birds were promptly put out for display, reporters took note, and his sisters back in Shrewsbury were able to read about their brother's collections in the *Morning Herald* (Desmond and Moore 1991, 210). And it fed his theorizing. Within six days of depositing them, the ornithologist John Gould informed Darwin that what he had taken for "Gross-beaks" and "black birds" from the Galapagos were, in fact, all true finches and were "an entirely new group, containing 12 species" (Desmond and Moore 1991, 209).

Thanks in no small part to their being placed in Owen's hands, Darwin's South American fossils were major scientific news in the first three months of 1837, as he discovered in case after case that Darwin's fossils were ancient representatives of contemporary forms. Two of the most spectacular and puzzling were an animal Owen dubbed *Macrauchenia,* which he at first thought was an extinct guanaco and then a relative of the camel—related to the living South American Camel or Llama—and a very big sloth, the Glyptodont. Owen later rethought the macrauchenia further and concluded it was intermediate between tapirs and llamas—but the idea that it was a predecessor to living llamas stuck in Darwin's mind. Darwin was also surprised to learn that a fossil horse tooth he had brought back really was a horse tooth—since horses were known not to be native to South America. Why should it be that in South America there was evidence of a natural law by which fossil remains anticipated living forms? Darwin, of course, placed this new information in the context of Lyell's world, in which continuity held the key to change. But still that did not explain either the change or the continuity (Browne 1995, 351).

Inference to Transmutation

The clarity with which the inference of transmutation occurred to Darwin should not detract from our appreciation of the gradual manner, the unlikely instruments, and the thoroughly polemical context through which this realization was achieved. Lyell, as always, knew everything about scientific London, and was thoroughly conversant with Owen's findings. And with good reason. Following up on his young disciple's interpretation of the geology of South America in January, Lyell used his farewell address as President of the Geological Society (William Whewell would take over the helm for 1837) to underscore the superiority of his own methods and his own assumptions. Lyell made a point of inviting Darwin to attend his speech (Darwin 1986, 2: 4). Darwin duly came up from Cambridge for the occasion and, in the company of his cousin, Hensleigh Wedgwood, and his brother, Erasmus, he could not help but be honored by the formal attention Lyell gave to his collections. The polemical use Lyell made of Darwin's material is well forecast in his letter to him shortly before the address when he speaks of "your new Llama, Armadillos, gigantic rodents, & other glorious additions to that new continent, which was heaved up *a un seul jet,* Anno mundi 1656." (Darwin 1986, 2: 4). Lyell's sarcasm was unmistakable and had to have been delicious to Darwin's ears, for it magnified the importance of Darwin's fossils—not only his rocks—for the gradualist case against the catastrophic geology of Elie de Beaumont. Even as Darwin while on the voyage had had to work through the de Beaumont view of geology in trying to make sense of South America, now Lyell, with studied and polished presentational skill, was making plain the philosophic and methodological bankruptcy of his old opponent by drawing out the implications of Darwin's research. Though he was familiar with the facts already, it was through Lyell's lecture that the full meaning and larger implications of Owen's interpretations began to register. Why indeed this remarkable continuity over time? If the older school of geology was clearly no longer tenable, what implication did his own findings have for the continuity of life?

Gould's surprising account of Darwin's birds, coming as it did hard upon Owen's equally surprising account of his South American giants—and Lyell's explication of their meaning—put Darwin over the edge. Darwin began a systematic search for an underlying natural law.

Ever one to economize his material, in the remaining pages of a red pocket notebook he had used on the voyage sometime around mid-March of 1837, Darwin recorded the following observation: "Speculate on neutral ground of the two ostriches, bigger one encroached on smaller. change not progressif<e>: produced at one blow" (Darwin 1987, 61). Succession in time and succession in space, paleontology and biogeography, had convinced Darwin, persuaded as he was already of the essential correctness of Lyell's actualism, that transformism was creation's law. Though his theory of saltation was to be short-lived, soon giving way to a series of gradualist theories that culminated in natural selection, the decisive step had been taken.

In June, his *Journal* at the printers, and in the lull before the endless proof sheets descended on him, Darwin opened a new set of notebooks starting with "B" ("A" having been devoted to geology), which by 1839 would eventually run through E and flow over into his M, N and finally what became his "torn-apart" notebook (Darwin 1986, 2: 433). What to Darwin's scientific superiors were mere technical reports or descriptions of his collections, or to Lyell a confirmation of his own theory, to Darwin seemed an unprecedented explanatory opportunity—a chance to establish for himself a distinctive place in science by producing an alternative discourse. By seeing a strategy whereby Lyell's rhetoric and hermeneutic could be conscripted into a new argument and thus into a new science, Darwin became a scientific revolutionary. What had begun as imitation was leading Darwin to ever more radical innovation.

The Matrix of Invention: Lyell's Last Gift to the Emergence of Darwin's Mature Theories

It might seem strange that in the summer of 1838, in the midst of the single most busy and, arguably, most fateful year of his life, aside from 1859, Darwin would take time off to visit the valley of Glen Roy in Scotland and spend six additional weeks on what he described as "one of the most difficult and instructive tasks I was ever employed on" (Darwin 1986, 2: 432). The focus of Darwin's life—from the time of his return from South America in October of 1836, up to his marriage in January, 1839, to Emma Wedgwood, and their eventual retreat, in 1842, to the Kentish countryside not far from London, where he and Emma would make their permanent home, in Down House (his thoroughly urban, dilettantish and sardonic brother called the place "Down in the mouth"—Darwin 1986, 2: 336)—was concentrated on establishing his reputation.

By mid-summer 1837, through a process of in-group, old-boy politicking, he obtained a government grant of £1,000 and proceeded, having established himself as their general editor, to shepherd through the press a series of technical volumes that would save his Beagle collections from oblivion and establish his reputation among men of science (Darwin 1986, 2: 17). Throughout 1838 and beyond, volumes of his government-backed series, Owen on *Fossil Mammalia,* Gould on *Birds,* Waterhouse on *Mammalia* issued from the press (Darwin 1986, 2: xv). While Darwin oversaw the works of his newfound colleagues and collaborators, he completed his *Journal of Researches* and raced against the calendar to make sure his personal work—none of which touched on the biology of the Beagle voyage, but on its geology—would appear on time. The Glen Roy paper, written during a period of frenetic overwork, in which the mysterious disease that would dog his career first became evident, was published in the *Philosophical Transactions of the Royal Society of London* in February, 1839, the first month of his marriage (Darwin 1977, 87–136). On the strength of this paper, he became Charles Darwin, F.R.S. (Darwin 1986, 2: 433; Rudwick 1974, 129).

The Strategic Context of Darwin's Paper on the Parallel Roads of Glen Roy

Darwin's decision to tackle the parallel-roads problem Glen Roy was a brilliant example of his ability to seize the moment and thereby consolidate several agendas. On the most elemental level, as Darwin had now, after a year of excuses which Lyell had advised him to make, acceded to Whewell's wish that he serve as Secretary to the geological society. He needed credibility. One of the reasons he had given for his earlier refusal of this office was his ignorance of British geology (Darwin 1986, 2: 9–10, 50–52). But Darwin's Glen Roy visit was motivated by an agenda deeper than merely his need for a British geological credential. As his notes prior to the trip and those taken in the field illustrate, Darwin saw Glen Roy as an opportunity to annex British geology to the actualist, uniformitarian project that he shared with Lyell. Considered simply in geological terms, his paper on the parallel roads of Glen Roy was a natural extension of the themes he had developed in each of his geological papers since his return (Rudwick 1974, 161–165).

Darwin's debut paper before the Geological Society of London on January 4 of 1837, "Observations of Proofs of Recent Elevation on the Coast of Chili," set the tone for his subsequent work. In his "Observations," Darwin argues not just that there has been widespread and recent uplift along the coast of Chili, but that South America itself has emerged from the sea by the same process by which it now continues to rise, and that the entire earth, geologically speaking, can be divided into zones of elevation and subsidence. His second paper, "A Sketch of the Deposits Containing Extinct Mammalia in the Neighbourhood of the Plata," given on 31 May, 1837, continues the theme of recent elevation in its account of how the bones arrived at where they were found and how this fact ties in with Lyell's theory of the succession of similar forms in the same region. His third paper, "On Certain Areas of Elevation and Subsidence in the Pacific and Indian Oceans, as Deduced from the Study of Coral Formations," on 7 March, 1838, rounds out the central theme of his South American geology while demonstrating the continuing suggestiveness of focusing on the interconnectedness of geological and biological phenomena (Darwin 1986, 2: 430–432). The project of developing a worldwide geological theory of uplift and subsidence was substantively advanced by his coral paper.[9]

The connection of Glen Roy to his working partnership with Lyell could hardly be more evident. And this makes all the more interesting the ultimately incommensurate nature of the argument Darwin perfected in this essay with the arguments and worldview of his mentor (which emerges not in the Glen Roy essay itself, but in the transfer and adaptation of its reasoning to his species work). Darwin had first heard about Glen Roy through Lyell's *Principles,* the third volume of which had reached him in Valparaiso in 1834. In this volume, Lyell had remarked on the "Parallel Roads of Coquimbo" and summarized the account of the Englishman Basil Hall who had first observed them (Lyell 1990, 3: 131). Hall had interpreted the Coquimbo roads as the beaches of a lake using, as an analogy, the work of John MacCullough and Sir Lauder Dick on the parallel roads of Glen Roy (Rudwick 1974, 114–115). Lyell also noted that Hall later had doubts and suspected that the roads of Coquimbo were of marine origin. When Darwin studied the Coquimbo roads himself in May of 1835, he observed not only five or six major terraces stretching mile after mile in a valley east and inland from the coast but found several intermediate terraces and abundant shells which settled all doubt that they had been formed by the sea (Rudwick 1974, 114). The facts on the ground provided immediate support for the thesis Darwin was then developing on the elevation of continents "along linear 'axes of elevation,' one of which was presumed to lie some way inland, parallel to the coast" (Rudwick 1974, 115). Darwin concluded that the same process had taken place in Scotland. Even on its face, and without Darwin's continent-elevating and sea-sculpting hypothesis, MacCullough and Dick's freshwater hypothesis had major problems. The Scottish lakes would have to have been impounded by barriers, and massive ones at that, yet there was no evidence in the Glen Roy area of the remains of the required barriers. Darwin had every reason to suspect they had never existed—and his continent-elevating, sea-sculpting hypothesis would explain why. Later Louis Agassiz's glacier hypothesis was to explain the roads, and Darwin would ultimately abandon his Glen Roy work as "a great failure" (de Beer 1983, 48). But the Agassiz position was still only a rough suggestion and at the time not carefully developed. Besides, Agassiz was associated with catastrophism and thus his whole line of inquiry was viewed with suspicion by Darwin and by his peers in the British scientific establishment (Herbert 1968, 28).

Glen Roy then, was a rich prize, simply in geological terms. But his Glen Roy speculations spilled also over beneficially into his transmutation notebooks. Darwin visited Glen Roy for eight days, from 28 June to 5 July, 1838. After a couple of weeks at Shrewsbury, where he rested and visited with his father and sisters, he began his Glen Roy paper on the first of August and completed it on the sixth of September. Darwin opened his notebooks D and M ("M" signaling the first of his separate notebooks on Man) at Shrewsbury during his visit to his father following his trip (Darwin 1986, 2: 432). Further, he copied into notebook D from his Glen Roy notebook several observations on inheritance that had emerged from his conversations with shepherds during his walks in the highlands (Darwin 1987, 345, D: 43). Not least significant, the opening section of D, preceding his Glen Roy passages, contains various reflections on strategy. "So far is there any appearance of animals being created. It is probable if created at once. <wd> according to ordinary laws, the character of offspring would vary." (Darwin 1987, 336, D: 19). Or again, "In my speculations. Must not go back to first stock of all animals, but merely to classes where types exist for if so, it will be necessary to show how the first eye is formed.—how one nerve becomes sensitive to light" (Darwin 1987, 337, D: 21). The fruit of Darwin's evident thought on how to reframe recalcitrant questions and reposition the direction of the argument and the burden of proof are abundantly evident in his Glen Roy paper.

Arrangement and Argument in Darwin's Paper on Glen Roy

Darwin's earlier papers were comparatively data driven. The elevation of the South American content on which two of his earlier papers had focused was something Darwin had documented profusely through his numerous first-hand observations. Even his more speculative theory of coral reefs rested on the known processes of subsidence and uplift and made more evident sense, on Lyellian grounds, than Lyell's own explanation of the sinking of extinguished volcanoes. The Glen Roy paper is unique in Darwin's early writings in the degree to which it relies upon a speculative or overtly philosophic form of argument grounded in Herschel's *Preliminary Discourse* (Herschel 1987) and exemplified in Lyell's *Principles,* to tip the scale of likelihood in his favor when the factual evidence he desired (or needed) was absent or ambiguous (Rudwick 1974, 167–170). The combination of theory and case—of case selected in the interest of theory and of fact carefully noted, but noted with an eye to its role in theory—and of the reception of that theory by an audience through a persuasive argument rooted in complex probabilities, signals the full emergence of Darwin, the self-conscious rhetorician of science long in preparation in his notebooks (Campbell 1990).

The argumentative spirit of Darwin's paper is manifest in his title, "Observations on the Parallel Roads of Glen Roy, and of Other Parts of Lochaber in Scotland, with an Attempt to Prove that They Are of Marine Origin" (Darwin 1977, 89–136). Darwin's title begins empirically with "Observations of the Parallel Roads of Glen Roy," a familiar, if contested object of geological reflection, the focus widens with "and of Other Parts of Lochaber in Scotland," the argumentative character appears with "an Attempt to Prove," and the last clause announces the thesis, "that They are of Marine Origin." As the title anticipates, every observation in the essay will be in the service of argument and every argument will draw on observation. Under color of description or the examination of a seemingly secondary point, again and again Darwin challenges fundamental premises of his opponents and substitutes for them assumptions favorable to his own case.[10] As with the *Origin* and his mature work in general, by the time Darwin mounts his principal argument many readers will have anticipated it and possibly shifted their opinions to match his (Campbell 1984; 1986).

Two points about that argument are worth stressing. First, important as is the role of description in Darwin's paper, underlying his argument is the *vera causa* logic advocated by Herschel and exemplified by Lyell (Rudwick 1974, 167–170).[11] Despite the medieval ring of the Latin expression, the vera causa principle was seen as, first, a reaction to scholasticism, and, second, British. Developing and summarizing a tradition which begins proximately with Newton, Herschel stressed that the method of science from Bacon through Newton placed importance on reasoning from *real causes* as opposed to the occult properties of Aristotelian substantial forms (Herschel 1987, 50 and especially chapters 5 and 6, Part II; Kavalovski 1974, 12–17). A real cause had to satisfy three criteria:

1. it had to have an existence independent of the phenomenon it was to be used to explain;
2. it had to be adequate to bring about the character of the change required; and
3. reasons had to be given to show it was responsible for the phenomenon (Kavalovski 1974, 40–41).

These tests—independence, adequacy, and responsibility—may sound stringent, and perhaps are when compared to "dormative properties" of quasi-Aristotelian science. But a certain facility, a happy gift, for analogy was required to make them work.[12]

As Herschel explained, though we do not see a sling when we observe the rotation of the planets around the sun, we may safely infer that something of the sort is in play to restrain them, however unseen and acting at a distance that something may be, for our experience tells us that a stone, but for the restraint of the sling, would otherwise fly off in a straight line (Herschel 1987, 148–149, 192–199). But how did one know if one's candidate for a true cause was real and not just a mere analogy? The answer was more analogies, particularly the analogies of similar phenomena that come unbidden from sometimes quite different regions of experience, and press in almost of their own will, to confirm the explanatory power of the analogy or cause.[13] Herschel's colleague Whewell famously called this part of Herschel's scientific logic "the consilience of inductions" (Whewell 1840, vol. 2; Butts 1973).

Darwin's Glen Roy paper illustrates at once the strengths and the possibilities for abuse in the role assigned to analogy in *vera causa* reasoning. It is here that we begin to see emerge the inventional path that would lead Darwin from the very point at which he was closest to Lyell to his development of a mode of reasoning that would separate Darwin from his mentor.

From the first, Darwin wanted to use the example of the parallel roads of Coquimbo to explain the similar-seeming phenomena of Glen Roy (Rudwick 1974, 153–157). To do this he could, and quite legitimately, note the number of similar features in the two formations that indicated not just the role of water but of a volume and force of water, which presumably only the sea could supply (Darwin 1977, 99–107). Further, in the absence of unequivocal confirming facts for the lake hypothesis, he could note the embarrassing assumptions it required and the comparative advantages and simplicity of his own assumptions (Darwin 1987, 95–99). The problem was that if only one set of causes were the true ones the other set risked not being regarded as scientific at all. What on the one hand, in the context of a comparative advantage argument, might seem a reasonable, if not an exclusive explanation, if framed as a choice between true causes vs. false ones could easily end up turning a contingent epistemological advantage into an outright ontological monopoly.[14] The origin of *vera causa* reasoning as a weapon of war, a demarcation device in Newton's attempt to discredit "the occult causes" of the Medievals as unscientific, made this possibility more than academic.

Recalling Lyell's attempt in *Principles* to associate sudden causes of unprecedented force with all that was retrograde in science, and thus have them dismissed *tout court* as unscientific, Darwin, on the basis of lengthy preliminary manoeuvring, does the same with the defenders of the lake hypothesis.[15] Using the absence of any remains of the necessarily massive retaining wall that would have been needed to have contained the supposed lakes as his starting point, Darwin proceeds to associate the lake hypothesis with unknown, even unknowable, causes

(Darwin 1977, 97–99). Working further at inconsistencies in his opponents' position as to how the failure of the barriers was to have proceeded and how the various breaches at various elevations at various times could have explained the roads, Darwin soon reduces the position of his opponents to a series of absurdities (Darwin 1987, 97, 99). To the apparently outrageous or extravagant assumptions of his opponents he contrasts the elegance of his own actualist assumptions based on the tidal action of an arm of the sea. Though he gives very plausible arguments for his own position, his argument finally rests on the view that only the marine explanation is scientific.[16]

The possibilities for abuse in establishing the analogies on which Darwin's *vera causa* argument rests are patent. Arthur Fine, in an introduction to *Preliminary Discourse,* an introduction that greatly supports Pera's thesis about the informal and argumentative character of the truly crucial turns in scientific reasoning, argues that Herschel's "neo-Baconian scheme for arriving at theories inductively is a rather looser and more open-ended scheme than talk of induction and 'the scientific method' might suggest" (Herschel 1987, xviii). That much in establishing what counts as a *vera causa* must rest upon skill in argument is indicated by the prominence Herschel gives to Cicero in the beginning of the volume.[17] Indeed, in his first note to Lyell about Glen Roy, Darwin said of his sea-water hypothesis, "I think I can explain away most, if not all the difficulties" (Darwin 1986, 2: 96). The key to Darwin's argument is in controlling the analogies that limit the range of what can count as a true cause.

As commentators have repeatedly noted, time and again in Darwin's mature work, particularly in *The Origin,* he is careful to air the objections to his theory.[18] He does the same thing in his Glen Roy paper. What merits closer attention is the way Darwin's concessions to his potential critics are often tied to his attempt to reverse the reader's expectation of the burden of proof. Having used his gradual elevation-plus-*stasis* thesis to explain how the roads could have been formed by the sea during a period of rising earth, Darwin notes, "Several objections [. . .] will occur to everyone. [. . . of which the strongest] is, that, as the upward movement probably affected a considerable area, or at least as it cannot be supposed to have been confined within a defined line, so ought the shelves to be continuous over an equal space" (Darwin 1977, 111). In no more than the space of the next line, the whole meaning of this admission is reversed: "I believe, however, from what I have seen in South America, that it would be more proper to consider the preservation of these ancient beaches as the anomaly, and their obliteration from meteoric agency the ordinary course of nature" (Darwin 1987, 111).

A similar line of argument emerges as Darwin explains an anomaly in his own position—the missing shells. Darwin begins with a seemingly devastating admission. When he first searched the Spean and Roy valleys, he says, he expected to find "fragments of sea-shells but I could not discover a particle, and the quarrymen assured me they had never observed any." The admission is necessary, but Darwin quickly turns it into an advantage, certainly away from being a threat. "This may at first be thought to be a strong objection against the theory of the marine origin of these deposits." What seems apparent at first will not survive the carefully worded, clause-punctuated particulars of his second thought.

> But having been led in consequence of Mr. Murchison's remarkable discovery of recent sea shells in the inland counties of Shropshire and Staffordshire, to examine many gravel pits there, and having observed how frequently it happens, that not the smallest particle can be discovered in vast accumulations of the rudely stratified matter, and that when found, the fragments are generally exceedingly few in number and partially decayed, I feel convinced that their preservation may be considered as a remarkable and not as an ordinary circumstance. (Darwin 1987, 113)

As Darwin's statement shows, in scientific reason as in law, establishing the appropriate precedent can be crucial. Read one way, a pattern of blanks where one had reason to expect evidence, might be understood as counting against the theory that required the expected evidence for its support. But filled out by analogy, or on the basis of probabilities, the conditions of which are difficult to establish except on the basis of further analogies, blanks can be read so that a factually underdetermined theory can be saved from rebuttal. "In some parts of South America I have found beds of gravel which did not contain a fragment of shell, and yet on the bare surface, nearly perfect ones were strewed in numbers" (Darwin 1987, 114). Darwin notes similar cases are known to have happened elsewhere, for both a Mr. Smith and Lyell himself have failed to find shells in upraised coastal areas known to have been of marine origin (Darwin 1987, 114).

From a rhetorical point of view, a great beauty in Darwin's argument is that it rests on a quasi *vera causa* of its own and is not, or at least can plausibly protect itself from the charge of being, a mere argument from absence. Not only does there exist, Darwin notes, a cause capable of destroying the evidence he needs, but, more convenient still, he shows this very cause is more active in Scotland than in England! "Thus in Shropshire, the gravel is covered in most parts by an earthy deposit, which contains a small proportion of lime; hence the rain water having absorbed carbonic acid gas in its descent, would find matter to dissolve before it reached the layers containing shells." This then explains the presence of shells in England—that is, it explains them as having not yet been destroyed or why they have escaped destruction. However, "in Lochaber the gravel and sand, being derived entirely from granite rocks, does not, as I ascertained, usually contain any free carbonate of lime, and consequently the fragments of shells would more readily be dissolved" (Darwin 1987, 114). Perhaps the greatest beauty of all in Darwin's argument is his restraint. Darwin would not have the reader confuse this argument with mere cleverness and warns the reader in methodologically grave terms about taking his argument too seriously. "I do not wish to assign this circumstance as the real cause of their disappearance," Darwin says, "but merely to indicate it, and other similar ones, as quite sufficient to show that the marine origin of the shelves cannot be controverted from the absence of organic remains" (Darwin 1987, 114). In the very act of denying that he means to give his argument any undue positive weight, Darwin blunts a potentially devastating (and devastatingly obvious) objection to it and solidifies the other probabilities that strengthen it.[19]

Darwin's "Path" to Natural Selection

The coincidence between Darwin's completion of his Glen Roy paper and his reading, or rather re-reading, of Malthus is striking. Darwin sent his Glen Roy paper to the Royal Society on 6 September, 1838 (Darwin 1986, 2: 432). On 28 September, three weeks and one day later, as he recorded in Notebook D, he read Malthus. In his *Autobiography,* Darwin claimed to have read Malthus for his "amusement" (de Beer 1983, 71). Given Darwin's proverbially omnivorous reading and the centrality of Malthus to the reform movement on the lips of everyone in the London world in which he and his celebrity-entertaining, dinner-party-giving brother moved, Darwin had to have been thoroughly familiar with Malthus's arguments well before September of 1838. What is far more likely than Darwin having read Malthus for amusement is that he read him following a hunch (Schweber 1977, esp. 231–234). In any event, when Darwin re-read Malthus, he saw something new. Here the incommensurate divide between Darwin and Lyell emerges in earnest, though paradoxically it was only Darwin's thorough assimilation of Lyell that prepared him for his supremely productive parting of the ways. The language of his entry in notebook D still communicates the shock of insight.

> <<I do not doubt, every one till he thinks deeply has assumed that increase of animals exactly proportiona[1] to the number that can live.—>> We ought to be far from wondering of changes in number of species, from small changes in nature of locality. Even the energetic language of <Malthus> <<Decandoelle>> does not convey the warring of the species as inference from Malthus.—<<increase of brutes, must be prevented solely by positive checks, excepting that famine may stop desire.—>>in Nature production does not increase, whilst no checks prevail, but the positive check of famine & consequently death.
>
> population increase at geometrical ratio in FAR SHORTER time than 25 years—yet until the one sentence of Malthus no one clearly perceived the great check amongst men.—<<Even a *few* years plenty, makes population in Men increase, & an *ordinary* crop. causes a dearth then in Spring, like food used for other purposes as wheat for making brandy. —>>take Europe on an average, every species must have same number killed, year with year, by hawks. by. cold &c—. even one species of hawk decreasing in number must effect instantaneously all the rest.— One may say there is a force like a hundred thousand

> wedges trying to force <into>every kind of adapted structure into the gaps <of> in the oeconomy of Nature, or rather forming gaps by thrusting out weaker ones.<<The final cause of all this wedgings, must be to sort out proper structure & adapt it to change.—to do that, for form, which Malthus shows, is the final effect, (by means however of volition) of this populousness, on the energy of Man>> (Darwin 1987, D134e-135e)

This somewhat breathless passage marks at once a moment of discovery and an intensification of the process of interpretation that always had been integral to Darwin's enterprise. Mara Beller notes that "scientific creation [. . .] is often characterized by the ingenious mingling and selective appropriation of ideas from different 'paradigms'" (Beller 1999, 301), and this passage marks an intensification of Darwin's process of interpretation—for in it he is realizing how differently Lyell's paradigm looks when read in light of Malthus. In David Kohn's haunting phrase, it is at this point, when Lyell and Malthus meet in his thought, that Darwin realized "the creative power of death" (Kohn 1980, 144).

As with his theory of coral reefs, Darwin's transition to his first "selection" theory emerges through a radicalization of the thought of his mentor. Lyell was well aware, of course, of the struggle for existence, but had thought of it as occurring at the level of the species, not at the level of the individual (Lyell 1990, 2: 172–175; Kohn 1980, 144–146; Herbert 1971, 216–217). As a result, he assumed that when geological change occurred, better-adapted forms in adjacent hitherto separate areas would out-compete native ones and drive them to extinction, stopping novelty before it could start (Lyell 1990, 2: 173–175). Darwin quickly saw how Malthus's population pressure, operating on individual organisms to produce an intra-specific struggle for existence, warranted the opposite interpretation—hence his emphasis on "wedging," "forming gaps" and "adapted structure." Darwin saw how Malthus supplied a biological principle—a "force" in the required Newtonian sense—that, operating on individual organisms just as relentlessly as vulcanism, wind, and water operated on regions in geology, would necessarily yield change (Schweber 1977, 231–234, 261, 199–200).

Though it would take Darwin months, even years, of thought and insight to realize fully what he had discovered, the focus of Darwin's thought after September 28th increasingly turned to the problem of exposition.[20] Where his old theory of generation had been rhetorical in fits and starts, the new theory was rhetorical from the ground up; it emerges, with but the sketchiest of evolutionary histories from the late entries of Notebooks D and E, drops the rich psycho-biological themes of notebooks M and N and presents its stripped down minimalist core as an exemplification of the logic of Herschel and Lyell—a logic, now that it takes up residence in biology, incommensurate with the views of the mentors who had unintentionally nurtured it (Herbert 1974). What would become the mature rhetorical strategy of the *Origin* emerges in full polemical form in the *Sketch* of 1842 and in the *Essay* of 1844 (Limoges 1970, esp. 105).[21]

The *Sketch* of 1842 and the Emergence of Darwin's Mature Strategy

The details by which Darwin moved from his initial Malthusian insight, centered on struggle, to his reinvented theory, centered in the analogy of the breeder, I have addressed elsewhere (Campbell 1990a, 58–90; 1990b). In this brief section, I wish to focus on how Darwin, by committing the "category mistake" of applying the Herschel/Lyell *vera causa* logic to biology, put the resources of this once conservative tradition in the service of radical change. If a "category mistake" can be understood socially as combining what convention seeks to separate, then it is through successfully committing such a mistake, however contested the success, that convention gives birth to innovation, and a strong writer in Bloom's sense overcomes tradition to become recognized for original thought.[22] It is through such processes that we see how radical difference emerges from imitation. I will confine my analysis just to the *Sketch* of 1842, though its implications extend to Darwin's *Origin* and beyond, since it is in this short and unpolished work that the inventional germ of Darwin's mature strategy first emerges.

The 1842 *Sketch* (as well as with lessening degrees of obviousness Darwin's 1844 *Essay* and the *Origin*) is organized into three sections united by a single three-part Herschelian logic (Hodge 1977, 237–245). An opening section establishes the existence of variation and selection as causes of organic change in domestication sufficient (were the organisms in a state of nature) to cross the division between a variety and a species. A second section establishes the existence of variation in nature and a struggle for existence that eventuates in "selection." A third section examines 'What then is the evidence in favour of it and what the evidence against it" (in Darwin and Wallace 1958, 60). As this three-part strategy makes clear, the central expository lesson of Glen Roy has found a new application.

If Darwin's Glen Roy paper, particularly in the subtle manner in which it prepares the reader to recognize the action of the sea in sculpting the land, can be characterized as description laced with argument, the *Sketch* and *Essay* can be characterized as argument laced with description. Given that the work of the revisionist can only occur when there is a sufficiently rich vocabulary to revise, the challenges Darwin faced in his Glen Roy and in his species work were parallel but distinct. In the Glen Roy paper, Darwin was reworking the already well-worked language of geology and addressing what everyone recognized as a classic problem in need of explanation. In his species work, Darwin's challenge was to get his reader to recognize biology as a field rich in problems begging to be addressed by the established methods of science—by the methods established for the sciences of motion (Kavalovski 1974, 41–42). Darwin addresses his persuasive problem not by a formal defense of extending the *vera causa* principle and its applicability to biology but, indirectly, almost in the manner of his notebooks, by engaging the reader in a series of intriguing puzzles he has assembled from his reading and has arranged in his argument.

Vera causa Argument and Expository Art

Under the unpretentious heading "On Variation Under Domestication, And on the Principles of Selection," Part I of Darwin's 1842 *Sketch* begins, "An individual organism placed under new conditions [often] sometimes varies in a small degree and in very trifling respects such as stature, fatness, sometimes colour, health, habits in animals and probably disposition. Also habits of life develop certain parts. Disuse atrophies. [Most of these slight variations tend to become hereditary]" (in Darwin and Wallace 1958, 41; Darwin's interpolations). And over the space of approximately two and one-half pages—as if in response to a reader asking "But what of it?"—Darwin offers a series of implications and extrapolations that do nothing if not tempt thought:

> A certain degree of variation (Muller's twins) seems inevitable effect of process of reproduction. But more important is that simple generation, especially under new conditions [when no crossing] causes infinite variation and not direct effect of external conditions, but only in as much as it affects reproductive functions. There seems to be no part (beau ideal of liver) of body, internal or external, or mind or habits, or instincts which does not vary in some small degree and [often] some to a great amount. (in Darwin and Wallace 1958, 41–42; Darwin's interpolations)

Into this mix of technically interesting, if diverse, nature lore, Darwin invites the reader to consider the lessons of ordinary experience. "But if man selects then new race rapidly formed—of late years systematically followed—in most ancient times often practically followed." This familiar idea, briefly introduced, Darwin turns yet again. "But man selects only what is useful and curious—has bad judgment, is capricious—grudges to destroy those that do not come up to his pattern." A brief run of development and Darwin's thought, and first section, ends, "This might all be otherwise" (in Darwin and Wallace 1958, 43). Darwin's tempting of the reader to think otherwise is, of course, the name of his enterprise. This enterprise is only strengthened by his constant fortifying of the reader with new material, little of it in itself controversial.

In "II. On Variation in a State of Nature, and On the Natural Means of Selection" (which eventually would become the second chapter of *Origin,* "Variation Under Nature") the opening assertion seems positive, "Let us see how far above principles of variation apply to wild animals." But the initial evidence is discouraging. "Wild animals vary exceedingly little—yet they are known as individuals." Still there is evidence on both sides, "Primrose and cowslip Wild animals from different [countries can be recognized]. Specific character gives some organs as varying" (in Darwin and Wallace 1958, 43–44). And then in a run of nearly continuous exposition, we see flashes of that capacity for argument developed in detail in the *Essay* and later perfected in *Origin.* What indeed, amidst all these possibilities and ambiguous signs, is the lesson of our experience?

> Our experience would lead us to expect that any and every one of these organisms would vary if the organism were taken away and placed under new conditions. Geology proclaims a constant round of change, by every possible change of climate and the death of pre-existing inhabitants, endless variations of new conditions. These generally very slow, doubtful though . . . how far the slowness would produce tendency to vary. But geologists show change in configuration which, together with the accidents of air and water and the means of transportal which every being possesses, must occasionally bring rather suddenly, organism to new conditions and expose it for several generations. Hence we should expect every now and then a wild form to vary; possibly this may be cause of some species varying more than others. (in Darwin and Wallace 1958, 44)

Though never himself a student of law, this passage is a perfect illustration of Darwin's mastery of all those habits of mind that Thomas Jefferson found so irritating in Chief Justice John Marshall when he declared him the enemy of well established distinctions who "sophisticates the law to his own mind, by the turn of his own reasoning" (qtd. in Simon 2002, 281). But Darwin had studied something better than law. He had studied *Principles of Geology.* Finding evidences potentially favorable to his case in the tangled apparently contradictory facts of his material, Darwin is here, to biology, applying lessons in prudential/adversarial argument written on every page of the geological work of Lyell, his professionally trained lawyer/mentor. In the spirit of a cross-examination, Darwin exhibits to his reader a series of evidences both for and against the permanence of organic forms. The evidence against the position for which he is apparently arguing garners the reader's trust while creating at the very same time an opening for anomalous or contrary evidence to register. The apparent weakness, or equivocalness of Darwin's argument, gives it a subtle and supple strength. As a reader continues, even a reader who finally will decide against Darwin, that reader cannot help but be struck by the sheer number of anomalies that could, were the evidence but stronger or conditions somewhat different, provide a starting point for the view of nature's pattern that he has rejected. In Darwin's mastery of how to translate technical knowledge into the enthymemes, topics and common sense idioms of his peers, we see him giving life to what otherwise would have been the blank, lifeless, and merely abstract form of *vera causa* reasoning.

Vera causa reasoning, as we have noted, required analogies to make it work. And, no less than in his Glen Roy work, Darwin was quick to supply them. Having established the fact of variation, he went on to ask

> But is there any means of selecting those offspring which vary. [. . .] But can varieties be produced adapted to end, which cannot possibly influence their structure and which it is absurd to look at as effects of chance. Can varieties like some vars of domesticated animals, like almost all wild species be produced [. . .] But if every part of a plant or animal was to vary [. . .] and if a being infinitely more sagacious than man (not an omniscient creator) during thousands and thousands of years were to select all the variations. (in Darwin and Wallace 1958, 44–45)

That Darwin was wary of tipping his hand to reveal how much farther he took the principle than would his reader is evident by his note to himself in mid paragraph, "(Good place to introduce, saying reasons hereafter to be given, how far I extend theory, so to all mammalia—reasons growing weaker and weaker)" (in Darwin and Wallace 1958, 46). Clearly Darwin wished his language to introduce just the right touch of social knowledge, while not encouraging (though he could not entirely block) wrong or inconveniently accurate, implications of human ancestry. In his demiurge figure, Darwin does not so much cancel traditional theism as make it familiarly strange, thereby dramatizing the possibility of convention being safely extended.

Sometimes Darwin is particularly cautious of not asking too much of his reader. Having introduced late in his first section the development of "some of the most difficult cases of instincts, whether they could be possibly acquired," he immediately cautions, "I do not say probably, for that belongs to our third part, I beg this may be remembered, nor do I mean to attempt to show exact method. I want only to show that whole theory ought not at once to be rejected on this score" (in Darwin and Wallace 1958, 56). At other times, when he wishes forcibly to impress upon the reader how the facts, premises and truths of convention can yield opposite conclusions, he breaks forth into what can only be described as a virtuoso performance—a performance, by the way, that reads and often sounds, particularly when read aloud, like oratory, the oratory befitting the closing argument in a trial and which was known to characterize the parliamentary style deliberations of the Geological Society (Secord 1986, 14–24). In the following passage, I have highlighted in contrasting **bold** and ***bold italics*** the parallel passages, to underscore the antiphonal character of Darwin's sentences.

> Summing up this division. **If variation be admitted** to occur occasionally in some wild animals, ***and how can we doubt it,*** when we see thousands of organisms, for whatever use taken by man, do vary. **If we admit** such variations tend to be hereditary, ***and how can we doubt it*** when we remember resemblances of features and character—diseases and monstrosities inherited and endless races produced (1200 cabbages). **If we admit** selection is steadily at work, ***and who will doubt it,*** when he considers [. . .] **If we admit** that external conditions vary, ***as all geology proclaims*** they have done and are now doing—then, if no law of nature be opposed, there must occasionally be formed races [slightly] differing from the parent races. [. . .] **Are not all the most varied species** [. . .] ***Take Dahlia and potato,*** who will pretend in 5000 years (that great changes might not be effected): [. . .] **Think what has been done in few last years,** ***look at pigeons, and cattle.*** **With the amount of food** [. . .] ***I conclude it is impossible to say we know the limit of variation.*** **And therefore with the [adapting] selecting power of nature, infinitely wise compared to those of man,** ***I conclude that it is impossible to say we know the limit*** of races, which would be true to their kind. (in Darwin and Wallace 1958, 57–58)

In each of the four opening *if*-clauses, and in the several declarative sentences that follow, Darwin does not simply identify a fact of nature, or class of evidence favoring his thesis, he appeals to the reader to consult and thus reconfigure his or her experience. The lengthy paragraph focuses the reader's attention not primarily on nature, but on topics, principles of reasoning that, however related to the facts to which Darwin alludes, are prior to them for they are rooted in everyday life. The paragraph can be read as a series of *a fortiori* analogies, the purpose of which is not to prove Darwin's thesis but to save it from extinction by making the reader confess—and *confess* is the precise term—that Darwin's claims are rooted in the reader's experience. But once merely saved from extinction, Darwin's thesis seems impossible to limit and begins to colonize the reader's experience. Grammatically, the passage is full of questions ("who will doubt ...", "are not the most ...", "who will pretend ..." and "how can we doubt ..." twice), with some mild imperatives ("Take Dahlia and potato...", "think what ..."), interspersed among inclusive assertions. The effect of those assertions is also, like the questions and imperatives, to invite his reader into the deliberative process. They are largely defined by first-person-plural subjects, alongside predicates of concession and agreement ("we admit," "we remember," "we see," ...). Overall, the rhetorical mood of the passage is defined by the verbs in these predicates, which build in epistemic strength to the even stronger verbs of cognition, paired up in the last two sentences ("conclude [. . .] know," twice).

Parallel to the arguments in his Glen Roy paper that countered the absence of shells or other marine remains, by establishing through a series of analogies that nothing prevented their having been there, here Darwin argues that if a phenomenon pertinent to the explanation of his thesis of organic change can be shown to occur in domestication, it must be assumed to have happened, and even more powerfully in nature. In each of the sentences of this passage Darwin uses what his reader does know to warrant by analogy the further extension of his thesis to what the reader does not know.

Once mundane experience can be shown to be consonant with Darwin's theory, then Darwin can seek to rule the alternative perspective out of court, by challenging it on methodological grounds just as in his Glen Roy paper he similarly sought to cancel the lake hypothesis. The final sentence of the first chapter in his sketch sets in place what appears to be the next step in the *vera causa* agenda that must be executed by chapter two: "But is there any evidence that species have been thus produced, this is a question wholly independent of all previous points, and which on examination of the kingdom of nature we ought to answer one way or another" (in Darwin and Wallace 1958, 58).

But before Darwin turns to the promised examination of the evidence, the change in the rules of the game, implicitly set in place in Chapter 1, is now made explicit. Lyell, it will be recalled, under cover of merely attempting to see how far actualistic explanations could go had sought to use the *vera causa* principle to rule out of geology any other kind of causes—forces different in kind than available to current inspection or operating at unprecedented intensities. In the beginning of Chapter 2, following closely in his mentor's footsteps, Darwin makes the question of whether there is "any evidence [. . .] wholly independent" of that of domestication entirely subordinate to what can count as evidence. First, he drives home to the reader how the mere possibility of the actualist thesis he has sketched in Chapter 1 is an enormous embarrassment to the established perspective.

> I may premise, that according to the view ordinarily received, the myriads of organisms peopling this world have been created by so many distinct acts of creation. As we know nothing of the [. . .] will of a Creator—we can see no reason why there should exist any relation between the organisms thus created; or again, they might be created according to any scheme. But it would be marvelous if this scheme should be the same as would result from the descent of groups of organisms from the same parents according to the circumstances, just attempted to be developed. (in Darwin and Wallace 1958, 59)

Then, interweaving his project with the narrative of the inevitable march of science toward actualist explanations as set forth by Lyell in the opening chapter of *Principles* (and driven home with force in Lyell's immediately following four-chapter history of the development of geology), Darwin shows the reader how the received view of species must similarly be placed on a path of extinction:

> With equal probability did old cosmogonists say fossils were created, as we now see them, with a false

> resemblance to living brings; what would the Astronomer say to the doctrine that the planets moved not according to the law of gravitation, but from the Creator having willed each separate planet to move in its particular orbit? I believe such a proposition (if we remove all prejudices) would be as legitimate as to admit that certain groups of living and extinct organisms, in their distribution, in their structure and in their relations one to another and to external conditions, agreed with the theory and showed signs of common descent, and yet were created distinct. As long as it was thought impossible that organisms should vary, or should anyhow become adapted to other organisms in a complicated manner, and yet be separated from them by an impassable barrier of sterility, it was justifiable, even with some appearance in favour of a common descent, to admit distinct creation according to the will of an Omniscient Creator; or, for it is the same thing, to say with Whewell that the beginnings of all things surpass the comprehension of man. In the former sections I have endeavoured to show that such variation or specification is not impossible, nay in many points of view is absolutely probable. What then is the evidence in favour of it and what the evidence against it. With our imperfect knowledge of past ages [surely there will be some] it would be strange if the imperfection did not create some unfavourable evidence. (in Darwin and Wallace 1958, 59–60; Darwin's interpolation)

In the concluding lines Darwin seems to pull his punches. The pull, however, is only apparent and is parallel with the similar root equivocation that runs throughout *Principles of Geology*. In effect Darwin is saying, "Having established the norm of actualist interpretation let us see, on this premise, whether any counter evidence occurs." Darwin admits there might be. But one must notice that little word *imperfection*. As in the Glen Roy paper, anomalous evidence, or the absence of evidence, is explained away in advance as an artifact of preservation. However great the challenge of explanation posed by any particular structure, or adaptation, the only principles of reasoning admissible on the rules Darwin has established are clear—actualist ones in principle rooted in the analogy, or *vera causa,* of the practice of domestic breeding and therefore referable to experience. The details are a mere matter for the hermeneutics and rhetoric of science to work out or imagine.

As a final illustration of how incommensurability is itself an artifact—indeed, in this case anyway, a principle—of invention, it is worth noting how Darwin suborns a favorite saying of Whewell as reluctant testimony, if not directly for transmutation, then for the special use of *vera causa* reasoning on which his argument is based. Whereas Whewell, who understood the *vera causa* principle just as Kant had, as limited in application to the world of inanimate objects, could declare the beginning of things as beyond science, Darwin, using the same principle with inventive transgression, in effect, says, "Exactly so! Therefore let us use it in my sense to show how the study of life might become scientific." One could explore Darwin's strategy in greater detail, and there is, I believe, much to be learned in doing so. But in sum, and in small, the germ of Darwin's entire mature strategy, beginning from a very few principles taken from convention and then almost deferentially turned against it, can be found in the *Sketch* of 1842. In an irreverent spirit parallel in playfulness, if not in profundity, to Darwin's own, we give to the master the final word. "Give sketch of the past—beginning with the facts appearing hostile under present knowledge—then proceed to geograph. distribution—order of appearance—affinities—morphology. Etc." (in Darwin and Wallace 1958, 60)

Rhetorical Invention and the Illusion of Incommensurability

Though Darwin's *Origin* did not immediately convince his scientific peers, the more progressive among them tended to accept the idea of evolution but ascribed to it a variety of different mechanisms, including use-inheritance and even occasional macro-mutation. Darwin's inventiveness was not in vain. *Origin* both changed and signaled the change in thought we associate with Darwin's name and with the theory of evolution. Darwin's skill in turning the beliefs, topics and conventions of scientific argument—what Harris phrases as "themes and practices" in the introduction to this volume—against themselves made possible an entirely new way of thinking, even if it took scientists until 1940 to integrate Mendel into the picture and to recognize the central role of natural selection in the completed Darwinian synthesis. Darwin changed history not by any single argument, important as were his individual lines of reasoning, but by showing what convention could authorize were it challenged, loosened, and the parts reconnected by the fragile skeins of probability and likelihood set forth in the *Origin*. The study of Darwin's actual forms of reasoning shows how the problem of incommensurability is itself an artifact of Cartesian anxiety—a prohibition imposed by formal logic on inventional ingenuity.

Whereas incommensurability prohibits any but either a rationalist or an irrationalist solution to the problem of difference, inventional thinkers such as Darwin or Cicero, or any practical reasoner, see difference less as a problem in formal logic and more as a challenge of negotiation. This is why any "rational reconstruction" of Darwin's thought, particularly any reconstruction grounded in his "method," is fated to end in caricature. Method, in science generally and in Darwin certainly, has always been a primary site for the study of rhetorical technique, historically speaking, because method is where one perspective is demarcated from another—usually with prejudice. In Darwin studies, two of these reconstructions have been particularly instructive.

The thesis of Michael T. Ghiselins's now-classic work is summarized in its title, *The Triumph of the Darwinian Method* (Ghiselin 1969). Though Ghiselin is astute in recognizing the plurality of Darwin's methods, appropriately sceptical of applying Hempel's covering-law model to so historical and contingent a science as geology, and apt in noting that science is "a living tongue," he tells an otherwise predictable—if exceptionally instructive—tale of why Darwin won (Ghiselin 1969, 27–31). Darwin won, and continues to hold the esteem of the scientific world, Ghiselin argues, because his reasoning was what we now call *hypothetico-deductive,* and exemplifies Popper's philosophy of falsification (Ghiselin 1969, 4, 15, 30, 62–63, 237).[23] According to Ghiselin, those who failed to recognize this in Darwin's time were Platonists, or what is scarcely better, Aristotelians, and both together, certainly not excepting those who have reservations about Darwin's overall logic and method today, are little more than enemies of the open society (Ghiselin 1969, 49–59, 122–123).[24] A key moment in the rationalism that informs Ghiselin's perspective throughout becomes clear when, evidently exasperated at the interpretations of

Darwin's doctrine made by scholars who have cited Darwin's texts in support of their interpretations, Ghiselin counsels intellectual historians "to abandon the study of words and to derive our understanding from concepts" (Ghiselin 1969, 75). His ire is particularly directed toward those who regard natural selection, which Darwin described as "daily scrutinizing, throughout the whole world, every variation" (Darwin 1964, 84), as though it were a conscious force (Ghiselin 1969, 74–75).[25] The value of Ghiselin's account of the overall logic and unity of Darwin's diverse projects is indisputable. But his reduction of the rationality of Darwin's triumph to method comes at a cost too high for Darwin's actual practice as an arguer to bear.

A far more historically nuanced interpretation of Darwin's method is Vincent Kavalovski's. Kavalovski supports Ghiselin's claim that Darwin, following Herschel and Whewell, made use of the hypothetico-deductive argument pattern (Kavalovski 1974, 123–124). Unlike Ghiselin, however, Kavalovski recognizes the polemical character of disputes over method in the history of science and does not regard Darwin's opponents simply as blinded by prejudice (Kavalovski 1974, 55–56). Kavalovski notes how Darwin's use of the *vera causa* principle, rather than uniting him with his fellow scientists, was one of their key objections to his theory (Kavalovski 1974, 2, 214–224). Herschel himself, most famously, called it "the theory of higgledy pigglety" (Darwin 1986, 7: 423). The problem, as Kavalovski notes, was the two-tiered structure of the traditional model of science. The laws of motion, of physics, the field in which Newton was supreme, or geology, where Lyell had made such a mark, could be explained *vera causally*. But no one, other than the methodologically unsound Frenchman, Lamarck, or the *outre* radical, Darwin's grandfather, Erasmus, or the rank amateur, Robert Chambers, supposed biology could be explained this way. Despite Hume's questioning of the argument for design, no less an authority than Kant recognized that students of biology had to proceed as though they were examining objects designed for a purpose.[26] Kavalovski notes that the *vera causa* principle understood in the context of an overtly theistic science had historically functioned as a buttress for a cosmos seen as designed. Given that one of the constituting ends of the scientific enterprise itself was to manifest the wisdom of God, Darwin's use of the *vera causa* argument to undermine evidence of design, in effect, established a different kind of science (Kavalovski 1974, 224–225). Strong providentialist that he was, Newton would have been the last to believe that the *vera causa* principle was a blank check to provide materialist explanations for everything (Kavalovski 1974, 36–37).

While Kavalovsky is at one with Ghiselin in insisting on the importance of the *vera causa* principle, his greater historical awareness of how this principle has been interpreted—and how it gained in meaning through use—prevents him from dismissing Darwin's opponents; that is, almost right up until his next-to-final chapter. In the end Kavalovski concludes that Darwin's peers ought to have recognized the anti-*vera causa* character of their own providentialist explanations and agreed with Darwin on methodological grounds (Kavalovski 1974, 204–205). It is here, where Kavalovski's admirable historiography gives way to his one lamentable lapse into rationalism, that we can see the wisdom of John Pocock's observation that however prestigious may be the language of any paradigm, like the grammar of a living language, that paradigm must submit itself to the relentless judgment of daily use (Pocock 1973). Rationalism and the openness of history, not excepting the history of science (as observable in the changing meaning not only of words but in the function of whole lines of reasoning), are indeed incommensurate, as incommensurate as the idea of real species and the persistence of Darwinian variation, inheritance, differential adaptation, endless time and changing ecologies. And in any contest between the two it is rationalism—not excepting the rationalism of science's pretence to "method" (one holy, catholic and universal taught at all times and places) that must give way.[27] But such necessary concessions to reality on the part of science and of historians and philosophers of science need not occasion a surrender to casuistry. As Wendy Raudenbush Olmsted has shown, the deliberation that accompanies the use of a scientific or technical term (a negotiation she also traces in law, practical thinking and literature) gives the term the specificity it needs for the work it must perform for the community that uses it (Olmstead 1997). As Olmsted makes clear, deliberation (what we have identified as Cicero's inventional process of *multiplex ratio disputandi*) explores the range of meanings of a term, how it relates to other terms and to the situation of its deployment, and how, despite the latitude in the forms of argument it allows, deliberation is an instance of reason rather than a substitute for it. The application of this method generally to science and particularly to the case of Darwin has been well argued by Marcello Pera.

Pera recognizes with Ghiselin and Kavalovski Darwin's use of hypothetico-deductive argument patterns, and he acknowledges with both of them the role of the *vera causa* principle in Darwin's thought (Pera 1994, 71–72, 74, 76, 78). Pera differs from Ghiselin and Kavalovski, however, in the way he sees these two modes of reasoning functioning in the larger economy of Darwin's thought. Pera finds in Darwin not just one mode of reason but a great variety, from hypothetico-deduction to induction to garden-variety rhetorical. Among the rhetorical modes of Darwin's reason he notes (all page numbers are for Pera 1994):

- arguments by analogy (73–77);
- arguments by double hierarchy (77–79);
- pragmatic arguments (79);
- appeals to the absurd and ridiculous (79–80);
- arguments by division (80–83);
- arguments *ad ignorantiam* (83);
- arguments *ad hominem* (83–85); and
- arguments of the possible (85–88).

For Pera, Darwin's reasoning cannot be rationally reconstructed according to some ideal of method without fundamentally falsifying its historical particularity. Nor, for Pera, does Darwin's use of any of these modes of argument count against his integrity as a thinker. For Pera, Darwin's use of fallacious arguments and personal attacks are not exceptions to reason but exemplifications of the factors of scientific rhetoric and dialectic always at play in scientific disputes—particularly when traditional ways of understanding are at risk.[28] Nor is it hardly to be wondered that modes of reasoning in disputes when major shifts in perspective are at stake should be varied. What is odd, indeed almost unseemly, is that defenders of Darwin—the man who, along with Wallace, understood the importance of adaptation better than any other—should blanche at noting situational adaptation in their analysis of him! It must be counted among the profound anomalies of neo-positivism that it does not hesitate to attribute the incredibly complex structures of the beings of the bio-physical world to directionless material self-sufficiency, and insists that science be governed (and at every step) by a rationalism that abhors, in the late Stephen Gould's memorable phrase, "odd arrangements and funny solutions" (Gould 1980, 20).

A rhetorical interpretation of Darwin's achievement, particularly his capacity for invention both in discovery and in justification, does not entail an abandonment of rational thought, but in the words of the title of Stephen Toulmin's book, involves rather *A Return to Reason* (Toulmin 2001). Darwin's use of reason is not to be measured simply by his use of a particular method but in his skill at negotiating the varied contingencies of his case. No small part of Darwin's power as a negotiator derives from his ability to take whole lines of argument, remove them from their initial contexts where they performed quite different functions and put them to new uses in ways that horrified many of the teachers from whom he had learned them.

Conclusion: the Triumph of Rhetorical Reason

In a comment that seems fitting to conclude an essay on the inventional evolution of Darwin's theory of evolution, Paul Ricoeur reflects that while we are not privileged to see the origin of a new order of meaning, we can perhaps gain a glimpse of how one originates by noting how an old one changes (Ricoeur 1975, 22). This observation leads Ricoeur not to concepts but to words and to their use. Ricoeur roots his interpretation of change in Aristotle's comment in the *Rhetoric* that to metaphorize—to see the similar in the different—is the signature of genius and cannot be taught (Ricoeur 1975, 23). Lest this observation seem to place radical invention on too elitist a footing, Ricoeur supplies a complementary definition by Nelson Goodman as an appropriate corrective. According to Goodman, "a metaphor is an affair between a predicate with a past and an object which yields while protesting" (Goodman 1976 [1968], 69; qtd. in Ricoeur 1975, 235). Goodman's definition aptly captures the tensional character of novelty's painful advent. As Goodman explains, a metaphor begins as a category mistake (an elegant if logicised view of an affair) and yet this mistake is not "mere" for it marks the beginning of a new relationship. The memory of the former or conventional meaning creates resistance to the new meaning, and realization of the new meaning, despite the cognitive jolt it delivers, marks the beginning of a new perspective (Ricoeur 1975, 231–239, 244–245, 305, 335, 349, 352–353).

Goodman's definition aptly captures and summarizes the sense of invention that we get from Bloom, Leff and Pera. The ground of invention, whether in poetry, political rhetoric, or science, is not in the first instance genius but the merest of errors. Genius enters with the intellectual courage of a specially favored individual, Bloom's "strong writer," to recognize the potential significance of the mistake—to insist that it was not an error—and to set forth the new intelligibility that it opens to best advantage. But even here the individual talent, though seemingly decisive, is not supreme. The individual whom, in the long sweep of history, seems but an accidental agent is also equipped by history to carry received tradition with her—for she is a representative of the tradition she overcomes. This is the sense in which T.S. Eliot argued in "Tradition and the Individual Talent" that the poet has, "not a 'personality' to express, but a particular medium" on which tradition is inscribed in particular ways (Eliot 1975, 42). The ground of invention whether in poetry, politics or science is tradition—thanks, as Darwin would have it, to the odd variation (and they seem to be copious, however ill fated and unhappy most of them turn out to be) that accompanies all descent. Despite appearances and its apparently forbidding safeguards, tradition is never safe. In the long sweep of history, rhetoric (encompassing both relative stability and change)—given contingency, semantic variation, slippage, *kairos*, and the pressure to adapt old terms and meanings to novel situations—rules.

Darwin is a prime illustration of how, through invention and imitation, incommensurate perspectives are assimilated. Even when Darwin failed to persuade his peers immediately, in the long run (and the run was not that long), by revealing a possible counter-hermeneutics embedded in convention, and by exploiting its rhetorical possibilities, he carried the day. Darwin won, not by decisively rebutting his opponents, but by arguing convention to a draw. Darwin's preferred scheme of reconciliation between the old natural theology paradigm and his new insights was (of course) extremely unstable and a close reader suspects, more than just a little, that Darwin was rather aware of this; that is why, unlike the more pugnacious Huxley, who is not buried in Westminster Abby, he encouraged them.[29] Darwin saw compromise or middle-of-the-road positions as the transitional forms necessary to lead to a view of the world very different than even he, perhaps, could fully imagine. The historical difference made by Darwin's rhetorical skill, though not to be measured in immediate direct audience results (Bowler 1988; Depew and Weber 1997, 159–160), is not therefore negligible; for it works, as does his theory. Once his alternative has appeared, clothed in the very language of convention, the old tradition can never regain, and must gradually lose, its former exclusive preoccupation of ground. The problem of incommensurability thus is the unconscious tribute positivist philosophy pays to the self-effacing inventional art of rhetorical reason.

Notes

[1] Lyell's statement of the case for species change and his rebuttal of it is given in volume 2 of *Principles* (Lyell 1990, 2: esp. chaps. 2 and 4).

[2] See Martin Rudwick's introduction to volume 1 of *Principles* (Lyell 1990, 1: xxix-xxxv). For a detailed discussion of Lyell's motives, see Bartholomew (1973).

[3] Wallace observed, "The succession of fossil remains throughout the whole geological series of rocks is the record of this change; and it became easy to see that the extreme slowness of these changes was such as to allow ample opportunity for the continuous automatic adjustment of the organic to the inorganic world, as well as of each organism to every other organism in the same area, by the simple process of 'variation and survival of the fittest.'" Cited by de Beer (1958, 8) .

[4] Leff's analysis illuminates how the hermeneutic turn cuts through the dichotomy between message reception and message production urged by late modernists (for instance, Gaonkar 1997) who wish to urge the irrelevance of classical rhetoric to problems of criticism because readers are no longer orators.

[5] For the difference between Beagle and post-Beagle annotations, contrast DiGregorio and Gill 1990, 1: 530–531 with 531–543. A large number of annotations are from the 5th edition of 1837.

[6] Lyell called this principle *modernism*. I have adopted the more common, and clearer, *uniformitarianism*.

[7] A good short general account is de Beer (1983).

[8] Only recently has Sulloway been able to rebut the thesis that there has been evolution on the Galapagos since Darwin's time by showing that the appearance of change resulted from the fact that Darwin put birds on the wrong islands to begin with (Sulloway 1982, 346–351).

[9] It was also one of his arguments against marriage, in a private memo (divided into columns for and against) he had written himself earlier in the year. Since Darwin had already decided to marry, and since by his own calculation marriage meant dropping geology in favor of his species work, Glen Roy provided him his last opportunity to pursue a new field study that would complete the theoretic thrust of his earlier geological work (Herbert 1968, 18–19).

[10] The definitive account of Darwin's Glen Roy paper, including a sophisticated and detailed account of Darwin's various strategies of argument, is Rudwick (1974).

[11] In his *Autobiography* Darwin says that von Humboldt's *Personal Narrative* and Herschel's *Preliminary Discourse,* "stirred up in me a burning zeal to add even the most humble contribution to the noble structure of Natural Science" (de Beer 1983, 38). That Darwin reread Herschel closely after the voyage is indicated by his "Reading List." See Charles Darwin, "Reading List," Cambridge University Darwin Collection, MSS Vol. 119, 120.

[12] On the relation between hypotheses and analogy, Herschel observes, "Hypotheses, with respect to theories, are what presumed proximate causes are with respect to particular inductions: they afford us motives for searching into analogies; grounds of citation to being before us all the cases which seem to bear upon them, for examination" (Herschel 1987, 196).

[13] "The surest and best characteristic of a well-founded and extensive induction, however, is when verifications of it spring up, as it were, spontaneously, into notice, from quarters where they might be least expected, or even among instances of that very kind which were at first considered hostile to them" (Herschel 1987, 170; see also 208–209, 272–273).

[14] Kavalovski states the problem precisely:

> [. . .] the avowed purpose of the principle is to eliminate arbitrary and fanciful hypotheses more or less peremptorily from any given scientific investigation, and indeed from science in general. This seems like a reasonable and innocent enough task. However, it is not uncommon in the history of science that the fanciful speculations of one era emerge as the accredited science of the next; conversely, the certified and credible concepts of an earlier era often become the "occult" ideas of a later one. On the basis of this one might be tempted to view the *vc* principle as simply an arbitrary and rather dogmatic way of excluding currently unfashionable ideas from scientific purview; the term *vc* or "true cause," would then be a purely honorific rubric, merely a polemical way of dignifying one's own concepts and reviling those of an opponent. (Kavalovski 1974, 55)

[15] Charles Lyell, *Principles,* Chs. 1–8. Invective most wry and delightful so regularly intersperses Lyell's first volume that an ungeological reader looking for nothing else never has to wait very long. "[. . .] but we shall dwell no longer on exploded errors but proceed at once to contend against weightier

objections, which require more attentive consideration" (Lyell 1990, 1: 91). For Lyell's strategy, see Rudwick (1970).

[16] Darwin writes, "the conclusion is inevitable, that no hypothesis founded on the supposed existence of a sheet of water confined by barriers, that is, a lake, can be admitted" (Darwin 1980, 99).

[17] A lengthy headnote from *De Officis* prefaces Ch.1 (Herschel 1987).

[18] See particularly the opening of Ch. 6 (Darwin1959).

[19] Rudwick (1974) provides the rhetorician a veritable catalogue of Darwin's suasive techniques: descriptions loaded in favor of the marine hypothesis (117); facility at explaining away (117); use of Biblical sounding language to lend dignity to his position (118); "making [. . .] explanatory virtue out of observational necessity" (120), which Rudwick later refers to as "his characteristic move" (128); use of the principle of exclusion to make his own position seem inevitable (118); "inverting the norm of expectation" (124); "evasion" of "central explanatory problems" (141); and the signature attempt "to make his own position unfalsifiable" (145).

[20] David Kohn: "it would also take him several years to develop the expository structures to argue his theory" (Kohn 1980, 149). In a larger sense Darwin's "discovery" kept on happening. Only in the 1850s (1854–1858) did he realize how population pressure entailed divergence of character. Darwin also, and notoriously, continued to tinker with his theory through all six editions of *The Origin.* On the development of Darwin's thought particularly after Malthus, see Ospovat (1981, esp. Ch. 7) ; on the differences in the various editions of *The Origin* see Peckham (1959) .

[21] Of the "Sketch," Limoges observes, "The order of exposition does not reproduce the order of research; it envisages demonstration, persuasion" (1970, 105; my translation).

[22] An excellent analysis of the relation of category mistake to creative insight is Ricoeur (1977). See my review (Campbell 1979).

[23] See also Manier (1978, 95): "the early drafts of the theory do not conform to the 'hypothetico-deductive model' of scientific explanation, though they indicate Darwin's intent to represent his view *as if* they did conform to that model. Darwin himself flatly states in the *Notebooks* that he found it impossible to distinguish the inductive and deductive phases of his investigation."

[24] That scientific purity does not line up neatly with any particular school or set of metaphysical commitments in science, and that polemicism in the censurable sense can be found also, and equally, in the winning party, is underscored by Kavalovski. "Vociferous champions of Darwinism such as Haeckel, and occasionally Huxley and Spencer, also indulged in rhetorical technique, substituting contemptuous dismissal of an opponent for careful

evaluation. They, too, often failed to grant that the opposition possessed a theory worthy of impartial examination. (Kavalovski 1974, 206).

[25] Ghiselin seems to take a more positive view of the study of words when he observes favorably of the Whorf-Sapir hypothesis, "the structure of our language determines the perception of reality." He seems to exempt the language of science from the scope of the hypothesis—for scientists have, he affirms, "no difficulty in constructing an artificial language [. . .] more suited to the unbiased apperception of the universe [. . .] and use [. . .] natural language only for communication" (Ghiselin 1969, 158–159).

[26] See Hume (1989); For Kant, see Lenoir (1982, 2, 4, 13–14, 22, 25, 31, 50–52, 61, 72) and Depew and Weber (1997). The implications of Kant for the design argument, however, were not always well understood.

[27] Kavalovski's study seems to end in a tantalizing ambiguity. On the one hand, he has been meticulous in stressing the integrity of Darwin's opponents, and in noting that part of their substantive picture of science was grounded in providentialism. On the other hand, he believes their commitment to *vera causa* reasoning should have convinced them of Darwin's thesis. There may be more than a tincture of methodological reasoning in Kavalovski's thinking—that there really was, after all, on some level, a clearly "rational" choice in the positivist sense, between Darwin's theory and the conventional one. It is here I would urge that the ordinarily censurable aspects of rhetoric, all those informal fallacies and points of disputable cogency that Pera and others have found in the *Origin* (and in science generally), have their reasonable force (see Pera 1994, Ch. 3). The issue may come down to this—which way of seeing best comports with our understanding of ourselves in relation to our other commitments? At a certain point "we"—whomever identifies with Darwin's reasoning—are no longer capable of taking providentialism seriously as part of our science. Hence, in this sense, the cogency of Darwin's appeal in the conclusion of *Origin* that accepting his theory entails that we no longer look at nature "as a savage looks at a ship" (Darwin 1959, 485). Darwin's argument may be an *ad hominem*. At a particular historical moment, and in a particular setting, the argument may cease to be a fallacy.

[28] Since Hamblin's influential *Fallacies,* of course, the traditional notion of fallacious argumentation has been under steady review in philosophy (Hamblin 1970; see van Eemeren *et al.* 1996, 50–92 for a good summarizing discussion of that book's wake); in rhetoric, there has always been more attention to particular situations such that blanket "all arguments of Pattern X are evil" declarations have been relatively rare. See, for instance, Perelman and Olbrechts-Tyteca (1969 [1958]) for a typically rhetorical attitude—such as their reanalysis of a Bentham fallacy in terms of Presumption: "What Bentham calls the fallacy of the fear of innovation or also the fallacy of universal veto, which consists of opposing any new measure just because it is new, is

not a fallacy at all, but the effect of inertia which operates in favor of the existing state of affairs (Perelman and Olbrechts-Tyteca 1969, 171; see also 202, 205, 283, 306, 326, 328, 414, 485).

[29] See, for instance, his letter to E. N. Aveling, in Feuer (1975). See also Ceccarelli (2001), for her systematic exploration of multiguity in close readings of several highly influential scientific texts whose success was largely a function of supporting different interpretations by different audiences—for instance, Shrödinger's *What Is Life?,* which appealed in distinct ways to biologists and physicists.

9

Cell and Membrane: The Rhetorical Strategies of a Marginalized View

Jeanne Fahnestock

The discovery of cells, the gradual realization that all plant and animal tissues are composed of cells, the explanation of cell reproduction and the detailed description of cell structure and function—all these steps follow a good Whiggish script in the history of science. Along the way, this history features many episodes of rival theories contending until the best account emerges, one with a better fit to observed facts and greater explanatory power. In the mid nineteenth-century, for example, opinions divided over whether new cells were formed within existing cells or grew from seeds on their surfaces (H. Harris 1999, 112–114). Improved microscopes eventually settled the issue.

The specific case of competing views about the nature of the cell discussed in this essay has not been settled. Beginning in the 1960s, a small group of researchers developed what they considered an innovative and comprehensive theory of cell physiology stemming from their criticism of the prevailing direction of research. They published their studies in mainstream journals like *Science* and the *Proceedings of the National Academy of Sciences,* received continued grant support, and trained students in their methods. They claim MRI medical imaging technology as a byproduct of their approach. The original proponents have persisted for forty years and recently a new generation of adherents has reasserted their views. Yet their perspective is not mentioned in current textbooks in cell or molecular biology, nor is it featured in histories of the field found in reviews or in Nobel Prize speeches. One account of this exclusion comes from the dissidents in published texts from small presses and on personal websites. These offer an alternate history claiming that their theories were marginalized, if not altogether suppressed, by misguided colleagues invested in an approach incommensurable with theirs. What are the sources of these divisions and, specifically, what explanation can be provided by a rhetorical perspective on the problem of incommensurability represented by this case?

Incommensurability in scientific theories is often described as the inevitable outcome of different conceptual vocabularies, different methodologies and research traditions, and different systems of accountability or even worldviews.[1] Any of these perspectives can be applied productively in a rhetorical account of the competing paradigms explored here. There are indeed disjointed conceptions, discrete terminologies, differing experimental procedures and even contrary disciplinary allegiances that can account for the variant views. As Randy Harris suggests in the introduction to this volume, such differences can often be overcome when the parties involved want or need to overcome them; and not just Harris. Paul Feyerabend puts it this way in his posthumous *Conquest of Abundance:* people who want to get along, such as "astronomers interested in unity," who "face new situations, products, challenges" can deal with them, "often successfully." If they want to get along, he says, "they do not excuse themselves by saying 'this is beyond the semantic boundaries of the language I am speaking'" (Feyerabend 1999, 32).

But incommensurability can also be an active rhetorical investment. As Mara Beller says, "[I]ncommensurability logically dictates total, unquestioning, dogmatic commitment," one that "excludes the possibility of being suspended between two different, incompatible worlds, of creatively participating in both" (Beller 1999, 292). When one buys into incommensurability, one declines negotiation. A particular research group can deliberately and self-consciously present its work as incommensurable with other views. It can set an either/or choice for a disciplinary audience by claiming that its truths make those of others impossible. This is a very high stakes position to take and it is, not surprisingly, usually avoided. Instead, tactics for minimizing differences or managing disagreements within the normal course of an advancing research front are more typical and even fruitful; in this volume, the study by Bazerman and De los Santos provides a particularly compelling example of just such a situation. A research area with seriously debated open questions is a research area that deserves funding. But a research program that claims all other approaches are misguided and a waste of money risks catastrophe.

The school of cell physiology explored in this paper made a heavy rhetorical investment in incommensurability. The term *rhetorical investment* here is used deliberately instead of the more familiar term, *rhetorical strategy.* A "strategy" suggests calculating deliberation and seems incompatible with sincerity. The proponents of the unorthodox view of cell physiology described below were and are passionately sincere in their convictions. They did not adopt incommensurability as a device. In the first generation, as the account below demonstrates, they coalesced as a view based on shared disagreements with the prevailing line of research. When they lost funding and footing in journals, they turned to a public campaign without effect. Second generation adherents have, however, reasserted their theories, but in doing so they have had to make explicit several appeals that a mainstream tradition can take for granted. Leah Ceccarelli's essay in this volume explores a case in which adopting a stance of incommensurability has contributed to widespread rejection of one theoretical program, E. O. Wilson's sociobiology. The present essay explores a case in which a very similar stance has contributed to the marginalization of a theoretical programme, Gilbert N. Ling's "Association Induction," or AI, model of cell physiology.

Dr. Gilbert N. Ling

The source of the alternate view claiming to be the "the first and only unifying theory of cell physiology in existence" (Ling 1998) is Gilbert N. Ling. Born (1919) and educated in China, Ling won the last offered Boxer Scholarship to pursue graduate work in the United States and completed a Ph.D. at the University of Chicago in 1948. During a long and active career, Ling has published four books and over 200 articles (Ling 2001, 371–372).[2]

A private history of this career appears in the introduction to Ling's third book, published in 1992, in an article he published in 1997 in the journal he edits, and, in its frankest form, embedded in a long article entitled "Why Science Cannot Cure Cancer and AIDS without Your Help" on Ling's website (Ling 1998). These sources offer a narrative of breakthroughs, harassment and persistence; they tell the story of a courageous researcher and a handful of followers who developed an innovative and comprehensive theory of cell physiology that has been suppressed by rivals with a virtual monopoly on institutional power.

Ling's theories grew from his attempts to explain the well established observation, known from the early twentieth century, that living cells tend to exclude sodium and retain potassium. When the availability of radioactive isotopes in the thirties led to the discovery that sodium was not excluded by the cell's membrane but could in fact pass through it, a "pump" was proposed that could actively remove sodium from the cell. Ling, however, concluded from experiments he performed in the 1950s on frog muscle that cells simply do not have the energy available to power the pumps needed to maintain the observed gradient of sodium ions, let alone the unequal distribution of other ions or small molecules between the outside and inside of a cell. He believed that another mechanism maintaining the different concentrations was needed, and that led him to his Association Induction (AI) model of cell physiology. Ling introduced his new hypothesis and published his experimental evidence against the pump in a 1962 book, *A Physical Theory of the Living State: The Association-Induction Hypothesis.*

According to Ling's AI hypothesis, the cell is not a dilute aqueous solution but is instead a somewhat structured latticework of water, ions and proteins. Water molecules, which are polar, line themselves up along the charged sites on the cell's proteins, that typically exist in extended states in Ling's view. In this structured water environment, described as an intermediate state between liquid water and ice, potassium ions (K^+) are preferentially bound while sodium ions (Na^+) are excluded. In short, the typical ion distributions observed between a cell and its extracellular environment can be explained without the necessity of a membrane, let alone of channels and pumps to shuttle ions and other molecules in and out of the cell. In still another departure from canonical belief, ATP (adenosine triphosphate), the metabolic powerhouse of the cell, is only needed in Ling's theory as a "cardinal adherent," to maintain the extended configurations of the macromolecules that promote water adsorption, the layering of structured water and the ion gradients. Polarization and depolarization are the sources of energy transfer in the cell: hence the term *induction.*

Not everything in this account is problematic to mainstream molecular biology; most molecular biologists would agree that there is some structuring of water in the cell due in part to the polarity of water molecules and charges on macromolecules. But the view that membranes, pumps, and channels are unnecessary certainly contradicts the view held by most cell physiologists for the last fifty years.

Ling's views attracted adherents in the sixties and seventies, who then built their own work on assumptions about the structuring of cell water. As a group, these researchers all used nuclear magnetic resonance (NMR) techniques to study cell physiology. Foremost among them was Freeman W. Cope, a physician with a background in physics working in a Navy-sponsored research lab in Pennsylvania in the sixties. Cope had an unusual view of the cell based on an analogy with semiconductors (Cope 1965, 227). In a 1965 publication considered interesting and valid enough to appear in the *Proceedings of the National Academy of Sciences,* Cope reported on an NMR analysis of the content of sodium ions in frog muscle compared to the sodium content based on other methods of analysis. He concluded that much of the sodium ion in the cell was invisible to NMR analysis because it was "complexed" to macromolecules in the cell, thereby supporting Ling's theories. Another proponent was Carleton F. Hazlewood, formerly at Baylor Medical College in Texas, a prolific researcher who also worked on ion concentrations in different tissues and NMR analysis in the seventies.

But Ling's most famous supporter was and is Raymond Damadian, whose life story has been the subject of numerous profiles (see Mattson and Simon 1996, Wakefield 2000, Schneider 1997). An MD with an interest in basic research, Damadian tried as a postdoctoral fellow to isolate a pump protein. He met Ling in 1969 (Mattson and Simon 1996, 643) and subsequently adopted a structured-water view of the cell, but he went several steps further. He theorized that the water in cancer cells had to be organized differently than water in normal cells and that that difference could be detected by NMR analysis and used as the basis of a screening or imaging device. Damadian used a then less common method of NMR analysis, namely transient or induction NMR, measuring the time between the excitation of hydrogen nuclei by transient bursts of energy in a constant magnetic field and their subsequent return to a normal state by emitting recordable pulses of energy (Mattson and Simon 1996, 644). Comparing samples of healthy and malignant tissue in rats, Damadian found significant differences in the "relaxation" times of the hydrogen nuclei. In a 1971 paper in *Science* that has become a "citation classic," he interpreted his results as indicating "a significant decrease in ordering of intracellular water in malignant tissue," and he concluded by observing that NMR might prove useful as a screening technique for tumors (Damadian 1971, 1152). In 1972, he filed a patent for an "Apparatus and method for detecting cancer in tissue," and finally, in 1977, achieved the first NMR image of the human body. In 1978 he formed a company to produce his new invention—now widely in use as Magnetic Resonance Imaging, or MRI (Mattson and Simon 1996, 668–685). According to Ling, Damadian wrote to him several months after the first successful trial in 1977, "The achievement originated in the modern concepts of salt water biophysics [introduced by] your treatise, the association-induction hypothesis" (Ling 1984, vii). Damadian's rationale for his results is no longer considered correct, not even by Ling himself (Ling 1992, 94); the tissue differentiation achieved in MRI imaging is the product of complex features including the presence of different ions in different tissues. But Damadian's basic faith in the usefulness of NMR imaging for tissue discrimination was more than justified.

In the years following his 1962 book, Ling continued to perform experiments reinforcing his theories and refuting those whose research centered on membranes and pumps. He had an active lab with graduate students and postdoctoral fellows and enjoyed continuous grant support. By the early seventies he could point to the publications of a handful of researchers who characterized their work as supporting his theories. On his web site, however, he describes pressure "building up here and there against my science" as early as 1960 and to a 1966 reference to him by Richard Keynes, Professor of Physiology at Cambridge, as someone "responsible for a major heresy in this field" (Ling 1998).

Then in 1973, according to his description of events, the Physiology Study Section at the National Institute of Health (NIH) reviewing his grant renewal under the chairmanship of a man whose work he had criticized, recommended "not only that my specific proposal submitted to NIH for renewal funding be refused (which was within their prerogative) but that NIH should *stop funding my future work* altogether (which was beyond their prerogative and thus illegal)" (Ling 1998). It is of course not unheard of for NIH to withdraw support, but the usual mechanism is not an outright refusal but rather a competitive score that places a proposal below the cut-off point for funding. Scientists so refused can submit a rebuttal of the criticisms leading to their score, which Ling did. He also complained to NIH officials. In a career retrospective in his 1992 book, Ling thanks the then Associate Director of the Division of Research Grants at NIH, Dr. Stephen Schiaffino, who referred his proposal to an *ad hoc* "Special Study Section" comprised not of membrane pump supporters, who were, in Ling's words, "by definition, my scientific enemies," but of "neutral scientists" who approved continued funding (Ling 1992, xiv). Ling had other sources of support, notably from an administrator at the Office of Naval Research where peer review procedures were not in place. When this administrator resigned and when a new Director of Research Grants at NIH abolished the Special Study Section, Ling finally lost all funding in 1988. His lab was, in his words, "forcibly closed" after 27 years (Ling 1997, 128).[3] At this point Raymond Damadian came to his rescue, sending seven trucks to move him and his equipment to a new lab established for him at the headquarters of Damadian's MRI company, Fonar, on Long Island. Ling is still there (as of 2005).

During Ling's funding problems in the mid-seventies another crisis occurred. As he describes these events on his website, "Aware of all the terrifying power of coercion, I was nonetheless deeply shaken when (virtually) all my graduate and postdoctoral students suddenly left en masse" (Ling 1998). This exodus could have been caused by the precariousness of support, but it seems that these were defections rather than merely departures and that what also occurred was a loss of confidence in Ling's research paradigm and experimental methods. As Ling says in the website version of events,

> Perhaps it was to offset the impact of their earlier published work confirming and extending the AI Hypothesis, and to convince others that they left my laboratory for purely scientific reasons, [but, whatever the reason,] each departing student from my laboratory soon published a paper, a short note, or merely circulated some unpublished pamphlets, all aimed at throwing doubts on the AI hypothesis and at rehabilitating, even glorifying the membrane pump theory. (Ling 1998)

The Public Campaign

Ling did not suffer his losses in silence. He began a letter-writing campaign to NIH and government officials complaining not only about his unfair treatment but also about the corrupt funding system that gave his opponents the power to suppress his views.[4] A more public airing of the complaints of AI researchers also took place at this time, and details of this confrontation can be gathered from sources other than Ling himself. In the summer of 1975, Representative John B. Conlan (Republican, from Arizona), a member of the Committee on Science and Technology, held standard oversight hearings on National Science Foundation (NSF) funding. At these hearings, Freeman Cope gave public testimony against the bias in funding decisions. He called for a halt to all grant support on the pump theory and for the resignation of the current administrators of the biomedical division of NSF, saying that they had been "wasting tax money funding research on a scientific hypothesis which has been disproved, and which has not and is unlikely ever to be of any practical use to anybody" (qtd.in Kolata 1976, 1221).[5] Next, Cope wrote to all the members of the Biophysical Society complaining that the editor of the society's *Biophysical Journal,* Frederick Dodge, engineered the rejection of papers by structured water advocates and published criticisms of their theory without giving them space to reply. He asked for their votes to put another Ling supporter, Carleton Hazlewood, on the Society's council, a vote that he lost by a narrow margin. Two months later, Cope wrote another letter, calling for Dodge's resignation. He complained of "a misuse of editorial power by Dr. Dodge to suppress publication of work which threatened his self-esteem, professional reputation and grants, which depend on maintaining the illusion of validity of old concepts of salt and water biophysics upon which the professional reputations of Dr. Dodge and his friends are built" (Hildenbrand 1979). Cope asked members to join a new society, the International Society for Supramolecular Biology, and announced that an existing journal, *Physiological Chemistry and Physics,* would now be the official organ of the new group. (This journal was then co-edited by Ling, and is now exclusively edited by him, but it has since changed its name to *Physiological Chemistry Physics and Medicine NMR.*) On his part, Ling also wrote to members of the council of the Biophysical Society, evidently early in 1976, accusing Dodge of censorship. But obviously Freeman Cope was the point person at the hearings, the author of the letters to members of the Biophysical Society, and very active in the attempts to extend the structured water view of the cell to cancer therapies.[6]

Many of these details were brought to light in an important watershed event in the public airing of this debate, a review article on the controversy that appeared in *Science* in June of 1976. The author of this review, Gina Kolata, who has since become a well-known science writer, obviously interviewed both sides about the issues. Dodge refused to be quoted but in summarizing his position, Kolata reported that he wished to "stay aloof." Dodge claimed that unlike his predecessors, he did not allow proponents of structured water to review one another's papers, so they were having more difficulty getting published. But he denied any conspiracy among the editors. Kolata sums up the situation in late 1976 in the following way:

> The bitterness and frustration expressed by members of the structured water group have made peaceful coexistence with their opponents impossible, and they are becoming increasingly isolated from other physiologists and biophysicists. As one observer notes, "there is paranoia on one side [the structured water advocates] and contempt on the other [the pump advocates]." Thus the communications gap between the structured water advocates and mainstream scientists seems likely to continue in the foreseeable future. (Kolata 1976, 1221)

It has.

The Review Articles

In the personal account of his trials on both his website and in the largely repetitive 1997 article published in the journal he edits, Ling engages is a very lengthy complaint about a supposed impropriety in the 1976 Kolata article,[7] and about the "fraud" perpetrated in a review article published in 1975. The second complaint raises interesting questions about how scientific controversies are represented in professional publications and presented to the public.

Volume 37 of the *Annual Review of Physiology* in 1975 contains the first review article devoted exclusively to "The sodium pump." According to Ling, the authors, I. M. Glynn and S. J. D. Karlish from the Physiological Laboratory at Cambridge, cite 245 references to articles on the sodium pump published over twenty years but never refer to his own numerous publications in this same period containing evidence against the pump (Ling 1997, 130). Indeed the review, he claims, contains no evidence against the pump at all. The significance of such reviews is well known: they canonize experimental conclusions into knowledge, assign credit for achievements, and set out the interesting issues facing a field. The absence of criticisms or alternatives from Ling and his cohorts in this, and all subsequent reviews, almost assured the absence of their ideas when textbooks were written. Ling was still protesting to NIH in 1986 that Glynn and Karlish were never chastised for their misleading piece (www.gilbertling.org/lp15.htm).[8]

The method of dealing with Ling's theory of the cell then, so far as mainstream researchers were concerned, was simply to ignore it—not to cite his work or continue to refute it.[9] The last explicit refutation of Ling's work apparently appeared in *Science* in 1977. In October of the preceding year, Ling's former graduate student Jagdish Gulati, along with another University of Pennsylvania professor, published a "test of the binding [AI] hypothesis" (Palmer and Gulati 1976, 521). They demonstrated that frog muscle cells can continue to accumulate K^+ in proportion to increasing external concentrations, contrary to the AI prediction that all the available K^+ adsorption sites should be saturated. In December of the following year, *Science* published Ling's lengthy answer, along with a response criticizing his new data from Palmer and Gulati. In that response, Palmer and Gulati point out the *ad hoc* assumptions and arbitrary parameters that Ling has to invoke to support an explanation of the evidence according to the AI hypothesis while "the membrane theory gave the simplest explanation of the data" (Ling, Gulati and Palmer 1977, 1284). They conclude that "even though the ideas contained in the association-induction hypothesis have, in the past, provided a stimulating challenge to the membrane theory, the new findings show that the binding hypothesis is no longer a useful alternative" (Ling, Gulati and Palmer 1977, 1284). Thus by the mid-1970s and perhaps even ten years before, the profession considered Ling's work on the AI hypothesis and his refutations of pumps and channels largely irrelevant. Ling's experiments suggesting that cells did not have sufficient energy to maintain the observed ion gradients were never answered to the satisfaction of AI adherents, as Carelton Hazlewood complained in a short *Science* piece, "Pumps or No Pumps?" in 1972. Their emphasis on the lack of specific refutation may reflect a Popperian assumption that every contradictory detail or anomalous result challenging a productive theory has to be answered.[10] But pump proponents simply moved on, convinced that ion partitioning was achieved by membrane-embedded proteins functioning as pumps and channels. Their paradigm was proving enormously productive, when measured by research activity and published results.

The Other Side

If in 1966 Ling was referred to as the source of a "heresy" and in 1976, in the Kolata article, the structured water proponents were labeled a "minority view," what views were held by the majority? Why did most people find the research of Ling and his circle uncompelling or irrelevant and eventually ignore it?

One way to answer this question is to look briefly at the history of research on pumps and channels, amply available in the venues of orthodoxy—Nobel Prize speeches, congratulatory retrospectives and textbook highlights. All begin from the known ion concentration gradients and from speculative papers in the 1940s postulating the existence of a pump specifically for sodium ions. In the early fifties two British researchers, A. I. Hodgkin and A. F. Huxley, using huge squid axons, showed that there was in fact a coupled movement of both Na^+ and K^+ ions across the cell membrane and that this movement created the electric discharge of this large nerve cell, work that earned them the Nobel Prize in 1963 (Hodgkin 1964, 1150). Channels and pumps were the inferred mechanisms for the movement of ions and hence the change in electrochemical potential across the membrane, but these inferred mechanisms were not linked with actual membrane structures until the work of Jens C. Skou in the late fifties. Skou discovered an enzyme that hydrolyzed ATP in the presence of potassium and sodium ions. By the mid-1960s, three years after Ling's first book, researchers knew that Skou's molecule was located in the membrane, had different ion affinities on its extracellular and cytoplasmic sides and was found in all cells that have coupled active transport of Na^+ and K^+. As a membrane-bound protein, this NaKATP-ase was extremely difficult to purify. But eventually it was isolated, sequenced, identified as having two subunits, and shown to be capable of working in reverse. Isoforms (that is, slightly different forms in the same tissue) were identified with different drug sensitivities (Skou 1989, 166). It was not until 1979 that experiments were published showing that the reconstituted enzyme had transport activity (Skou 1989, 164), but by then there was little doubt that Skou's enzyme was indeed the sodium pump predicted in 1941. In 1997 Skou received the Nobel Prize for his work, now cited as a "Classic Experiment" in a widely used molecular biology text (Lodish *et al.* 2000, CD-ROM).

There does not seem to be a single, critical experiment in the early seventies, shortly before the departures of Ling's students, that one could point to and say, "Before this research the sodium pump was only a promising hypothesis, but after this no one doubted it." Most researchers bet on active transport from the beginning. Ling's attempts to refute the sodium pump were perhaps discounted not so much because of definitive evidence for the pump, but because of the enormous amount of parallel work being produced at the same time on the nature of the membrane and on other trans-membrane devices controlling the flow of ions, and other molecules, into as well as out of the cell. By 1972, to judge by a review in *Science,* the still reigning model of the membrane was largely in place (Culliton 1972, 1350). In addition to multiple pumps, membranes also contain many channels discovered in the last forty years that open to admit molecules in response to voltage changes across the membrane or to the binding of specific ligands, like the well known channels in neurons that respond to specific neurotransmitters. The sodium pump, in other words, did not remain an isolated phenomenon; it became a comprehensible part of a larger picture revealing the basic mechanisms not only of nerve and muscle cells, but of all cells.

Moreover, the accumulation of knowledge about the structure and function of pumps and channels has drawn on every emerging technique in molecular biology in the last four decades: the purification of the proteins constituting the pumps or channels; the determination of their amino acid sequences leading to inferences about their secondary and tertiary structures, their functional reconstitution in synthetic lipid bilayers; the patch clamp technique for measuring voltage gating across a small segment of membrane and the use of cloning, mutagenesis and recombinant DNA methods where particular residues of pump or channel proteins are altered and the effects of these alterations investigated. Work on different organisms has also revealed homologous families of channels and pumps, supplying evidence for the evolutionary conservation of these critical structures. The recent determination of several bacterial genomes has revealed that a large proportion of an organism's genes (roughly 30 percent in *E. coli*) code for membrane proteins.

Undergraduate, graduate and medical students now learn the structure and functions of families of pumps and channels. The most popular textbook in this field has been *The Molecular Biology of the Cell,* one of whose authors is James Watson. The third edition of this text (Alberts *et al.* 1994) contains several chapters devoted to membrane transport. A more recent text is the fourth edition of *Molecular Cell Biology* (Lodish *et al.* 2000). It contains a comparably extensive treatment. There has been no diminution in the enthusiasm for pumps and channels between 1994 and 2005. In fact, in the last few years, there have been significant breakthroughs in the field in the form of the first determinations of the structure of one-pump and two-channel proteins by x-ray crystallography (see Doyle *et al.* 1998; Toyoshima *et al.* 2000; and Dutzler *et al.* 2002). The description of a potassium channel in 1998 was called a "dream come true for biophysicists" after the work of Hodgkin and Huxley fifty years earlier (Armstrong 1998, 56). When the structure of a chloride ion channel was announced in early 2002 in *Nature,* an image of the protein itself appeared on the cover, evidence of the extremely well developed visualization practices that lend reality to these molecular devices. This last report also explains how the channel might admit some ions while excluding others; the absence of such a detailed mechanism for over fifty years, though another source of criticism, was never perceived as a fatal impediment to advancement in the field. In fact it would be no exaggeration to say that membranes and their embedded proteins, both the cell's plasma membrane and the membranes of organelles like the mitochondria within the cell, are the sites of "normal science" in the field. Not only excitation and inhibition by neurotransmitters in the brain, but also photosynthesis, viral binding, signal transduction and regulatory pathways are all understood in terms of how the cell responds to its environment at membrane interfaces.

The Next Generation

In the face of this enormous enterprise and its results, Ling's views would seem to have lapsed into deserved obscurity. But they have not disappeared. Ling himself continued to campaign for his views in a series of books: *In Search of the Physical Basis of Life* (Ling 1984), *A Revolution in the Physiology of the Living Cell* (Ling 1992), and *Life at the Cell and Below-Cell Level* (Ling 2001). And his views have received a new stimulus from the efforts of Gerald H. Pollack, currently a Professor of Bioengineering at the University of Washington. In 2001 Pollack published *Cells, Gels and the Engines Of Life,* a book which reasserts the structured-water view of cell physiology and the essential features of Ling's association-induction hypothesis. There is no doubt that this work is meant to connect to the earlier research. Pollack writes in his opening,

> I warmly acknowledge the influence of Gilbert Ling. I met Gilbert at a meeting in Hungary in the mid-1980s. There I discovered a new conceptual world. It became clear that an approach to cell physiology orthogonal to current wisdom had considerable merit and enjoyed appreciable experimental support from a cadre of intellectually independent scientists. Ling has been a continuing force not only among that group but also for myself. Without his influence, this work would never have begun. (Pollack 2001, vi)

In the preface Pollack also testifies to the heroic dimensions of Ling's persona: "Ling in particular has spent his long career courageously advancing scientific frontiers in the face of frequent derision for the apparent extremity of his views" (Pollack 2001, xiii).

Reviving an ignored or marginalized position—especially one that looks to fit the standard criteria of incommensurability with respect to the dominant position—poses a significant rhetorical challenge; the tactics for attempting it are worth investigating. Unlike Ling, Pollack has one readily available source of rhetorical capital to draw on. In terms of chronological and career age, he is a second generation believer, and the ideas of a predecessor are often better represented by a disciple or convert, particularly a younger follower who replaces an entrenched previous generation in a version of what Randy Allen Harris calls the "Max Planck Effect" (R. A. Harris 1998, 86). Furthermore, the follower does not inherit whatever personal baggage may have been acquired by the source of challenging ideas. And finally, as Alan Gross pointed out in the case of Rheticus in relation to Copernicus, a disciple's conversion can itself become testimony, witness and model for audience members (Gross 1990, 101–105).

In his book reasserting the AI position, Pollack also made a strategic rhetorical decision that Ling did not make in his works: Pollack accommodates his text to non-experts and particularly to students. As he explains in his preface, he writes "for those with a minimal background in biology. It [his text] could have been written for experts, rich with theory and dense with detail. My experience, however, is that too-deep immersion often obscures the underlying logic—or lack of it" (2001, xii). The resulting book is a hybrid, a work that looks in some ways like a textbook, but that is structured as a unified argument, beginning with a refutation of current views, a return to the basics of the AI theory, and an application of that theory to various cellular processes. It is highly readable (Pollack won an award from the Society for Technical Communication for a 1990 book), but it is uneven as an accommodation, addressing as it does an implied reader (in the sense of Booth 1974b, 1983, 1989; see also Ewald 1988) who does not know the basics of mitosis but is relatively familiar with the structure of the proteins actin and myosin. Given this chosen audience of the uncommitted and uninitiated, Pollack is freer to speculate without the substantiation that would be required in a publication for experts. Despite its advocacy of an unorthodox view and its nature as a popularization, the book was reviewed last year in *Science* by Thomas P. Stossel, a researcher in the Hematology Division of Brigham and Women's Hospital affiliated with Harvard Medical School, who characterizes himself as initially receptive (Stossel 2001, 611). The final assessment of the book is not entirely favorable, but the fact that the review exists is a warrant of legitimacy in itself.

Writing as outsiders representing a marginalized view, both Pollack and Ling use appeals that are not necessary for those within the accepted tradition. These appeals are peripheral to the scientific arguments from both sides. The scientific arguments are the same in their general form, if not in their content, for insiders and outsiders. These basic arguments are typically constructed by antitheses (e.g., when proteins are in the longer state, water is ordered and when they are in the shorter state, water is disordered; Pollack 2001, 126), by reversals (e.g., structured water drives proteins to their extended state or their extended state drives watering structuring; Pollack 2001, 264) and by graded sequences (e.g. permeability in the cytoplasm is proportional to the size of a hydrated ion; Pollack 2001, 81).[11] But the marginalized view has to stage or surround these material arguments with other lines of support to gain a hearing for them. Among these enabling devices are the construction of a group of supporters, the use of history to legitimate a position, and the attempt to create an impression of newness. A mainstream view, by definition, has supporters, controls the history of a field, and has current research activity to report. It has presumption on its side; the burden of proof falls on marginal views (Whately 1963, 112–115).

Group Creation

In 1984 Ling published a book defending his views that includes, among its many visuals, a picture of nine people arranged in a row outside a building (Ling 1984, 329). The caption provides only the names of the individuals, but this group, including Ling on the far right, provides evidence of a cadre of supporters. As such, this photograph amounts to an argument in favor of Ling's position. It can be taken as evidence that he has supporters.

In intellectual movements there is no permanent merit in being a single voice crying in the wilderness. Movements need followers with real bodies and brains, and this support must also be represented textually. Both Ling and Pollack work at producing an impression of the agreement of others. For example, in the prefaces to both his 1984 and 1992 books, Ling lists the names of many people in the act of thanking them. Some of these are people who supported his ideas, some were perhaps loyal to him personally, and others may have had only institutional affiliations with him. Yet everyone listed in a preface, *unless* their agreement is specifically disclaimed, can count as a supporter.

Similarly in the Acknowledgements section of his 2001 book, Pollack mentions those who read his manuscript, including undergraduate and graduate students, peers, and experts on various subjects. He reports, "one or two [colleagues] responded with hostility" to his overtures, but he nevertheless thanks them for their willingness to answer follow-up questions "that helped me understand why this approach might be seen by some as unpalatable" (Pollack 2001, vii). There follows a list of twenty-one names in alphabetical order of those "whose help has made a real difference." The comments or observations of these solicited colleagues may have ranged from the wholeheartedly endorsing to the vaguely polite. Pollack does not represent them as a set of adherents, but he does present them as a supportive group; they have been willing to read and make comments. They do not reject his ideas out of hand, the fate of those who cannot even gain a hearing.

Even more notable is the apparent consensus-building on the back cover of Pollack's book and on the inside flaps where numerous approving readers are quoted, including students and those having institutional affiliations. Only one of these comments apparently comes from a researcher in the field: James Clegg, a professor of molecular and cell biology at UC Davis. His remarks are complimentary, and evoke Kuhn's ubiquitous notions of scientific change, but they are not an endorsement of AI theories:

> If the thesis advanced here is proven to be sound, there will be a revolution in cell biology. This book offers a dramatic alternative to the 'conventional wisdom' of contemporary cell biology. Some of the material is deeply embedded in controversy, and although I did not always agree with the details, the overall message comes across clearly in this stimulating and enjoyable book. Pollack writes in a pleasant, often humorous style. If you wish to explore the spirit of qualified scientific dissent, this book is for you.

From such mild good will and collegiality, both Ling and Pollack construct the appearance of a positive reception and hence a group of supporters for their ideas.

Constructing a History

Ling's first book, introducing much (if not all) of his AI hypothesis in 1962, proceeds from basic assumptions about charges on proteins and electron sharing to build up his theory. It includes data from his extensive ion concentration experiments and his refutation of the sodium pump hypothesis on the basis of the lack of sufficient energy. But his 1984 book, written after the crises of the seventies, though also containing a defense on experimental grounds, is quite different. It devotes several opening chapters to a history of cell physiology, tracing the rise and fall of competing membrane and "bulk phase" theories. A similar but shorter historical section opens both the 1992 and 2001 books. Historical sections and links are also featured on his web site. The textbook view of the membrane with pumps and channels has its history as well, of course, but its proponents typically separate historical accounts from research reports and reviews. Why did Ling become a historian in these later works? One answer is that these historical sketches serve as potential support for his views. Given the several centuries and the many actors involved in cell research and related areas, any historical account can select from a large number of potential items, but those items selected have to serve a narrative useful for the marginalized view.

The historical account that Ling is after is one that can give his views a lineage. The warrant for this rhetorical use of history might be expressed as follows: if a theory has better explanatory power, competent researchers should have glimpsed it before, supposing that at least some evidence supporting it were technically accessible; hence giving a view historical roots or placing it in a tradition over time can argue for its truthfulness. Darwin used this tactic when he opened later editions of *The Origin* with a historical sketch placing himself in a line of thinkers with similar ideas. With a plausible history in place, current supporters of an unpopular view can then reach back over the heads of their peers and immediate predecessors to earlier forebears who glimpsed the truth.[12] Ling creates a line of predecessors including those who identified and emphasized protoplasm as an inherently living substance in the nineteenth century–Lorenz Oken, Felix Dujardin, Hugo von Mohl, Max Schultze, Thomas Henry Huxley (Ling 1984, 6–8). But more important to him is a little known group in the twentieth century who maintained "bulk phase" theories of the cytoplasm even after the membrane hypothesis was in place. These predecessors include the Germans Martin Fischer and Otto Butschli and the Russian Dimitri Nasanov (Ling 1984, 37–41). While the relevance of their work and ideas can be challenged, the fact that they held certain views cannot be. They suggest the continuity of an alternate view. Historical accounts such as this one creating predecessors can be effective, if indirect, sources of support.

Another argumentative use of history occurs in Pollack's text when he tries to explain why Ling's ideas did not catch on. He first creates a lineage for the idea of "structured" water, citing a 1907 paper, a 1956 monograph by a Russian scientist Aphanasij Troshin, and another written in 1972 by Nobel Prize winner Albert Szent-Gyorgyi on how macromolecules structure water and vice versa (Polack 2001, 53–54). These publications, according to Pollack, all support Ling's 1962 book and his 1965 paper, offering evidence that the water inside a living cell is largely structured. However, according to Pollack, reception of ideas about structured water in the cell was damaged by the "polywater debacle" of the late sixties. Russian scientists announced the discovery of water with unusual properties, a claim that was thoroughly debunked within a few years. Ling's case for structured water in cells was presumably tainted by association with what many came to see as a scientific fraud, or at least that is the implication of Pollack's narrative, which also hints at western bias against Eastern bloc science (Polack 2001, 55). Thus while Ling historically supports his views by giving them predecessors, Pollack additionally supports them by explaining their dismissal as the result of historical accident.

However, perhaps the most obvious use of history comes from the conception of the history of science itself that the anti-pump group used and uses: it is the notion, explicitly derived from Kuhn, that science progresses by revolutionary change. Ling congratulates himself in the opening of his 1992 book, titled appropriately *A Revolution in the Physiology of the Living Cell,* in having been part of a scientific revolution (xi, xv), and in the concluding chapter he lists the five criteria his theories have satisfied to qualify as revolutionary (Ling 1992, 324).[13] He does, however, amend Kuhn by distinguishing revolutions that are constituted by the individual scientist's work (what he calls the "scientist's scientific revolution") from those revolutions represented by widespread acceptance of a new paradigm (Ling 1992, 319).

Ling cites many exemplars of revolutionary science, including Lavoisier, Mendel and Semmelweis, but the favorite model for the anti-pump group is Galileo, whose story in this tradition appears perhaps for the first time in Raymond Damadian's letter responding to Kolata's 1976 *Science* review. Galileo is also mentioned on Ling's website and in the opening of Pollack's *Cells,* which features Galileo conquering the epicycles, simplifying anomalous orbital complexities that resemble the current "bewildering" array of pumps and channels (Pollack 2001,

3). A story of revolutionary science, like Galileo's, has as potential subplots the persecution of the scientist and the corruption of official institutions. This master narrative can be so compelling as to invite the construction of impediments in accounts of scientific breakthroughs so that they can conform to this story grammar of insight, misunderstanding, persecution, and (one hopes in one's lifetime) eventual triumph. So strong is this plot structure that minority status and presumed persecution can be taken as signs of being right.

Newness and Age

The passage of decades works against this confidence. Scientific revolutions should replace older views, not be them. If newness usually wins out, then claims to newness will also have some persuasive appeal, even as historical precedent does. An arguer will claim that new instruments, new approaches and the resulting new evidence have produced a new perspective or revived an old one. The *Science* reviewer of Pollack's book, Thomas P. Stossel, endangers Pollack's case when he describes the book as "an eloquent and accessible statement of a heresy that has smoldered at the fringe of orthodox biology for about 30 years" (Stossel 2001, 611), and when he comments further that "Defenders of pumps, channels, and molecular details will notice that most of the references Pollack cites to support his claims are elderly" (Stossel 2001, 611). The implication here is that a view that has been around for a while and that is still a heresy is not going anywhere. How long does an alternate view have? The Copernican world system was not widely accepted for over a century, but perhaps the optimal time frame for serious consideration of an alternative has decreased. It is possible to simply "wait out" a heterodox position. If over time it receives no substantial reinforcement and there are no substantial converts to it, in number or prestige, it loses ground. Passage of time is especially fatal if a view receives no mention in annual reviews or textbooks as a possible alternative.

Another assumption, an implicit part of the rhetorical staging in scientific arguments, is revealed in Stossel's charge that Pollack's citations are "elderly." One would expect that the results of a well-conducted set of experiments should simply stand and not "age," though ten, fifty or even one hundred years have passed. Why is it then a refutation to describe sources as "elderly"? One answer is that experimental or observational facts, as rhetorical counters, are not impervious to context. Rhetorically, all evidence decays over time. Even if it is unchallenged, it ceases to have the same probative power in newer contexts. Pollack shows his awareness of the problematic of aging data when he brings up the energy refutation in Ling's 1962 book but prefaces the discussion with the admission that these tests were carried out "long ago" (Pollack 2001, xx). By this concession, Pollack attempts to defuse the refutation on the basis of age. Obscuring the datedness, he also cites a 1997 source for the data demonstrating that the cell has insufficient energy for the pumps, but this piece only reprints the data from the then thirty-five year old publication (Ling 1997,126).

Pollack, who in fact is reviving an older view, does of course easily achieve "newness" simply by the 2001 imprint on his book. Another writer citing his text can create the legitimacy of recency. But though Pollack does depend extensively on Ling's early arguments, he also constructs an appeal to "newness" by drawing support from a field that saw if not its beginning then a significant expansion in the nineties: namely research on polymer gels.[14]

In the first half of his book, Pollack introduces the basics about structured water and ion adsorption that are features of Ling's AI theory. But cells do not just sit there, and Ling's views must also account for how cells function. Pollack explains that he is looking for a functional mechanism based on cytoplasmic structure, and the framework for his search, he announces, lies in "the cytoplasm's similarity to the polymer gel" (Pollack 2001, 113). Pollack lists several common attributes that justify this analogy, but immediately he makes it clear that he means something more: "recognizing this correspondence, we treat the cell as a gel and ask whether such treatment can bring us toward a common paradigm for cellular action" (Pollack 2001, 113). In other words, the analogy functions as an identity though the exact nature of this relationship remains somewhat obscure, an ambiguity underscored in the title with its list format, *Cells, Gels and the Engines of Life.*

The cell/gel melding is extremely useful for Pollack. It allows him to take research results, largely about synthetic gels, and apply them by implication to living cells. Ling had actually drawn on the gel analogy in both his 1984 (175–178) and 1992 (81–85) books, emphasizing the well known "colloid" gelatin (or Jello, denatured collagen). But since a great deal of research on synthetic gels was done in the 1990s, many of the references to the properties of gels in Pollack's eighth chapter, which introduces the gel model of phase transitions, are not "elderly." This new material, though it concerns inorganic gels, is presented as a new source of legitimacy for the Ling view. Hence the demand for newness, for some new method, new input, new perspective, is satisfied to justify reviving the AI view of the cell.

Sources of Incommensurability

Overall, an account of Ling and his supporters provides ample evidence of the institutional and professional fallout that can accompany intellectual conflict. One explanation of the conflict might focus on sociological dimensions such as the institutional locations of the players, finding a narrative of outsiders challenging the work of insiders. But there is no question that genuine intellectual differences characterize the opposing positions. These differences both come from and lead toward different experimental practices and views on what a theory of the cell has to explain. These differences can be looked at from the AI perspective, focusing on their attempts at refutation, on the incompatible research tradition they represent, and, finally, on their rhetorical investment in incommensurability.

Though Ling has an overarching theory of the cell, much of his long career has been devoted to refuting features of the canonical view. In addition to his frequently referenced experiments that presumably proved the energetic impossibility of the pumps, Ling did experiments with sliced cells trying to show that a membrane is not needed for ion partitioning (Ling 1978, 1105–123), that red blood cell ghosts, the favorite model for studying membranes, are not free of protein as supposed (Ling 1984, 128–133),[15] that the potassium in cells is adsorbed and not in free solution (Ling, Gulati and Palmer 1977, 1281–1283), and that water in cells is structured and therefore not easily separated in centrifugation (Ling 1976, 293–295). Pollack also piles up presumably challenging problems for the membrane/pump tradition in the opening of his book: that patch clamp results showing the gating of the sodium channel can be generated from non-biological samples (Pollack 2001,13), that the mechanisms of pumping are unknown (Pollack 2001,10), that the exclusion of smaller ions from pores intended for larger ones has never been explained (Pollack 2001,14–15), that the energy requirements for all the newly postulated pumps are beyond the cell's capacity (Pollack 2001,17–19), and that furthermore the cell has insufficient surface space for the many pumps and channels (Pollack 2001,17). None of these observations or refutations has been compelling to membrane proponents; the gaps in knowledge cited by Ling and his supporters are research goals for the canonical view, not reasons for abandoning it. Like a Popperite with a rigid standard of falsification, Ling himself insists that the membrane pump view was once and for all "unequivocally disproved" (Ling 1997, 126), while the oblivious majority continues to define and describe ever more complex pumps and channels. Since the seventies, there have been no explicit refutations of Ling's views. Mainstream proponents are not defending early experiments on transport; they have found other sources of evidence.

These two groups then are not engaged with one another; they are not competing on the same field. This lack of engagement comes in part from different disciplinary allegiances and hence different explanatory demands and research methods. To begin with the disciplinary differences, Ling's first generation of supporters, Cope, Hazlewood and Damadian, were all MDs who shared a background in physics and a preference for NMR techniques. Ling himself, influenced by his fellow Boxer scholar, the physicist C. N. Yang, sought a model of cell physiology based on an analogy with the magnetic "cooperative phenomena" that won Yang the Nobel Prize (Ling 1984, 208–225). It is not accidental that the title of Ling's first book emphasizes that he offers a "physical" theory of the living state. Thirty years later he was equally explicit about this grounding: "a biologist," he writes, "must begin with a hypothesis that is, first, a logical deduction from the existing laws of physics and, secondly, verifiable on a relevant inanimate model system. A failure to recognize one or both of these steps would lead back to eighteenth-century *vitalism,* according to which life phenomena fall outside of the realm of physical laws governing the inanimate world, and are mystical in nature" (Ling 1992, 320).

The Ling group has been explicit in its preference for physics-based theories. In a 1982 interview, Damadian cited the difference in orientation: "I think the real reason for the delay in acceptance [of structured water views of the cell] is that most of the people doing this kind of work are biologists. The kind of stuff we're talking about gets into the depths of physical chemistry. Now an old-line organic chemist or organic biochemist who's not a terribly quantitative fellow—if he attempts to follow the trail—instantly runs into this quantitative barrier that he's completely unfamiliar with. Whereas the concept of the 'sodium pump' is sort of qualitative and vitalistic. [. . .]" (Hildenbrand 1979). In the second generation, Pollack is an electrical engineer who was, in his own words, "flabbergasted to learn in my first physiology class that the very same charges and potential gradients that drove induction motors, surrounded power lines, and gated semiconductor junctions also pervaded the living cell. The charm of this revelation has yet to evaporate" (Pollack 2001, 99). Pollack also talks frequently about what chemists and physicists are familiar with that "biologists" do not know (e.g. phase transitions; Pollack 2001, 143).

The underlying disciplinary differences are more subtle than a quantitative versus qualitative contrast, which is based on an incorrect characterization of molecular biology anyway. Ultimately, the Ling group wants a simplified, unified physical model of the cell. For them, complexity like that displayed by the current account of pumps and channels is prima facie a sign of error. A true account must be a simple one. Hence the attraction of the structured water ion-partitioning model. As Pollack puts it, "No special widgets are needed to keep potassium inside the cell or sodium outside the cell—all of this follows directly from the cytoplasm's basic physical chemical features" (Pollack 2001, 97). In his 1992 book Ling also cites simplicity as the achievement of his AI hypothesis (Ling 1992, 336), and Pollack could not be clearer that simplicity, or parsimony, is his explanatory goal: "Such quests [he writes] are driven by the notion of parsimony—the presumption that once the proper underlying principle has been identified, a seemingly complex array of phenomena will fall into place simply. It is this quest for simplicity that drives the search for a common cell-action paradigm" (Pollack 2001, 113; see also 129).

The search for underlying explanation of cellular structure and function in terms of a simplified physical model both comes from and produces an essentially mechanistic view of the cell: "If cellular mechanisms are to work reliably," Pollack writes, "should they not be based on simple—even beautiful—principles?" (Pollack 2001, 283). Given this mind set, it is not surprising that Pollack's argument is studded with analogies between cells and simple physical phenomena and machines—magnets (Pollack 2001, 262), zippers and jack-in-the-boxes (130), wedges and diesel engines (205) and batteries (263). To explain how phase transitions could move solutes, for example, he compares the process to the "zone refining" used by engineers to make transistors. In his ultimate definition, the cell is an "entropy machine" creating potential energy by ordering water and then releasing that energy through the disordering achieved by phase transitions (Pollack 2001, 265). Hence there is special force in the title of his book: *Cells, Gels and the Engines of Life.*

But molecular biologists, the pump and channel researchers, are arguably not driven by a quest for simplicity. If they were to label their approach, it might be called a quest for specificity. They seek the detailed mechanisms that have evolved to respond to unique environmental conditions in cells with unique functions. Thus the transmitter-gated chloride ion channels in postsynaptic neurons that inhibit firing or the chloride channel in epithelial cells where a single mutation leads to cystic fibrosis—these and many other highly specific and complex regulatory pathways are the goals of research, and this pursuit of complexity is inherently a more successful strategy for a research establishment.[16] Of course the complexity in biological systems has been used by intelligent designs proponents like Michael Behe as an opening for nonscientific explanations; hence the warning that the lack of reduction to simple physical processes represents a resurgence of a suspicious vitalism.

Because their goal is a simplified physical model of the cell, the AI researchers, especially Ling himself and those of the first generation, use rather stark single difference experiments on whole tissue preparations, modifying a condition and seeing what results.[17] As a group, they do not use the dazzling armamentarium of techniques now routinely exploited by molecular biologists. It is not technical ineptness on their part but rather that they are not interested in the results that these methods achieve. Ling for instance only demands that a "complete" theory of the cell explain in general solute distribution, permeability, cell volume and electrical potential (Ling 1992, 336); cell reproduction and the "subtle differences in molecular interactions" from organism to organism or tissue to tissue (Stossel 2001, 611) are not a central interest.

According to the explanatory hierarchy of the sciences, there should be no deep-seated incommensurability based on these disciplinary allegiances. Biological phenomena should translate into chemical phenomena and these in turn into physical; physics explains chemistry and chemistry explains molecular biology and molecular biology explains observable biological processes (digestion, infection, cancer).

But this ultimate reduction may be inconsequential or unproductive when it comes to studying biological systems that are the result of genetic change over time. Nature need not follow the principle of parsimony in biological systems.

There are then deep-seated differences in explanatory allegiances, but one would still expect that these differing views could be made compatible and here is where the rhetorical differences return. The anti-pump advocates, especially those of the first generation promoted a suite of radical views that had to be accepted altogether. They invested rhetorically in the incommensurability of their views with those of membrane/pump researchers. It was not just that the proposed pumps in the membrane could not operate as they were described; rather there were no pumps and no membrane of any physiological significance (see esp. Hazlewood 1972).[18] In other words, a cell with both active transport and structured water was deemed impossible. Their theories also required radically different notions about the role of ATP and the behavior of proteins in the cell. When scientific recognition of their work declined, they complained in public about the peer review infrastructure. Meanwhile the research methods of molecular biologists, using purification, sequencing and cloning, offered new evidence of the existence of the pump proteins, and altering pump or channel genes with site-directed mutagenesis demonstrated their functions. The AI edifice was discredited, including, temporarily at least, what were perhaps its less contestable parts.

In his 2001 book, Pollack maintains the rhetorical investment in incommensurability, though with a bit more nuance. He concedes the existence of "proteins with pump-like or channel-like features" (Pollack 2001, 5). He admits that these membrane-situated proteins may function as receptors initiating changes in the cytoplasm's permeability though their actual "pumping and channelling" might remain "functionally insignificant" (Pollack 2001, 21, see also 47). He never mentions pumps, channels or membranes again until the last ten pages where he says there is no evidence that NaK-ATPase pumps anything, though he allows that, "conceptualizing the process within a framework of pumping could remain a convenient expedient so long as the underlying reality is borne in mind" (Pollack 2001, 275). This view is certainly, as Pollack himself puts it, "orthogonal to convention" and "worlds apart from current views which place most of the action in the membrane instead of in the cytoplasm" (Pollack 2001, 109). Adopting it altogether would require not just minor adjustments but massive recantation of the ruling paradigm on the part of practicing molecular biologists.

The views originated by Ling and reasserted by Pollack may be, as a totality, growing further from mainstream research. In current descriptions even the cytoplasm of eukaryotic cells contains the membranes of the organelles, the mitochondria and the endoplasmic reticulum, so that many cellular processes are now understood in terms of events occurring across a definable membrane-mediated inside and outside. And in November of 2001, after Pollack's book appeared, *Nature* published three papers explaining the selectivity and favorable energetics of potassium channel proteins (Martindale 2002, 22).

But there are parts of the AI view that were never as contestable as the whole theory, namely the structuring or layering of some cell water and the possibility of gel-sol transitions in some cellular processes (Stossel 2001, 611). The incommensurability could be converted rhetorically into less serious differences of emphasis or perspective, grounds for potential assimilations. In other words, the elements might be the same while the emphases differed. Molecular biologists are used to invoking conformational changes as the agency for some cellular functions; Pollack would describe these same phenomena as phase transitions. They are not as interested in the intermediate processes as he is; he is not as interested as they are in what triggers a change such as the contraction of a muscle fiber. Water structuring and phase transition might be likely candidates for common ground between these approaches, but not if they are always part of the total view of the cell presented in the AI hypothesis.

A blending of the views would also require letting go of earlier claims to definitive refutation and a relinquishing of the "enclave mentality" that the first generation, given the painful personal histories involved, would not find easy.[19] Finally, to reconfigure their perspective within the mainstream approach, Ling and his supporters would have to give up their self-identification as a Kuhnian vanguard with a revolutionary theory of cell physiology. With these changes in emphasis and perspective, there may be another review in *Science* some day that describes the resurgence of a formerly discredited view. Incommensurability that is rhetorically constructed or enhanced over decades can be eventually be rhetorically diminished and even overcome.

Notes

[1] Hacking (1983, 67–73) has the most compact account; for more elaborate treatments, see, in particular, Sankey (1994), the introduction and papers in Hoyningen-Huene and Sankey (2001), a rhetorical approach in Heidlebaugh (2001)—as well, of course, as the introduction and the other essays in this volume.

[2] Ling's reputation was established when, as a graduate student, he perfected what came to be known as the Gerard-Graham-Ling microelectrode, which could measure the electrical potential in a single cell. Ling held positions at Johns Hopkins Medical School (1950–53), the University of Illinois Medical School (1953–57), the Eastern Pennsylvania Psychiatric Institute (1957–61), and the Department of Molecular Biology at the Pennsylvania Hospital in Philadelphia (1961–1988). He is now at Fonar Corporation.

[3] Ling writes of this period of sustained funding and final loss as follows in his 1992 book. "[U]nder the protective wings of Drs. Callahan [Office of Naval Research] and Schiaffino [NIH] our work was fairly judged periodically by neutral scientists who consistently recommended continued support. Unfortunately, Dr. Callahan and Schiaffino eventually left their posts. The new administrators held entirely different views in regard to my work. Thus the new Director of Division of Research Grants, Dr. Jerome Green and his subordinates believed strongly that no one applicant should be treated differently from all the others, and accordingly insisted that the new (in-name-only) Special Study Sections must include scientists from both sides. My repeated protests were to no avail. Strongly dominated by my scientific opponents in the panel—who have never openly defended their membrane-pump theory in public against my published refuting evidence, nor challenged in print the AI hypothesis and the confirmatory supportive evidence gathering in the literature for well over a quarter of a century—the panel 'massacred' all three renewal proposals and NIH withdrew all its support. In consequence, my laboratory was forced to close October 31, 1988 at the height of its productivity" (Ling 1992, xv).

[4] According to his web site, Ling wrote repeatedly to all seventy-five top NIH officials; and eventually sent a twenty-five-page letter to Presidents Carter and Reagan, to all representatives and senators of the 100th Congress, to twenty-four newspapers and magazines, and to ten prominent newsmen. One Congressional aide who bothered to answer him told him that letters like his went straight into the wastebasket, and that he should never write more than two paragraphs (Ling 1998).

[5] A similar charge concerning NIH's inability to fund unusual research or heterodox approaches has surfaced again recently in complaints from Craig Venter, decoder of the humane genome, about NIH's refusal to fund

his eventually successful "shotgun" approach to sequencing (Wade 2002, D5).

[6] After a career publishing in major journals, by the late seventies his papers, with only one exception, appeared solely in Ling's journal. Perhaps as a result of his frustrations or, according to Ling, his inability to find funding, Cope committed suicide in October 1982 (Ling 1984, ix).

[7] On his website and in the 1997 article, Ling claims that it was only in 1997 that Gerald Pollack drew his attention to a phrase in Kolata's 1976 mentioning "crucial experiments and calculations [. . .] that provide strong evidence for the existence of pumps" (Ling 1997, 122); the article also mentions criticisms by Jeffrey Freedman and Chris Miller (actually two of Ling's former graduate students) refuting Ling's assessment of sodium efflux from muscle cells. In Ling's reading, the objectionable phrase "crucial experiments and calculations" refers only to Freedman and Miller. Ling wrote to these two and to Kolata asking where their data had been published. Freedman never answered; Ling quotes Miller's answer as follows: "When I was a postdoc at Cornell and Jeff was one at Yale, the two of us, still fresh from the tumultuous experiences in your lab, wrote a manuscript on analyzing Na^+ efflux data in muscle—a sort of literature review—taking into account compartmentation effects" (Ling 1997, 129); Miller goes on to explain that this paper was rejected by the *Journal of Membrane Biology* and never published elsewhere. Ling also accused Kolata of making changes in the article after the version he had seen. Kolata said she could not remember the details behind this article but doubted that she would have made these attributions without good sources (Ling 1997, 128–129). Ling wrote her two more letters that she did not answer.

[8] I have adopted the mildly unconventional practice of citing URLs in the text because of the linking practices of Ling (1998), which is a long article with embedded links encoded as parts of the text. This one, for instance, is encoded as "Corruption also wastes a lot of money," (www.gilbertling.org/lp15.htm). My dating of the article, incidentally, is a bit of guesswork, based on a note that suggests the last alteration to the actual *text* was in March of 1998. A further note says the *site* was last updated in May of 2004. For the purposes of this essay, the site was last *accessed* in April of 2005.

In the page linked to "Corruption ..." (www.gilbertling.org/lp15.htm), Ling mentions a published exchange between Glynn and Freeman Cope in *Trends in Biochemical Sciences* for October 1977, pages 225–227. Ling is quite detailed about the number of his papers that Cope cited and that would, therefore, have been known to Glynn. However, no such exchange is published on those pages or anywhere else in that journal that I could find. Furthermore, an investigation of both Cope and Glynn's citations in the PubMed database revealed no published exchange.

[9] Ling claims to have found "at least five other reviews and large published symposiums" between 1975 and 1986 on the sodium pump or active sodium transport that also cited only supporting evidence and no work of his or his supporters (www.gilbertling.org/lp15.htm). He does not cite them. It seems unusual that, of all the possible reviews consolidating knowledge about the pump, he should have focused on the Glynn and Karlish one. However, the date of this review (1975) coincides with the defections of his graduate students, and Ling believes that his students left because they became aware that, "jobs and research grants were all totally under the control of my scientific opponents" (Ling 1997, 128).

[10] Whether writing in 1962, 1976, 1984, 1992, or 1997, Ling is unusual in the level of certainty he claims for his experimental conclusions. He firmly believes that he disproved the feasibility of sodium pumps on the grounds that cells do not have the energy available to drive these pumps. "Thus, on energy grounds alone, the sodium pump hypothesis has been unequivocally disproved. The time was 1956, if one considers the time the definitive experiments were completed. It would be 1962, if one relies on time of publication" (Ling 1997, 126).

[11] The patterns of reasoning with the scientific evidence are taken from my *Rhetorical Figures in Science* (Fahnestock 1999).

[12] In addition to citing predecessors, Ling's historical sections also record early mistakes that were never rectified. So, for instance, Ling claims that the early focus on plant cells, which have large central fluid-filled vacuoles, set up a conception of the cell as a sac filled with water; "The long erroneous usage," he writes on his web page, "and the work of early cell physiologists drove deep into the psyche of biologists the belief that cells are hollow chambers filled with a watery liquid—which lasted in the text books [sic] version of cell physiology to this very day." Nineteenth-century researchers, he claims, were also misled because their light microscopes, which could never have resolved the cell membrane, showed them a layer of protoplasm (the cytoskeleton) that they mistook for the membrane (www.gilbertling.org/lpb.htm).

[13] The following passage demonstrates Ling's explicit use of Kuhn and his five criteria:

> The starting point for a scientific revolution is the existence of a time-honored scientific theory which underlies the explanations of a spectrum of related natural phenomena. A revolution begins with the introduction of a contending new theory which deals either with the same or with a broader set of natural phenomena. The new theory is often built upon diametrically opposite basic postulates from those on which the old theory was built. I believe that a suc-

cessful revolution depends upon meeting a set of five criteria (for background information, see Kuhn 1962, 152 [1996, 152]).

Criterion 1: The accomplishment of one (or more) crucial set(s) of experiments disproving the old theory, which has (or have) successfully stood the test of (a prolonged period of) time.

Criterion 2: The demonstration that all of the key evidence once considered to unequivocally support the old theory is incorrect or equivocal.

Criterion 3: The disproof of the fundamental postulates of the old theory and the verification of the postulates of the new theory.

Criterion 4: The demonstration that the new theory can predict and explain significant experimental findings that can be explained by the old theory, *as well* as other significant experimental findings which the old theory *cannot* explain.

Criterion 5: Establishing that the new theory, cast in the rigorous form of equations, can quantitatively explain the experimental findings that can also be quantitatively explained by the old theory, *as well as* those that *cannot* be quantitatively explained by the old theory (Ling 1984, 324).

[14] Pollack makes another case for legitimacy based on new work. After describing the defeat by association with polywater, Pollack claims, "stunning breakthroughs by physical scientists have reopened the door of consideration" (Pollack 2001, 55). These stunning breakthroughs, in papers dated 1981, 1988, 1991, and 1996, concern evidence about the force required to squeeze thin films of water from between metal plates, and the discovery that the force-separation relation oscillates, suggesting that the remaining water is layered. By invoking research on inorganic gels and on water as an inorganic solvent, Pollack changes the scope of the field under consideration. The cell becomes part of wider disciplinary investigations; it is pulled toward the physical sciences.

[15] Techniques with red blood cell "ghosts" proved especially important in early studies of membrane composition. Red blood cells are already simplified by nature since they lack a nucleus and other internal membrane-bound organelles. They can be ruptured and their hemoglobin removed; they can then be studied either unsealed or resealed either normally or inside out. In these conformations they offered the first evidence that the lipid composition of the membrane bilayers differed and that some proteins extend through the membrane, while some are bound internally (e.g. spectrin, the cytoskeleten

under the membrane) and some externally. One of Ling's defecting students, Jeffrey Freedman, used red blood cell ghosts to show that the Na^+ and K^+ gradients were maintained even in the absence of the cell's cytoplasm, the agent of maintaining the ion gradient in Ling's model. Ling sought to specifically counter this work by refuting the assumption that red blood cell ghosts were in fact protein-free, or even relatively protein free.

[16] My thanks to Professor Catherine Schryer, University of Waterloo, for pointing out to me that an appeal to complexity in a field makes it possible to justify a larger research enterprise involving a greater number of people and greater grant support.

[17] One of the links on Ling's website records an affidavit that he had one of his graduate students write and sign in 1971 in which the student records a conversation with one of his professors at University of Pennsylvania. The Professor had explained in a lecture that there were two schools of thought in cell physiology, the membrane theory and the association-induction hypothesis held by a small but vocal minority. The course he was teaching would follow the membrane theory. The student went to talk to the faculty member afterwards, and the faculty member reportedly commented that he could not consider Ling's ideas because he had a position to protect and had to think of his wife and children. But the professor also said, "Ling does his experiments one way and everyone else does them another. The conditions are different, so you can't compare anything" (www.gilbertling.org/lp18.htm). Ling's accounts of his various lines research do show that complex enabling assumptions and interpretations are required in order to make the results favorable to his theories.

[18] This "all or nothing" attitude occurs in Carlton Hazlewood's letter in response to Kolata's review of the controversy. Kolata had reported that many pump advocates agree with the more ordered structure of cell water; but Hazlewood disallows any partial concession: "If this is so, then the proponents of pumps are stating that the fundamental premise (that is dilute solution theory) that gave rise to their concept is incorrect. Such contradictions between fundamental assumptions and experimental findings should provoke considerable doubt as to the validity of the hypothesis" (Hazlewood 1976, 532).

[19] Another painful personal history reported on Ling's website is that of Ludwig Edelman whose testimonial is quoted at length:

> In 1971 I showed that the influence of K (potassium ion), Rb (rubidium ion), and Cs (cesium ion) on the resting potential of guinea-pig heart muscle cells could be precisely predicted from flux measurements when following the assumptions of the association-induction hypothesis. On the other hand, the results contradicted

> the predictions of the membrane pump theory. Since that time several professors of our faculty tried to convince me that the membrane pump theory is basically correct and that Ling's theory has been disproved. In July 1972, four referees were invited to discuss my scientific results and conclusions together with the members of our Electrophysiology Department. During this discussion my conclusion could not be refuted. Furthermore, I put forward several arguments against the membrane pump theory and asked the invited referees to write down their opinion against these arguments. I never got an answer. In October 1973 I lost my job.

Edelman retrained as an electron microscopist and continued to publish in support of the association-induction hypothesis (http://www.gilbertling.org/lp12.htm).

10

Measuring Incommensurability: Are Toxicology and Ecotoxicology Blind to What the Other Sees?

Charles Bazerman and René Agustín De los Santos

In the account of incommensurability Thomas Kuhn introduced in *The Structure of Scientific Revolutions,* scientists who inhabit one theory-based perspective are unable to recognize, understand, or accept entities revealed through observations made from an alternative theoretical perspective.* The difference of ontologies stands as a fundamental roadblock to communication. Scientists can move from one perspective to the other only through a gestalt switch, which makes the accounts of nature from the previous theory incoherent and lacking reference to the world.[1] In the switch the old gestalt has evaporated. Later, when Kuhn reformulated the cognitive divide as a matter of a communal switch of a taxonomic lexicon rather than an individual psychological switch of gestalt, incommensurability remained. The terms in the old taxonomy do not exist in the same world of relations—of similitudes, contrasts, subordinations—as the terms in the new. The new and old terms are incoherent in each other's presence, and are only intelligible when each is viewed within its own taxonomic world. In his later for-

*We would like to thank Doug Bright, Scott Frickel, Randy Harris, Greg Kelly, Wayne Landis, Michael Osbourne, and Mike Truscello for their comments on an earlier draft of this essay.

mulations, Kuhn does think understanding across boundaries is possible, but only by a process akin to learning a new language. Translation is inevitably misleading (Kuhn 2000, 33–57).

This difficulty of communication goes well beyond Norwood Hanson's recognition that all observations are theory-laden (Hanson 1958). Kuhn suggests it is impossible to adopt an alternative framework to see the entities and relationships posited by alternative theories while keeping one's initial entities and relationships in mind. Even the measurements made from one theoretical perspective would find little credibility in the other, for the methods of measurement and the entities measured from one framework would have little standing in the other, because the methods and the entities of interest are both warranted by the theory, "paradigm-determined" (Kuhn 1996, 126).[2] Even more suspect would be the theoretical or second-order entities that cannot be directly observed and measured, but are conceived only as the theoretical consequences of the observable and measurable phenomena—which would constitute higher order terms in a taxonomic hierarchy. Hanson's observation suggests a degree of generosity and empathy in cross-theory discussion, while Kuhn's anticipates mutual incomprehension, if not hostility.

This paper examines the comprehension and incomprehension between an existing field (toxicology) and a newer one (ecotoxicology), which appeared to engage in paradigmatic conflict with the prior field to establish its own meaning and value. Examination of the shifting relations between the two fields, along with an intermediary field of environmental toxicology, suggests that neither incomprehension nor generosity ruled the day. Rather the practical problems and interests that sponsored the fields prompted an accommodation that respected the value of the work of each. Each of the three fields needed the intellectual and evidentiary resources of the others.[3]

Habits of Thought in Pure, Bounded Science Versus Socially Saturated Sciences

Incommensurability gives philosophic warrant to a kind of intellectual stubbornness within a knowledge space wholly occupied by rational considerations.[4] One theory, one perspective on the world, in Kuhn's view, replaces another because anomalies become intellectually intolerable to a group of scientists, particularly newer ones, who are less fully habituated into the previous perspective. Scientists are not led to abandon the old theory because their problem changes or their interests shift to new concerns or they bring new resources and cultural dispositions. After a period of confusion, a new theory emerges that better matches the data and gains the adherence of scientists who find the theory solves their intellectual problem. The Kuhnian scientific world assumes a bounded rational world of science influenced only by its own internal logic, uninfluenced by what would be considered external to its investigative and reasoning procedures. Even as he moved from an individualist mental view of cognition to one of group cognition supported through a shared lexicon, the commitment remained to a science moving by its own logic, separate from worldly concerns and interests. Similarly, as he simultaneously shifted his focus from revolutionary scientific change to specialization, he continued to locate incommensurability at the boundaries of specialties (Kuhn 2000).

In a self-contained world of science, fundamental problems do not shift greatly; more adequate accounts of phenomena are just sought and gain ascendancy. To Kuhn the problems of Anaximander, Aristotle, Ptolemy, Copernicus, Kepler, Galileo, and Newton were all much the same (Kuhn 1957): explaining the motions of the heavenly bodies. The problem solved by phlogiston theory was not far from the problem solved by oxygen (Kuhn 1996). Stephen Toulmin (1972), in presenting a survival-of-the-fittest theory refinement of Kuhn's rationalist theory, says the continuity between generations of scientists and their theories (what he calls the *transmit*) is the problem formulation.

Much of the history and sociology of science dating back to Robert Merton's 1938 *Science, Technology, and Society in Seventeenth Century England* suggests that in many, if not all, instances science is not fully bounded from other socio-cultural domains, that the processes of investigation and adjudication are not purely or simply rational, and that problem choice changes substantially and sometimes rapidly, as people are motivated by different concerns and interests (in both the intellectual and self/group-seeking senses of the term). Further, even the notion of a bounded rational science is a new historical construction, maintained with considerable rhetorical energy (Gieryn 1999; Shapin 1994; Shapin and Schaffer 1985; Bazerman 1988). From a philosophy-of-science perspective, this would suggest there is no essential criterion demarcating a boundary between science and non-science, but only a contingent historical definition serving local practical purposes. This position was suggested in a broad way by Feyerabend (1975) and was more fully articulated by Taylor (1996).

If there were a bounded space where scientific problems were held stable and action were simplified to a pure intellectual pursuit of best answers, theory change perhaps might be adequately described in Kuhnian terms of revolutionary gestalt shifts or new specializations with novel taxonomic lexicons. Within a world where there are no other forces beyond intellectual conviction to motivate change in concerns, problems and points of view, one can be stubborn in sticking by one's intellectual guns and can dismiss what is beyond one's vision as faulted, chimerical, unsubstantiated and unsubstantiatable. There is no exigency for change beyond what you and others think of your thought—and for those purposes it may be better to be a champion of a cause, even if it is losing, than to be a late convert to another's creed.

The case we examine in this paper does indeed in some respects resemble Kuhnian normal science replaced by a revolutionary paradigm shift. Even more it might be seen to resemble his later model of proliferation of specializations, each with different hierarchies and taxonomies of conceptual terms. Some actors in this case did resist and dismiss the new, particularly with respect to measurement outside the practices of the older normal science. Ultimately, however, there was little problem of incommensurability between the disciplinary perspectives of the two specializations. Further, while a number of pioneers of the new field describe the new endeavor in Kuhnian revolutionary terms, they never dismiss or displace the older discipline. That field continues going about its traditional business "normally." The new field never stops accepting the findings, measurements, or methods of the old, although placing them within a new framework, and the old gradually comes to recognize the findings, work, and methods of the new. The holders of alternative hierarchies and taxonomies find effective enough ways of communicating with each other to carry on their respective businesses, with the help of each other.

An "Impure" Case

But then, the case we will be looking at is not bounded in the pure space of rational science imagined by Kuhn—it is pervaded by the interests of industries, professions, politics, government regulations, the daily lives of people, and the health of the planet. One might think that those varieties of "nonscientific" interests might lead to an even greater intransigence and refusal to recognize the importance of alternative points of view, methods, measurements, and ontologies. One might expect that "rational science" might be more open-minded than applied sciences caught up in the ideological, economic, and political push and pull of life. But at least in this case, it is the complexity of nonscientific life that creates changing exigencies of concerns, changing definitions of problems and changing domains of interest, and complex multiple areas of engagement and activity. These complexities leave seemingly overlapping sciences and theoretical perspectives alive, side by side, each accomplishing their work and respecting the work of the other insofar as it fits their needs and interests. Each set of interests focuses empirical inquiries as well as provides boundaries around empirical attention. As interests change or divide, attention shifts. Loyalty to theory and theory-warranted methods evokes some defensive reaction, but as interests shift or expand attention, the need for empirical knowledge about new domains leads to acceptance of phenomena and methods, even if they are not fully within the toolkit of prior theory. As intellectual interests are so integrated by the interests of other domains of action, it is hardly possible to talk about these sciences as having hard boundaries dividing the internal and the external. The sciences here provide epistemic means for carrying out various human ends in as empirically grounded a way as current theory and methods will allow. The sciences we are looking at here are means of knowing the world so as to act effectively within this world while avoiding untoward effects of human action. What some might call "external factors" set the agenda for work and attention. It is then up to the applied fields to find and develop what they consider the most empirically grounded and intellectually justifiable way to carry out that work.

The case we are examining here is that of the relations of the recent field of toxicology to the even newer field of ecotoxicology, along with an intermediate field of environmental toxicology. Toxicology itself emerged as a distinct field from its parent, pharmacology, only in mid-twentieth century, but soon normalized in theory and practical method—even as both deepened with advances in biology, medicine, and chemistry. From its beginning it was a field that served a diverse set of needs—from the fields of medicine and pharmacology; from drug, chemical, and cosmetic corporations; from government regulatory bodies; from the criminal justice system; and from various other political concerns. Those interests directly motivated toxicology's projects and methods: identifying the dosages at which chemical substances had toxic effects on humans and, secondarily, on economically valuable domesticated animals, describing and understanding those effects, uncovering the mechanisms of those effects, and detecting the presence and levels of toxicants in individuals. Its studies were conducted under laboratory conditions where individual organisms were exposed to controlled amounts of toxicants. In the 1960s, with increasing public awareness of environmental pollution, toxicologists had a new range of substances to examine and a new set of government regulations, industries, and public interest groups to motivate and or/sponsor their work.

Soon some scientists began arguing that environmental pollution problems could not be fully understood in the laboratories by studying dose effects on individual organisms nor on individual species. They argued for gathering data and studying effects in the field, for studying populations rather than individuals, and for studying dynamic and complex effects in the ecosystem overall rather than isolated effects on separate species regarded independently of each other. Further concern turned from protecting humans, and species economically valuable to humans, to protecting all species participating in ecosystems. Although borrowing some techniques and knowledge from toxicology, these new concerns required a fundamentally different form of knowledge, with a radically different theoretical perspective, one that expanded the scope of entities and systems of interest to the field, employed new theories to describe relations among the entities, and new ranges of measurement to gather data about these entities. These incipient ecotoxicologists measured things that were not part of the ontology of toxicology and they measured them in ways that were not accepted and had no standing within toxicology.

As ecotoxicology emerged just when Kuhn's work was becoming widely known, the founders of the field adopted Kuhnian terminology to describe the novelty of their enterprise and to assess its maturity. Yet when we look at this story in greater detail we do not find the hard-edged verbal battles of opponents with different worldviews—those who see the rabbit versus those who see the duck, Unitarians versus Trinitarians, small-enders versus big-enders. Nor do we see opponents refusing to talk. Over time we see a complex pattern of communication, an emerging division of labor, continuity and evolution of practices in both the traditional field and the revolutionary one, and a deepening accommodation between the two. Not only is mutual communication taken seriously, their respective entities and measures are taken seriously and used as needed by each other. Although there was some early dismissal of each other's approach, suspicion of the methodological clarity and direction of attention motivated by alternative approaches, over time there seems little that fits the Kuhnian profile of incommensurability, although foci of attention do remain largely distinctive. While the two disciplines for the most part look in different directions, when they need to, their eyes wander to the other side, and they each accept what the other finds.

We draw evidence for the degree of commensurability between specialties by looking at the historical development of publications that present and codify knowledge in the specialties. Our measure of commensurability is whether each field accepts the phenomena and results the other sees through its methods and theory. We examine field-defining statements of leading researchers of toxicology and ecotoxicology to see how they explicitly connect or divide fields and how they evaluate methods and evidence relevant to their field definitions. We also examine the leading journals of toxicology, environmental toxicology, and ecotoxicology, at their founding and in their current manifestations, in order to see the range of phenomena, methods, and references to journals of other fields. Finally, we examine textbooks and other compendia to see how the three specialties of toxicology, environmental toxicology, and ecotoxicology characterize their phenomena of interest and methods to newcomers, as well as how they position the value of the work and findings of the other fields to their own. While textbooks typically lag several years behind the research front, they represent widely held views and so can be taken to indicate consensus views. The historically grounded analysis we present here examines how texts form relations with particular ways of viewing the world, empirical practices, and observed phenomena, as well as with other bodies of texts and groups of practitioners. Such an analysis can be seen as rhetorical in its close attention to what texts do, even though our particular analytical concepts are not drawn from classical rhetoric and we examine patterns across large numbers of texts rather than closely examining extended passages from individual texts.[5]

Classic Toxicology

Toxicology has its roots in the earliest human selection of food, avoidance of poisonous animals, and use of poisons on weapons. The word *toxin* comes from the Greek word *toxon,* meaning 'bow.' Papyri dating from the second millennium BCE indicate Egyptians had extensive knowledge of poisons and curatives; Nicander of Colophon (204–138 BCE) wrote two early treatises on snake and plant poisons; medieval and renaissance herbal compendia and other documents attest to a wide knowledge of toxic substances. However, the principles of modern toxicology rest on the insight of the sixteenth century Swiss Physician Paracelsus who pointed to the importance of the dose. "Everything is a poison," he wrote, "it is only the dose that makes it not a poison." In the early nineteenth century the Spanish chemist Bonaventura Orfila, began the "systematic use of test animals" and developed "methods of chemical analysis to identify poisons in tissue and body fluids" (Niesink *et al.* 1996, 5).

Standard procedures for the investigation of toxic effects of substances soon emerged, and in the first half of the twentieth century development of the pharmacological, processed-food, and industrial-chemical industries increased the pressure for standardized toxicological tests, particularly in the wake of government regulation. In the United States the Food and Drugs Act of 1906 made mislabeling and adulteration crimes, but did not provide for prior regulation. The replacement Food, Drug and Cosmetic Act of 1938 mandated proof of the safety of drugs before their marketing, established safe tolerances of unavoidable poisons, and extended regulation to the cosmetics industry. World War II military needs created the influential Toxicity Laboratory at the University of Chicago (Doull 2001). In the United Kingdom the 1925 Therapeutic Substances Act had requirements for labeling and record keeping, but safety testing was not required until the 1960s, with the 1963 creation of the Committee on Safety of Drugs, and the Medicines Act of 1968. The first journal of experimental toxicology was *Archives fuer Toxikologie* (*Archives of Toxicology*) founded in 1930, and continuing until today.

In both the U.S. and the U.K., the primary disciplinary sponsor of toxicological studies has been and remains pharmacology. From 1961 to 1975, toxicological research was covered within the *Annual Review of Pharmacology*; in 1976 toxicology gained status but remained paired with pharmacology in the *Annual Review of Pharmacology and Toxicology*. This serial has an impact factor three to four times greater that of any other serial in the field (based on 1997 and 1998 Journal Citation Reports). The longest continuously published U.S. toxicological journal is entitled *Toxicology and Applied Pharmacology,* founded in 1959. The Society of Toxicology was formed in the U.S. in 1960 to distinguish itself from pharmacology, but many of the articles in the journal relate to pharmacological substances. According to the *Society of Toxicology Resource Guide to Careers in Toxicology* (1989) the Chemical, Pharmaceutical and Support Industries provided 37 percent of the jobs in the field, followed by academic institutions (33 percent)—mostly in schools of medicine or public health. The overwhelming majority of graduate programs listed in the guide are affiliated with schools of pharmacology or medicine. Government is next with 15 percent—mostly in federal regulatory agencies. Nowhere is environmental work mentioned in the *Resource Guide*, although it is likely that some of the regulation involves environmental related work.

As described in the standard textbook, *Essentials of Toxicology,* the field concerns "the study of quantitative effects of chemicals on biological tissues" with a particular focus on "harmful actions" (Loomis 1968, 2). Following Paracelsus, Loomis notes "the single factor that determines the degree of harmfulness of a compound is the dose of the compound." Thus the cornerstone method of toxicology is dose-response studies done under laboratory conditions by exposing organisms to measured amounts of toxicants by inhalation, oral ingestion, injection, or cutaneous administration and measuring the degree of uptake of the toxicant and effect in each organism. As the largest interest is on toxic effects on humans, studies typically use surrogate species. Test species are bred for laboratory uniformity. These are the principles and methods reviewed with little change in the four editions (1968, 1974, 1978 and 1996) of Loomis' *Essentials.* The major difference among the editions is the increasingly detailed treatment of laboratory principles and methods along with an increasing number of laboratory tests presented. The fourth edition also adds a new chapter on clinical toxicology.

Method and Ontology in Journals in Toxicology

These well-defined and reasonably stable principles, practices, and procedures are reflected in the articles appearing in the top-line journals of the field. The journal *Toxicology,* the official journal of the British Toxicological Society, defined its concerns in its first volume in 1973 as

> the biological effects on tissues arising from the administration of chemical compounds, principally to animals, but also to man. Such compounds include food additives, pesticides, drugs, additives to animal feed, chemical contaminants, industrial chemicals and residues consequent upon their use. (*Toxicology* 1973)

It notes however, "A section devoted to brief reports on toxicological evidence related to the environment will be included." Research articles were limited to reporting quantitative studies. Almost all the articles in volume one report or review laboratory studies, mostly of dose-response (with three exceptions, to be discussed shortly). Representative titles include "The Influence of Dichlorvos from Strips or Sprays on Cholinesterase Activity in Chickens" (Rauws and van Logten 1973), "Tryptophan Pyrrolase in Rat Liver after Phenobarbitol Administration" (Seifert 1973), and "Short-Term Peroral Toxicity of the Food Colour Orange RN in Pigs" (Olsen *et al.* 1973). There are three exceptions to this pattern. One looks for an "objective measure of environmental effects through use of plants" (Berge 1973, 79). A second reports on the regulatory limits for toxicants in water in the Soviet Union (Stofen 1973), a concern closely related to the concept of laboratory-established threshold levels of toxic dose. The third exception, the first article in the first volume, to be discussed shortly, strongly reflects commitment to the laboratory dose-effect model of toxicological science (Worden 1973).

The opening purpose statement of *Toxicology* remained in 2001 much the same as it did at its founding, with only some reordering of terms and some additions to the list of chemical compounds of "consumer products, metals, cosmetics" and as

> the biological effects arising from the administration of chemical compounds, principally to animals, tissues or cell, but also to man. Such compounds include industrial chemicals and residues, chemical contaminants, consumer products, drugs, metals,

> pesticides, food additives, cosmetics and additives to animal feeding stuffs. (*Toxicology* 2001)

Again there is a limitation to quantitative studies: "Preference will be given to investigations dealing with the mechanisms of action of toxic agents. Papers describing molecular interactions with cellular and genetic processes will be welcomed" (*Toxicology* 2001). All articles in volume 159 (2001), for example, were laboratory dose-response studies with titles such as "109 CD Accumulation in the Calcified Parts of Rat Bones" (Hunder *et al.* 2001) and "The Effect of Polychlorinated Biphenyls on the High Affinity Uptake of the Neurotransmitters, Dopamine, Serotonin, Glutamate and GABA into Rat Brain Synaptosomes" (Mariussen and Fonnum 2001).

The longest continuingly published toxicology journal in the United States (founded 1959) is *Toxicology and Applied Pharmacology,* and it remains a leading journal of the toxicological field. It is one of two journals published by the Society of Toxicology.[6] As its name suggests it is closely tied to the testing of pharmaceutical agents and other medical applications of toxicology. The work published from the beginning has been and currently remains entirely laboratory based, primarily of dose-response (and associated mechanisms) in live organisms and, increasingly, in vitro tissues. The official statement of the journal says

> *Toxicology and Applied Pharmacology* publishes original scientific research pertaining to action on tissue structure or function resulting from administration of chemicals, drugs, or natural products to animals or humans. Articles address mechanistic approaches to physiological, biochemical, cellular, or molecular understanding of toxicologic/pathologic lesions and to methods used to describe these responses. Papers concerned with alternatives to the use of experimental animals are encouraged. (*Toxicology and Applied Pharmacology* 2005)

The first sentence has barely changed from the purpose statement of the earliest years of the journal.[7] The second sentence (added in the 1980s) adds a particular interest in the mechanisms of the induced effect, and the last sentence, added in the last few years, reflects increasing sensitivities to the use of test animals.[8]

All sixty-one research reports in *Toxicology and Applied Pharmacology* in a representative early volume (Volume 8, January-June 1966), concerned laboratory dose-response studies on animals, with fifty involving pharmaceutical and medical-related substances. Typical titles include "Acute Toxicity of Cephaloridine, an Antibiotic Derived from Cephalorisporin C" (Atkinson, *et al.* 1966) and "On the Mechanism of Sulfide Inactivation by Methemoglobin" (Smith and Gosselin 1966). Of the research reports, five were directed towards food treatments and additives (e.g., "Feeding of Irradiated Beef to Rats"—Blood *et al.* 1966), and six were related to agricultural and industrial environmental pollutants. The environmentally polluting toxicants were studied in exactly the same manner as the pharmacological agents, and the articles are of the same genre and read similarly (e.g., "Foot Deformity in Ducks from Injection of EPN During Embryogenesis" —Khera *et al.* 1966). Two of the articles are particularly interesting in terms of bringing environmental issues into the normalized practice of toxicology. "Cholinesterase Inhibition and Toxicological Evaluation of Two Organophosphate Pesticides in Japanese Quail" (Schellenberger *et al.* 1966) has as its aim evaluating the usefulness of Japanese quail as a laboratory species to test toxicity of pollutants on all game birds. In order for efficient and normal toxicity studies to proceed a single species needs to be identified that will "provide representative data" and is suitable for laboratory breeding. The second article attempts to develop new laboratory measurement methods to record the effect of air pollution, and thereby make this environmental problem more amenable to study by laboratory toxicology: "Application of the Evoked Response Technique in Air Pollution Toxicology" (Xintaras *et al.* 1966).

Just a few years later, in 1969, volume fourteen reveals the effect of increased environmental concern with at least twenty research reports on the toxicity of various agricultural and industrial environmental pollutants, with the majority concerned with pesticides such as DDT, dieldrin, and parathion. We also see the effect of other changing social concerns, with four studies directed towards the effects of alcohol, another towards the effects of marijuana, and a sixth towards amphetamine. All articles follow the normalized practices of the field. One, however, again shows an interesting attempt to bring complex environmental problems into the normalized science: "An Exploration of Joint Toxic Action: Twenty-Seven Industrial Chemicals Intubated in Rats in all Possible Pairs" (Smyth *et al.* 1969). As the report explains

"In the occupational, domestic or ambient environments, encounters with mixtures of chemicals far outnumber encounters with individual chemicals" (Smyth *et al.* 1969, 340). The study develops laboratory methods for measuring and analyzing joint toxicity and points to instances where the effect of joint exposure was greater or lesser than one would expect by additive effects of the separate exposures. It then notes patterns in these variances.

Current issues of the journal continue with the laboratory studies, but with a particular focus on uncovering mechanisms of action, as suggested by the current statement. The selection of topics has tended to be more narrowly focused on substances of pharmaceutical and medical interest, perhaps because environmental issues now have a range of journals in which they may be pursued, as we will discuss below. However, when they are discussed they follow the current normalized laboratory practices of the field, as in "Induction and Inhibition of Aromatase (CYP19) Activity by Various Classes of Pesticides in H295R Human Adrenocortical Carcinoma Cells" (Sanderson, *et al.* 2001).

The Social and Political Birth of Ecotoxicology

During the 1960s and 1970s, as toxicology was settling into its modern "normal science" themes and practices, a new social concern was developing in the environment, catalyzed by Rachel Carson's 1962 *Silent Spring,* publicizing the dangers of DDT and other insecticides and herbicides.[9] In some of her chapters, Carson adopted an ecological perspective that cast the threat not just to individual organisms and species but to the entire balance of nature. And while she documents the direct toxic effects of insecticides on humans, as well as the cumulative effects as they pass up the food chain, she also considers the effect of ecosystem change on humans. A report of the President's Science Advisory Council on "The Use of Pesticides" soon followed, taking a much stronger stand on the dangers of pesticide use than it and other government agencies (such as the Department of Agriculture and the Food and Drug Administration) previously had (Wang 1997). One consequence of the heightened concern for the effect of chemical poisons was a 1972 strengthening of the Federal Insecticide, Fungicide, and Rodenticide Act. This law provided for registration, data submission, and approval of pesticides for particular uses, as well as monitoring of production facilities and seizure and criminal penalties for illegal distribution and use. Other regulatory legislation soon followed. But unlike most previous toxicological problems, the exposure was not direct (as through ingested food or medicine). Exposure was ambient and required field monitoring of toxicant levels, as well as complex estimates of exposures of individuals. Further, it was unclear whether the exposure within natural contingencies was adequately modeled by laboratory tests of controlled exposures under controlled conditions. Rather than being concerned about thresholds for a well-defined toxic effect, demographic studies of effects of lower level entered in, as did the concept of risk assessment. Nonetheless most regulation at least followed the notion of regulating the level of exposure or dose. Other scientists, however, saw the issues in ecological terms and called for ecological studies aimed at protecting the biosphere as a whole.

Truhaut's Revolutionary Vision

René Truhaut in 1969 coined the term *ecotoxicology* for this new approach in a speech before the International Council of Scientific Unions, and he is generally recognized internationally as the founder of the field.[10] In 1977 the first journal of this new field appeared, *Ecotoxicology and Environmental Safety.* The editors' foreword to the first issue of this flagship journal of ecotoxicology draws a sharp distinction between the new field of ecotoxicology and classical toxicology:

> Ecotoxicology can be defined. [. . .] as the study of the adverse effects of chemicals, natural products, and physical agents on populations and communities of species of plants, animals, and microorganisms as they occur and are organized in nature. In contrast to classical toxicology, which deals predominantly with the toxic effects of chemicals on individual organisms, Ecotoxicology is essentially the study of the toxic effects of environmental chemicals on naturally occurring populations in various ecosystems, including Man. (Foreword 1977, iii.)

Although there is no overt opposition here, the statement by clear implication ("In contrast to classical toxicology ... ") draws dividing lines on a major conceptual issue and a major methodological issue. Conceptually, the units of analysis are populations and communities within ecosystems rather than individual organisms or isolated species. This position further implies that ecological systems theory is to play an important role, and not just organic chemistry, biochemistry, pharmacology and other studies that isolate mechanisms within individual organisms. Methodologically effects are to be studied *in situ*—in the field, as organized in nature rather than under the controlled and abstracted conditions of the laboratory. The guide-for-authors statement in 2005 goes significantly further in making ecosystems the unit of analysis, rather than organisms or populations in ecosystems:

> [. . .] Novel technologies, techniques and methods such as the biomedical photonic technologies, biomarkers, biosensors and bioanalytical systems, QSARs and QSPRs, advanced high-performance computational methods, models and storage systems, and their applications in the obtaining and processing of interdisciplinary ecotoxicological information, are [...] addressed in the journal. We welcome the applied outcome of the complex ecotoxicological research such as developing the science-based Environmental Quality Criteria (EQC), standard toxicity tests, techniques and methods for ecotoxicological evaluation of the environment, as well as developing ecotoxicologically proven methods and technologies for prevention, interception and remediation of human-induced damage to ecosystems. (Guide for authors 2005)

Qualitative studies are now welcomed, as are papers using the techniques of a wide range of disciplines.

Truhaut, in an article in the premier issue, takes on Paracelsus's founding principle of toxicology, cited by almost every overview of toxicology, that the dose makes the poison. In "Can Permissible Levels of Carcinogenic Compounds in the Environment be Envisaged? Critical Remarks" Truhaut argues that some substances being released in the environment are toxic at any level. He frames the challenge to classical toxicology baldly in the opening sentence and a single-sentence contrasting third paragraph.

> The golden rule in toxicological evaluations of environmental pollutants is to establish dose-effect (exposure-effect) relationships in order to be able to set toxicity threshholds and, consequently, permissible limits. [. . .]
>
> But in the case of carcinogens, a current view is that it is impossible to establish safe levels, because there are no thresholds for action. (Truhaut 1977a, 31)

After reviewing the evidence and arguments on both sides, he supports a World Health Organization conclusion that while thresholds might be envisioned, they have not been determined and are exceedingly hard if not impossible to determine. Thus rather than seeking no-effect thresholds we must calculate risks and adopt "socially acceptable risk" levels (Truhaut 1977a, 31).

In the second issue of this journal, a long article by Truhaut spells out "Ecotoxicology: Objectives, Principles, and Perspectives." In this article, which is often cited as the founding document of ecotoxicology, he associates the development of toxicology with the chemical age and the consequent need to study the effect of the industrial chemicals upon organisms, particularly man. However, he points out that man is part of an ecosystem and that chemicals affect the entire biosphere. It is, therefore, important to study toxic effects "in the context of biologic equilibria, the study of the harmful effects on the various constituents of ecosystems of chemical pollution of the environment, for which man is to a large extent responsible"(Truhaut 1977b, 152). An earlier 1974 definition of the field talked of the effect of pollutants "to the constituents of ecosystems, animal (including humans), vegetable and microbial, in an integrated context" (Truhaut 1974). But already the distinction was being reframed as the impact not on organisms but on "populations of living organisms [. . .] constituting ecosystems" (Truhaut 1977b, 152). In light of this redefinition, Truhaut identifies three principal sequences of ecotoxicological studies.

1. Study of the emission and entry of pollutants into the abiotic environment, with their distribution and fate.

2. Study of the entry into and fate of pollutants in the biosphere, with the very important problem of contami-

nation of biological chains, in the first place of food chains.

3. Study, qualitative and quantitative, of the toxic effects of chemical pollutants, at a certain level, to ecosystems, with investigation of the impact on man. (Truhaut 1977b, 153)

Even though he casts ecotoxicology as a subfield of toxicology, he clearly marks the field as different in object of analysis—the ecosystem and populations within ecosystems. He also marks the field as different in conceptual frame and different in method. Ecotoxicology is not limited to quantitative studies in the laboratory; it studies effects on populations and ecosytems in the field, as they are organized in nature. In his detailed comments on these three areas he follows disciplinary orientations that are outside the normalized world of toxicology. He does incorporate traditional toxicological work in talking about the importance of establishing quantitative dose-effect relationships in order to determine toxicity thresholds and allowable limits—for which laboratory studies are useful. But then he talks about the limitations of laboratory studies and surrogate animal studies for humans, pointing out the necessity of qualitatively comparing laboratory effects to populations of actually exposed humans, and calls attention to species differences and selective toxicity (Truhaut 1977b, 165- 166), stage of life, and interactions within complex ecosystems.

It is in the final section on "Perspectives and Prospectives in Ecotoxicology" that Truhaut takes the most transformative stand in calling for the multidisciplinary collaboration of ecology and toxicology. Ecology is anchored in the field-based study of systems over time, requiring statistical studies as well as qualitative observation. Toxicology is anchored in the laboratory-based study of individual organisms demonstrating specific concrete effects. While Truhaut retains a central place for dose-effect studies and what he calls the routine tests of toxicology, they need to be reassessed in light of the actual life, habitat, conditions of exposure and other field-based considerations:

> Joining efforts, toxicologists and ecologists should not forget to pay attention to the possible consequences for an ecosystem in its totality. To this end, *models with predictive value* should be established

> (Metcalf 1974), and laboratory studies should be complemented as much as possible by field studies on a much larger scale, using data from chemical analysis in a *continuous surveillance* of the environment in adequately programmed "monitoring." (Truhaut 1977b, 171)

Truhaut takes the stance, common at the beginning of widespread environmental concern, that we do not know enough to act and we need new studies. This was indeed the political attitude and stance that led to National Environmental Policy Act of 1969, which mandated Environmental Impact Statements for projects involving federal lands. The EIS, which soon became widely imitated, was a mechanism for increasing information to lead to wiser decisions rather than directly regulating behavior. The need for more knowledge about the environment was central to the rhetoric of the Impact statement's great advocate, Lynton Keith Caldwell, as well as the congressional discussion preceding the passage of NEPA (see Bazerman *et al.* 2002). Just as Caldwell felt that new governmental mechanisms were needed to generate the needed knowledge, which led to NEPA and the EIS, so Truhaut finds a new discipline necessary.

In Kuhnian terms, this is revolutionary science, despite Truhaut's meliorative language. While the work of toxicology is embedded in ecotoxicology as a contributing element, it is clearly cast as limited and inadequate in theory and method to deal with the full scope of problems. Specifically, in terms of incommensurability, one must measure and take into theoretical account entities that are outside of the scope or vision of classical toxicology, using methods and devices that are not recognized within the traditional field.

First issues of the journal *Ecotoxicology and Environmental Safety* included some articles in the dose-response laboratory paradigm, but concerning environmentally important chemicals, such as the first article of volume one, issue one, "Studies on the Interactions of Dieldrin with Mammalial Liver Cells at the Subcellular Level" (Wright *et al.* 1977). However, it includes a number of broad ranging essays, some of which use field, demographic, economic, and public health data, as in "The Importance of Chlorinated Hydrocarbons in World Agriculture" (Snelson 1977), and "Organicochlorine Pesticides and Liver Cancer Deaths in the United States, 1930–1972" (Deichmann and

MacDonald 1977). Other articles propose and evaluate methods for monitoring environment, such as "Assessment of the Trace Organic Molecular Composition of Industrial and Municipal Wastewater Effluents by Capillary Gas Chromatography/Real-Time High-Resolution Mass Spectrometry: A Preliminary Report" (Burlingame 1977).

Contrasting Disciplinary Visions

While the two flagship articles in the 1977 first volume of *Ecotoxicology and Environmental Safety* by the founder of the field of ecotoxicology call for new methods and new ways of thinking, the flagship article of the journal *Toxicology,* the first one in the first volume ("Toxicology and the Environment" by Alistair Worden) shows methodological caution and skepticism about non-traditional evidence suggesting environmental damage, as well as potentially risky products. The article is the text of the "Annual Lecture at the Royal College of Physicians, London, Sponsored by Merck, Sharp and Dohme Research Laboratories." As he reviews the state of evidence on a number of high-profile issues concerning pharmaceuticals, industrial chemicals and the environment, his attention is as much on those raising concerns as on the science. The spirit of his talk is captured by his introductory statement:

> It is possible to speak from experience of some of the approaches that are being made to tackle toxicological and environmental problems, but an evaluation of these problems is complicated beyond the comprehension of most of us by the genetic make-up, the innate or early-acquired behavioural patterns and the varying social and political motivation of our own species. [. . .]
>
> It is not surprising that the reaction among so many of us is alarmist, or that the allegiance of the cranks, the do-gooders or the somewhat unkindly designated lunatic fringe is so frequently transferred to an anti-pollution campaign or to stressing the potential dangers of new therapeutic agents. (Worden 1977, 4)

Worden does grant some grudging respect to Rachel Carson for making the public aware of toxic issues "whatever the special pleading or the technical shortcomings of [*Silent Spring* . . .] its intended message was a stimulus to research and thought that will not, in the long run, have proved damaging either to the general public or to the agricultural and chemical industries" (Worden 1977, 5). He, however, consistently shows skepticism about research outside laboratory dose-effect studies on animals, which remains his gold standard. He argues for maintaining animal laboratory studies as central even if satisfactory tissue culture methods are developed. Methods and calculations that go beyond the lab are dismissed by him, such as the systems models used for the Club of Rome's *The Limits to Growth* (Meadows *et al.* 1972). He does allow some measures of environmental contamination but he is also ready to state that evidence of higher concentration of mercury in fish and humans in contiguous coastal areas can be traced to the eating of canned tuna, and that higher concentrations apparent in England in comparison to other countries appears "no cause for alarm" (Worden 1977, 13). Similarly, he discounts heavy concentrations of chemicals and fish mortality in Hungary as being confounded by river flooding. Real-world measures seem for him to be too multicausal and situationally complex to draw conclusions from. He does express concern for pollutants in industrial waste discharged into sea water, and does guardedly note that manufacturers are not fully forthcoming. "I have tried to pay tributes to industry and to its co-operative attitude," he says, "but it is difficult to escape the conclusion that not all the information that could, and should, be provided on the subject of the industrial effluents discharged into rivers, estuaries and the sea has been forthcoming" (Worden 1977, 19). This guarded criticism is immediately cushioned by praise of industrial cooperation in decreasing PCBs. He ends with some jokes implying environmental criticism is the moral equivalent of adultery (tempted to be faithless to true principles) that suggest he misses the point of toxicological science by a rather wide margin. He then reaffirms his methodological faith: "We are facing quantitative problems and should try to quantify the help we try to give to their solution" (Worden 1977, 24). Given his clear distrust of numbers gathered in the field, this can only mean laboratory measured quantities under controlled conditions.

The division between the stances of Truhaut and Worden is sharp. One argues that since we do not know enough about the effect of human activity on the environment we need new theories and new methods which gather data widely from the complex world where effects take place, in order to be able to understand and model the total picture and understand the risks we are putting ourselves in. The other argues that we should be cautious in identifying toxic effects, holding ourselves to the highest standards of scientific data, with controlled laboratory conditions and precision of measurement of dose and response defining that standard. If we restrict industry and economic growth without that level of certainty we do more harm to human development than good. It is not just their theoretical and epistemic commitments that differ. Truhaut ascribes no overt commitment to those who stand on the opposite side from him, but he is clearly an advocate of environmental protection who believes that we are in great difficulty and must know more so we can act. Worden clearly sees human development and interests advanced through industrial growth, particularly through the introduction of agricultural chemicals and new pharmaceuticals, but unlike Truhaut he is willing to ascribe unscientific political, social and psychological motives to those who take the position opposite from him, representing his own motives, and those of traditional toxicologists generally, as simply pursuing the interests of true science.

This case in many ways appears to be a sharply drawn example of the Kuhnian conflict between normal science and revolutionary science, with a strong dose of incommensurability thrown in. The themes and practices are clearly at odds, and the incommensurability has most clearly to do with the willingness to take the other side's measurements and phenomena seriously. In this particular case incommensurability appears to be asymmetrical, even unilaterial. For Truhaut's ecotoxicology, toxicological methods of laboratory study of dose-response are accepted as valid and important, although they must be viewed in relation to the complicating factors of real, complex, *in situ* processes, and then are recontextualized in an ecologically based account of the systemic effects. For Worden's toxicology, field data are questionable because of confounding effects and multicausality. Models, similarly, are necessarily incomplete, so that it is very difficult to talk with confidence about anything outside the laboratory—the numbers are uncertain and the phenomena those numbers are supposed to indicate are hazy. So for Truhaut the real object of knowledge is *in situ* systematic processes—lab studies are useful but simplified contributors to unpacking the complex big picture. For Worden the complex big picture is too messy to be seriously knowable, so only lab studies provide true, warrantable knowledge. The laboratory is where the action is.

Textbooks and Disciplinary Definition

The views presented by these two leaders were reflected in the words and actions of many in their field during the seventies, and later. While all were cognizant of the fact that toxicants in the environment were an issue, traditional toxicology at first treated this contamination just as it treated other toxicological problems. The mid-seventies textbook *Toxicology: The Basic Science of Poisons* (Casarett and Doull 1975), while it has chapters on Clinical Toxicology, Forensic Toxicology, Industrial Toxicology, and Veterinary Toxicology, has no chapter on environmental toxicology (or ecotoxiology). Among its various chapters on pollutants it does, however, have chapters on air pollutants and pesticides. Thus it treats environmental toxicants as nothing special—simply substances needing to undergo standard toxicological measurement of toxicity.

The first edition of Loomis's *Elements of Toxicology* includes two pages in the introduction defining environmental toxicology as "that branch of toxicology which deals with incidental exposure of biologic tissue, and more specifically in man, to chemicals that are basically contaminants of his environment, food, or water" (Loomis 1968, 7). The section describes the various sources and substances that comprise toxic pollution, with no indication that different methods or theories are needed for the problem. The substance of the book provides no mention of any theories or methods that are specifically aimed at environmental issues. No field methods are discussed. The section remained essentially unchanged in the second (1974) and third (1978) editions. This is particularly interesting as in those same three editions Loomis includes a chart indicating the resource disciplines and the areas of specialization within Toxicology. It lists environmental as equal to economic and forensic. Thus it subsumes environmental issues into normal toxicological practice, based in methods developed largely for pharmaceutical and secondarily for medical, industrial and agricultural purposes.

The ecotoxicology textbooks (all of a more recent vintage), on the other hand, treat ecotoxicology as making a substantial break from toxicology and adopting a new "broad conceptual framework for evaluating chemicals in the environment," as it is phrased in *Introduction to Ecotoxicology* by Connell, Lam, Richardson, and Wu (Connell *et al.* 1999, v). It identifies ecotoxicology as having its roots in both the sciences of ecology and toxicology, and drawing on chemistry, pharmacology, and epidemiology as well. In this respect it is much like Loomis's and other toxicologists' view of toxicology as multidisciplinary—except that ecology is more a field-and-systems science than a lab-and-organism one. The authors make one further addition: "a managerial aspect, resulting from the increasing need to regulate industrial and human activities" (Connell *et al.* 1999, 1). This managerial aspect serves as the warrant for adding "risk assessment and risk management into the ecotoxicological equation." Toxicologists viewed themselves as scientists meeting social needs set by others, but did not view themselves engaged in management and regulation. The book cites Truhaut's invention of the term *ecotoxicology*, and defines the goal of the field (following Moriarty 1988) as "to assess, monitor and predict the fate of foreign substances in the environment" (Connell *et al.* 1999, 1). Note how this definition shifts attention from the effect on humans and other organisms to chemicals and to the environment. Also it includes all substances foreign to the environment under its purview and not just toxicants. Although citing the Paracelsian principle that all substances are potentially poisons, depending on the dose, the authors invert its use to identify potential harm at all dosages rather than to suggest safety at lower dosages. Since ecotoxicology is concerned with persistent chemicals that may accumulate, all substances have the potential of being toxic, even if the immediate exposure is low.

Rombke and Moltmann's *Applied Toxicology* defines its field as "concerned with the toxic effects of chemical and physical agents on living organisms, especially on populations and communities within defined ecosystems" and as "the science which seeks to predict the impacts of chemicals on ecosystems" (Rombke and Moltmann 1996, 3). Consequently, the textbook identifies several ways ecotoxicology is distinct from toxicology. In contrast to toxicology's attempt to identify the dose which determines toxic effect, in ecotoxicology "no single measurement of the concentration of a substance in the environment is sufficient in itself to evaluate the stress on the ecosystem" (Rombke and Moltmann 1996, 5). Ecosystems are more complex in their operations than even the most complex single organism. And the environmental impact of various foreign substances is not always measurable in terms of acute responses. Ecosystems need to be studied through ecological theory.

Similarly, *Fundamentals of Ecotoxicology* defines the field as "the science of contaminants in the biosphere and their effects on constituents of the biosphere, including humans" (Newman 1998, 130). In seeing the field as synthetic and multidisciplinary, the author sees effects of concern occurring at every level of a hierarchy "from the molecular (e.g., enzyme inactivation by a contaminant) to the population (e.g., local extinction) to the biosphere (e.g., global worming) levels of biological organization" (Newman 1998, 14). However, he also sees "questions dealing with the lower levels of the conceptual hierachy, e.g., biochemical effects of toxicants, are more tractable and have more potential for linkage to a specific cause than effects at higher levels such as the biosphere" (Newman 1998, 14–15). A bit later he describes the field in the explicitly Kuhnian terms of *paradigm* and *normal science*, the latter in partial contrast with "innovative science"(Newman 1998, 19). He suggests that while some parts of this synthetic field are appropriately mature to proceed in normal ways, others still require innovation and should not be saddled by a "preoccupation with details [. . .] or measurement" (Newman 1998, 19–20). The study of the lower levels of hierarchy of biological systems (those that are encompassed within traditional toxicology) is mature and normalized with specific expectations for detailed measurement. The study of higher orders of biological organization (those added by the ecological sciences) needs to be taken seriously, but is pre-paradigmatic. It needs license to innovate before it can produce mature, normal science.

While the tables of contents of toxicology textbooks and other overview works are organized by kinds of toxicants, kinds of effects, domains of practice, and methods of laboratory measurement (see Loomis 1968, 1974, 1978; Loomis and Hayes 1996; Cassarett and Doull 1975; Kent 1995; and Niesink, deVries and Hollinger 1996), the ecotoxicology books have very different chapter organizations, including chapters on ecological theory, on processes of chemical transformations in the environment, on populations, communities and ecosystems, and other topics. There is no fixed set of organizing principles, suggesting that the field is, in Kuhnian terms, pre-paradigmatic. However, each of the ecotoxicology books does have chapters on laboratory dose-response studies, indicating that the work of toxicology has a respected, though limited, place within the work of ecotoxicology to establish the biological risk of specific levels of exposure to toxicants.

Symposium volumes, and other overview statements of ecotoxicology reveal the same sense of a creative, innovative science that has not yet reached maturity or a normalization of practices, theories, and methods. Smeets, in a late-seventies article, "New Challenges to Ecotoxicology," states that ecotoxicology has yet "to develop adequate and representative test methodology techniques [...] to simplify test procedures" and thus to define "standard methods" (Smeets 1979, 120). Public concern and regulatory interest now made it imperative that the field mature into more normalized practices. In the first chapter of *Ecotoxicology: A Hierarchical Treatment,* Newman similarly represents ecotoxicology as a creative field not yet developed into a mature science, specifically citing Kuhn. It is not yet ready to be normalized despite strong practical and technological pressures that necessitate standard methods and immediate guidelines for specific actions (Newman 1998, 3). But he does urge the development of theoretical and methodological tools to soon make "the transition into a mature science" (Newman 1998, 7) where there is a balance between normal and innovative work (Newman 1998, 5). These views are echoed by a number of the authors of other chapters in the volume. *Ecotoxicology in Theory and Practice* (Forbes and Forbes 1994) also sees the struggle of ecotoxicology in becoming a fully independent science, with a key site of definition being a deeper engagement with ecological theory.

But the most radical statements of a deep division between ecotoxicology and toxicology accompanied by limited communication and lack of cooperation appear in the volume *Ecological Toxicity Testing* (Cairns and Niederlehner 1995). Cairns in his opening chapter on the genesis of ecotoxicology sees a deep division in the field between toxicological testing and an ecosystem approach, which is an aspiration of ecotoxicology (Cairns 1995, 7). A follow-up chapter by the sociologist Halffman documents and examines the tall boundary that exists between ecology and toxicology. He finds the boundary realized in discontinuities in organizational structure, funding agencies, journals, theories, and research programs (Halffman 1995, 16). He supports his observations with evidence from the citation structure of international journals through 1988, an analysis of Dutch scientific funding agencies, the testimony of practitioners, and an historical analysis of the theories and research programs in both the U.S. and Europe.

But when we look further into the histories and analyses documented by Cairns and Halffman, we find a more complex story about the relationship of toxicology and ecotoxicology, intertwined with their relations to industry, government regulation, university disciplines, and academic research. While the face of their stories is very much like the previously recounted story about an older normalized field with institutional infrastructure and a new revolutionary field attempting to displace the older as authoritative, the complexities expose more dynamics at play than just intellectual commitments. These complexities provide a way of approaching another level of materials that do not fit so easily into the dichotomous Kuhnian story adopted by the founders of ecotoxicology.

Cairns casts his own professional position in the dichotomy in a curious way. Cairns adheres to an ecological view, has criticized the limitations of normal toxicological studies for understanding environmental problems, has numerous publications in ecotoxicological journals, and is viewed as a leader in the field; but he does not yet consider himself to deserve the name of ecotoxicologist.

> I have stated that I do not yet feel that I deserve to call myself an ecotoxicologist although I had been involved for two decades in multispecies and community–level testing. The reason that this remains an aspiration rather than an accomplished fact is that attributes recognized as fundamentally important to ecologists are not yet routine endpoints in the field of environmental toxicology. My assumption is that people label themselves ecotoxicologists to indicate that they reject the sole use of single species laboratory toxicity tests low in environmental realism as the primary means of estimating the effects of toxicants on ecosystems. A concomitant intent in the use of the term is almost always certainly to relate toxicological testing more closely to ecosystem responses even though it is more of an aspiration than a reality at this time. If this is the case, then the term ecotoxicologist is used to indicate an expansion of the toxicity testing array now available without denigrating either single species tests or the events leading to the present

> evolutionary stage of the development of the field. Without doubt, there will be the usual lag of two or three decades before regulatory agencies embrace the new position. (Cairns 1995, 7–8)

Ecotoxicology seeks to realize ecological theory in toxicological tests, according to Cairns. Importantly, he operationalizes the maturity of the discipline in its ability to develop tests that match its theoretical ambitions. Further he notes the relationship between those tests and government regulation. Also even as he seeks the maturity of a new field he respects the well-established measures and methods of toxicology and its evolving specialty of environmental toxicology. Each of these elements points us to a more complicated and enriched story of disciplinary evolution in a complex social, political and scientific environment.

Halffman's follow-up essay places the development of toxicological studies of environmental pollution and ecotoxicology in the context of governmental need to legitimate decisions in an emerging arena of policy and politics. Part of that story involves the support of the American Chemical Society in the late 1960s while seeking a place in the monitoring of pollution called for more ecological research within chemistry. But when regulation was established in the 1970s the monitoring tasks went largely to the well-established procedures of toxicology and its new subspecialty environmental toxicology, which adapted traditional toxicological procedures by finding critical species for testing chemical pollutants in the laboratory. Ecotoxicology, then, did not develop within the context of government regulation. To some extent it was a university-based endeavor, although disciplinary differences in the location of ecological and toxicological sciences created some barriers to integration.

Because of the political charisma of the term *ecology*, people doing conventional toxicological and chemical work at times attempted to adopt the term, but few people at the time of Halffman's study had been able to develop new techniques to study and monitor an ecological system as their primary object of analysis. Government regulatory measures—particularly as administered by the EPA in the U.S., which set the terms for approvals of listed chemicals to be released in the environment—remained tied to traditional toxicity testing, despite the ecologists' argument that such tests lacked consonance with the complexity of real ecosystems. Without factoring in the complexities, judgments of chemical toxicity would not reflect the actual toxicity in real situations. One attempt to establish microecosystems as an experimental standard for regulation failed in the mid 1980s, which left single species testing as the dominant force at the end of the 1980s where Halffmann's study ends. Significantly, though, Halffman does note that just at the end of his study period toxicological and ecological journals began to be more tightly linked through citations. We will examine this change more closely when we look at recent publications in toxicology.

The story Halffman tells is of the important role of institutional sponsorship, and the necessity of producing regularizable tests that would meet the needs of sponsors. The contention over methods is in part a debate for control of regulatory regimes, but without a commodifiable procedure that can win arguments for regulatory use, ecotoxicology is left only with an unfulfilled moral ambition and a marginal institutional role. The argument over alternative measures and thus the commensurability or incommensurability of alternative views is an interested argument. Indeed, recounting the changing political and regulatory environment is a standard introductory move in most textbooks and other overviews of the field, even if they typically do not continue to keep the economics and politics at the forefront of their presentations. This is equally true of presentations of toxicology (for examples, see Loomis 1968; 1974; 1978, 4–9; Loomis and Hayes 1996, 4–12; Cassarett and Doull 1975, 9; Kent 1998, 3–4), environmental toxicology (Cockerham and Shane 1994, 5–7; Yu 2001, 1–5; Landis and Yu 1995, 4–5; Zakrzewski 1997, 1–14; Hughes 1996, 10), and ecotoxicology (Rombke and Moltmann 1996, 5–6; Forbes and Forbes 1994, viii, 185–6; Newman 1998, 1–7; Levin *et al.* 1989, 497–540). These are scientific specialties that grew directly in response to industrial developments, perceived public harm and political reaction, and government regulatory regimes. These fields aim to provide the knowledge and measurement practices that meet the public concern and government need for scientifically warranted decisions and regulation. And as fields their continued prosperity and research agendas depend on the sponsorship that comes with being percieved as delivering the goods on public regulation, corporate liability, public concern, and the many other sites where toxicity and pollution are adjudicated and negotiated.

Environmental Toxicology in the Shifting Middle

This changing view of what must be studied in order to regulate the environment can be traced in the changing definition of the specialty of environmental toxicology, which was originally very closely linked to toxicology, as a means of expanding its traditional methods to the new market of environmental concern. As mentioned earlier, the first edition of Ted Loomis' standard *Essentials of Toxicology* had two pages describing environmental toxicology in terms entirely normalized into toxicological procedures with no environmentally specific discussion in the remainder of the book. This was unchanged across the first three editions (1968, 1974, 1978.) However, the fourth edition, co-written with A. Wallace Hayes (who is employed in the office of Corporate Product Integrity of the Gillette Company) entirely revises the introductory material on environmental toxicology. The new discussion describes the complexities in understanding pollution in the environment, including issues of biological, solar, and mechanical transformations, interactions within the environment, transformations, dilution and accumulation and fates of populations rather than individuals. The discussion takes something of a systems perspective, even including a diagram that approximates an ecological system, though never using the word *ecology*. And at the end of the volume is a new chapter on Risk Assessment.

Environmental toxicology undergoes an even more major redefinition in other publications. In 1975, a 750-page reference on *Toxicology: The Basic Science of Poisons* mentions ecological terms in only two paragraphs on DDT affecting ecological balance (431). Such a mention would be hard to ignore in any discussion of DDT in the wake of Carson's explicitly ecological discussion. In comparison, a 1996 compendium of similar genre, the 1,250-page *Toxicology Principles and Applications,* has a 28-page chapter on ecotoxicology (and not environmental toxicology), plus another chapter on biotransformations.

The transformation of environmental toxicology into the equivalent of ecotoxicology can be traced in the various textbook definitions of the field. One 1980 textbook *Environmental Toxicology* (Duffus 1980) defines the field as "the study of the effects of toxic substances occurring in both natural and man-made environments." This definition does not move far from traditional toxicology doing dose-effect lab studies of pollutants that appear in the environment. Another textbook from the period, *Introduction to Environmental Toxicology* (Guthrie and Perry 1980), however, while maintaining a strong grounding in traditional toxicology moves further into the field and considers ecosystem issues. While the first four chapters consider a range of toxicants, measurable in classical toxicological ways, chapters five and six consider the social and economic processes that generate pollutants in the actual world, and chapters seven through ten consider the impact of pollution on various organisms as they exist in the world. Chapters eleven and twelve develop a dynamic view of the movement of toxicants and the role of behavior; chapters thirteen and fourteen begin to develop an ecosystem approach to aquatic and estuarine ecologies; and chapters fifteen and sixteen consider the effects of oil and pulp industries in ecological terms (localizing the social and economic themes of chapters five and six to specific industries). Chapters seventeen through twenty-one are a series of case studies of pollutants in ecosystems, and twenty-two examines "Pesticide Effects on the Agroecosystem." Beginning with traditional toxicological tools, then, the book gradually leads into complex problems as they are found in the field, and thus moves to an ecological perspective.

The mid-1990s collection, *Basic Environmental Toxicology* (Cockerham and Shane 1994), opens with two chapters introducing ecotoxicology and its principles. While the chapters in section two cover the laboratory treatment of toxicity of environmentally sensitive materials, section three gives a systems treatment of air, soils, aquatics, estuarine ecosystems and wildlife based on field studies. Section four on methods and areas of work includes ecological risk assessment.

The 1995 *Introduction to Environmental Toxicology: Impacts of Chemicals upon Ecological Systems* redefines environmental toxicology to be virtually equivalent to ecotoxicology. The book's opening sentence states "Environmental toxicology is the study of pollutants upon the structure and function of ecosystems. For the purpose of this text, the emphasis will be on ecological systems, at every level of organization, from molecular to ecosystem" (Landis and Yu 1995, 1). The substance of the book follows suit. While it includes chapters on "Typical Toxicity Test Methods," it also includes chapters on field measures and risk assessment as well as on routes of exposure and modes of action, and the factors that effect toxicity in field. Yu's 2001 textbook *Environmental Toxicology: Impacts of Environmental Toxicants on Living Systems* also takes a largely systems approach.

By the late 1990s, Newman could state that the definitions of environmental toxicology and ecotoxicology "are rapidly converging" (Newman 1998, 12). What this history of convergence reflects is the inability of traditional toxicology to adequately speak to public and scientific concerns about changes to the environment. In order to maintain a central role in providing service to the various sponsors of toxicology, the field had to recognize the kinds of phenomena and problems manifest in the ecology, even if measures and tests had not yet developed that could be administered in a normal, commodified way.

While a number of environmental toxicology books of recent vintage maintain a more traditional toxicological orientation, they all make a nod towards ecology, such as *Essentials of Environmental Toxicology: The Effects of Environmentally Hazardous Substances on Human Health,* which stays methodologically quite close to traditional toxicology lab dose-response studies (Hughes 1996). But it does have a couple introductory pages on ecological concepts. Interesting is Sigmund Zakrzewski's 1997 second edition of his 1991 *Principles of Environmental Toxicology.* In the preface to the second edition he comments that in response to criticism of the first edition he now adds "a section on wetlands and estuaries" and includes a description of the fate of one ecosystem. He then explains,

> Despite these changes, this book is primarily a toxicology, and not an ecology, text. Thus, certain important areas of interest to environmentalists have been omitted. To remedy these shortcomings, a list of subjects for student research and seminars have been included. (Zakrzewski 1997)

Thus while he bounds his area of presentation, he also legitimates the value of ecological study.

So Who Recognizes Whose Work?

There seems to have developed a division of labor that leaves the core field of toxicology with only limited attention to environmental issues and ecological theory. This leaves toxicology largely free to serve its primary clients of medicine, pharmacology, cosmetics, and other industries—and the regulation of them. Core toxicological journals such as the two sponsored by the Society of Toxicology (*Toxicology and Applied Pharmacology* and *Toxicological Sciences*) currently carry little that directly speaks to environmental or ecological issues. However, when they do run articles in that area, those articles cite appropriate ecologically based literature.

Toxicology and Applied Pharmacology has particularly devoted itself to mechanisms by which toxicants have effect within organisms, a focus well suited to laboratory organism-based studies. An inspection of the contents in volumes 170 to 183, covering the period from January, 2001, to September, 2002, reveals that only a handful of articles concern environmentally related toxicants. The journal currently appears twice monthly, with each volume constituted of three issues for a total of 42 issues in the covered period, containing 343 articles. Of those 343 articles, only eight appear to treat environmentally related toxicants. Those few articles treat the toxicant primarily within the organism, and thus cite primarily health, medical, biochemical, and toxicological literature. However, the first few sentences or paragraphs typically gesture towards the environmental presence and/or the epidemiological consequences of the toxicant, and may cite field studies of contamination levels or population studies of effects. Thus they use environmental and ecological literatures and data to identify the importance of the problem to be studied through the laboratory, organism-based methods of toxicology. The first paragraph, for example of "Percutaneous Absorption of Explosives and Related Compounds: An Empirical Model of Bioavailability of Organic Nitro Compounds from Soil" (Reifenrath *et al.* 2002, 160) details the history of military use of TNT and other explosives, and the mechanisms of soil contamination. The second paragraph then identifies contamination levels, citing *in situ* studies, and establishes research problems concerning uptake. Then the article moves into laboratory experimental studies for its remainder. In these introductory comments, such articles show respect for the findings and methods of field, demographic, and even ecologically-based studies, even though the laboratory accounts withdraw into laboratory-based literature. One article, however, enters more fully into field ecological studies: "Evidence for Endocrine Disruption in Perch (*Perca fluviatilis*) and Roach (*Rutilus rutilus*) in a Remote Swedish Lake in the Vicinity of a Public Refuse Dump" (Noakson *et al.* 2001). This *in situ* study of entire populations of fish uses many kinds of field data, including statistical samples, and cites several articles from journals in aquatic science, environmental science, marine environmental science, and ecotoxicology. Thus, while publications in this journal currently raise issues far from ecotoxicology, when the study enters into problems or issues that intersect with the interests of ecotoxicology there is no difficulty in accepting the findings and methods of environmental science, ecology, or ecotoxicology.

A similar pattern holds for *Toxicological Sciences,* the other journal of the Society of Toxicology, but because the journal is not restricted to studies of mechanisms of effect on organisms, it is freer to range into the field and into environmental problems. This monthly journal categorizes articles by topic and lists environmental toxicology as a topic of interest. However, only nine of the 225 articles appearing in calendar year 2001 were under this heading with perhaps an equal number of environmentally related articles appearing under the categories of Risk Assessment and Forums. Most of the articles, in contrast, appeared under topics such as carcinogenicity, neurotoxicology, respiratory toxicology, biotransformation, and toxikinetics. While the articles focusing on environmental issues are few, there seems no stricture against citing ecotoxicological findings and even using ecological theory and ecotoxicological measures.

An example of the acceptance of ecotoxicology within this toxicological forum is "Fitness Paramaters and DNA Effects Are Sensitive Indicators of Copper-Induced Toxicity in Daphnia Magna" (Atienzar 2001). This article compares effects occurring at the molecular and the population levels. While the population studies of water fleas here are within the laboratory, the article situates the work within literatures from aquatic science, aquatic toxicology, environmental science, ecology, environmental toxicology and ecotoxicology. Another study, "Acquired Resistance to Ah Receptor Agonists in a Population of Atlantic Killifish (*Fundulus heteroclitus*) Inhabiting a Marine Superfund Site: *In Vivo* and *In Vitro* Studies of Inducibility of Xenobiotic Metabolizing Enzymes" (Bello *et al.* 2001), rather than using laboratory-bred specimens, takes the experimental species directly from a contaminated superfund site, in order to study the particular adaptations of the fish that have allowed them to survive in a polluted environment. Again the article includes in its citations and intellectual context multiple articles from environmental, ecological, marine, and ecotoxicological sciences. "Masculinization of Female Mosquitofish in Kraft Mill Effluent-Contaminated Fenholloway River Water is Associated with Androgen Receptor Agonist Activity" (Parks *et al.* 2001) similarly collects both specimens and water samples from the field to be studied further in the laboratory. This article includes environmental, ecological and ecotoxicological literature in its discussion.

The journal *Toxicology* seems to have an even broader mandate, but except for the special articles described below, only ten of the almost 230 articles appearing in 2001 were on environmental toxicants. In seven studies the toxicant (in several cases diesel exhaust particulates) was brought into the laboratory for study on test animals. Another article was a longitudinal public health study of chromosome damage in Croatian workers producing pesticides (Garaj-Vhrovac and Zeljezic). Two articles, however, demonstrated significant respect for the measures and methods of ecotoxicology. The first, "Pesticide Use in Developing Countries" (Echobichon 2001) is a wide-ranging review article that includes considerations of economics, politics, regulation, ecological and environmental transmission and transformations, complexities of real-life exposures, and the value of long-term population studies to determine chronic exposure effects. The second, "Toxicological Profile of Pollutants in Surface Water from an Area in Taihu Lake, Yangtze Delta" describes findings from "monitoring the toxicological profile of aquatic ecosystems" (Shen *et al.* 2001). The study revealed "significant mutagenic activity" (71) and identified aspects of the hormones that were affecting the ecosystem. Among the very interdisciplinary resources cited were several articles from ecological and environmental journals. So we do again have a pattern of division of labor so that toxicology largely takes up industrial, pharmacological and medical issues and leaves most environmental issues to other specialties. When however, the subject warrants ecological theory and complex field measures, these are accepted as appropriate, intelligible and commensurable, revealing valuable and reliable data.

Most revealing about the respectful division of labor that now exists between toxicology and ecotoxicology is the special 2001 double issue of *Toxicology* (Volume 157: 1–2) devoted to "digital information and tools" (1). In this double issue, eleven articles catalog and describe the various information sources that have appeared on the internet, particularly the Web. While some articles predominantly focus on medical and pharmaceutical toxicological databases containing largely laboratory based information, the majority of the articles cover some environmental issues using field-based data and information about ecosystems. The report on "Toxicology information from US government agencies," for example, describes a wide range of agencies making toxicological data of various sorts available to the public—including environmentally focused agencies gathering field data from an ecological perspective (Brinkhuis 2001).

Several of the articles are focused particularly on environmental and ecological issues. "Toxicology Information Resources at the Environmental Protection Agency" (Poore *et al.* 2001) describes the EPA website, which contains basic public education and more professional data on environmental management and ecosystems, as well as pesticides and other pollutants. The article particularly mentions the Office of Research and Development's National Health and Environmental Effects Research Laboratory, which is concerned with "the effects of contaminants and environmental stressors on human health and ecosystem integrity" (Poore *et al.* 2001, 17).[11]

"Toxicology and Environmental Digital Resources from and for Citizen Groups" gives a rationale for attending to data from groups usually not considered scientifically significant by toxicologists: citizen groups "have provided an 'early warning' network" for "emerging problems" and "have initiated and advocated public policy initiatives" (Montague and Pellerano 2001, 77). The article reviews this history of citizens' movements in bringing attention to and regulation of environmental issues. Many of the databases described take an ecological perspective and provide field data and reports on the state of ecosystems, particularly as they have been affected by toxicants. "Online Resources for News about Toxicology and other Environmental Topics" lists in detail the sites devoted to environmental news (South 2001). Overall this special issue makes evident the many environmental information sources that toxicology needs to attend to if it is to contribute to environmental issues, even if toxicology's contribution is to be made primarily through traditional methods and theories.

Incommensurability Lost and Complex Practice Found

This catholic representation of the data resources available to toxicologists suggests that toxicologists must pay attention to a wide range of data and dynamics that reach far beyond the kinds of laboratory studies they focus their work on. It makes visible and legitimizes the work of colleagues taking ecological perspectives and gathering population and system data from field sites. The economic, political, and regulatory clients of environmental knowledge and the complexity of toxicants in the environment do not leave toxicologists the luxury of ignoring areas where their traditional methods fail to look, if they are to maintain authority within environmental spheres. This is especially true once the internet distributed widely the information generated by many organizations of different sorts. Because toxicology does have increasing amounts of work in areas that do not challenge the traditional laboratory methods and organism orientation, such as pharmacology, medicine, and consumer products, the field did have the option of simply withdrawing from environmental issues. Since, however, ecotoxicology has reserved a substantial though limited role for traditional toxicology, it seems in the field's interests to maintain a working synthetic alliance. Holding by methodological purity and ontological narrowness—in short acting as though ecological and field-based work were not commensurable with toxicology's ideals—seems to run counter to the field's practical goals, social responsibilities, and economic sponsorship.

Unlike Kuhn's vision of science, where practitioners seem free to follow the theoretical and practical commitments of their paradigms, fields like toxicology and ecotoxicology must be responsive to the complexity of applied problems, and changing economic, political, and regulatory climates. They do not have the insulation from practical concerns that would allow the luxury of incommensurability with the accompanying methodological intolerance and ontological blinders. The pressures are great to attend to all data and phenemoena that might be construed as relevant by the social, economic, and political sponsors. Although a science insulated from practical concerns is often thought to be more creative, flexible, and truth seeking because it claims not to be driven by interests, in this case we find the opposite: practical concerns of applications and interests *foster* the creativity and flexibility. Here application broadens the vision and mitigates methodological obstinacy of a field to allow a fuller understanding of the issues, acceptance of a greater range of data and phenomena, and tolerance of more methodological tools.

Even in the late 1980s when Halffman notices a great divide between toxicology and ecotoxicology, the Society of Toxicology's 1989 *Resource Guide to Careers inToxicology* lists over twenty-three programs with an environmental focus, even though the majority of the almost ninety programs listed have a traditional medical and pharmacological focus. The environmentally focused programs often include faculty with ecological and field-study orientations, according to faculty biographies that are part of the program descriptions.[12] When it comes to preparing students for jobs, providing students with relevant and useful perspectives may trump disciplinary purity.

Here continuity and difference within research programs is driven by the needs, interests and perceptions of varying sponsors that are more concerned with policy, politics, profits, and life-styles than with pure knowledge for its own sake. Perhaps incommensurability requires a purer science, driven only by its internal intellectual dynamics, with theory change being a response to anomalies. Kuhn's sociology is limited to recruitment and induction into paradigmatic camps. But as Fuller (2000) and others have noticed this ideology of a pure science driven only by its internal dynamics had its heyday in the post World War II period (when, in fact, physics and other sciences were heavily sponsored by governmental interest). It may be that all of science is more responsive to the complex exigencies of practical problems and interests of various human communities than the believers in pure science would have it. If that is the case the exigencies and complexities of the world militate against the motivated blindness of those who are strongly attached and ingrained to an insulated way of viewing the science. New specialties and new theories arise not just because anomalies make former accounts increasingly untenable, but because there are new problems to address in the world, and new groups bringing new interests to bear on scientific inquiry. It may be the problems of living in the world lead one to keep opening one's eyes wider, to counteract the psychic ease and sociological comfort that comes from the paradigmatic security of communally held gestalts or tightly structured taxonomic lexicons. Such would be at least the pragmatist hope for increasing knowledge in a world without foreordained correct ways of knowing it.[13]

Such a form of pragmatic relativism—roughly, the inverse rhetorical stance to what Harris diagnoses as pragmatic incommensurability in the introduction—rather than being a threat to the epistemic grounds of science, may be the means of holding accountable our intellectually proud ways of knowing to the world that we are trying to live in. The practicality and multiplicity of interests of science may help keep its spirit of open inquiry alive. At moments of change, the readjustment to be made, the tensions of alternative views, may seem to create great divides between different worldviews. The exigencies of coming to knowledgeable solutions for practical problems experienced by people in all walks of life, however, constantly humble our certitudes and keep us seeking better, more comprehensive, and more practically successful truths. For that, we can keep our eyes narrowed for only so long, no matter how much we are committed to our favored ideas and habituated perspectives.

Notes

[1] See, for instance, Kuhn (1996, 122–125); in the introduction to this volume, Randy Harris discusses these elements of Kuhn's conception under the label *cosmic incommensurability.* In "Second Thoughts on Paradigms" (Kuhn 1977, 293–319), Kuhn explicitly recognizes that the cognitive commitments are embedded with the complex of practices he calls the "disciplinary matrix." The embedding in practice makes the cognitive commitments even more resistant to change, as a cognitive change would disrupt an entire way of life. In an even later formulation, "The Road since *Structure,*" he sees disciplinary practices rather than individual cognition maintaining the disciplinary way of thought and perception. As Alan Gross takes up in his contribution to this volume, Kuhn here particularly points to the role of the taxonomically structured lexicon as the vehicle that structures disciplinary thought (Kuhn 2000, 90–104). Gross explicitly disagrees in that essay with our characterization of Kuhnian incommensurability here, but, as we see it, while he moved from an individualist model of cognition to a group model, based in the publicly displayed thought of communal vocabulary, Kuhn's model remains one of cognition, with incommensurability across cognitive boundaries. A switch of taxonomies is the group equivalent of an individual's switch of gestalt.

[2] Further: "operations and measurements that a scientist undertakes in the laboratory are not 'the given' of experience but rather 'the collected with difficulty.' [. . .] they are selected for the close scrutiny of normal research only because they promise opportunity for the fruitful elaboration of an accepted paradigm" (Kuhn 1996, 126).

[3] Carolyn R. Miller's essay on a controversy over non-thermal EMF effects, which follows ours in this volume, also looks at incommensurability through an interdisciplinary case study. The developments she explores are quite different, however, in that the two parties are steadfastly recalcitrant.

[4] For the depth of paradigmatic commitment based on solved problems, see Kuhn (1996, 169).

[5] The methods and theoretical approach of this analysis are an extension of an approach presented by one of us in Bazerman (1988) and elaborated most recently in Bazerman and Prior (2004).

[6] The second journal of the Society of Toxicology, *Toxicological Sciences,* has a somewhat broader mandate, but also stays close to laboratory dose-effect studies on organisms or tissues, as its current purpose statement indicates:

> *Toxicological Sciences* publishes research articles 12 times a year that are broadly relevant to assessing the potential adverse health effects resulting from exposure of human or animals to chemicals, drugs, natural products, or synthetic materials. Manuscripts are published in all areas of toxicology, both descriptive and mechanistic, as well as interpretive or theoretical investigations that elucidate the risk assessment implications of exposure to toxic agents alone or in combination. Studies may involve experimental animals or human subjects, or they may focus on *in vitro* methods or alternatives to the use of experimental animals. Other articles include historical topics, contemporary issues in toxicology, scientific and regulatory reviews, and international perspectives. (Society of Toxicology 2005)

[7] See, for example, January 1969 14:1 p. 205, Information for Authors: "*The Journal of Toxicology and Applied Pharmacology* publishes original scientific research pertaining to effects on tissue structure or functions resulting from administration of chemicals, drugs or natural products to animals or man." The title of the journal has changed, *effects* has become *action*, *function* shows up as a mass noun, rather than a pluralized count noun, and *man* has become *human.*

[8] We can't date the sentence exactly, but it showed up between 2002 when we first accessed the description, on the Society for Toxicology site, and 2005, when we accessed it from the Elsevier Science site, which publishes the journal.

[9] See Waddell (2000) for several rhetorical analyses of *Silent Spring.* These analyses, as do most other analyses of environmental rhetoric, examine strategies of arguing environmental issues in various public spheres, and are based on rhetorical readings of texts or recounting particular episodes of controversy or advocacy. These studies differ in character from this current study, which considers the rhetorical organization of knowledge fields arising in the wake of environmental concerns.

[10] Although French, he is widely cited by Americans as the founder and first theorist of ecotoxicology. Thus his statements have importance for the development of ecotoxicology in the United States.

[11] Also mentioned and described are the databases of the mid-continent ecology division AQUIRE, Phytotox, and TERRETOX (which has field data on the toxicity of exposures on wildlife). These three databases are gathered in the ECOTOX database. Also described is the ENVIROFATE database that has data on "the environmental fate or behavior of chemicals released in the environment" (Poore *et al.* 2001, 20).

[12] See, for example, in the *Resource Guide*, the program profiles on Clemson 18–19; Duke 22–23, Iowa State University. 30–31, University of Illinois Urbana-Champaign 112–113; University of Wisconsin-Madison 172–173.

[13] For related historically grounded philosophic accounts of the hybridity of scientific practice and communication see Pickering (1995) and Galison (1997), whose results are highly consonant with ours. The first author of this paper, however, grounds his view of hybridity and contingency in an utterance-based view of language in the tradition of Volosinov, a sociocultural, interpersonalist view of psychological development in the traditions of Vygotsky and Sullivan, and a pragmatist historical account of social relations in the tradition of G. H. Mead. See, for examples, Bazerman (2000, 2001, 2004); Bazerman and Prior (2004).

11

Novelty and Heresy in the Debate on Nonthermal Effects of Electromagnetic Fields

Carolyn R. Miller

The Controversy about EMF Bioeffects

Do cell phones cause brain cancer?* Are electric blankets safe? Is childhood leukemia, or breast cancer, caused by electric power lines? Should we worry about all the radio and television broadcast transmissions that surround us? Among the many risks that seem to plague contemporary life are electromagnetic fields (EMFs), forms of radiation that

* *Disclosure*: I am married to one of the proponents in the bioeffects debate, a founder of the Bioelectromagnetics Society; his research is not cited or discussed in this chapter. *Abbreviations:* A complete list of abbreviations used in this essay (and there are many) is given at the end. *Acknowledgements:* This chapter was initially conceived as a collaborative effort with Dale Sullivan, and I have benefited greatly from many conversations and exchange of ideas with him. Unfortunately, he was not able to continue contributing to the project. I am also grateful to Dr. Louis Slesin, editor and publisher of *Microwave News*, and to Dr. Carl Blackman, research biologist at the U.S. Environmental Protection Agency, for many helpful comments that helped ensure that the scientific details are accurately represented. Any remaining errors are my responsibility.

are emitted by many commonplace technologies. Several government reports, multiple liability lawsuits, frequent Congressional testimony, and extensive coverage in the mass media have helped alert us to the possible health effects of this invisible part of our environment.[1] The questions about safety have been controversial because the scientific research into biological and health effects of EMFs has been inconclusive. A 1992 "News and Comment" item in the journal *Science* summarized the state of EMF research as "deeply tortured," calling it a world "where deeply held points of view function like opposite poles, and between them stretches a powerful force field that whips the lay media into a frenzy. What keeps the force field strong is a lack of conclusive data to settle the EMF–cancer question" (Stone 1992, 1724). Little has changed in the intervening years.

The scientific polarization is strong enough that several observers have attributed it to a Kuhnian "paradigm shift" (Becker 1991; Morgan 1990; Nair 1990). Indeed, the polarization is characterized by seemingly incommensurable understandings of the interaction of fields with biological systems. My purpose in this essay is to examine these understandings and the debates about them from a rhetorical perspective, to give a rhetorical account of what has been called *incommensurability.* My examination is in two major parts: first, a topical analysis of the preferred lines of argument of the two sides in the scientific debate illustrates in some detail the rhetorical constituents of incommensurability; and second, an analysis of the defensive strategies used by those resisting paradigm change highlights the socio-political dynamics that arise from and reinforce failures of comprehension. The EMF case involves a number of issues that complicate Kuhn's and Feyerabend's discussions of scientific change, including the conduct of interdisciplinary science and the influence of social and political interests, such as the military and industry. Together these issues focus our attention on the forums in which debate occurs and judgments are made.

In EMF research, the conventional scientific understanding has been that fields of the kinds represented by microwaves, communications bands, and electric power could not have effects on biological systems other than heating, because they do not have sufficient energy to break chemical bonds and create the charged particles called *ions.* These fields in the frequency range below visible light in the electromagnetic spectrum, with long wavelengths and low frequencies, are therefore classified as non-ionizing radiation (see Figure 1).

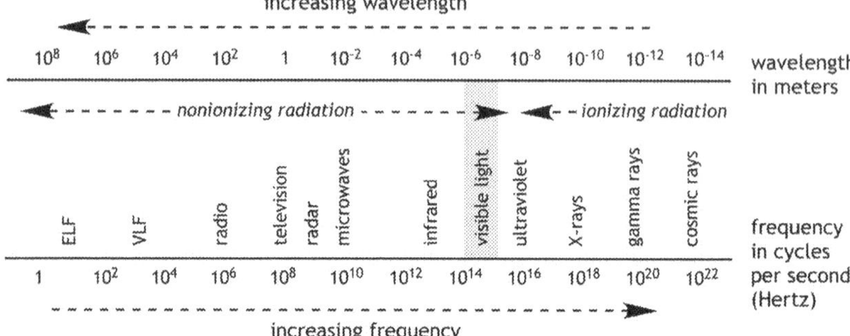

Figure 1. Electromagnetic spectrum. Electromagnetic fields at frequencies in the ultraviolet region and above are classified as ionizing radiation because they have sufficient energy to create charged particles (ions) by breaking chemical bonds, which is a non-thermal effect. Nonionizing radiation includes visible light as well as other conventionally named frequency bands with common applications. The conventional scientific understanding is that fields in these frequency bands can have only thermal effects, but this understanding has been unsettled since the 1970s by an accumulation of anomalous epidemiological and laboratory research.

In contrast, EMFs in the range above visible light, such as x-rays and gamma rays, with short wavelengths and high frequencies, are classified as ionizing radiation, and it is well accepted that they can cause nonthermal biological effects, including genetic mutations and cancer. Another reason underlying the assumption that nonionizing radiation is safe is that the electric fields that already exist within biological systems are much larger than those that can be created by external sources like power lines. Thus, the addition of relatively weak external fields to these internal background fields, or "noise," should not have detectable effects.

Bioeffects research is complicated by three facts. First, various frequency bands have different properties that make them not only useful for different applications (as suggested by their ordinary-language names) but also potentially capable of having different biological effects. Second, EMFs can vary not only in their frequency (and corresponding wavelength) and intensity ("brightness") but also in their wave forms: they can be "pulsed" and "modulated" in ways that change their properties, multiplying the number of variables that must be controlled and examined in research designs. And third, the relationship between the electric and magnetic components of EMFs changes with frequency and with distance from the transmitting source, so that at lower frequencies these components must be considered and controlled separately.

The accepted understanding that nonionizing EMFs are harmless has been perturbed by anomalous research findings. These findings are of two basic kinds, epidemiological studies of human health and laboratory studies of cell and animal systems. One early sign that the conventional understanding might be inadequate was a 1979 epidemiological study of childhood leukemia in Denver that showed a correlation with high EMF exposure from power lines (Wertheimer and Leeper 1979). As a 1990 *Science* "News and Comment" item noted about this study, "almost no one believed it" (Pool 1990, 1096). There were problems with the study, which had been done with essentially no funding support, including the fact that exposure could not be directly measured but was inferred from inspection of power lines outside residences, and the fact that the coding of exposures was not done blind. But in 1988, the study was replicated with an improved methodology and showed a risk factor of 1.5 (children with high exposures were 1.5 times as likely to develop cancer as those with low exposures) (Savitz *et al.* 1988).

Another set of anomalous results came from laboratory cell- and tissue-culture studies also beginning in the 1970s. This work showed that weak fields at extremely low frequencies (ELF) as well as at radio frequencies (RF) affected the rate at which calcium ions flow out of the cell membrane in brain tissue; because calcium ions are involved in the functioning of the cell membrane and specifically in neuron function, these findings had possible biological significance. One study in 1976 reported an unusual pattern of response: rather than an effect that continually increased with increasing amplitude (power), the effect increased and then decreased at higher amplitudes; a similar "window" effect was observed for increasing frequencies, with a "maximal field sensitivity at 'biological' frequencies" (Bawin 1976, 2001). This window effect was replicated at a range of frequencies and powers. These experiments, as well as others on DNA synthesis and RNA transcription, kinetics of some cellular biochemistry, and responses to hormones and other signaling molecules, produced results that were difficult to replicate, often transient, and sensitive to apparently small changes in experimental conditions, all of which provoked "considerable skepticism in established circles" (Morgan 1990, 119).

The anomalous nature of many of the scientific findings in this area and the resistance and skepticism with which they have been met are consistent with Kuhn's description of the early stages of a paradigm change (granting all the ambiguities of that phrase), and the persistence of incompatible interpretations of many research findings suggests incommensurable understandings of the phenomena being observed. Kuhn noted that "The reception of a new paradigm often necessitates a redefinition of the corresponding science" (Kuhn 1996, 103), and such a redefinition began in this case with the formation of a new scientific society devoted to the study of biological effects of EMFs. The society was founded in 1978, just as these early findings started coming in. The following year, it sponsored its first annual meeting and published the first issue of its journal, *Bioelectromagnetics.* In 1993 Ross Adey, the scientist responsible for some of the earliest calcium ion work, claimed confidently that "there is a reasonable prospect that bioelectromagnetics may emerge as a separate biological discipline, having developed unique tools and experimental approaches in a search for essential order in living systems" (Adey 1993, 415).

Those who remain skeptical, however, have continued to see the accumulation of anomalous results as evidence of error, incompetence, vested interest, and even fraud. The continuing "polarization" described in the 1992 *Science* article is well illustrated by two statements released by scientific societies in the mid-1990s. The first, from the American Physical Society, reflects the traditional view of EMFs:

> The scientific literature and the reports of reviews by other panels show no consistent, significant link between cancer and power line fields. This literature includes epidemiological studies, research on biological systems, and analyses of theoretical interaction mechanisms. No plausible biophysical mechanisms for the systematic initiation of promotion of cancer by these power line fields have been identified. Furthermore, the preponderance of the epidemiological and biophysical/biological research findings [... has] failed to substantiate those studies which have reported specific adverse health effects from exposure to such fields. [. . .T]he conjectures relating cancer to

> power line fields have not been scientifically substantiated. (American Physical Society 1995)

The second, a letter to Congress from the Bioelectromagnetics Society, was in part a response to the first:

> [E]lectric and magnetic fields, unlike many other environmental agents, are not characterized by a single quantity but involve many different factors. [. . .] A wealth of published, peer-reviewed scientific evidence indicates that exposure to different combinations of electric and magnetic fields consistently affects biological systems in the living body as well as in laboratories [. . .] Major strides have been made in the past 20 years of research in this area. The program has only recently expanded to a critical mass of interdisciplinary and multi-laboratory effort that, in our opinion, must be continued. In this still emerging area of scientific research, controversy about reported results is a natural and healthy part of the scientific process. (Bioelectromagnetics Society Presidents 1996)

Further evidence of the profound disagreements involved is displayed by a series of major reports issued by several governmental agencies and other organizations over the years and the often contentious responses to them. A timeline of these reports follows.

1990 The Environmental Protection Agency released for comment a draft report that classified extremely low-frequency EMFs as "probable human carcinogens" and radio and microwave frequencies as "possible" carcinogens. However, these recommended classifications were removed from the draft almost immediately by EPA administrators, and speculation was that pressure from the White House was involved (*Microwave News* 1990; Pool 1990; Park 2000, 154). This report was never completed. The draft was submitted to the EPA's Science Advisory Board in 1991, which criticized it as unsupported by evidence, and in 1996, the EPA announced that the final version would be indefinitely delayed (*Microwave News* 1996a).

1992 The Committee on Interagency Radiation Research and Policy Coordination (CIRRPC), a unit of the White House Office of Science and Technology Policy, issued a report that had been requested by the Department of Labor in 1989 after a stir created by the publication of a series of *New Yorker* articles by Paul Brodeur alleging a cover-up of serious human health effects. The CIRRPC report concluded that "there is no convincing evidence in the published literature to support the contention that exposures to extremely low-frequency electric and magnetic fields generated by sources such as household appliances, video display terminals, and local power lines are demonstrable health hazards" (Committee on Interagency Radiation Research and Policy Coordination 1992, ES-11).

1995 The conclusion of a report by the National Council on Radiation Protection and Measurements, was unofficially released in draft form. It stated that some health effects of EMFs were well enough documented to warrant preventive steps: "Although incomplete, available epidemiological and laboratory data share certain consistencies that would link ELF environmental EMFs with increased health risks. These findings appear to warrant a substantive national commitment to further research, and the serious attention of cognate regulatory agencies and of the general public" (qtd. in *Microwave News* 1995b, 13). The panel, chaired by Adey, endorsed an exposure limit for new day care centers, schools, and playgrounds and for new transmission lines near residential areas. "It took us nine years but we finally reached agreement," Adey commented (*Microwave News* 1995b, 1). This report was never completed because of changes in personnel and agency priorities.

1996 The National Research Council weighed in with a report that "no conclusive and consistent evidence shows that exposures to residential electric and magnetic fields produce cancer, adverse neurobehavioral effects, or reproductive and developmental effects" (U.S. National Research Council 1997, 2). However, three of the sixteen panel members released a statement through the Bioelectromagnetics Society indicating that the report's most important finding was "a reliable, though low, statistical association between power lines and at least one form of cancer" and highlighting the report's call for more research (Kaiser 1996). And panelists quoted in *Microwave News* indicated that the report set a very high standard in seeking "conclusive" proof; according to Savitz, the vice-chair of the panel, "Only by setting the threshold that high could we come to a consensus" (*Microwave News* 1996b, 5). Other comments quoted in this article indicate that the panel was deeply divided in its interpretation of the evidence.

1999 The National Institute of Environmental Health Science (NIEHS) completed a six-year review of research relevant to powerline health effects, concluding that there is "weak" evidence of health risk from exposure to ELF-EMFs: "The lack of positive findings in animal or mechanistic studies weakens the belief that this association [between exposure and leukemia] is actually due to ELF-EMF, but it cannot completely discount the epidemiological findings" (National Institute of Environmental Health Sciences 1999, Executive Summary). The report, which recommended "passive regulatory action" and continued research, was based on an earlier NIEHS report that classified EMFs as a "possible" human carcinogen, according to criteria defined by the International Agency for Research on Cancer, part of the World Health Organization (*Microwave News* 1998). There was virtually no press coverage of the final NIEHS report, according to *Microwave News* (1999b).

2002 A committee of the World Health Organization issued a report that classified extremely low-frequency magnetic fields as "possibly carcinogenic to humans" on the basis of limited evidence for the link to childhood leukemia (World Health

Organization 2002, 338). This report has received virtually no press coverage in the U.S.

Kuhn introduced the term *incommensurability* to describe several related aspects of the differences between successive paradigms, or explanatory frameworks: that they present different problems for science to solve, invoke different standards or solutions, use concepts and apparatus in different and sometimes incompatible ways, and (thereby) constitute "different worlds" within which scientists work.[2] This case study of the research on EMF bioeffects[3] will complicate Kuhn's account of incommensurability in three areas—the disciplinary locus of scientific change, the relevance of the public forum in the process of change, and the adequacy of the metaphors used to describe that process.

First, Kuhn (and Feyerabend) describe scientific change over time largely as a phenomenon internal to a single scientific discipline, but the EMF case involves differences between disciplines in an interdisciplinary enterprise. Different understandings of EMFs derive not only from the differential acceptance of a newly proposed explanatory framework within a scientific community but also from different long-standing and deeply engrained commitments in multiple disciplines.[4] The Bioelectromagnetics Society, though it formed by splitting off from several primarily engineering groups[5] to provide more attention to biological matters, was conceived as fundamentally interdisciplinary; for example, its constitution stipulates that the Board of Directors reflect the basic structure of the field of bioelectromagnetic research, with three members elected from the engineering and physical sciences, six from the biological and medical sciences, and three elected at large. But many of the difficulties this field has experienced derive from differences between the ways that those in the biological and medical sciences and those in the physical and engineering sciences understand the problems and phenomena in EMF research. I will make the case that the different worlds inhabited by researchers in this debate are to some extent defined and limited by their different disciplinary matrices, to use the Kuhnian term—their intellectual commitments, their habits of mind and of argument. Thus, the different worlds illustrated by this particular case are not the mutually inaccessible perceptual systems implied in the critiques of cosmic incommensurability described by Harris in his introduction to this volume; rather, they are the contextual differences of values, practices, and rhetorical themes that he calls *pragmatic incommensurability.*

Second, the EMF controversy must be understood as taking place not only within the restricted forum of the scientific enterprise but also in the public forum. Kuhn's discussion of scientific change and paradigm choice focuses exclusively on the explanatory commitments of scientists to their evolving state of knowledge about the world. The debate about EMF research has played out not only within scientific forums but also in governmental and policy forums. What is at stake in the debate are health and safety standards for exposure of human populations to electromagnetic fields of various strengths and frequencies—in specific workplaces (such as with radar and electric power distribution), for specific consumer products (such as electric blankets and cell phones), and in the ambient environment. The results of the debate are of interest not only to the general public but also to workers in specific electrical occupations, to electronics manufacturers, to the military, and to the broadcast and communications industries. I treat both scientific and public forums here, because the research has often been motivated (and funded) by groups with a policy interest and because research results are used routinely to justify policy claims. As Jeanne Fahnestock has found, the most polarizing statements by parties on both sides of a controversy are made in public forums, though often with a keen sense of audience in the scientific world as well (Fahnestock 1986; Fahnestock 1989). A complication of the EMF case is that because the debate, to at least some extent, takes place *between* existing disciplines, locating and controlling an appropriate forum becomes an issue in itself.

Third, with others in this volume, I will be taking the road taken by neither Kuhn nor Feyerabend, though gestured at by both, in explicating the types of impasse they included under their term, *incommensurability.* In addition to his overarching metaphor of political revolution, Kuhn invokes the models of gestalt psychology, religious conversion, and linguistic translation in an effort to describe the mutual incomprehension that can develop as scientific beliefs and commitments change. In fact, Kuhn seems somewhat at a loss to describe what he well characterizes as a rhetorical process. He asks how conversion is induced and how resisted, and answers, "Just because it is asked about techniques of persuasion, or about argument and counterargument in a situation in which there can be no proof, our question is a new one, demanding a sort of study that has not previously been undertaken" (Kuhn 1996, 152). He goes on to provide an "impressionistic" survey of the kinds of arguments that seem most persuasive: superior puzzle-solving ability, improved quantitative precision, prediction of new phenomena, aesthetic and social considerations (1996, 153–159). Kuhn's question was not new, of course, and his survey is much like a list of Aristotelian special *topoi* (*Rhetoric* 1358[a]). But the metaphor of incommensurability conflicts with this approach to the problem, since it implies that mutual incomprehension makes argument (as well as proof) impossible. And in subsequent work, Kuhn does not pursue his inquiry about argument and persuasion or seem much interested in the work of people who might have helped him do so. Instead, beginning as early as the "Postscript" to *Structure,* he pursues a philosophical approach to meaning change and translatability (Hoyningen-Huene 1993, 252–258), a trajectory that diverges significantly from his original rhetorical characterization. Meanwhile, though Feyerabend's trajectory, which began not far from where Kuhn's ended up, propels incommensurability deeper and deeper into a territory populated with values, diverse argumentations, and persuasion, in his actual analyses of specific incommensurabilities and in his choice of descriptive terminology, he no more followed a rhetorical brief than did Kuhn.

I will argue, as does Harris in his introduction, that the mathematical notion of incommensurability is not a particularly productive way to characterize differences between disciplinary commitments, being more descriptively tautological than explanatory. Instead, I will explore the explanatory power of two related tensions operating in this case, the tension between novelty and tradition and the tension between heresy and orthodoxy. The general claim I will make is that anomaly can be welcomed as productive scientific novelty that contributes to the growth of scientific knowledge, or it can be rejected as scientific heresy that must be corrected or expunged in order to preserve scientific knowledge. What Kuhn and Feyerabend called *incommensurability* can be understood alternatively as the result of these rhetorical dynamics, which involve a complex combination of both intellectual commitments and socio-political relations.

Disciplinary Differences in the Response to Anomaly

> In science, [. . .] novelty emerges only with difficulty, manifested by resistance, against a background provided by expectation.
>
> —Thomas S. Kuhn, *The Structure of Scientific Revolutions*

EMF research requires the cooperative contributions of multiple disciplinary perspectives, including those of engineers, physicists, epidemiologists, physicians, and biologists of many kinds, but these multiple perspectives have not always produced cooperation—rather, they have helped create the controversy. A public sign of disciplinary discord appeared in 1991 when *Science* reported that physicists had been excluded from a panel set up by the Environmental Protection Agency to review the controversial draft report that had identified EMFs as a "probable, but not proven, cause of cancer." The official reason for the selection of panelists was that they were to review epidemiological and laboratory animal data that did not seem to call for a physicist. An insider, however, reported that physicists were considered too skeptical of EMF bioeffects and that they "have trouble accepting what's going on in the field" (*Science* 1991). An analogous flare-up, with reverse polarity, occurred in 1995 over the constitution of two committees of the National Council on Radiation Protection and Measurements (NCRP), one established at the request of the EPA to address possible health standards for modulated radio frequency (RF) and microwave (MW) radiation (this would include cell phones), and the other to revise a 1986 report on biological effects and exposure criteria for RF fields. Originally chartered to focus on dosimetry (i.e., measuring doses of radiation), the first committee was made up of engineers and physicists, but the scope of the work was enlarged to include biological effects without any change in the membership. "Where are the biologists?" one federal official was quoted as asking (*Microwave News* 1995a, 14). And in explaining why neither the chair nor the vice-chair of the second committee was a biologist, an NCRP official said, "The best understanding in the field is more on the physical science–engineering end than on the biology end" (*Microwave News* 1995d, 13).

Although disciplinary differences can't account for all dimensions of the controversy over bioeffects research, they are central to the way the debate has played out. In particular, they reveal different ways that physical and biological scientists have dealt with certain anomalies. In what follows, I examine some of the prominent statements of difference, focusing on exchanges of comment and response where participants are ostensibly discussing the "same" data or research issues; these exchanges are where evidence of incommensurability would most likely appear in the written record.[6] What is striking in these statements are differences in the favored rhetorical *topoi* used by the physicists and engineers who argued against nonthermal bioeffects and by the biological scientists who argued in favor of such effects.[7]

One of the earliest programmatic statements of resistance to the mounting number of laboratory and epidemiological studies was by Robert K. Adair, a physicist at Yale, in *Physical Review A* in 1991. His argument illustrates in both substance and strategy many subsequent statements of resistance. The conclusion argues from the topic of impossibility that, "It does not appear to be possible for weak external ELF electromagnetic fields to affect biological processes significantly at the cell level" (Adair 1991a, 1047). The paper reasons from the topic of more and less in the cases of both electric fields and magnetic fields:

> ELF electric fields are so completely shielded by the conductivity of the body tissues that the interaction of external fields with a strength less than 300V/m with cells is far weaker than fundamental thermal noise.
>
> ELF magnetic fields may act through static interactions with magnetic dipole moments of biological material or through the induced electric fields generated by changes in the magnetic fields. Since the static effects of ELF fields of 50 μT are no greater than the earth's field, it is difficult to believe that the intensity is harmful. Since the maximum induced electric field in the body induced by 60-Hz 4-μT magnetic fields is no greater than the electric field induced by walking through the earth's field, it is difficult to believe that such changing ELF magnetic fields are harmful. (Adair 1991a, 1047–1048)

This is also an *a fortiori* argument: if the fields to which the body is normally subject (thermal noise, the earth's magnetic field, induced electric fields) are not harmful, then these far weaker fields can't possibly be harmful. There's a tone of incredulity that accompanies the argument, as well ("it is difficult to believe"), connecting the *a fortiori* to impossibility. In subsequent discussions, the *a fortiori* argument becomes a premise, often abbreviated as a point about signal-to-noise ratio (S/N).

Adair dismisses the experimental record, arguing from (lack of) agreement and contradiction: "There is no near-consensus among those who work in the field to the effect that any of the reports of effects in any of these areas is valid; none have been satisfactorily replicated, many of the more substantial results have been contradicted" (Adair 1991a, 1047). He specifically calls the "window" effects into question:

> It is, perhaps, the intensity windows that are [. . .] most difficult to accept. [. . .] It is an almost firm rule of the behavior of systems that, above an action threshold, the response to a perturbing signal increases at least linearly with the incremental signal. The linear effect will generally be terminated only when the signal is so large that it can no longer be considered a perturbation. Since it is very difficult to consider that the small signals in question are sufficiently large to have any effect at all, the view that they can be so large as to dampen out a response is even more troubling. (Adair 1991a, 1047)

Here Adair also combines an appeal to physical impossibility with a kind of inverted *a fortiori* argument.

Another form of argument that is characteristic of the physicists is from cause, which often takes the form of an appeal to "mechanism."[8] Indeed, Adair's entire article is an argument from cause; it examines the mechanisms of interaction between fields and biological tissues, reasoning from well accepted laws about the characteristics of electric and magnetic fields and from assumed values for parameters such as the specific resistance of a cell membrane, the radius of a cell, and the thickness of the membrane. Once these mechanisms are understood, Adair can claim that EMFs cannot cause effects that can have biological significance. It is Adair's confidence about the lack of a causal mechanism that supports the overall argument from impossibility.

The statement quoted earlier from the American Physical Society provides another example of causal argument: "No plausible biophysical mechanisms for the systematic initiation or promotion of cancer by these power line fields have been identified" (American Physical Society 1995). Similarly, the CIRRPC report concludes, "Epidemiologic findings of an association between electric and magnetic fields and childhood leukemia or adult cancers are inconsistent and inconclusive. No plausible biological mechanism is presented that would explain causality," and in the discussion of the calcium ion research, it notes that "Although a statistically significant effect has been reported in these experiments, no adequate explanation of the interaction mechanism has been provided" (Committee on Interagency Radiation Research and Policy Coordination 1992, ES-11, II-83). The NRC report notes, "A serious barrier to acceptance of a possible weak connection between human health and exposure to extremely-low-frequency electric and magnetic fields in residences is the absence of a plausible physical mechanism to account for such a connection" (U.S. National Research Council 1997, 206).

The epidemiological work is also subject to refutation by causal arguments on the general ground that epidemiology can show only correlation, not causation. For example, the NRC's Executive Summary summarizes its review of the epidemiology in this way: "An association between residential wiring configurations [. . .] and childhood leukemia persists in multiple studies, although the causative factor responsible for that statistical association has not been identified" (U.S. National Research Council 1997, 2). One of the members of the CIRRPC panel, William Bennett, another Yale physicist, challenged the epidemiology on the grounds that it had not accounted for all possible confounding factors, that is, alternative causes (Bennett 1994, 6).[9] And others charged that no consistent dose-response relationship had been found—that is, there is no linear causal relationship in which increased exposures produce increased effects (U.S. National Research Council 1997, 201; Adair 1992a, 1869).

Challenges to the anomalous bioeffects research also take the argument one *topos* further. Based on the premise that a causal mechanism determines the possibility or impossibility of a phenomenon, physicists can then define what is real and what is not and invoke the appearance/reality dissociation. These appeals tend to show up not in formal scientific reports but in public or semi-technical forums. Adair, who has been a frequent and outspoken respondent in letters and commentaries, has charged that the bioeffects research "has created an imaginary problem where no real problem exists [. . .] experimental errors have been accepted as real effects" (Adair 1992a, 1869). The APS statement quoted above concludes that "the conjectures relating cancer to power line fields have not been scientifically substantiated," contrasting the appearance of conjectures with the reality of scientific substantiation.

Another common argumentative theme follows from this one, building on the topical sequence of cause ⇒ impossibility ⇒ reality. This theme follows from what we can call the *physics-trumps-biology argument*, an appeal to the "fundamental laws of physics" as necessarily constraining any and all scientific knowledge. For example, the CIRRPC report states, "The lack of converging epidemiological and biological support for the occasionally reported adverse health effects is consistent with calculations of quantities based on fundamental laws of physics for describing electric or magnetic fields" (Committee on Interagency Radiation Research and Policy Coordination 1992, 12). Similarly, Adair's abstract concludes, "any biological effects of weak ELF fields on the cellular level must be found outside the scope of conventional physics" (Adair 1991a, 1039). And Bennett's book claims that

> Much of the speculation on biological interactions with extremely low-frequency fields has tacitly ignored the fundamental physical laws involved. The physical properties of electromagnetic fields were well described by the work of Maxwell and his predecessors in the nineteenth century, and the solution of Maxwell's equations for most cases of interest is straightforward. Although biological mechanisms can be complicated, they cannot violate the fundamental laws of physics. (Bennett 1994, 14)

Since physics trumps biology, biologists are not qualified to speak about physics, but with their asymmetrical privilege physicists feel qualified to speak about biology.[10] This point is illustrated best by an exchange initiated by Berkeley particle physicist J. David Jackson with a 1992 article in the prestigious *Proceedings of the National Academy of Sciences*. Jackson argued that the "putative causal relation" between ELFs and cancer could be tested "by examining historical data on the growth of the generation and consumption of electric power since 1900 and corresponding data on cancer death and incidence rates" (Jackson 1992, 3508). Such data, he claims, "are *prima facie* evidence that the stray low-frequency electromagnetic fields associated with the generation, distribution, and use of electricity in the home and office are no significant cause of cancer deaths" or of incidence rates (Jackson 1992, 3509). Jackson's argument became quite influential, being relied on substantially in the CIRRPC report and elsewhere. Several epidemiologists, including Wertheimer and Leeper (1992) and Savitz (1993), criticized it severely as amateur epidemiology that ignored confounding factors, and an article in the *Los Angeles Times* about Jackson's publication quoted epidemiologist Richard Stevens calling it "a fatuous piece of work." In response, Jackson is reported as saying, "It takes a physicist to know how to deal with the power data. [. . .] We're looking at historic facts, and I am as competent as anyone to deal with that" (Maugh 1992).

Why does physics trump biology? One reason is the topical sequence we have traced, from cause to reality. This sequence underlies the assumption that all other scientific disciplines can ultimately be reduced to the terms of physics.[11] For example, Adair describes the advantages of reasoning from physics: "the statistical significances of some of the biological work and many of the epidemiological reports have been seriously overstated. Such analyses are usually subjective; experience with simpler, falsifiable, physical science experiments has shown that significance levels are generally exaggerated" (Adair 1992a, 1869). Reducing biology to physics is authorized by a philosophy of science based on the conduct and methods of the physical sciences; only in the past 30 years or so has biology had serious influence on the philosophy of science. A classic statement of reductionism is that of Rudolf Carnap from 1938, in the first volume of the *International Encyclopaedia of Unified Science:* "Biology presupposes physics, but not vice-versa." Carnap goes on to claim that "a biological law contains only terms which are reducible to physical terms" (Carnap 1938, 46, 60).[12] In his useful essay "How Biology Differs from the Physical Sciences," Ernst Mayr quotes Nobel laureate physicist Steven Weinberg with a more contemporary example of reductionism: "the closest we can come to a unified view of nature is a description in terms of elementary particles and their mutual interactions." Mayr comments: "By contrast, every biologist would insist that to dissect complex biological systems into elementary particles would be by all odds the worst way to study nature" (Mayr 1985, 44).

We have seen that physicists rely on arguments from cause, impossibility, appearance/reality, a recurrent *a fortiori* argument, and an implicit reductionism. What sorts of arguments do the biologists use? If we return to the statement by the Bioelectromagnetics Society Presidents, we see arguments from agreement ("a wealth of [. . .] evidence"; "a critical mass of [. . .] effort"), from consistency, and from past fact: the strong appeal is that there are many research findings that have to be accounted for, anomalous or not. Past fact is an important appeal throughout the statements and responses from biologists. For example, Adey's 1993 article draws on past fact and agreement: "Laboratory studies have tested a spectrum of EM fields for bioeffects at cell and molecular levels, focusing on exposures at athermal levels. A clear emerging conclusion is that many observed interactions are not based on tissue heating" (Adey 1993, 410). It goes on to project future fact from past fact: "As evidence has mounted confirming occurrence of bioeffects of EM fields that are not only dwarfed by much larger intrinsic bioelectric processes, but may also be substantially below the level of tissue thermal noise, there is a mainstream of theoretical and experimental studies seeking the first transductive steps" (Adey 1993, 411). In a response to Adair's 1991a article, Kirschvink contrasts past fact with impossibility: "the credibility of weak ELF magnetic effects on living systems must stand or fall mainly on the merits and reproducibility of the biological or epidemiological experiments that suggest them, rather than on dogma about physical implausibility" (Kirschvink 1992, 2178).

In his response to the CIRRPC report, Savitz also argues from agreement (providing a copia of evidence, if not of expression) and past fact: "Epidemiologic evidence linking power lines near residences and elevated magnetic fields to childhood cancer continues to accrue; employment in selected electrical occupations seems to confer an increased risk of leukemia, brain cancer, and perhaps breast cancer; and there have been numerous laboratory studies indicative of influences on circadian rhythms, calcium efflux from nerve cells, and a hypothesized pathway linking such exposures to cancer" (Savitz 1993, 52). A companion response by biophysicist and radiation biologist Thomas Tenforde uses the same strategies:

> Epidemiological evidence for an elevated cancer risk in children residing near high-current distributions lines has been reinforced by the positive findings in recent studies conducted in Los Angeles County and in Sweden, which add credibility to the results of earlier studies [. . .] Similarly, there is now a substantial body of epidemiological evidence [. . .] indicating that individuals in the general class of 'electrical workers' exhibit an elevated cancer risk. (Tenforde 1993, 56)

A second major strategy the biologists use is to emphasize the complexity of biological systems, an implicit comparison with the simplifying assumptions of physical analysis. Mayr's discussion reinforces the centrality of this strategy to biological thinking, again by contrast with Weinberg's statements about physics: "One of man's enduring hopes," says Weinberg, "has been to find a few simple laws that would explain why nature with all of its seeming complexity and variety is the way it is." Mayr responds, "Surely no biologist would ever express such a hope" (Mayr 1985, 44). The argument from complexity counters the physicists' reductionism and stakes out the autonomy of biology. In an earlier essay on the distinct nature of causality in biology, Mayr noted that causal prediction of the sort expected in classical mechanics is impossible for biological systems because of their extreme complexity, the uniqueness of biological entities, and the emergence of new qualities at higher levels of integration (Mayr 1961, 1505–1506). Similarly, philosopher David Hull notes that the central issues in a distinct philosophy of biology are "the complexity and consequent uniqueness of biological systems, their stratified, teleological organization, and the central role of historical considerations in biology" (Hull 1974, 142).

There are many arguments from complexity in this debate, but I'll focus on those that appear in, or respond to, examples already used here. First, the Bioelectromagnetics Society's response to the APS notes that "electric and magnetic fields, unlike many other environmental agents, are not characterized by a single quantity but involve many different factors" (Bioelectromagnetics Society Presidents 1996). A letter in *Science* responding to Adair's 1992a letter (quoted above) uses the complexity argument, as well as agreement and an argument from authority:

> Many credentialed observers believe that the chance that EMF hazards are real is far from negligible. A large number of biologists and epidemiologists contend that an equally plausible explanation of the record is that EMF bioeffects are simply more subtle than those of many other environmental agents. (Florig 1992, 1869, 1960)

Several defenses of the epidemiology work rely on the complexity *topos.* Savitz's invited commentary on the CIRRPC report makes this point at length (Savitz 1993, 54), and in the *Los Angeles Times* article on the Jackson work, Savitz is quoted to the same effect but in pithier form: "Intuitively, it sounds logical to look at such data. But there are so many things going on over that period—new drugs, new treatments, new causes of cancer—that it really doesn't tell us a thing" (Savitz qtd. in Maugh 1992).

Responses to Adair about the mechanism of interaction also argue from complexity. For example, Robert Becker, a physician, responds to Adair's letter criticizing a review of Becker's 1990 book, *Cross Currents,* by refuting Adair's reductionist *a fortiori* signal-to-noise ratio argument and introducing an appeal to future fact:

> Adair's rejection of any biological effects from low-level electromagnetic fields rests entirely on the outmoded concept that *kT* [thermal noise] must be exceeded for such effects to occur. This concept in turn rests upon the also outmoded biological concept that living things are simply chemical machines all of whose functions result from chemical reactions in an aqueous medium. [. . .] The new biological paradigm is far richer than the old and offers great opportunities for medical therapies as well as cautions for our ever expanding use of electromagnetic energy. Both urgently require full explanation. (Becker 1991, 103–104)

Adey makes a similar, though more general point regarding the calcium ion window results: "these phenomena are in the realm of non-equilibrium thermodynamics, and are thus far removed from traditional equilibrium models of cellular excitation based on depolarization of the membrane potential and on associated massive changes in ionic equilibria across the cell membrane" (Adey 1993, 411). And Tenforde's response to the CIRRPC report draws on the complexity *topos* to refute reductionism:

> The [. . .] panel discusses the fact that [. . .] the signal-to-noise ratio (S/N) is expected to be much less than unity for ELF fields from power lines and other common sources. Although this is true at the level of a single cell, most biological tissues are composed of large cell aggregates that are electrically coupled via junctions between adjacent cell membranes. As a result, the S/N ratio for an imposed ELF field improves roughly in proportion to the number of cells in the junctionally coupled aggregate. [. . .] Hence the argument that ELF field signals would be 'drowned' in a sea of background electrical noise is incorrect. (Tenforde 1993, 56–57)

To summarize then, in comparison with the physical scientists, the biological scientists rely on a very different cluster of characteristic *topoi:* past fact, future fact, and complexity, with some attention to agreement and credibility. The emphasis on both past and future fact suggests an empirical preference for "appearance" over the rationalist preference for "reality" that we saw among the physicists. Future fact is related to possibility, in contrast to the physicists' arguments from impossibility. These differing patterns of argument can help us see, in this particular case, "how conversion is induced and how resisted" (Kuhn 1996, 152). The resisters, the physical scientists, rely on an ontology that physics has confidence in, a rational structure of reality that is comprehensible and predictable by established causal law and to which observational fact must conform. This ontology produces an axiomatic world from which phenomena can be deduced, a world in which logical relations (more and less, consistency) control and limit the acceptance of anomaly. The proselytizers, the biological scientists who attempt to induce conversion, rely on an inductive epistemology, in which anomalous phenomena are granted a status that demands attention: their complexity gives them an empirical weight that requires explanation. Note that biologists do not offer a new dogma to which they urge conversion, for there is yet no new structure of commitments that they can offer; they merely point to the inadequacy of existing tradition and to the hope that future fact will resolve the problems that past fact has created.

This analysis of topical preferences gives a detailed picture of how these two groups of scientists struggle with anomalies—and specifically what sorts of intellectual commitments and presuppositions may shape their willingness to see anomaly as scientific novelty. Agreement, cause, impossibility as an entailment of reality, and reductionism are hostile to innovation; deductive reasoning has long been understood as tautological, unable to produce novelty. In contrast, past and future fact, possibility, and complexity all can produce innovation; inductive reasoning can incorporate new facts and may produce novel generalizations (though these are less reliable than the conclusions of deductive arguments). These patterns of thinking, revealed as clusters of rhetorical *topoi,* characterize what one close observer has called the "gulf between the physicists and the biologists" (*Microwave News* 2000, 11). This rhetorical characterization can provide one way of understanding what Kuhn meant (or should have meant) when he said that "the proponents of competing paradigms practice their trades in different worlds" (Kuhn 1996, 150). The "different worlds" they inhabit are in an important way rhetorical worlds.

Orthodox Resistance to Anomaly

> The transfer of allegiance from paradigm to paradigm is a conversion experience that cannot be forced. Lifelong resistance, particularly from those whose productive careers have committed them to an older tradition [. . .], is not a violation of scientific standards but an index to the nature of scientific research itself.
>
> —Thomas S. Kuhn, *The Structure of Scientific Revolutions*

In an essay that predates *Structure* by three years, Kuhn described an "essential tension" between tradition and innovation in science, calling our attention to the curious fact that in science, "work within a well-defined and deeply ingrained tradition" is "productive of tradition-shattering novelties" (Kuhn 1977b, 234). Indeed, he argued, it is the strength and definition of the tradition that makes it possible for anomalies to be identified, evaluated, and explored. Anomalies must negotiate the essential tension—are they errors that the tradition is justified in rejecting, or are they novel discoveries that will revise or remake the tradition? In his essay Kuhn emphasized the "central role" of tradition (Kuhn 1977b, 236), as in *Structure* he emphasized the necessity of the "paradigm" to normal science. As a "strong network of commitments—conceptual, theoretical, instrumental, and methodological" (Kuhn 1996, 42)—the paradigm defines the science, requires the allegiance of its practitioners, and accounts for the "relative unanimity of their professional judgments" (Kuhn 1996, 182). These commitments produce an "assurance that the older paradigm will solve all its problems, that nature can be shoved into the box the paradigm provides. Inevitably, [. . .] that assurance seems stubborn and pigheaded as indeed it sometimes becomes. But it is also something more. That same assurance is what makes normal [. . .] science possible" (Kuhn 1970, 151–152). It does this in part by focusing scientific attention, distinguishing the significant and interesting from the trivial and mistaken (Kuhn 1977b, 236; Kuhn 1996, 35). The paradigm is a powerful exclusionary system.

The power of tradition in science is such that it functions as an orthodoxy, as Thomas Lessl and Dale Sullivan have both noted (Lessl 1988; Sullivan 2000). An orthodoxy is a body of doctrine and practices adhered to by a group that protects the doctrine and defends its own authority to determine both the doctrine and group membership. Deviance from orthodox doctrine is heresy, an "affront to an institution's authority" (Lessl 1988, 23), and group members who challenge the correctness of the doctrine and the authority of the adherents are rejected as heretics. Kuhn himself noted that a scientific tradition rewrites history to demonstrate its own inevitable correctness (Kuhn 1996, 137) and described initiation into a scientific tradition as "dogmatic" (Kuhn 1977b, 229). The tradition inspires not only commitment and "assurance" among its adherents but also "faith" (Kuhn 1977b, 236). Kuhn's later term, *disciplinary matrix,* captures this aspect of scientific traditions well, implying the mutual embeddedness of the intellectual and socio-political components of a tradition. The need for intellectual exclusion can involve a parallel need for social exclusion, and a paradigm must constrain not only ideas but also those who hold them.

As we saw in the previous section, some of the rhetorical differences between physicists and biologists in the debate on EMF bioeffects derive from the specific intellectual commitments and preferences of these two broad disciplinary traditions.[13] However, other differences derive from the relative positioning of the fields in this specific debate, whether defending or challenging a tradition. Thus, those who do not believe there can be nonthermal bioeffects can be seen as defending an orthodoxy against heretical proponents of nonthermal effects, reaffirming strong conceptual commitments in the form of symbolic generalizations about mechanisms of interaction between matter and electromagnetic energy and attempting to reduce the authority of those who do not adhere to the doctrine.[14] The response to heresy, as Lessl explains it, often strengthens orthodoxy, creating internal clarity and solidarity. The challenge of heresy makes it important to clarify not only doctrine but also the boundary between doctrine and heresy. Thus, there may be much attention to what sociologists call boundary-work, and what philosophers of science have called the demarcation problem: heresy is often driven out by defining it as nonscientific. The heretic may also be driven out by rituals that remove tension from the community under challenge (Lessl 1988, 20–22). To this picture, Sullivan adds the notion that a successful response to heresy must control the forums of the doctrinal community, that is, must supervise official communication by authorizing and de-authorizing speakers and writers. He describes four specific means of "forum control" available to most scientific communities: peer review, denial of forum, public correction, and public ridicule (Sullivan 2000, 128). With these methods of control, scientific communities enact the general social processes that characterize responses to heresy. In what follows, I examine the EMF debate for evidence of the latter two forms of control. The first two, peer review and denial of forum, are not strategies of argument that typically appear in the published record, although in the previous section we did see attempts to deny access to the forum in situations where the composition of review committees was at issue; these efforts to gain or retain authority, we noted, occurred on both sides. The analysis here focuses on the response of the physicists to see whether it is consistent with the hypothesis that they are engaged in defending a scientific orthodoxy, but with attention also to the biologists to determine whether the argument is symmetrical.

Public correction occurs quite frequently, on both sides, and in both scientific and public forums. It often takes the form of impersonal correction of ideas but can also be more personally directed, correcting those responsible for the ideas. A great many of Adair's contributions can be seen as efforts at correction. In his letter to *Science,* he notes that "the statistical significances of some of the biological work and many of the epidemiological reports have been seriously overstated" (Adair 1992a, 1869). And in a technical article criticizing a physical model put forth by biophysicists, which proposes a theoretical mechanism for the window effects of EMFs, Adair claims, "Arguably, neither the physical character of the IPR model, nor the required initial conditions nor the effects generated by ELF fields according to the model, nor its severe defects are clearly understood. Thus I attempt here to provide a critical description of the model" (Adair 1998, 181). Note Adair's polite use of the passive voice in both instances, declining to attribute agency to those who have "overstated" or misunderstood. Adair is not the only physicist to engage in correction: Allan Bromley, Science Advisor to the first President Bush, worked behind the scenes to delay the release of the EPA report because, as he told *Time* magazine, EPA's "findings of a 'positive association' between EMFs and childhood cancer are 'quite incorrect'" (qtd. in *Microwave News* 1991).[15]

Correction works in the other direction, as well: nonthermal-effects heretics correct the defenders of orthodoxy. In their comments on the CIRRPC report, both Savitz and Tenforde engage in extensive public correction of the report's reliance on Jackson's historical correlations by stating the argument, claiming that it's mistaken, and then explaining the flaws. Here is the beginning of Savitz's commentary on this point:

> A specific theme that is invoked [in the CIRRPC report] for both cancer and adverse reproductive outcomes is that whatever the research might suggest, electric or magnetic fields could not adversely affect health because the rise in electric power use over time has not produced epidemics of cancer, birth defects,

> or miscarriage. In fact, the two observations are true but unrelated. (Savitz 1993, 54)

Savitz goes on to explain what data would be needed to establish the correlation. Jackson's argument was also directly refuted, and Jackson himself corrected, in a letter to *Microwave News* by Wertheimer and Leeper, the authors of the earliest positive epidemiological study:

> Dr. Jackson unquestioningly assumes that power frequency EMF exposure increased proportionately with increasing power use over the last 50 years. The fact, however, is that numerous engineering changes have occurred over the years that markedly decreased the level of magnetic field exposure likely to accompany a given power use. Thus, in spite of increased power use, it is likely that magnetic field exposures associated with power distribution have decreased, if anything, over the years for which reliable cancer data are available. [. . .] (Wertheimer and Leeper 1992)

In a final example, Tenforde engages in correction in a review of the book that Bennett published based on his work for the CIRRPC committee, using a simple opposition:

> Bennett's arguments generally proceed from a simple fact: that the ELF field levels reported to elicit these biological effects are too small to do so, being less than the body's intrinsic electrical noise over the dimensions of a single cell. But there is ample reason to expect that biological systems may exhibit responses to ELF fields that cannot readily be modeled by treating an integrated organ or tissue as a noninteracting collection of individual cells. (Tenforde 1995, 10)

Correction is a means of forum control that parties use in their struggle over authority, specifically attempting to establish the boundary that defines their science. As the examples above illustrate, corrections tend to adhere to scientific decorum, being impersonal and often well supported with detail, whether appearing in a narrow scientific forum or a semitechnical or public forum like a book review in *IEEE Spectrum* or an interview in *Time* magazine. The strategy of correction addresses the substance of the science, the propositional content of the orthodoxy being contested and only indirectly attempts to control those making statements about it.[16] That the biologists engage in correction suggests that they do not view themselves as heretics but, perhaps, as guardians of a different orthodoxy, one that is jeopardized by the interfering involvement of physicists. The situation thus epitomizes the difficulties of interdisciplinary work: without a jointly owned orthodoxy, the separate disciplines simply contest the other's right to govern the forum.

Other forms of boundary work address the authority and legitimacy of those who make heretical statements, by calling into question their competence and motives. These strategies usually appear in nontechnical forums and often depart from scientific decorum—the tone is more clearly adversarial and sometimes quite personal. Notably, in these more public forums the boundary being defended is not that of a particular scientific discipline but that of science itself. One politically important strategy on the part of the defenders of orthodoxy has been to draw a distinction between "sound science" and "junk science" (or "voodoo science," or "pathological science," etc.), in effect denying the scientific competence of those who are challenging the orthodoxy.[17] For example, in his response to a review of one of Brodeur's books, Adair says, "good scientists hold these very weak 60-Hz fields harmless;" the bioeffects work, he says, is not a "paradigm shift" but "illegitimate science" (Adair 1991b). Similarly, Bromley commented in an interview about his role in the EPA report, "My role as science advisor to the president is to be sure that statements that come out of this administration are based on sound science" (*Microwave News* 1991). Robert Park's book *Voodoo Science* has an entire chapter on the EMF bioeffects work, other chapters being devoted to cold fusion and perpetual motion machines (Park 2000).[18] Bennett's 1994 *Wall Street Journal* column concludes by characterizing the entire debate as "the electromagnetic hoax." On the way to that conclusion, he accuses the bioeffects research community of "scaremongering," "alarmism," "marginal statistical accuracy and extreme susceptibility to systematic error," and "far-fetched hypotheses" (Bennett 1994).

It is only a short additional step from questioning the competence of heretics to questioning their motives. Bennett's article in *Physics Today,* published at about the same time as the *Wall Street Journal* column, opens with a description of a "growth industry" that includes "a land-office business" in gaussmeters (a measurement instrument for magnetic fields), a "bonanza" for researchers, and "sensational articles" in the media (Bennett 1994, 23). In his column for the *New York Times* commenting on the recently released National Research Council report, Park similarly charges that "An industry has grown up around the power line controversy. Tort lawyers, engineers who measure electric fields, some journalists and newsletter publishers have an interest in the issue never being quite settled" (Park 1996). In his book, he makes similar attributions, adding to the list of those with a vested interest the scientists whose research is "thriving" (Park 2000, 158). And in 1992, Adair also raised the issue of motive in a letter to *Physics World:* "the self-interest that leads to scientists grossly exaggerating the linkage between their researches and health effects is also enormous" (Adair 1992b, 17). Occasionally a bioeffects researcher resorts to this kind of argument; for example, questioning the impartiality of authors who have been paid consultants to the manufacturers of microwave equipment (Frey 1986), or noting that much of the funding for EMF research comes from agencies with a connection to the electric power industry (notably, the Department of Energy) or to the military (Pool 1990, 24). There have also been claims that the NCRP committee on exposure criteria for RF fields had conflicts of interest, since five of the seven members had been consultants to the communications industry (*Microwave News* 1995a).

Another way of questioning the authority of other scientists is to cast doubt on their simple common sense. This is a muted form of ridicule that operates by *reductio ad absurdum.* For example, Bennett's *Wall Street Journal* column points out that "If fields of two milligauss really are a serious threat in Denver, Los Angeles and Sweden, then commuters on East coast electric trains—where the fields at power line frequencies can be hundreds of times larger—ought to be dying like flies" (Bennett 1994). Adair uses a similar argument; reasoning that we all know that falling leaves will not cause fractured skulls, he notes, "few of us understand magnetic fields as we do tree parts. Are the minute magnetic fields from our power distribution systems [. . .] leaves or tree limbs? I answer, '[. . .] it is no more possible that they cause cancer than that real leaves crack skulls'" (Adair 1999). And in his book, Park ends the chapter on EMFs by suggesting that EMF bioeffects are as likely as a stegosaurus running down Fifth Avenue

(Park 2000, 161). Interestingly, the heretics fight back using much the same technique, suggesting again that some of them, at least, do not cede authority for policing the boundary to the physicists. In a response to Bennett's *Physics Today* article, one notes that "Accepting the article by [Bennett] as guidance on the question of health effects of [EMFs] seems to me analogous to accepting the advice of the village blacksmith on how to fix your Swiss watch" (qtd. in *Microwave News* 1995c). And Stevens's comment on the Jackson *PNAS* paper claims that "based on this data, you couldn't even show any significant effect of smoking on cancer, much less a more subtle effect like this one" (qtd. in Maugh 1992).

Much of the public commentary by Adair, Bennett, and Park[19] can be seen as attempts to assert authority within the field, to control the boundary between sound and junk science, and to eject from any position of scientific authority those who refuse to recognize and practice what they consider orthodox science. All three are physicists, as noted earlier, and as such they come by their presumption of authority by training and tradition. Physics is considered the oldest "mature" science (in Kuhn's terms), and it predates paradigmatic biology by centuries. Partly because of its historical status, physics has also enjoyed a heightened social and political status. Perhaps the most telling illustrative fact is that all nineteen Presidential science advisors (dating back to Vannevar Bush during World War II) have been physical scientists (mostly physicists with a few chemists and engineers); none has been a biological scientist (Halber 2001).[20] Physicists have an authority in the public and political realm that is part of their disciplinary identity, and the sometimes presumptuous tone of their comments in the EMF debate is probably related to this identity.[21]

A final means of defending the orthodoxy extends ridicule into ritual through the process of scapegoating. Neither Lessl nor Sullivan mentions scapegoating by name, though Lessl's discussion of the heretic as a "ritual object upon which anxieties can be heaped" clearly invokes it (Lessl 1988, 24), and Sullivan's discussion of public ridicule and exposé suggests it less directly (Sullivan 2000). In the EMF case the orthodox had a scapegoat delivered to them in 1999 by the NIH Office of Research Integrity (ORI). The ORI made an official finding of scientific misconduct against Robert Liburdy, then at Lawrence Berkeley National Laboratory (LBL), who had published two papers in 1992 about EMF effects on calcium flow into rat lymphocytes. Liburdy had, the ORI charged, "intentionally falsif[ied] and fabricat[ed] data and claims about the purported cellular effects of electric and magnetic fields" (*Federal Register* 1999). At issue were three graphs, in which, the ORI contended, Liburdy presented only selected data and fabricated parts of the curves. Liburdy admitted no scientific wrongdoing but agreed to retract the graphs from the published sources. In a letter to *Science,* he emphasized that both the raw data and the scientific conclusions "stand as published" and were supported by other results that had not been challenged, as well as by subsequent replication. He did admit to poor presentation of his data and failure to explain how he processed the data, though the methods were standard ones. He said he had agreed to settle with ORI because he could not afford a legal battle (Liburdy 1999).[22] Liburdy resigned his position at LBL and agreed not to receive any federal funds for a period of three years.

The ORI finding stirred up what *Microwave News* called a "media storm" (*Microwave News* 1999a). There were accounts in *Science,* in the *San Francisco Chronicle,* and then on page one of the *New York Times*, with the headline "Data Tying Cancer to Electric Power Lines Found to be False." The *Times*'s story was picked up in many other papers, and the *Wall Street Journal* and the *Washington Post* ran an Associated Press story on Liburdy. *Science* claimed that "Liburdy's findings were among the first to offer a plausible mechanism for a possible link between EMF exposure and cancer or other diseases" (Vergano 1999). The *New York Times* carried the reasoning to the next step in its lead paragraph: "A Federal investigation has concluded that a scientist at the Lawrence Berkeley Laboratory in Berkeley, Calif., faked what had been considered crucial evidence of a tie between electric power lines and cancer. The disclosure appears to strengthen the case that electric power is safe" (Broad 1999). The *Times* interviewed Park, who said, "Liburdy's deception was probably typical for the field." Several days later, the *Wall Street Journal* published a column by the president of the American Council on Science and Health, an industry-funded group that has defended Alar, DDT, and endocrine disruptors, among other environmental agents (American Council on Science and Health 2002), saying that the Liburdy "deception did much to keep the myth [about powerline dangers] alive"; she called the research a "taxpayer-subsidized fraud" and a "new low in the annals of junk science," and questioned the "objectivity of environmental health researchers"

(Whelan 1999). Researchers in bioelectromagnetics, however, noted that Liburdy's disputed findings were only indirectly related to the EMF–cancer debate and certainly did not play the central role that was portrayed in the press coverage. *Microwave News* criticized both the *New York Times* and the AP for failing to interview any biological scientist for their coverage of the Liburdy story and for failing to cover the nearly simultaneous NIEHS report that found evidence (albeit weak) for health risks (*Microwave News* 1999b).

Once the scapegoat is killed, the community is purified (partly through moral indignation) and reunified (Burke 1969a, 406ff), reaffirming the authority of the orthodox. In this case, Liburdy's ejection from the lab and denial of funding is a kind of scientific death that allows the critics to reject all heretical research findings as "probably typical for the field." The orthodox can claim victory, reaffirm their authority, and declare the matter closed; as Park told the *New York Times,* the Liburdy affair would "aid the growing consensus on safety" (Broad 1999). And it was precisely this aspect of the *Times*'s coverage to which the officers of the Bioelectromagnetics Society objected in a letter sent to the *Times* but never published: "The story [. . .] is unfairly misleading to the general public by presenting as finally resolved and settled a scientific issue that is still controversial" (Bioelectromagnetics Society 1999).

And this points up a final rhetorical difference between the opponents and proponents of bioeffects research—the use of a rhetoric of closure versus a rhetoric of open-endedness.[23] The rhetoric of closure is consistent with the *topos* of impossibility, and the rhetoric of open-endedness with the *topos* of possibility. Closure would cut off research funding, halt the endless series of review reports, and permit the setting of standards. Although the impulse toward closure was particularly strong after the Liburdy affair, it was a theme throughout the debate we have been examining, strengthening when there was an occasion for the exercise of orthodox authority. After the NRC report was released, for example, Park sounded this call multiple times. In the *New York Times* news coverage, he said, "Scientifically, it's essentially over now. I hope this puts the whole issue to rest" (Leary 1996). In his *New York Times* op-ed column several days later, he said, "After 17 years, the scientific debate over possible adverse health effects of electromagnetic fields from power lines has finally been put to rest by experts at the National Academy of Sciences" (Park 1996). And in *Voo-*

doo Science, he calls the final section of his EMF chapter, "Slamming the Door Shut," referring again to the NRC report (Park 2000, 158). The *Science* coverage of the NRC report suggested, however, that the debate wouldn't be so easily concluded, since three NRC panel members in a separate statement to the press indicated that "it's still an open question whether EMFs threaten health" (Kaiser 1996). A final pair of examples illustrates the extreme differences on closure. A *Technology Review* article quoted Adair saying that "There is probably nothing on earth, or in the universe, that we understand as well as electromagnetic fields and the interaction of electromagnetic fields with matter, including biological matter" (Palfreman 1996, 26). At about the same time the Bioelectromagnetics Society stated that "A great deal is yet to be discovered about the interaction of EMFs with biological systems" (Bioelectromagnetics Society Board of Directors 1996).

Both closure and openness are efforts to determine the future of the community: closure returns the community to a status quo *ante bellum*, a condition of orthodoxy defined by traditional authority; and openness looks toward a condition in the future, to an orthodoxy not yet achieved. As the Bioelectromagnetics Society Presidents said in their letter to Congress, "In this still emerging area of scientific research, controversy about reported results is a natural and healthy part of the scientific process. Such controversy should not be the basis for discarding programs of research before the important questions are answered conclusively" (Bioelectromagnetics Society Presidents 1996). What is at stake in the controversy over closure or openness is control of the orthodoxy, and of the community: Who are the authorities? What is the research agenda? What is its future?

Community, Interests, and Incommensurability

> The very existence of science depends upon vesting the power to choose between paradigms in the members of a special kind of community.
>
> —Thomas S. Kuhn, *The Structure of Scientific Revolutions*

A complication of the EMF debate is its interdisciplinary character, which calls into question the locus of the relevant community. Is it the case that bioelectromagnetics, although it is an interdisciplinary enterprise, functions as the special kind of community capable of choosing between paradigms? Or, do the conflicting intellectual traditions noted in our earlier discussion of topical preferences preclude there being a single operative community whose members can do the choosing? Is intellectual conflict inevitable in interdisciplinary work? Is mutual incomprehension? Incommensurability? While I can't answer these questions about science in general (and there may well be no general answer), I believe the interdisciplinary nature of EMF bioeffects research raises issues highly relevant to a critique of incommensurability.

It is noteworthy that in the most vociferous boundary-work recounted in the previous section, the boundary being policed is that of science-in-general, not that of a particular discipline or subdiscipline, whether physics, biophysics, molecular biology, epidemiology, or bioelectromagnetics. A debate between equal disciplines or subdisciplines cannot be decided by either one of them but must be referred to some superordinate forum. Thus, the translation of the debate to a broader policy or public forum is a natural and perhaps necessary move when interdisciplinary debate does not get resolved within the scientific forum. In the examples above, appeals are addressed to a public outside the territory being contested—those who read *Science, Scientific American, Technology Review, Microwave News, Physics Today,* the *New York Times,* the *Wall Street Journal,* and the like—scientists and engineers in any discipline, policy-makers, research administrators, journalists, and industry analysts. This pattern of expanding the scene of argument has also been noted by Charles Alan Taylor and by David Mercer in similar cases. In describing the cold fusion controversy as initially a conflict between physicists and chemists, Taylor notes the construction of an "internal demarcation line between the competing research communities of fusion physics and electrochemistry" (Taylor 1996, 211). Ultimately, however, the debate was played out in public forums, notably in Congressional testimony, and decided, to some extent, by federal funding decisions (Taylor 1996, 211–221). Mercer points out that the Australian debate about EMFs that he followed in the public forum of a governmental inquiry relied on competing images of scientific method. Such "method discourses," which address the questions of "how to, and who could, do legitimate EMF science" (Mercer 2002 208), are based on normative definitions of science at large.

Transferring the debate to a broader forum brings several rhetorical advantages to the defenders of electromagnetic orthodoxy. Challenging a heretic within the small subdiscipline of bioelectromagnetics, whose very existence is being contested, offers little rhetorical payoff, but if the scene can be expanded and the heresy shown to be an affront to science itself, the errant subdiscipline can then, *a fortiori,* be made to conform to the orthodoxy. Further, since physics trumps biology, the transfer of the EMF debate undoes the assumption that the debate is between disciplinary equals, and physics gains the right to dictate the decision, at the same time identifying itself with the larger forum of judgment. The orthodox represent science itself in the public forum and to the public forum, speaking both as plaintiff and as judge. As Richard Whately saw it, the orthodox always begin with the presumption in their favor (Whately 1963, 112–115), and in this case they gain another layer of presumption from the socio-historical status of the discipline of physics.

The scenes played out in the public forum direct our attention to yet another dimension of the EMF debate: that it involves not only differing disciplinary commitments and the asymmetrical positionings of heresy and orthodoxy but also multiple *interests*—social, economic, health, and policy interests that properly exert their influences in the public realm. The effects and uses of EMFs are of interest to industry and the military, as well as to citizens and consumers, and have extensive economic and national security implications. In the public forum, the debate cannot be purely scientific. A glance back at Figure 1 reminds us that EMFs in the nonionizing ranges have applications in the electric power infrastructure, broadcast media, and telecommunications. Electric power in the U.S. uses 60 Hz, in the ELF band at the extreme left of Figure 1. Long-range communication, such as submarine and other marine communication, uses the lower frequencies, in the ELF and VLF bands (300Hz to 3 MHz). Other communication media, including radio, television, satellite communications, radar, cell phones, and other forms of broadcast and point-to-point communication use a range of frequencies called RF, from 0.5 MHz to 100 GHz.

Public attention to EMFs has been affected by a series of events since the first uses of radar in World War II. The earliest research and standard-setting efforts were conducted by the military to protect servicemen working with radar, which uses pulsed microwaves. Little of this work attracted public attention until the Moscow Embassy crisis in the 1960s when it became known that the U.S. Embassy in the Soviet Union was receiving targeted RF radiation (Steneck 1984). A second wave of public concern followed the Wertheimer study on powerline effects in 1979, and another followed in the 1980s, after a series of epidemiological studies of occupational hazards to electrical and electronic workers (Slesin 1987). In the 1990s, a series of lawsuits seeking damages from cell phone manufacturers for brain tumors brought another wave of media attention (Parascandola 2001).

Two military projects also became public issues.[24] A Navy project, called successively Project Sanguine, Project Seafarer, and Project ELF, was designed to communicate with submerged submarines by means of extremely long-wavelength, low-frequency signals (76 Hz). The massive antenna system needed was to be buried in bedrock in northern Michigan and Wisconsin over 22,000 square miles. The Navy's 1973 Environmental Impact Statement prompted resistance from the public and criticism from many scientists, causing delays, relocations, downsizing, and name changes, but the system was built and became fully operational in 1985 (FAS Weapons of Mass Destruction 1998). In 1976 the Air Force issued an environmental assessment for a project called PAVE PAWS,[25] a huge radar array to be installed on Cape Cod to detect incoming ballistic missiles. A similar installation was to be built in California. The environmental assessment concluded that there were no risks to human health or the natural environment from the 420–450 MHz pulsed fields that would be used. Despite intense public resistance to both projects, they were built and became operational in 1980, and there is now a third site, in Alaska (FAS Space Policy Project 2000). Controversy about PAVE PAWS continues, and the National Research Council initiated a study funded by the Air Force to evaluate recent claims about the effects of pulsed fields on biological systems (*Microwave News* 2002a, 10). Electronic warfare in the future will utilize EMFs in communication and detection systems like these two systems from the 1970s, as well as in offensive systems to disable enemy infrastructure and possibly to cause human injury.

The complexity of debate in the public forum is demonstrated in a discussion of the origin of the 1966 exposure standard for microwave radiation by historian Nicholas Steneck and his co-authors. They enumerate the multiple interest groups involved in the standard-setting process: military operations, military research, defense contractors, medical device manufacturers, clinical researchers, biological researchers, engineering-physics researchers, and the civilian-consumer public (Steneck *et al.* 1980, 1234). Because the question of microwave exposure standards first came up in the context of World War II, the discussion was initiated, and continued, primarily in the context of national security: "Those who set the standard in 1966 still viewed microwaves as radar and radar as a military and industrial problem" (Steneck *et al.* 1980, 1234). Steneck and his co-authors note that "in the push to set the standard, there can be no doubt that possible evidence against its safety was ignored and that research that might have clarified certain details was not undertaken"(Steneck *et al.* 1980, 1235). In a booklength study that covers the development of the 1966 standard in more detail, as well as its revision in 1982, Steneck concludes that public health interests have not been adequately represented in the "microwave debate" and that military influence on bioeffects research must be eliminated (Steneck 1984, 240). However, in the time since then, the public interest has been even less well represented, as most federal funding for bioeffects research was eliminated by the Reagan administration in the mid-1980s. Research in the past fifteen years has been funded primarily by those interested in expanding the use and applications of EMFs: the electric power industry, the communications and defense industries, and the military.

By and large, active researchers have left the job of pointing out the potential for conflicts of interest to gadflies in the public realm, particularly Paul Brodeur, the author of three mass-market books, and Louis Slesin, editor and publisher of *Microwave News,* an independent bi-monthly review of research and policy. Brodeur's main thesis has been that powerful interests have covered up potential hazardous effects of EMFs, as the titles and subtitles of his three books indicate: *The Zapping of America: Microwaves, Their Deadly Risk, and the Cover-Up* (Brodeur 1977), *Currents of Death: Power Lines, Computer Terminals, and the Attempt to Cover Up Their Threat to Your Health* (Brodeur 1989), and *the Great Power-Line Cover-Up: How the Utilities and the Government Are Trying to Hide the Cancer Hazard Posed by Electromagnetic Fields* (Brodeur 1993). His work has been sternly reviewed by scientists, including those who are sympathetic to the nonthermal effects hypothesis; one review, for example, says that he "gets far out ahead of the evidence" and "deliberately oversimplifies and misrepresents the complexity of the scientific process and the evidence it has produced" (Morgan 1990, 118).

Slesin, in addition to exerting editorial control in his newsletter, has also been an active public commentator in the science and technology press with letters to the editor and columns in *Scientific American, Technology Review,* and elsewhere. In 1986, he identified a basic feature of "the microwave problem" as the "domination of military funding for biomedical research on nonionizing radiation and the reliance on engineers rather than biologists to do the research" (Slesin 1986). In 1990, he charged that "with only a very few exceptions, all the [EMF] research in the United States is paid for by commercial interests and the Departments of Defense and Energy" (Slesin 1990, 17). In 1994 he criticized *Technology Review* for a column on cell phones by pointing out that it had been written by a consultant to the industry: "It's one thing to let the fox guard the henhouse, but it's quite another to let him lecture us on the best way to stand guard" (Slesin 1994).

In his editorials in *Microwave News,* Slesin has provided a running commentary on the role of interests in EMF research. Regarding press coverage of the Liburdy affair in 1999, he charged that "at times the influence of corporate power in both science and the media is so overwhelming that it starts to resemble [a conspiracy]" (*Microwave News* 1999b). Several months later, he charged that "industry is firmly in control of decisions on wireless phones and public health" (*Microwave News* 1999c). In 2001, commenting on the recent announcement of a new Air Force microwave weapon for crowd control, he noted that the exposure standards for RF and microwaves had been loosened in the 1980s in a way that permitted the development of this weapon. Pointing out that a majority of the IEEE committee in charge of the standard worked either for the Air Force or for Raytheon, the contractor that developed the new weapon, he said, "It seems obvious, but it's worth repeating: Health standards should be written by medical and public health professionals, not those who make weapons for the military-industrial complex" (*Microwave News* 2001). In a final example, he claimed that "Motorola is the single most important force in bioelectromagnetics today. It is the largest sponsor of health research, [. . .] and it controls key positions on standard-setting committees and professional society such as BEMS" (*Microwave News* 2002b).

In the public forum, the debate about EMF bioeffects seems to turn on the Ciceronian forensic question, *cui bono*? or "for whose benefit?" We saw in the previous section that the orthodox raised questions about competence and motives, in an attempt to control the intellectual content of EMF science by determining the boundary between science and non-science. Arguers in the public forum, especially those without standing in the scientific forum, focus on interests in order to challenge credibility (or *ethos*), since they usually do not have the qualifications to challenge technical competence. Although there can be little opportunity for these defenders of the public interest to engage in forum control or boundary work, and the public forum lacks the structure and procedures of scientific forums, commercial and military interests may exert control through exclusion and secrecy, versions of what Sullivan called *denial of forum*, by virtue of their access to sources of power.

The EMF debate is argumentatively complex, and there is much more to it than the selections I have been able to present here. However, we've seen enough to appreciate that the polarization of positions is long-standing, intransigent, and resistant to simple resolution. Within the relevant scientific community, the divergent understandings of the biological and epidemiological results fit Kuhn's characterization of a paradigm debate: they present different problems to be solved, different solutions or standards for obtaining solutions, and most importantly different visions of the scientific future. To attribute these conceptual differences to scientific incommensurability would imply that the specific nature of the ideas themselves is responsible for the "gulf" between the physical and biological scientists in this case and that the apparent mutual incomprehension that results makes argumentation difficult if not impossible. Incommensurability as an explanation leaves both the rhetorical actor and the rhetorical analyst with little to say.

But we've also seen that in the EMF debate the conceptual differences are exacerbated by disciplinary politics and by the operation of interests that originate in the public sphere but have ways of influencing the conduct of the debate within the disciplinary community. In addition, the interdisciplinary nature of the field has encouraged the transfer of the debate into the public sphere where those socio-economic-political interests can exert even greater influence. Thus it seems too simple to chalk up the EMF debate to incommensurability. My argument, then, is that incommensurability is an *impression*, which can be created not only by the differing intellectual commitments and habits that constitute a disciplinary matrix but also by argumentative positionings—by accusation or defense, presumptions of authority, expected alliances—and which can be magnified by socio-political interests. Argumentation is complexly socio-cognitive, and the continuing EMF controversy demonstrates multiple dynamics that the incommensurability model tends to obscure. A rhetorical description of the debate accounts for more of its multiple features, its twists and turns, its partisan strategies, its contested forums. A rhetorical description is also just more interesting. In the study of scientific change and controversy, incommensurability is an idea we can probably just do without.

Glossary of Abbreviations Used

APS	American Physical Society
CIRRPC	Committee on Interagency Radiation Research and Policy Coordination, a unit of the White House Office of Science and Technology Policy
ELF	Extremely low frequency band; fields with frequencies below 300 Hz; includes powerline frequency of 60 Hz and some submarine communications
EMF	Electromagnetic fields; a form of energy produced by the motion of a charged particle or object, consisting of separate, but related, electric fields and magnetic fields (for the electromagnetic spectrum, see Figure 1). Engineers have divided the spectrum into somewhat arbitrary bands, such ELF and VLF.
EPA	Environmental Protection Agency; an agency of the U.S. government
Hz	Hertz, a measure of frequency, or cycles per second
LBL	Lawrence Berkeley National Laboratory
MW	Microwave; fields with frequencies between 10^{10} and 10^{11} Hz
NIEHS	National Institute for Environmental Health Sciences, a division of the National Institutes of Health
NRC	National Research Council
NCRP	National Council on Radiation Protection and Measurements; an independent organization chartered by Congress in 1964 to formulate and disseminate informa-

	tion and recommendations on radiation protection and measurements
ORI	Office of Research Integrity, a division of the National Institutes of Health
RF	Radio frequency; generally refers to a broad band of fields in the range from 10^5 to 10^{11} Hz (0.5 MHz to 100 GHz); includes frequencies assigned to uses such as radar, satellite communications, CB radio, radio, television, marine communications, and cell phones.
S/N	Signal-to-noise ratio
VLF	Very low frequency band; these fields have frequencies from 10^3 to 10^4 Hz (3–30 kHz); used primarily for navigation

Notes

[1] Media coverage has included three controversial books by Paul Brodeur and the *New Yorker* articles on which they are based (Brodeur 1977; Brodeur 1989; Brodeur 1993).

[2] See Harris's introduction and the essays in this volume by Hoyningen-Huene and Gross for more detailed discussions of the proposal and development of the incommensurability notion by both Kuhn and Feyerabend. Here I focus on Kuhn's use of it, partly because I'm more familiar with it and partly because his notion, at least in its early form, presents a more widely known model.

[3] The EMF controversy has been discussed before in the science studies literature; one study, similar in some ways to the present one, examined the argumentative strategies of proponents and opponents of high voltage powerlines in the Australian state of New South Wales (Mercer 2002), and the other examined the use of dosimetry to construct risk (Mitchell 1997). Mercer focuses on differential representations about the nature of science and scientific method in public submissions to the Gibbs Inquiry in 1990–91.

[4] See Bazerman and De los Santos's contribution to this volume for another look at incommensurability through a case study of interdisciplinarity. Their findings are almost inverse to mine, suggesting that interdisciplinary developments can pursue assimilation, as well as (in my case) division.

[5] The major precursor groups include the International Union of Radio Science (URSI), the International Microwave Power Institute (IMPI), and the IEEE Microwave Theory and Techniques Society.

[6] A different way of examining disciplinary differences has been elaborated by Knorr-Cetina (Knorr-Cetina 1999), who develops the notion of "epistemic cultures" as a way of emphasizing the disunity of science. Her two cases are high-energy physics and molecular biology.

[7] This is not to say that all physicists argue against bioeffects or that all biological scientists argue in their favor. My point, rather, is that there are certain types of arguments favored by physical scientists who do not accept the biological effects research, and certain types of arguments favored by biological scientists who believe the bioeffects require further research and explanation.

[8] Mercer notes the use of this argument in the Australian case (Mercer 2002, 218).

[9] Bennett's book was based on the work he did for the CIRRPC panel.

[10] The cold fusion debate between physicists and chemists in 1989 demonstrates that physics trumps chemistry, as well (Taylor 1996).

[11] In their discussion of how the "frontiers of science" are defended, Grabner and Reiter attribute the trumping asymmetry to reductionism and the presumed unity of the scientific method. They note, sarcastically, that "physicists are expert in everything; other scientists only in those fields of science that are not more basic than their own. But in any case, physicists are the best experts" (Grabner 1979, 92). See Leah Ceccarelli's contribution to this volume for an account of how E. O. Wilson's uncomprommising reductionist program in sociobiology has run afoul of social scientists.

[12] He does admit that at the current state of development "it is certainly not possible to derive the biological laws from the physical ones" (Carnap 1938, 60).

[13] As I noted earlier, my characterization is necessarily general. Not all physicists resist the hypothesis of nonthermal effects, and not all biological and medical scientists adhere to it. Several physical scientists have proposed models that would explain the mechanism by which nonthermal effects could occur. But they represent a minority of those physicists involved in the debate, and they have been less vocal and less visible.

[14] Conceivably, the disciplinary positions could be reversed in another debate, with physicists championing novelty and biologists defending a doctrine, although, given other dimensions of disciplinary politics I discuss a bit later on, this seems unlikely.

[15] Bromley had been a member of the APS committee that approved the statement about EMFs quoted above.

[16] Sullivan's examples of correction are personal, including direct reprimand and censure by disciplinary or professional authorities (Sullivan 2000). However, I prefer in the present case to use this category of forum control for

the more routine dialectical exchange of scientific and public debate, partly because it's an important part of the struggle for forum control and partly because it's an open question just who the disciplinary or professional authorities are in this case.

[17] See Edmond and Mercer's discussion of the origin and political uses of the "junk science" model (Edmond 1998).

[18] Like Bromley, Park was also on the committee that approved the APS statement about EMFs. Interestingly, he does not take up creationism in his book.

[19] These three are the most adversarial and the most audible within the public forum, though by no means the only physical scientists who engage in boundary policing. Others include K. R. Foster, J. E. Moulder, and W. F. Pickard.

[20] The first eighteen science advisers are listed in Halber (2001), and I verified their disciplinary backgrounds in *American Men and Women of Science*; the current science adviser (as of 2005), John Marburger, has a Stanford degree in applied physics (see http://www.ostp.gov).

[21]Similarly, Taylor notes in his discussion of the cold fusion controversy, which at one level played out between physicists and chemists, that physicists acted with a proprietary presumption to the field of fusion, hot or cold (Taylor 1996, 213). The elite status of physics suggests that some of the argumentative strategies used might be explained by Andrew King's analysis of the rhetoric of power maintenance, and in fact, two of the seven strategies he discusses are clearly present in the examples above: ridicule and creating boundaries by setting impossible standards (sound science, for example) (King 1976). A closer examination might reveal more of these strategies in this case.

[22] The Office of Research Integrity was stripped of its ability to conduct investigations just four months after it made its judgment in the Liburdy case, in part because several of its decisions had been overturned on appeal (Kaiser 1999).

[23] I am indebted to Dale Sullivan for noticing this particular point. Mercer also notes this strategy in passing (Mercer 2002).

[24] These accounts are taken from Brodeur (1989). I am not able to include here many details of research, technology, public reaction, and policy decisions that complicate both these stories.

[25] Precision Acquisition of Vehicle Entry Phased Array Warning System.

References

Adair, R. K. (1991a). Constraints on biological effects of weak extremely-low-frequency electromagnetic-fields. *Physical Review A*, 43 (2), 1039–1048.

—. (1991b). ELF effects—paradigm shift or fabric rip? *Physics Today,* 44 (12), 103.

—. (1992a, December 18). EMF research. *Science,* 258, 1868–1869.

—. (1992b). Magnetic effects. *Physics World,* 5 (9), 17.

—. (1998). A physical analysis of the Ion Parametric Resonance Model. *Bioelectromagnetics,* 19 (3), 181–191.

—. (1999). The fear of weak electromagnetic fields. *Scientific Review of Alternative Medicine,* 3 (1). Retrieved 27 February 2005 from http://www.sram.org/0301/electromagnetic-fields.html.

Adey, W. R. (1993). Biological effects of electromagnetic fields. *Journal of Cellular Biochemistry* 51 (4), 410–416.

Ahlbom, A., & M. Feychting. (1993). EMF and cancer: letters. *Science,* 260, 14, 16.

Alberts, B., D. Bray, J. Lewis, M. Raff, K. Roberts, & J. D. Watson. (1994). *Molecular biology of the cell.* 3rd ed. New York: Garland Publishing.

Alcock, J. (2001). *The triumph of sociobiology.* New York: Oxford University Press.

Alford, D. M. (2000). *Linguistics and nominalising languages.* Retrieved 21 March 2005 from http://www.enformy.com/ll02.htm

Allen, E., B. Beckwith, J. Beckwith, S. Chorover, D. Culver, M. Duncan, *et al.* (1975, November 13). Against sociobiology. *New York Review of Books* 22 (18), 43-44.

Allén, S., ed. (1989). *Possible worlds in humanities, arts, and science.* Berlin: Walter de Gruyter.

American Council on Science and Health. (2002). *American Council on Science and Health.* Retrieved 1 July 2002 from http://www.acsh.org/

American Physical Society. (1995). *Statement on power line fields and public health.* American Physical Society. Retrieved 20 February 2005 from http://www.aps.org/statements/95.2.cfm

Aristotle. (1991). *On rhetoric: a theory of civic discourse.* (G. Kennedy, Trans.). Oxford: Oxford University Press.

Armstrong, C. (1998, April 3). The vision of the pore. *Science,* 280, 56–57.

Asquith, P. D., & T. Nickles, eds. (1983). *PSA 1982. Proceedings of the 1982 Biennial Meeting of the Philosophy of Science Association.* Vol. 2. East Lansing: Philosophy of Science Association.

Astumian, D. (1997). Body works. *New Scientist,* 156 (2112), 38–41.

Atienzar, F. A., V. V. Cheung, A. N. Jha, & M. H. Depledge. (2001). Fitness parameters and DNA effects are sensitive indicators of copper-induced toxicity in dauphnia magna. *Toxicological Sciences,* 59, 241–250.

Atkinson, R., J. P. Currie, B. Davis, D. A. H. Pratt, H. M. Sharpe, & E. G. Tomich. (1966). Acute toxicity of Cephaloridine, an antibiotic derived from Cephalorisporin C. *Journal of Toxicology and Applied Pharmacology,* 8, 398–406.

Avis, J. M. (1992). Where are all the family therapists? Abuse and violence within families and family therapy's response. *Journal of Marital and Family Therapy,* 18, 225–232.

Bacon, F. (1968). *The works of Francis Bacon.* J. Spedding (Ed.). New York: Garrett Press.

Barash, D. P. (1982). *Sociobiology and behavior.* 2nd ed. Oxford, UK: Elsevier.

Barker, P. (2001). Kuhn, incommensurability, and cognitive science. *Perspectives on Science,* 9 (4), 433–462.

Barlow, N., ed. (1963). Darwin's ornithological notes. *Bulletin of the British Museum, Natural History,* Historical Series 2 (7), 200-278.

—, ed. (1987). *Diary of the voyage of H.M.S. Beagle.* Vol. 1. New York and London: New York University Press.

Barnes, B. (1982). T. S. Kuhn and social science. New York: Columbia University Press.

—, & D. MacKenzie. (1979). On the role of interests in scientific change. In R. Wallis (Ed.), *On the margins of science: the social construction of rejected knowledge.* Staffordshire, Keele: University of Keele.

Barthes, R. (1957). *Mythologies.* Paris: Editions du Seuil.

—. (1975). The pleasure of the text. (R. Miller, Trans.). New York: Hill and Wang.

Bartholomew, M. (1973). Lyell and evolution: an account of Lyell's response to the prospect of an evolutionary ancestry for man. *The British Journal for the History of Science,* 6, 261–303.

—. (1975). Huxley's defence of Darwin. *Annals of Science,* 32, 525–535.

Bawin, S. M., & W. R. Adey. (1976). Sensitivity of calcium binding in cerebral tissue to weak environmental electric fields oscillating at low frequency. *Proceedings of the National Academy of Sciences of the United States of America,* 73 (6), 1999–2003.

Bazerman, C. (1988). *Shaping written knowledge: the genre and activity of the experimental article in science.* Madison: University of Wisconsin Press.

—. (1993). Forums of validation and forms of knowledge: the magical rhetoric of Otto von Guericke's sulfur globe. *Configurations,* 1 (2), 201–227.

—. (1999). *The languages of Edison's light.* Cambridge, MA: The MIT Press.

—. (2000). A rhetoric for literate society: the tension between expanding practices and restricted theories. In M. Goggin (Ed.), *Inventing a discipline: rhetoric and composition in action.* Urbana, IL: NCTE.

—. (2001). Anxiety in action: Sullivan's interpersonal psychiatry as a supplement to Vygotskian psychology. *Mind, Culture and Activity,* 8 (2), 174–186.

—. (2004). Intertextualities: Volosinov, Bakhtin, literary theory, and literacy studies. In A. Ball and S. Freedman (Eds.), *Bakhtinian perspectives on language, literacy and learning.* Cambridge, UK: Cambridge University Press, 53-65.

—. (2005). Participating in emergent socio-literate worlds: genre, disciplinarity, interdisciplinarity. In R. Beach, J. Green, M. Michael, & T. Shanahan (Eds.), *Multidisciplinary perspectives on literacy research.* 2nd ed. Cresskill, NJ: Hampton Press.

—, J. Little, & T. Chavkin. (2003). The production of information for genred activity spaces: informational motives and consequences of the Environmental Impact Statement. *Written Communication,* 20 (4), 455-477.

—, & P. Prior. (2004). *What texts do and how they do it.* Mahwah, NJ: Lawrence Erlbaum Associates.

Beaumont, J. (1665.) *Some observations upon the apologie of Dr. Henry More for his mystery of godliness.* Cambridge: John Field.

Becker, R. O. (1990). *Cross currents: the promise of electromedicine, the perils of electropollution.* Los Angeles, CA: J. P. Tarcher.

—. (1991). Becker replies. *Physics Today,* 44 (12), 103–104.

Beller, M. (1993). Einstein and Bohr's rhetoric of complementarity. *Science in Context,* 6 (1), 241–255.

—. (1999). *Quantum dialogue: the making of a revolution.* Chicago: University of Chicago Press.

Bello, S. M., D. G. Frans, J. J. Stegman, & M. E. Hahn. (2001). Acquired resistance to Ah receptor agonists in a population of Atlantic Killifish (*Fundulus heteroclitus*) inhabiting a marine superfund site: in vivo and in vitro studies of inducibility of xenobiotic metabolizing enzymes. *Toxicological Sciences,* 60, 77–91.

Bennett, W. R., Jr. (1994a). *Health and low-frequency electromagnetic fields.* New Haven: Yale University Press.

—. (1994b). Cancer and power lines. *Physics Today* 47 (3), 23–29.

—. (1994c, August 10). Power lines are homely, not hazardous. *Wall Street Journal,* A8.

Berge, H. (1973). Plants as indicators of air pollution. *Toxicology,* 1, 79–89.

Berk, R. A., S. F. Berk, D. R. Loseke, & D. Rauma. (1983). Mutual combat and other family violence myths. In D. Finkelhor, R. Gelles, G. T. Hotaling & M. A. Straus (Eds.), *The dark side of families: current family violence research.* Newbury Park, CA: Sage.

Berlin, I. (1969). *Four essays on liberty.* New York: Oxford University Press.

—. (1979). *Against the current.* London: Hogarth Press.

—. (1998, April 3). My intellectual path. *New York Review of Books,* 53–60.

Berman, P. L., ed. (1985). *The courage of conviction.* New York: Dodd, Mead & Company.

Bernstein, R. (1983). *Beyond objectivism and relativism.* Philadelphia: University of Pennsylvania Press.

Best, J. (1990). *Threatened children: rhetoric and concern about child-victims.* Chicago: University of Chicago Press.

Biagioli, M. (1993). *Galileo, courtier: the practice of science in the culture of absolutism.* Chicago: University of Chicago Press.

Billig, M. (1987). *Arguing and thinking: a rhetorical approach to social psychology.* Cambridge, UK: Cambridge University Press.

Bioelectromagnetics Society Board of Directors. (1996). Statement on EMF issues. *Microwave News,* 16 (4), 13.

Bioelectromagnetics Society Presidents. (1996). *BEMS Presidents' letter to the Congress* Retrieved March 2005 from http://www.bioelectromagnetics.org/newsletter/news131.html#BM2

Bioelectromagnetics Society Public Policy Committee. (1999). *Response to New York Times article by William Broad.* Retrieved 5 March 2005 from http://www.bioelectromagnetics.org/newsletter/news149.html#response

Bird, Alexander. (2000). *Thomas Kuhn.* Princeton: Princeton University Press.

Black, Edwin. (1976). The sentimental style as escapism, or the devil with Daniel Webster. In K. K. Campbell & K. H. Jamieson (Eds.), *Form and genre.* Falls Church, VA: Speech Communication Association.

Bloom, H. (1975). *The anxiety of influence: a theory of poetry.* New York: Oxford University Press.

Bonvillain, N. (1997). *Language, culture, and communication.* 2nd ed. Upper Saddle River, NJ: Prentice Hall.

Booth, W. C. (1974). *Modern dogma and the rhetoric of assent.* Chicago: University of Chicago Press.

—. (1979). *Critical understanding: the power and limits of pluralism.* Chicago: University of Chicago Press.

Borradori, G. (1994). *The American philosopher: conversations with Quine, Davidson, Putnam, Nozick, Danto, Rorty, Cavell, Macintyre, and Kuhn.* (R. Crocitto, Trans.). Chicago: University of Chicago Press.

Bostock, D. (1988). *Plato's Theaetetus.* Oxford: Clarendon Press.

Bowler, P. J. (1988). *The non-Darwinian revolution.* Baltimore: The Johns Hopkins University Press.

—. (1989). *Evolution: the history of an idea.* Rev. ed. Berkeley: University of California Press.

Brainard, G. C., R. Kavet, & I. Kheifets. (1999). The relationship between electromagnetic field and light exposures to melatonin and breast cancer risk: a review of the relevant literature. *Journal of Pineal Research,* 26 (2), 65–100.

Brienes, W, & L. Gordon. (1983). The new scholarship on family violence. *Signs: Journal of Women in Culture and Society* (8), 490–531.

Brinkhuis, R. P. (2001). Toxicology information from US government agencies. *Toxicology,* 157 (1–2), 25–49.

Broad, W. J. (1999, July 24). Data tying cancer to electric power found to be false. *New York Times,* A-1.

Brodeur, Paul. (1977). *The zapping of America: microwaves, their deadly risk, and the cover-up.* New York: W. W. Norton & Company.

—. (1989). *Currents of death: power lines, computer terminals, and the attempt to cover up their threat to your health.* New York: Simon and Schuster.

—. (1993). *The great power-line cover-up: how the utilities and the government are trying to hide the cancer hazard posed by electromagnetic fields.* Boston: Little, Brown and Co.

Brown, H. I. (2005). Incommensurability reconsidered. *Studies in History and Philosophy of Science Part A,* 36 (1), 149-169.

Browne, Janet. (1995). *Charles Darwin, voyaging: a biography.* Princeton: Princeton University Press.

Bryant, D. C. (1953). Rhetoric: its functions and its scope. *Quarterly Journal of Speech* 39, 401–24.

Buchwald, J. Z. (1989). *The rise of the wave theory: optical theory and experiment in the early nineteenth century.* Chicago: University of Chicago Press.

—. (1992). Kinds and the wave theory of light. *Studies in the History and Philosophy of Science* 23, 39–74.

Buckley, M. J. (1970). Philosophic method in Cicero. *Journal of the History of Philosophy* 8, 143–154.

Burian, R. M. (1984). Scientific realism and incommensurability: some criticisms of Kuhn and Feyerabend. In R. S. Cohen & M. W. Wartofsky (Eds.), *Methodology, metaphysics and the history of science: in memory of Benjamin Nelson.* Dordrecht: Reidel, 1-31.

Burke, K. (1966). *Language as symbolic action: essays on life, literature, and method.* Berkeley: University of California Press.

—. (1969a). *A grammar of motives.* Berkeley: University of California Press.

—. (1969b). *A rhetoric of motives.* Berkeley: University of California Press.

—. (1984a). *Attitudes toward history.* 3rd edition. Berkeley: University of California Press.

—. (1984b). *Permanence and change: an anatomy of purpose.* 3rd edition. Berkeley: University of California Press.

Burlingame, A. L. (1977). Assessment of the trace organic molecular composition of industrial and municipal wastewater effluents by capillary gas chromatography/real-time high-resolution mass spectrometry: a preliminary report. *Ecotoxicology and Environmental Health*, 1, 111–140.

Butterfield, H. (1957). *The origins of modern science.* Rev. ed. New York: Macmillan.

Butts, R. E. (1973). Whewell's logic of induction. In R. N. Giere & R. S. Westfall (Eds.), *Foundations of scientific method: the nineteenth century.* Bloomington: Indiana University Press, 53-85.

Cairns J., Jr., & B. R. Niederlehner. (1995). *Ecological toxicity testing: scale, complexity, and relevance.* Boca Raton: CRC Lewis Publishers.

Campbell, G. (1873). *The philosophy of rhetoric.* New York: Harper & Brothers.

Campbell, J. A. (1975). The polemical Mr. Darwin. *Quarterly Journal of Speech*, 61, 375–390.

—. (1979). The rule of metaphor. *Quarterly Journal of Speech*, 65, 335–338.

—. (1984). Of orchids, insects and natural theology: timing, tactics and cultural critique in Darwin's post-*Origin* strategy. *Argumentation*, 8, 63–80.

—. (1986). Scientific revolution and the grammar of culture: the case of Darwin's *Origin*. *The Quarterly Journal of Speech*, 72, 351–376.

—. (1987). Charles Darwin: rhetorician of science. In J. S. Nelson, A. Megill & D. N. McCloskey (Eds.), *The rhetoric of the human sciences: language and argument in scholarship and public affairs.* Madison: University of Wisconsin Press.

—. (1989). The invisible rhetorician: Charles Darwin's third-party strategy. *Rhetorica*, 7, 55–85.

—. (1990a). *On the way to the* Origin*: Darwin's evolutionary insight and its rhetorical transformation.* Evanston: Northwestern University School of Speech.

—. (1990b). Scientific discovery and rhetorical invention: the path to Darwin's *Origin*. In H. W. Simons (Ed.), *The rhetorical turn: invention and persuasion in the conduct of inquiry.* Chicago: University of Chicago Press.

—. (1998). Rhetorical theory in the twenty-first century: a neo-classical perspective. *Southern Journal of Communication*, 63, 291–308.

—. (2002). Rhetoriography: an essay in invention. Conference paper. New Orleans: National Communication Association.

—. (2003). Intelligent design, Darwinism, and the philosophy of public education. In J. A. Campbell & S. C. Meyer (Eds.), *Darwinism, design, and public education.* Ann Arbor: Michigan State University Press.

—, & S. C. Meyer, eds. (2003). *Darwinism, design, and public education.* East Lansing, MI: Michigan Statue University Press.

Carnap, R. (1938). *Logical foundations of the unity of science.* Encyclopedia of Unified Science. Chicago: University of Chicago Press.

—. (1956). *The methodological character of theoretical concepts.* Minnesota Studies in the Philosophy of Science. Minneapolis: University of Minnesota Press.

Carrier, M. (2001). Changing laws and shifting concepts: On the nature and impact of incommensurability. In P. Hoyningen-Huene & H. Sankey (Eds.), *Incommensurability and related matters.* Dordrecht: Kluwer, 65-90.

Carson, R. (1962). *Silent spring.* Boston: Houghton Mifflin.

—. (2002). *Silent spring, 40th anniversary edition.* Boston: Houghton-Mifflin.

Carstensen, E. L. (1995). Magnetic-fields and cancer. *IEEE Engineering in Medicine and Biology Magazine,* 14 (4), 362–369.

—. (1997). Biological effects of power frequency electric fields. *Journal of Electrostatics,* 39 (3), 157–174.

—, A. Buettner, V. L. Genberg, & M. W. Miller. (1985). Sensitivity of the human-eye to power frequency electric-fields. *IEEE Transactions on Biomedical Engineering* 32 (8), 561–565.

Casarett, L. J., & J. D. Doull, eds. (1975). *Toxicology: the basic science of poisons.* New York: Macmillan.

Ceccarelli, L. (2001a). *Shaping science with rhetoric: the cases of Dobzhansky, Schrodinger, and Wilson.* Chicago: University of Chicago Press.

—. (2001b). Uniting biology and the social sciences: a rhetorical comparison of E. O. Wilson's *Consilience* and Theodosius Dobzhansky's *Mankind evolving. Poroi* 1 (1) Retrieved 2 July 2005 from http://inpress.lib.uiowa.edu/poroi/papers/ceccarelli010101.html.

Cedarbaum, D. G. (1983). Paradigms. *Studies in History and Philosophy of Science,* 14, 173–213.

Chalmers, A. F. (1999). *What is this thing called science?* 3rd ed. St. Lucia: University of Queensland Press.

Chambers, R. (1994). *Vestiges of the natural history of creation and other evolutionary writings.* Chicago: University of Chicago Press.

Chang, R., ed. (1997). *Incommensurability, incomparability, and practical reason.* Cambridge, MA: Harvard University Press.

Charatan, F. (1999). Research claiming link between electromagnetic fields and cancer deemed fraudulent. *British Medical Journal* 319 (7206), 337.

Charland, M. (2001). Constitutive rhetoric. In T. Sloane (Ed.), *Encyclopedia of rhetoric.* New York: Oxford University Press, 616-619.

—. (2003). The incommensurability thesis and the status of knowledge. *Philosophy and Rhetoric,* 36, 248–263.

Chasin, R., M. Herzig, S. Roth, L. Chasin, C. Becker, & C. R. Stains. (1996). From diatribe to dialogue. *Mediation Quarterly,* 13, 323-44.

Chen, X. (1997). Thomas Kuhn's latest notion of incommensurability. *Journal for General Philosophy of Science,* 28, 257–273.

Child, S. Z., C. L. Hartman, L. Schery, & E. L. Carstensen. (1992). A test for increased lethality in land snails exposed to 60 Hz magnetic-fields. *Neuroscience Letters,* 134 (2), 169–170.

Chomsky, N. (1965). *Aspects of the theory of syntax.* Cambridge, MA: The MIT Press.

Cicero. (1942). *De oratore.* (E. W. Sutton & H. Rackham, Trans.). Cambridge, MA: Harvard University Press.

—. (1949a). *De inventione.* (H. M. Hubbell, Trans.). Cambridge, MA: Harvard University Press.

—. (1949b). *Topica.* (H. M. Hubbell, Trans.). Cambridge, MA: Harvard University Press.

—. (1969). *Selected political speeches.* (M. Grant, Trans.). New York: Penguin Books.

Cline, B. L. (1965). *The men who made a new physics.* Chicago: University of Chicago Press.

Cockerham, L., & B. S. Shane, eds. (1994). *Basic environmental toxicology.* Boca Raton: CRC Lewis Publishers.

Cohen, R. S., & M. W. Wartofsky, eds. (1965). *In honor of Philipp Frank.* Vol. 2, Boston Studies the Philosophy of Science. New York: Humanities Press.

—, eds. (1984). *Methodology, metaphysics and the history of science: in memory of Benjamin Nelson.* Dordrecht: Reidel.

Coke, R. (1660). *Justice vindicated from the false fucus put upon it by Thomas White, gent., Mr. Thomas Hobbs, and Hugo Grotius. As also Elements of power & subjection, wherein is demonstrated the cause of all humane, Christian, and legal society. And as previous introduction to these is shewed the method by which men must necessarily attain arts & sciences.* London: Tho. Newcomb for G. Bedell and T. Collins.

Cole, T. (1991). *The origins of rhetoric in ancient Greece.* Baltimore: The Johns Hopkins University Press.

—. (1992). Who was Corax? *Illinois Classical Studies* 16, 65–84.

Collier, J. (1984). Pragmatic incommensurability. In P. D. Asquith & P. Kitcher (Eds.), *PSA 1984. Proceedings of the 1984 Biennial Meeting of the Philosophy of Science Association.* East Lansing: Philosophy of Science Association, 146-153.

Collins, R. (1998). *The sociology of philosophies: a global theory of intellectual change.* Cambridge, MA: The Belknap Press of Harvard University Press.

Colodny, R. G., ed. (1965). *Beyond the edge of certainty.* Englewood Cliffs: Prentice-Hall.

Committee 89–3, See National Council on Radiation Protection and Measurements.

Committee on Interagency Radiation Research and Policy Coordination. (1992). Health effects of low-frequency electric and magnetic fields. Washington, DC: Oak Ridge Associated Universities.

Comte, A. (1893). *The positive philosophy.* (H. Martineau, Trans.). 3rd ed. London: Kegan Paul.

—. (1953). *Republic of the West order and progress: a general view of positivism.* (J. H. Bridges, Trans.). Stanford: Academic Reprints. Original edition, 1848.

—. (1973). *The catechism of positive religion.* (R. Congreve, Trans.). 3rd ed. London: Kegan Paul, Trench, Trubner & Co.

Condit, C. M. (1999). *The meanings of the gene: public debates about human heredity.* Madison: University of Wisconsin Press.

—. (1996). How bad science stays that way: brain, sex, demarcation, and the status of truth in the rhetoric of science. *Rhetoric Society Quarterly* 26 (4), 83–110.

Conley, T. M. (1985). The virtues of controversy: in memoriam, Richard P. McKeon. *Quarterly Journal of Speech,* 71, 470–475.

—. (1990). *Rhetoric in the European tradition.* New York and London: Longman.

Connell, D., P. Lam, B. Richardson, & R. Wu. (1999). *Introduction to exotoxicology.* Oxford: Blackwell Science.

Cook, H. J., N. H. Steneck, A. J. Vander, & G. L. Kane. (1980). Early research on the biological effects of microwave-radiation—1940–1960. *Annals of Science,* 37 (3), 323–351.

Cook, T., & R. Seamon. (1980). Ein Feyerabenteur:Who is Feyerabend and where can he go from here: rhetoric and skepticism in Feyerabend's philosophy of science *PRE/TEXT* 1, 124-60.

Cope, F. W. (1965). Nuclear magnetic resonance evidence for complexing of sodium ions in muscle. *Proceedings of the National Academy of Sciences of the United States of America,* 54 (1), 225–227.

Cox, C. F., L. J. Brewer, C. H. Raeman, C. A. Schryver, S. Z. Child, & E. L. Carstensen. (1993). A test for teratological effects of power frequency magnetic- fields on chick-embryos. *IEEE Transactions on Biomedical Engineering* 40 (7), 605–610.

Crew, Richard. (2000). *Pythagoras, Hippasos, and the square root of two.* Retrieved 21 March 2005 from http://www.math.ufl.edu/~crew/texts/pythagoras.html

Crismore, A. & R. Farnsworth. (1989). Scientific rhetoric, metadiscourse, and power: Darwin's *Origin.* In C. W. Kneupper (Ed.), *Rhetoric and ide-*

ology: compositions and criticism of power. Arlington: Rhetoric Society of America, 174-188.

Crusius, T. W. (1999). *Kenneth Burke and the conversation after philosophy.* Carbondale: Southern Illinois University Press.

Culler, J. (1982). *On deconstruction: theory and criticism after structuralism.* Ithaca: Cornell University Press.

Culliton, B. J. (1972, March 24). Cell membranes: a new look at how they work. *Science,* 175, 1348–1350.

Cushman, D. P., & P. K. Tompkins. (1980). A theory of rhetoric for contemporary society. *Philosophy and Rhetoric,* 13, 43–67.

Damadian, R. (1971, March 19). Tumor detection by nuclear magnetic resonance. *Science,* 171, 1151–1153.

—, G. N. Ling, C. F. Hazlewood, F. W. Cope, & J. W. Woodbury. (1976, August). Structured water or pumps? *Science,* 193, 528, 530, 532, 608.

Darwin, C. (1952). *Journal of researches into the geology and natural history of the various countries visited by H.M.S. Beagle.* New York and London: Hafner.

—. (1958). *The autobiography of Charles Darwin and selected letters.* New York: Dover. Original edition, 1892.

—. (1959). *The origin of species by Charles Darwin: a variorum text.* (M. Peckham, ed.) Philadelphia: University of Pennsylvania Press.

—. (1964). *On the origin of species: a facsimile of the first edition.* Cambridge, MA: Harvard University Press.

—. (1977). *The collected papers of Charles Darwin.* (P. H. Barrett, ed.). 2 vols. Chicago: University of Chicago Press.

—. (1986). *The correspondence of Charles Darwin.* (F. Burkhardt *et al.* eds.). 13 vols. Cambridge, MA: Cambridge University Press.

—. (1987). *Charles Darwin's notebooks, 1836–1844: geology, transmutation of species, metaphysical enquiries.* (P. H. Barrett, P. J. Gautrey, S. Herbert, D. Kohn & S. Smith, eds.). Ithaca: Cornell University Press.

—., and A. Wallace. (1958). *Evolution by natural selection.* [Including Darwin's 1842 *Sketch*, Darwin's 1844 *Essay*, and Darwin and Wallace's "On the tendency of species to form varieties; and on the perpetuation of varieties and species by natural means of selection."] (G. de Beer, ed.). Cambridge, UK: Cambridge University Press.

Darwin, F., ed. (1887). *The life and letters of Charles Darwin.* 2 vols. London: Murray.

—, ed. (1904). *The life and letters of Charles Darwin.* 3 vols. London: Appleton.

Davidson, D. (1984). *Inquiries into truth and interpretation.* Oxford: Clarendon Press.

Davis, D. L., D. Axelrod, L. Bailey, M. Gaynor, & A. J. Sasco. (1998). Rethinking breast cancer risk and the environment: the case for the precautionary principle. *Environmental Health Perspectives*, 106 (9), 523–529.

Dear, P., ed. (1991). *The literary structure of scientific argument.* Philadelphia: University of Pennsylvania Press.

de Beer, G. (1958). Foreword. In C. Darwin and A. R. Wallace, *Evolution by natural selection.* Cambridge, UK: Cambridge University Press.

—. (1963). *Charles Darwin: a scientific biography.* Garden City: Doubleday Anchor Books.

—, ed. (1983). *Charles Darwin and Thomas Henry Huxley: autobiographies.* New York: Oxford University Press.

Deichmann, W. B., & W. E. MacDonald. (1977). Organicochlorine pesticides and liver cancer deaths in the United States, 1930–1972. *Ecotoxicology and Environmental Health,* 1, 89–110.

de Man, P. (1978). The epistemology of metaphor. *Critical Inquiry*, 5, 13-30.

—. (1979). *Allegories of reading.* New Haven: Yale University Press.

Depew, D. J., & B. H. Weber. (1997). *Darwinism evolving: systems dynamics and the genealogy of natural selection.* Cambridge, MA: The MIT Press.

Derrida, J. (1976). *Of grammatology.* Baltimore: The Johns Hopkins University Press.

—. (1988). Letter to a Japanese friend. In D. Wood & R. Bernasconi (Eds.), *Derrida and difference.* Evanston, IL: Northwestern University Press, 1-5.

Desmond, A. (1994). *Huxley: from devil's disciple to evolution's high priest.* Reading, MA: Addison-Wesley.

—, & R.Moore. (1991). *Darwin.* London: Michael Joseph.

Dewey, J. (1934). Philosophy. In E. R. A. Seligman (Ed.), *Encyclopedia of the social sciences.* New York: Macmillan, 118-129. Reprinted in Dewey, J. *The later works, 1925-1953.* (J. Boydston, ed.). Carbondale, IL: Southern Illinois University Press, 1981-1990, 8: 19-39.

Dieter, O. A. L. (1950). Stasis. *Speech Monographs,* 17, 345–369.

Dieterich, R. (1995). The body electric. *The Sciences,* 35 (3), 9–11.

—. (1995). Electric scaremongering. *The Sciences,* 35 (6), 7.

DiGregorio, M. A. (1984). *T.H. Huxley's place in natural science.* New Haven and London: Yale University Press.

—, Mario A., & N.W. Gill, eds. (1990). *Charles Darwin's marginalia.* 2 vols. New York and London: Garland Publishing.

Diogenes Laërtius. (1853). *The lives and opinions of eminent philosophers.* (C. D. Yonge, Trans.). London: Henry G. Bohn.

Dobash, R. E., & R. P. Dobash. (1979). *Violence against wives: a case against the patriarchy.* New York: Free Press.

—. (1988). Research as social action: the struggle for battered women. In M. Bograd & K. Yllo (Eds.), *Feminist perspectives.* Newbury Park, CA: Sage, 51-74.

—. (1992). *Women, violence and social change.* London: Routledge.

Dobson, R. (1999). US research scientist found guilty of fraud. *British Medical Journal* 319 (7218), 1156–1156.

Dobzhansky, T. G. (1937). *Genetics and the origin of species.* New York: Columbia University Press.

Doida, Y., M. W. Miller, A. A. Brayman, & E. L. Carstensen. (1996). A test of the hypothesis that ELF magnetic fields affect calcium uptake in rat thymocytes in vitro. *Biochemical and Biophysical Research Communications,* 227 (3), 834–838.

Doppelt, G. (1982). Kuhn's epistemological relativism: an interpretation and defense. In Krausz, M., & J. W. Meiland (Eds.), *Relativism: cognitive and moral.* Notre Dame: Notre Dame University Press, 113-146.

Doull, John. (2001). Toxicology comes of age. *Annual Reviews of Pharmacology and Toxicology* (41), 1–21.

Doyle, D. A., J. M. Cabral, R. A. Pfuetzner, A. Kuo, J. M. Gulbis, *et al.* (1998, April 3). The structure of the potassium channel: molecular basis of K+ conduction and selectivity. *Science,* 280, 69–77.

Duffus, J. H. (1980). *Environmental toxicology.* New York: John Wiley and Sons.

Dutzler, R., E. B. Campbell, M. Cadene, B. T. Chalt, & R. MacKinnon. (2002, January 17). X-ray structure of a CIC chloride channel at 3.0 Å reveals the molecular basis of anion selectivity. *Nature,* 415, 287–294.

Earman, J. (1993). Carnap, Kuhn, and the philosophy of scientific methodology. In P. Horwich (Ed.), *World changes: Thomas Kuhn and the nature of science.* Cambridge, MA: The MIT Press, 9-36.

Echobichon, D. (2001). Pesticide use in developing countries. *Toxicology,* 160, 27–33.

Edmonds, G., & D. Mercer. (1998). Trashing 'junk science.' *Stanford Technology Law Review* 3. Retrieved 20 February 2005 from http://stlr.stanford.edu/STLR/Articles/98_STLR_3/index.htm

Edwards, D., M. Ashmore, & J. Potter. (1995). Death and furniture: the rhetoric, politics and theology of bottom line arguments against relativism. *History of the Human Sciences,* 8, 25–49.

Eisen, S. (1964). Huxley and the positivists. *Victorian Studies,* 7, 337–358.

Elgin, S. H. (2000). *Native tongue.* New York: The Feminist Press. (Original edition, 1984, DAW Books.)

Eliot, T. S. (1975). Tradition and the individual talent. In F. Kermode (Ed.), *Selected prose of T.S. Eliot.* London: Faber & Faber. Original edition, 1919.

Engels, F. (1975). *Socialism utopian and scientific.* (E. Aveling, Trans.). New York: International Publishers.

Erickson, B. M. (1992). Feminist fundamentalism: reactions to Avis, Kaufman, and Bograd. *Journal of Marriage and Family Therapy,* 18, 263–267.

Ewald, H. R. (1988). The implied reader in persuasive discourse. *Journal of Advanced Composition* 8, 167-179.

Fahnestock, J. (1986). Accommodating science: the rhetorical life of scientific facts. *Written Communication,* 3, 275–296.

—. (1989). Arguing in different forums: the Bering crossover controversy. *Science, Technology, and Human Values,* 14 (1), 26–42.

—. (1999). *Rhetorical figures in science.* New York: Oxford University Press.

FAS Space Policy Project. (2000). *AN/FPS-115 PAVE PAWS radar* Federation of American Scientists. Retrieved 21 February 2003 from http://www.fas.org/spp/military/program/track/pavepaws.htm

FAS Weapons of Mass Destruction. (1998). *Extremely low frequency communications program.* Federation of American Scientists. Retrieved 21 February 2003 from http://www.fas.org/nuke/guide/usa/c3i/elf.htm

Favretti, R. R., G. Sandri, & R. Scazzieri, eds. (1999). *Incommensurability and translation: Kuhnian perspectives on scientific communication and theory change.* Cheltenham: Edward Elgar.

Federal Register—Staff. (1999). Findings of scientific misconduct. *Federal Register,* 64 (116), 32503–32504.

Feigl, H., & G. Maxwell, eds. (1962). *Scientific explanation, space, and time.* Vol. 3. Minnesota Studies in the Philosophy of Science. Minneapolis: University of Minnesota Press.

Feuer, L. S. (1975). Is the 'Darwin-Marx correspondence' authentic? *Annals of Science,* 32, 11-12.

Feyerabend, P. K. (1958a). An attempt at a realistic interpretation of experience. *Proceedings of the Aristotelian Society,* 58, 143–170.

—. (1958b). Complementarity. *Proceedings of the Aristotelian Society, suppl.,* 32, 75–104.

—. (1962). Explanation, reduction and empiricism. In H. Feigl & G. Maxwell (Eds.), *Scientific explanation, space, and time.* Minnesota Studies in the Philosophy of Science. Minneapolis: University of Minnesota Press, 28-97.

—. (1965a). On the 'meaning' of scientific terms. *Journal of Philosophy,* 62, 266–274.

—. (1965b). Problems of empiricism. In R. G. Colodny (Ed.), *Beyond the edge of certainty.* Englewood Cliffs: Prentice-Hall, 275-353.

—. (1965c). Reply to criticism: comments on Smart, Sellars and Putnam. In *Boston studies in the philosophy of science, vol. 2: In honor of Philipp Frank.*

(R. S. Cohen & M. W. Wartofsky, eds.). New York: Humanities Press, 223-261.

—. (1970). Consolations for the specialist. In I. Lakatos & A. Musgrave (Eds.), *Criticism and the growth of knowledge.* Cambridge, UK: Cambridge University Press, 197-230.

—. (1975). *Against method: outline of an anarchistic theory of knowledge.* London: New Left Books.

—. (1976). *Wider den methodenzwang. Skizze einer anarchistischen erkenntnistheorie.* Frankfurt: Suhrkamp.

—. (1978a). *Against method: outline of an anarchistic theory of knowledge.* 2nd ed. London: Verso.

—. (1978b). *Science in a free society.* London: New Left Books.

—. (1981a). *Realism, rationalism and scientific method. Philosophical papers 1.* Cambridge, UK: Cambridge University Press.

—. (1981b). *Problems of empiricism. Philosophical papers 2.* Cambridge, UK: Cambridge University Press.

—. (1981c). More clothes from the emperor's bargain basement: a review of Laudan's *Progress and its problems.* In *Problems of empiricism.* Cambridge, UK: Cambridge University Press, 57-71.

—. (1981d). Consolations for the specialist. In *Problems of empiricism.* Cambridge, UK: Cambridge University Press, 131-161.

—. (1981e). Introduction: scientific realism and philosophical realism. In *Realism, rationalism and scientific method.* Cambridge, UK: Cambridge University Press, 3-16.

—. (1981f). Introduction: proliferation and realism as methodological principles. In *Realism, rationalism, and scientific method.* Cambridge, UK: Cambridge University Press.

—. (1983). *Wider den Methodenzwang* 2nd ed. Frankfurt: Suhrkamp.

—. (1987a). *Farewell to reason.* London: Verso.

—. (1987b). Putnam on incommensurability. *British Journal for the Philosophy of Science,* 38, 75–81.

—. 1988. Response. In J. Agassi, *The gentle art of philosophical polemics.* La Salle, IL: Open Court, 405-15.

—. (1989). Realism and the historicity of knowledge. *Journal of Philosophy* 86, 393–406.

—. (1993). *Against method.* 3rd ed. London: Verso.

—. (1995). *Killing time: the autobiography of Paul Feyerabend.* Chicago: University of Chicago Press.

—. (1999). *Conquest of abundance: a tale of abstraction versus the richness of being.* Chicago: University of Chicago Press.

Fields, M. J., & R. M. Kirchner. (1978). Battered women are still in need. *Victimology,* 3, 216–222.

Finnis, J. (1997). Commensuration and public reason. In R. Chang (Ed.), *Incommensurability, incomparability, and practical reason.* Cambridge, MA: Harvard University Press, 215-233.

Finocchiaro, M. A. (1977). Logic and rhetoric in Lavoisier's sealed note: toward a rhetoric of science. *Philosophy and Rhetoric,* 10, 111–122.

Fish, S. E. (1989). *Doing what comes naturally: change, rhetoric, and the practice of theory in literary and legal studies.* Oxford: Clarendon Press.

—. (1994). *There's no such thing as free speech.* New York: Oxford University Press.

Fleck, L. (1979). *Genesis and development of a scientific fact.* (F. B. T. Trenn, Trans.). Chicago: University of Chicago.

Florig, H. K. (1992, December 18). EMF research. *Science* 258, 1869, 1960.

—. (1996, November 29). EMF report: is there consensus? [letter]. *Science* 274, 1449–1450.

Forbes, V. E., & T. L. Forbes. (1994). *Ecotoxicology in theory and practice.* London: Chapman & Hall.

Foreword. (1977). *Ecotoxicology and Environmental Health,* 1, iii.

Forster, M. R. (2000). Hard problems in the philosophy of science: idealisation and commensurability. In R. Nola & H. Sankey (Eds.), *After Popper, Kuhn and Feyerabend: recent issues in theories of scientific method.* Dordrecht: Kluwer.

Freeman, S. A., S. W. Littlejohn, & B. Pearce. (1992). Communication and moral conflict. *Western Journal of Communication,* 56, 311–29.

Frey, A. H. (1986). Letter. *Scientific American* 255 (6), 4–6.

Fritz, K. von. (1945). The discovery of incommensurability by Hippasus of Metapontum. *Annals of Mathematics,* 46, 242–264.

Fuller, S. (1995). The strong program in the rhetoric of science. In H. Krips, J. E. McGuire & T. Melia (Eds.), *Science, reason, and rhetoric.* Pittsburgh: University of Pittsburgh Press.

—. (2000). *Thomas Kuhn: a philosophical history for our times.* Chicago: University of Chicago Press.

—. (2002). *Social epistemology.* 2nd ed. Bloomington: Indiana University Press.

Gadamer, H. G. (1982). *Truth and method.* New York: Crossroad.

Galison, P. (1997). *Image and logic: a material culture of microphysics.* Chicago: University of Chicago Press.

Gallie, W. B. (1968). *Philosophy and historical understanding.* 2nd ed. New York: Schoken Books.

Gaonkar, D. P. (1990). Rhetoric and its double: reflections on the rhetorical turn in the human sciences. In Simons, H. W. (Ed.), *The rhetorical turn: invention and persuasion in the conduct of inquiry.* Chicago: University of Chicago Press, 341-366.

—. (1997). The idea of rhetoric in the rhetoric of science. In A. G. Gross & W. Keith (Eds.), *Rhetorical hermeneutics: invention and interpretation in the age of science*. Albany: SUNY Press.

Garaj-Vhrovac, V., & D. Zeljezic. (2001). Cytogenetic monitoring of Croation population occupationally exposed to a complex mixture of pesticides. *Toxicology,* 165, 153–162.

Garver, E. (1990). Arguing over incommensurable values: the case of Machiavelli. In H. W. Simons (Ed.), *Rhetoric in the human sciences*. Chicago: University of Chicago Press.

—. (2003). Rhetorical arguments and scientific arguments: do my children have to listen to more arguments against evolution? In J. A. Campbell & S. C. Meyer (Eds.), *Darwinism, design, and public education*. East Lansing, MI: Michigan State University Press, 487-497.

Gee, J. P. (1993). *An introduction to human language.* Upper Saddle River, NJ: Prentice-Hall.

Geertz, C. (1980). Blurred genres: the refiguration of social thought. *American Scholar*, 49, 165-79.

Geller, J. L. (1997). The stalemate of reason: Barbara Herrnstein Smith on the problems of circularity and self-contradiction. *Philosophy and Rhetoric*, 30, 376-394.

Ghiselin, M. T. (1969). *The triumph of the Darwinian method.* Los Angeles: University of California Press.

Gieryn, T. F. (1983). Boundary-work and the demarcation of science from non-science: strains and interests in professional ideologies of scientists. *American Sociological Review*, 48, 781–795.

—. (1999). *Cultural boundaries of science: credibility on the line.* Chicago: University of Chicago Press.

Gladwell, M. (2002, December 2). Group think. *The New Yorker*, 102–107.

Glynn, I. M. (1993). All hands to the sodium pump: Annual Review Prize lecture. *Journal of Physiology,* 462, 1–30.

—, & S. J. D. Karlish. (1975). The sodium pump. In J. H. Comroe Jr., R. R. Sonnenschein & I. S. Edelman (Eds.), *Annual Review of Physiology*. Palo Alta, CA: Annual Reviews.

Golinski, Jan V. (1987). Robert Boyle: scepticism and authority in seventeenth-century chemical discourse. In A. E. Benjamin, G. N. Cantor & J. R. R. Christie (Eds.), *The figural and the literal: problems of language in the history of science and philosophy, 1630–1800*. Manchester: Manchester University Press.

Goodman, N. (1976). *Languages of art: an approach to a theory of symbols.* 2nd ed. Indianapolis: Hackett.

—. (1978). *Ways of world making.* Indianapolis: Hackett.

Goodman, R., & M. Blank. (1998). Magnetic field stress induces expression of hsp70. *Cell Stress & Chaperones,* 3 (2), 79–88.

—, J. Bumann, L. X. Wei, & A. S. Henderson. (1992). Exposure of human-cells to electromagnetic-fields—effect of time and field-strength on transcript levels. *Electro- and Magnetobiology* 11 (1), 19–28.

—, & A. S. Henderson. (1986a). Some biological effects of electromagnetic-fields. *Bioelectrochemistry and Bioenergetics,* 15 (1), 39–55.

—. (1986b). Sine waves enhance cellular transcription. *Bioelectromagnetics,* 7 (1), 23–29.

—. (1987a). Stimulation of RNA-synthesis in the salivary-gland cells of Sciara-Coprophila by an electromagnetic signal used for treatment of skeletal problems in horses. *Journal of Bioelectricity,* 6 (1), 37–47.

—. (1987b). Effect of low-frequency nonionizing radiation on transcription/translation. *Abstracts of Papers of the American Chemical Society,* 193, 152.

—. (1988). Exposure of salivary-gland cells to low-frequency electromagnetic-fields alters polypeptide-synthesis. *Proceedings of the National Academy of Sciences of the United States of America* 85 (11), 3928–3932.

—. (1990). Exposure of cells to extremely low-frequency electromagnetic-fields—relationship to malignancy. *Cancer Cells,* 2 (11), 355–359.

—. (1991). Transcription and translation in cells exposed to extremely low-frequency electromagnetic-fields. *Bioelectrochemistry and Bioenergetics,* 25 (3), 335–355.

—, L. X. Wei, J. C. Xu, & A. S. Henderson. (1989). Exposure of human-cells to low-frequency electromagnetic-fields results in quantitative changes in transcripts. *Biochimica et biophysica acta,* 1009 (3), 216–220.

Gorgias of Leontini. (1982). *Encomium of Helen.* Bristol: Bristol Classical Press.

Gould, S. J. (1980). *The panda's thumb: more reflections in natural history.* New York: W.W. Norton and Company.

—. (2000). More things in heaven and earth. In H. Rose & S. Rose (Eds.), *Alas, poor Darwin: arguments against evolutionary psychology.* New York: Harmony Books.

Grabner, I., & W. Reiter. (1979). Guardians at the frontiers of science. In H. Nowotny & H. Rose. (Eds.), *Counter-movements in the sciences.* Dordrecht: D. Reidel.

Grant, C. K. (1956). Akrasia and the criteria of assent to practical principles. *Mind, New Series,* 65 (259), 400–407.

Gray, J. (2000). *Two faces of liberalism.* New York: The New Press.

Greene, J. C. (1981). *Science, ideology and world view: essays in the history of evolutionary ideas.* Berkeley: University of California Press.

Grice, H. P. (1989). *Studies in the way of words.* Cambridge, MA: Harvard University Press.

Griffin, J. (1986). *Well-being: its meaning, measurement, and moral importance.* Oxford: Clarendon Press.

—. (1997). Incommensurability: what's the problem? In R. Chang (Ed.), *Incommensurability, incomparability, and practical reason*. Cambridge, MA: Harvard University Press, 35-51.

Gross, A. G. (1990). The origin of species: evolutionary taxonomy as an example of the rhetoric of science. In H. W. Simons, (Ed.), *The rhetorical turn: invention and persuasion in the conduct of inquiry*. Chicago: University of Chicago Press, 91-115.

—. (1996). *The rhetoric of science*. 2nd ed. Cambridge, MA: Harvard University Press.

—, J. E. Harmon, & M. Reidy. (2002). *Communicating science: the scientific article from the 17th century to the present*. New York: Oxford University Press.

Grotevant, H. D., & C. I. Carlson, eds. (1989). *Family assessment: a guide to methods and measures*. New York: Guilford.

Gudernatsch, J. F. (1912). Feeding experiments on tadpoles. *Archiv fuer Entwicklungsmechanik* 35, 475.

Guide for authors. (2005). Ecotoxicology and Environmental Safety. Retrieved 12 March 2005 from http://authors.elsevier.com/JournalDetail.html?PubID=622819&Precis=DESC

Gusfield, J.R. (1989). Constructing the ownership of social problems: fun and profit in the welfare state. *Social Problems*, 36, 431-441.

Guthrie, F., & J. J. Perry, eds. (1980). *Introduction to environmental toxicology*. New York: Elsevier.

Habermas, J. (1987). *The philosophical discourse of modernity*. (F. G. Lawrence, Trans.). Cambridge, MA: The MIT Press.

Hacking, I. (1983). *Representing and intervening: introductory topics in the philosophy of natural science*. Cambridge, UK: Cambridge University Press.

—. (1993). Working in a new world: the taxonomic solution. In P. Horwich (Ed.), *World changes: Thomas Kuhn and the nature of science*. Cambridge, MA: The MIT Press, 275-310.

—. (1996). The disunities of the sciences. In P. Galison & D. J. Stump (Eds.), *The disunity of science: boundaries, context, and power*. Stanford: Stanford University Press, 37-74.

Halber, D. (2001). *Eight of 18 Presidential Advisors on Science Have MIT Ties* [News release]. MIT News Office, 1 May 2001. Retrieved 28 February 2005 from http://web.mit.edu/newsoffice/2001/ostpside.html.

Halffman, W. (1995). The boundary between ecology and toxicology: a sociologist's perspective. In J. Cairns Jr. & B. R. Niederlehner (Eds.), *Ecological toxicity testing: scale, complexity, and relevance*. Boca Raton: CRC Lewis Publishers, 11-34.

Halliday, M. A. K., & J. R. Martin, eds. (1993). *Writing science: literacy and discursive power*. Pittsburgh: Pittsburgh University Press.

Hamblin, C. L. (1970). *Fallacies*. London: Methuen.

Hanson, N. (1958). *Patterns of discovery: an inquiry into the conceptual foundations of science.* Cambridge, UK: Cambridge University Press.

Hardy, H. (2004). *Writings about pluralism before/independently of Isaiah Berlin.* The Isaiah Berlin Virtual Library (2001). Retrieved 21 March 2005 from http://berlin.wolf.ox.ac.uk/lists/pluralism/onpluralism.htm.

Harré, R. and Krausz, M. (1996). *Varieties of relativism.* Oxford: Blackwell.

Harris, H. (1999). *The birth of the cell.* New Haven: Yale University Press.

Harris, J. F. (1992). *Against relativism: a philosophical defense of method.* La-Salle, IL: Open court.

Harris, R. A. (1990). Assent, dissent, and rhetoric in science. *Rhetoric Society Quarterly* 20 (1), 13-37.

—. (1993). Generative semantics: secret handshakes, anarchy notes, and the implosion of *ethos. Rhetoric Review,* 23, 125–60.

—. (1994). The Chomskyan revolution 2: Sturm und Drang. *Perspectives on Science,* 2, 176–230.

—, ed. (1997). *Landmark essays in rhetoric of science: case studies.* Mahwah, NJ: Lawrence Erlbaum and Associates.

—. (1998). A note on the Max Planck effect. *Rhetoric Society Quarterly,* 28, 85–89.

Harvey, D. (1989). *The condition of postmodernism: an inquiry into the origin of cultural change.* Cambridge, MA: Blackwells.

Hasian Jr., M. A. (1996). *The rhetoric of eugenics in Anglo-American thought.* Athens: University of Georgia Press.

Hauser, G., & D. P. Cushman. (1973). McKeon's philosophy of communication: architectonic and interdisciplinary arts. *Philosophy and Rhetoric,* 6, 211–234.

Havas, Magda. (2000). Biological effects of non-ionizing electromagnetic energy: a critical review of the reports by the US national research council and the US national institute of environmental health sciences as they relate to the broad range of EMF bioeffects. *Environmental Review,* 8, 173–253.

Hazelwood, C. F. (1972, November 29). Pumps or no pumps. *Science,* 177, 815–816.

—. (1976). Letter to the editor [Response to Kolata (1976)]. *Science,* 193, 528.

Heidlebaugh, N. J. (2001). *Judgment, rhetoric, and the problem of incommensurability.* Studies in Rhetoric/Communication. Columbia, SC: University of South Carolina Press.

Heilbron, J. L. (1998). Thomas Samuel Kuhn. *Isis,* 89, 505–515.

Heisenberg, W. (1971). *Physics and beyond: encounters and conversations.* (A. J. Pomeranz, Trans.). New York: Harper and Row.

Hempel, C. G. (1966). *Philosophy of natural science.* Foundations of Philosophy. Englewood Cliffs, NJ: Prentice-Hall.

Herbert, S. S. (1968). The logic of Darwin's discovery. Unpublished doctoral dissertation, Brandeis University.

—. (1971). Darwin, Malthus and selection. *Journal of the History of Biology* 4, 209-217.

—. (1974). The place of man in the development of Darwin's theory of transmutation pt. 1. *Journal of the History of Biology,* 7, 217–258.

—. (1977). The place of man in the development of Darwin's theory of transmutation pt. 2. *Journal of the History of Biology,* 10, 155–227.

Herschel, J. F. W. (1987). *Preliminary discourse on the study of natural philosophy.* Chicago: University of Chicago Press.

Hesse, M. (1983). Comment on Kuhn's "Commensurability, comparability, communicability." In P. D. Asquith & T. Nickles (Eds.), *PSA 1982: Proceedings of the 1982 Biennial Meeting of the Philosophy of Science Association.* East Lansing: Philosophy of Science Association, 704-711.

Higashikubo, R., V. O. Culbreth, D. R. Spitz, M. C. LaRegina, W. F. Pickard, W. L. Straube, *et al.* (1999). Radiofrequency electromagnetic fields have no effect on the in vivo proliferation of the 9L brain tumor. *Radiation Research,* 152 (6), 665–671.

Hildenbrand, G. (1979). The American revolution in cellular biology. *Healing Journal* 2(1). Retrieved 12 March 2005 from http://www.gerson-research.org/docs/HildenbrandGLG-1979-1/

Hinks, D. A. G. (1940). Tisias and Corax and the invention of rhetoric. *Classical Quarterly* 34, 61–6.

Hintikka, J. (1988). On the incommensurability of theories. *Philosophy of Science,* 55, 25–38.

Hodge, M. J. S. (1977). The structure and strategy of Darwin's 'long argument.' *The British Journal for the History of Science,* 10, 237–245.

—. (1982). Darwin and the laws of the animate part of the terrestrial system (1835–1837): on the Lyellian origins of his zoonomical explanatory program. *Studies in the History of Biology,* 7, 1-106.

Hodgkin, A. L. (1964, September 11). The ionic basis of nervous conduction. *Science,* 145, 1148–1154.

Holmquest, A. (1990). The rhetorical strategy of boundary-work. *Argumentation,* 4, 235–258.

Holton, G. (1988). *Thematic origins of scientific thought: Kepler to Einstein.* Rev. ed. Cambridge, MA: Harvard University Press. Original edition, 1973.

—. (1998). *Scientific imagination.* Rev. ed. Cambridge, MA: Harvard University Press. Original edition, 1978.

Hooykaas, R. (1966). Geological uniformitarianism and evolutionism. *Archives internationales d'histoire des sciences,* 19, 3–19.

Hoyningen-Huene, P. (1989). Idealist elements in Thomas Kuhn's philosophy of science. *History of Philosophy Quarterly,* 6, 393–401.

—. (1990). Kuhn's conception of incommensurability. *Studies in History and Philosophy of Science*, 21, 481–492.

—. (1992). The interrelations between the philosophy, history and sociology of science in Thomas Kuhn's theory of scientific development. *British Journal for the Philosophy of Science*, 43, 487–501.

—. (1993). *Reconstructing scientific revolutions. Thomas S. Kuhn's philosophy of science.* Chicago: University of Chicago Press.

—. (1994). Obituary of Paul K. Feyerabend (1924–1994). *Erkenntnis* 40, 289–292.

—. (1995). Two letters of Paul Feyerabend to Thomas S. Kuhn on a draft of *The structure of scientific revolutions. Studies in History and Philosophy of Science* 26 (3), 353–387.

—. (1997a). Obituary of Thomas S. Kuhn (1922–1996). *Erkenntnis* 45, v-viii.

—. (1997b). Thomas S. Kuhn. *Journal for General Philosophy of Science,* 28 (2), 235–256.

—. (2000a). Paul Feyerabend and Thomas Kuhn. In J. Preston, G. Munevar & D. Lamb (Eds.), *The worst enemy of science? Essays in memory of Paul Feyerabend.* New York: Oxford University Press.

—. (2000b). Paul K. Feyerabend: an obituary. In J. Preston, G. Munevar & D. Lamb (Eds.), *The worst enemy of science? Essays in memory of Paul Feyerabend.* New York: Oxford University Press.

—. (2002). Paul Feyerabend und Thomas Kuhn. *Journal for General Philosophy of Science,* 33 (1), 61–83.

—, E. Oberheim, & H. Andersen. (1996). On incommensurability. *Studies in History and Philosophy of Science,* 27, 131–141.

—, & H. Sankey, eds. (2001). *Incommensurability and related matters.* Dordrecht: Kluwer.

Hubble, E. (1954). *The nature of science and other lectures.* San Marino: The Huntington Library.

Hughes, W. W. (1996). *Essentials of environmental toxicology.* Washington DC: Taylor & Francis.

Hull, D. L. (1974). *Philosophy of biological science.* Prentice-Hall Foundations of Philosophy. Englewood Cliffs, NJ: Prentice-Hall.

—. (1978). Scientific bandwagon or traveling medicine show? In M. S. Gregory, A. Silvers & D. Sutch (Eds.), *Sociobiology and human nature: an interdisciplinary critique and defense.* San Francisco: Jossey-Bass, 50-59.

—. (1980). Sociobiology: another new synthesis. In G. W. Barlow & J. Silverberg (Eds.), *Beyond nature/nurture: reports, definitions, and debate.* Boulder, Colorado: Westview Press, 77-96.

—. (1988). *Science as a process.* Chicago: University of Chicago Press.

—, P. D. Tessner, & A. M. Diamond. (1978, November 17). Planck's principle: do younger scientists accept new scientific ideas with greater alacrity than older scientists? *Science* 202, 717-723,

Hume, D. (1989). *Dialogues concerning natural religion.* Buffalo: Prometheus Books.

Hunder, G. , J. Javdani, B. Elsenhaus, & Klaus Schuemann. (2001). 109 CD accumulation in the calcified parts of rat bones. *Toxicology,* 159, 1–10.

Hutten, E. H. (1956). *The language of modern physics: an introduction to the philosophy of science.* London: Allen & Unwin.

Huxley, L, ed. (1900). *Life and letters of Thomas Henry Huxley.* 2 vols. London: Macmillan.

Huxley, T. H. (1873). On the methods of studying zoology. In E. L. Youmans (Ed.), *The culture demanded by modern life: a series of addresses and arguments on the claims of scientific education.* New York: Appleton & Co.

—. (1893). *Selected works of Thomas H. Huxley.* 9 vols. New York: Appleton.

—. (1897). *Darwiniana.* New York: Appleton.

Isocrates. (1928–1954). *Works.* (G. Norlin & L. V. Hook, eds.). 3 vols. Loeb Classical Library. Cambridge, MA: Harvard University Press.

Jackson, J. D. (1992). Are the stray 60-Hz electromagnetic fields associated with the distribution and use of electric power a significant cause of cancer? *Proceedings of the National Academy of Sciences of the United States of America,* 89, 3508–3510.

James, H. (1948). *The art of fiction and other essays.* (M. Roberts, ed.) New York: Oxford University Press.

James, W. (1911). *Some problems of philosophy: a beginning of an introduction to philosophy.* New York: Longmans, Green, & Co.

Jameson, F. (1972). *The prison-house of language.* Princeton: Princeton University Press.

Jauss, H. R. (1969). Paradigmwechsel in der Literaturwissenschaft. *Linguistische Berichte,* 1, 44–56.

Johnson, M. P. (1995). Patriarchal terrorism and common couple violence: two forms of violence against women. *Journal of Marriage and the Family,* 57, 283–294.

—, & K. J. Ferraro. (2000). Research on domestic violence in the 1990s: making distinctions. *Journal of Marriage and the Family,* 62, 948–963.

Johnstone, C. (2001). Philosophy: perennial topics and terms. In T. Sloane (Ed.), *Encyclopedia of rhetoric.* New York: Oxford, 592-601.

Johnstone Jr., H. W. (1959). *Philosophy and argument.* University Park, PA: Pennsylvania State University Press.

Jonsen, A.E. & S. Toulmin. (1988). *The abuse of casuistry.* Berkeley: University of California Press.

Journet, D. (1993). Biological explanation, political ideology, and "blurred genres": a Bakhtinian reading of the science essays of J. B. S. Haldane. *Technical Communication Quarterly,* 2, 185–204.

Kaiser, J. (1996, November 8). Panel finds EMFs pose no threat. *Science,* 274, 910.

—. (1999, October 29). Shalala takes watchdog office out of the hunt. *Science,* 286, 883.

Kaufman Jr., G. (1992). The mysterious disappearance of battered women in family therapists' offices: male privilege colluding with male violence. *Journal of Marriage and Family Therapy,* 18, 233–243.

Kavalovski, V. (1974). The *vera causa* principle: a historico-philosophical study of a metatheoretical concept from Newton through Darwin. Unpublished doctoral thesis, University of Chicago.

Kenshur, O. (1984). The rhetoric of incommensurability. *Journal of Aesthetics and Art Criticism*, 42, 375-82.

Kent, Chris. (1998). *Basics of toxicology.* New York: Wiley.

Kenyon, D., & G. Steinman, 1969. *Biochemical predestination.* New York: McGraw-Hill.

Kerferd, G.B. (1981). *The sophistic movement.* Cambridge, UK: Cambridge University Press.

Khera, K. S., Q. N. LaHam, C. F. G. Ellis, Z. Z. Zawidzka, & H. C. Grice. (1966). Foot deformity in ducks from injection of EPN during embryogenesis. *Journal of Toxicology and applied Pharmacology,* 8, 540–549.

King, A. (1976). The rhetoric of power maintenance: elites at the precipice. *Quarterly Journal of Speech,* 62, 127–134.

King, R. W. P. (1998). The Interaction of power-line electromagnetic fields with the human body. *IEEE Engineering in Medicine and Biology Magazine,* 17 (6), 67–73.

Kirschvink, J. L. (1992). Comment on 'Constraints on biological effects of weak extremely-low-frequency electromagnetic fields.' *Physical Review A* 46 (4), 2178–2184.

Kitcher, P. (1983). Implications of incommensurability. In P. D. Asquith & T. Nickles (Eds.), *PSA 1982. Proceedings of the 1982 Biennial Meeting of the Philosophy of Science Association*. East Lansing: Philosophy of Science Association, 689-703.

Kloss, D. A., & E. L. Carstensen. (1983). Effects of ELF electric-fields on the isolated frog-heart. *IEEE Transactions on Biomedical Engineering* 30 (6), 347–348.

Knorr-Cetina, K. (1999). *Epistemic cultures: how the sciences make knowledge.* Cambridge, MA: Harvard University Press.

Koertge, N. (1980). [Review of Feyerabend 1978b]. *The British Journal for the Philosophy of Science*, 31 (4), 385-390.

Koestler, A. (1959. *The sleep walkers: a history of man's changing vision of the universe.* New York: Macmillan.

Kohn, D. (1980). Theories to work by: rejected theories, reproduction, and Darwin's path to natural selection. *Studies in the History of Biology,* 4, 67-70.

Kolakowski, L. (1968). *The alienation of reason: a history of positivist thought.* (N. Guterman, Trans.). Garden City: Doubleday & Co.

Kolata, G. B. (1976, June 18). Water structure and ion binding: a role in cell physiology? *Science,* 192, 1220–1222.

Krips, H. (1995). Rhetoric, ideology, and desire in von Neumann's Gründlagen. In H. Krips, J. E. McGuire & T. Melia (Eds.), *Science, reason, and rhetoric.* Pittsburgh: University of Pittsburgh Press.

Kuhn, T. S. (1952). Robert Boyle and structural chemistry in the seventeenth century. *Isis,* 43, 12–36.

—. (1957). *The Copernican revolution: planetary astronomy in the development of western thought.* Cambridge, MA: Harvard University Press.

—. (1959). Energy conservation as an example of simultaneous discovery. In M. Clagett (Ed.), *Critical problems the history of science.* Madison: University of Wisconsin Press, 321-356.

—. (1960). [Proto-Structure] *The structure of scientific revolutions.* Unpublished manuscript.

—. (1962). *The structure of scientific revolutions.* Chicago: University of Chicago Press.

—. (1970a). *The structure of scientific revolutions.* 2nd ed. Chicago: University of Chicago Press.

—. (1970b). Reflections on my critics. In I. Lakatos & A. Musgrave (Eds.), *Criticism and the growth of knowledge.* Cambridge, UK: Cambridge University Press, 231-278.

—. (1977a). *The essential tension: selected studies in scientific tradition and change.* Chicago: University of Chicago Press.

—. (1977b). The essential tension: tradition and innovation in scientific research. In *The essential tension: selected studies in scientific tradition and change.* Chicago: University of Chicago Press, 225-239.

—. (1977c). Objectivity, value judgment, and theory choice. In *The essential tension: selected studies in scientific tradition and change.* Chicago: University of Chicago Press, 320-339.

—. (1978). *Black body theory and the quantum discontinuity, 1894–1912.* Chicago: University of Chicago Press.

—. (1980). The halt and the blind: philosophy and the history of science. *British Journal of Philosophy of Science,* 31, 181–192.

—. (1981). What are scientific revolutions? Occasional paper #18. Cambridge, MA: Center for Cognitive Science, M.I.T.

—. (1983a). Commensurability, comparability, communicability. In P. D. Asquith & T. Nickles (Eds.), *PSA 1982. Proceedings of the 1982 Biennial Meeting of the Philosophy of Science Association*. East Lansing: Philosophy of Science Association, 669-688.

—. (1983b). Rationality and theory choice. *Journal of Philosophy*, 80, 563-70.

—. (1983c). Response to commentaries [on "Commensurability, comparability, communicability"]. In P. D. Asquith & T. Nickles (Eds.), *PSA 1982. Proceedings of the 1982 Biennial Meeting of the Philosophy of Science Association*. East Lansing: Philosophy of Science Association, 712-716.

—. (1993). Afterwords. In P. Horwich (Ed.), *World changes: Thomas Kuhn and the nature of science*. Cambridge, MA: The MIT Press, 311–41.

—. (1999). Remarks on incommensurability and translation. In R. R. Favretti, G. Sandri & R. Scazzieri (Eds.), *Incommensurability and translation: Kuhnian perspectives on scientific communication and theory change*. Cheltenham: Edward Elgar.

—. (2000). *The road since* Structure*: philosophical essays, 1970—1993, with an autobigraphical interview*. (J. Conant & J. Haugeland, eds.). Chicago: University of Chicago Press.

—, A. Baltas, K. Gavroglu, & V. Kindi. (1997). A discussion with Thomas S. Kuhn: a physicist who became a historian for philosophical purposes. *Neusis,* 6 (Spring-Summer), 145–200.

Kunz, F., & B. L. Whorf. (1941). Toward a higher mental world. *Main Currents in Modern Thought* 1 (7), 14–15.

Kurz, D. (1993). Physical assaults by husbands: a major social problem. In R. L. Gelles & D. R. Loseke (Eds.), *Current controversies in family violence*. London: Sage, 88-103.

Laërtius, Diogenes, see Diogenes Laërtius.

Lakatos, I. (1970). Falsification and the methodology of scientific research programmes. In I. Lakatos & A. Musgrave (Eds.), *Criticism and the growth of knowledge*. Cambridge, UK: Cambridge University Press, 91-196.

—, & A. Musgrave, eds. (1970). *Criticism and the growth of knowledge.* Cambridge, UK: Cambridge University Press.

Lakoff, G., & M. Johnson. (1980). *Metaphors we live by.* Chicago: University of Chicago Press.

Lamprecht, S. P. (1920). The need for a pluralistic emphasis in ethics. *Journal of Philosophy, Psychology and Scientific Methods,* 17, 561–72.

—. (1921). Some political implications of ethical pluralism. *Journal of Philosophy,* 18, 225–244.

—. (1955). *Our philosophical traditions: a brief history of philosophy in Western civilization.* New York: Appleton-Century-Crofts.

—. (1962). *The moral and political philosophy of John Locke.* New York: Russell and Russell.

Lanham, R. (1976). *The motives of eloquence.* New Haven: Yale University Press.

Landis, W. G., & M. Yu. (1995). *Introduction to environmental toxicology: impacts of chemicals upon ecological systems.* Boca Raton: CRC Lewis Publishers.

Larrain, J. (1994). The postmodern critique of ideology. *The Sociological Review,* 42, 289–314.

Latour, B. (1993). *We have never been modern.* (C. Porter, Trans.). Cambridge, MA: Harvard University Press

—, & P. Fabbri. (2000). The rhetoric of science: authority and duty in an article from the exact sciences. (S. Cummins, Trans.) *Technostyle* 15 (2), 119–138.

Laudan, L. (1990). *Science and relativism: some key controversies in the philosophy of science.* Chicago: University of Chicago Press.

Leary, W. E. (1996, November 1). Panel sees no proof of health hazards from power lines. *New York Times,* A1.

Leff, M.C. (1987). Modern sophistic and the unity of rhetoric. In J. Nelson, A. Megill & D. N. McCloskey (Eds.), *Rhetoric of the human sciences.* Madison: University of Wisconsin Press.

—. (1997). Hermeneutic rhetoric. In M. Hyde & W. Jost (Eds.), *Rhetorical hermeneutics in our time: a reader.* New Haven and London: Yale University Press, 196-214.

—. (2003). Tradition and agency in humanistic rhetoric. *Philosophy and Rhetoric* 36 (2), 135-147.

Lenin, V. I. (1970). *Materialism and empirio-criticism: critical comments on a reactionary philosophy.* Moscow: Progress.

Lenoir, T. (1982). *The strategy of life: teleology and mechanics in nineteenth-century german biology.* Chicago: University of Chicago Press.

Lenzer, G., ed. (1975). *Auguste Comte and positivism: the essential writings.* Chicago: University of Chicago Press.

Lessl, T. M. (1988). Heresy, orthodoxy, and the politics of science. *Quarterly Journal of Speech,* 74 (1), 18–34.

—. (1989). The priestly voice. *Quarterly Journal of Speech,* 75, 183–197.

—. (1996). Naturalizing science: two episodes in the evolution of a rhetoric of scientism. *Western Journal of Communication,* 60 (4), 379–396.

Levin, S. A., M. A. Harwell, J. R. Kelly, & K. Kimball, eds. (1989). *Ecotoxicology: problems and approaches.* New York: Springer-Verlag.

Levitt, B. B. (1995). *Electromagnetic fields: a consumer's guide to the issues and how to protect ourselves.* San Diego, CA: Harcourt, Brace, & Company.

Liburdy, R. P. (1992). Biological interactions of cellular-systems with time-varying magnetic-fields. *Annals of the New York Academy of Sciences,* 649, 74–95.

—. (1992). Calcium signaling in lymphocytes and ELF fields—evidence for an electric-field metric and a site of interaction involving the calcium-ion channel. *FEBS Letters* 301 (1), 53–59.

—. (1995). Cellular studies and interaction mechanisms of extremely-low-frequency fields. *Radio Science,* 30 (1), 179–203.

—. (1999, July 16). Calcium and EMFs: graphing the data. *Science,* 285 (5426), 337.

—, J. D. Harland, & S. M. J. Afzal. (1996). Inhibition of melatonin's action on MCf-7 cell proliferation by magnetic fields. *Molecular Biology of the Cell,* 7, 1057–1057.

Lightman, B. (1987). *The origins of agnosticism: Victorian unbelief and the limits of knowledge.* Baltimore: The Johns Hopkins University Press.

Limoges, C. (1970). *La selection naturelle: etude sur la premiere constitution d'un concept.* Paris: Presses Universitaires de France.

Ling, G. N. (1962). *A physical theory of the living state: the Association-Induction hypothesis.* New York: Blaisdell Publishing.

—. (1984). *In search of the physical basis of life.* New York: Plenum Press.

—. (1992). *A revolution in the physiology of the living cell.* Malabar, FL; Krieger Publishing Company.

—. (1997). Debunking the alleged resurrection of the sodium pump hypothesis. *Physiological Chemistry, Physics and Medicine NMR,* 29, 123–198.

—. (1998). *Why science cannot cure AIDS without your help.* Retrieved 21 March 2005 from http://www.gilbertling.org.

—. (2001). *Life at the cell and below-cell level.* New York: Pacific Press.

—, J. Gulati, & L. G. Palmer. (1977, December 23). Potassium accumulation in frog muscle: the Association-Induction hypothesis versus the membrane theory. *Science,* 198, 1281–1284.

—, & C. L. Walton. (1976). What retains water in living cells? *Science* 191, 293–295.

Livingston, E. (1986). *The ethnomethodological foundations of mathematics.* London: Routledge & Kegan Paul.

Lodish, H., A. Berk, S. L. Zipursky, P. Matsudaira, D. Baltimore, & J. Darnell. (2000). *Molecular cell biology.* 4th ed. New York: W. H. Freeman.

Loeb, J. (1916). *The organism as a whole from a physicochemical viewpoint.* New York: Knickerbocker Press.

Loomis, D. P., D. A. Savitz, & C. V. Ananth. (1994). Breast cancer mortality among female electrical workers in the United States. *Journal of the National Cancer Institute,* 86, 921–925.

Loomis, T. (1968). *Essentials of toxicology.* 1st ed. Philadelphia: Lea & Febiger.

—. (1974). *Essentials of toxicology.* 2nd ed. Philadelphia: Lea & Febiger.

—. (1978). *Essentials of toxicology.* 3rd ed. Philadelphia: Lea & Febiger.

—, & A. W. Hayes. (1996). *Essentials of toxicology.* 4th ed. San Diego, CA: Academic Press.

Lukes, S. (1997). Comparing the incomparable: trade-offs and sacrifices. In R. Chang (Ed.), *Incommensurability, incomparability, and practical reason.* Cambridge, MA: Harvard University Press, 184-195.

Lyell, C. (1990). *Principles of geology.* (M. J. S. Rudwick, ed.). 3 vols. Chicago: U of Chicago Press.

Lyne, J. (1990a). Bio-rhetorics: moralizing the life sciences. In H. W. Simons (Ed.), *The rhetorical turn: invention and persuasion in the conduct of inquiry.* Chicago: University of Chicago Press, 35-57.

—. (1990b). Rhetorics of expertise: E. O. Wilson and sociobiology. *Quarterly Journal of speech* 76 (2), 134–51.

—. (2002). Review of *Shaping science with rhetoric* [Ceccarelli 2001a]. *Argumentation and Advocacy,* 38, 76–77.

—, & H. F. Howe. (1986). Punctuated equilibria: rhetorical dynamics of a scientific controversy. *Quarterly Journal of Speech,* 72 (2), 132–147.

Lyons, S. (1999). *Thomas Henry Huxley: the evolution of a scientist.* Amherst, NY: Prometheus Books.

Lyotard, J.-F. (1984). *The postmodern condition: a report on knowledge.* (G. Bennington & B. Massumi, Trans.). Minneapolis: University of Minnesota Press.

—. (1988). *The differend: phrases in dispute.* (G. Van Den Abbeele, Trans.). Minneapolis: University of Minnesota Press.

—, W. van Reijen, & D. Veerman. (1988). An interview with Jean-François Lyotard. *Theory, Culture and Society,* 5, 277–309.

Machlup, S., C. F. Blackman, & J. P. Blanchard. (1996). Biological effects of extremely-low-frequency (ELF) magnetic fields may go linearly (not quadratically) with field amplitude. *Biophysical Journal,* 70 (2), MP419.

MacIntyre, A. (1981). *After virtue.* London: Duckworth.

Maffeo, S., A. A. Brayman, M. W. Miller, E. L. Carstensen, V. Ciaravino, & C. Cox. (1988). Weak low-frequency electromagnetic-fields and chick embryogenesis—failure to reproduce positive findings. *Journal of Anatomy,* 157, 101–104.

Malone, M.E. (1993). Kuhn reconstructed: incommensurability without relativism. *Studies in the History and Philosophy of Science,* 24, 69-93.

Malthus, T. R. (1992). *An essay on the principle of population.* Cambridge, UK: Cambridge University Press.

Manier, E. (1978). *The young Darwin and his cultural circle.* Dordrecht: D. Reidel.

Mann, R. M. (2000). *Who owns domestic violence? The local politics of a social problem.* Toronto: University of Toronto Press.

Mannheim, Karl. (1936). *Ideology and utopia: an introduction to the sociology of knowledge.* (L. Wirth & E. Shils, Trans.). New York: Harcourt, Brace & World.

Margolis, H. (1993). *Paradigms and barriers: how habits of mind govern scientific beliefs.* Chicago: University of Chicago Press.

Margolis, J. (1990). Reconciling realism and relativism. In H. W. Simons, (Ed.), *The rhetorical turn: invention and persuasion in the conduct of inquiry.* Chicago: University of Chicago Press, 308-319.

—. (1992). *The truth about relativism.* Oxford: Blackwell.

Marino, A. A., R. M. Wolcott, R. Chervenak, F. Jourd'heuil, E. Nilsen, & C. Frilot, II. (2000). Nonlinear response of the immune system to power-frequency magnetic fields. *American Journal of Physiology—Regulatory, Integrative and Comparative Physiology,* 279 (3), R761–R768.

Mariussen, E., & F. Fonnum. (2001). The effect of polychlorinated biphenyls on the high affinity uptake of the neurotransmitters, dopamine, serotonin, glutamate and GABA into rat brain synaptosomes. *Toxicology,* 159, 11–21.

Martin, L. (1986). Eskimo words for snow: a case study in the genesis and decay of an anthropological example. *American Anthropologist,* 88, 418–23.

Martindale, D. (2002). Channel crossings. *Scientific American* 286 (1), 22.

Mason, S. F. (1962). *A history of the sciences.* New York: Collier Books.

Masterman, M. (1970). The nature of a paradigm. In I. Lakatos & A. Musgrave (Eds.), *Criticism and the growth of knowledge.* Cambridge, UK: Cambridge University Press, 59-89.

Mattson, J., & M. Simon. (1996). *The pioneers of NMR and magnetic resonance in medicine: the story of MRI.* Jericho, NY: Dean Books.

Maugh, T. H., II. (1992, April 3). Study sparks debate on electromagnetic fields, cancer risk. *Los Angeles Times,* 15.

Mayr, E. (1961, November 10). Cause and effect in biology. *Science,* 134, 1501–1506.

—. (1985). How biology differs from the physical sciences. In D. J. Depew & B. H. Weber (Eds.), *Evolution at a crossroads: the new biology and the new philosophy of science.* Cambridge, MA: The MIT Press.

—. (1988). *Toward a new philosophy of biology: observations of an evolutionist.* Cambridge, MA: Harvard University Press.

McCarthy, J. T. (1987). *The rights of publicity and privacy.* New York: C. Boardman.

McCarthy, T. (1987). General introduction. In K. Baynes, J. Bohman, & T. McCarthy, (Eds.), *After philosophy: end or transformation?* Cambridge, MA: The MIT Press.

McCloskey, D.N. (1985). *The rhetoric of economics.* Madison: University of Wisconsin Press.

McGee, M. (1975). In search of the "people": a rhetorical alternative. *Quarterly Journal of Speech*, 1975, 235-249.

McKeon, R. (1957). Communication, truth, and society. *Ethics*, 67, 89–99.

—. (1966). The methods of rhetoric and philosophy: invention and judgment. In L. Wallach (Ed.), *The classical tradition*. Ithaca, NY: Cornell University Press, 365-373.

—. (1971). The uses of rhetoric in a technological age: architectonic productive arts. In L. F. Bitzer & E. Black (Eds.), *The prospect of rhetoric*. Englewood Cliffs, NJ: Prentice-Hall, 44-63.

McLauchlan, K. (1992). Are environmental magnetic fields dangerous? *Physics World*, 5 (1), 41–45.

McMullin, E. (1991). Rhetoric and theory choice in science. In M. Pera & W. R. Shea (Eds.), *Persuading science: the art of scientific rhetoric*. Canton, MA: Science History Publications, 55-76.

Mead, G. H. (1934). *Mind, self, and society*. Chicago: University of Chicago Press.

Meadows, D. H., D. I. Meadows, J. Randers, & W. W. Behrens III. (1972), *The limits to growth: a report to The Club of Rome*. New York: Universe Books.

Megill, A. (1989). What does the word "postmodern" mean? *Annals of Scholarship*, 6, 129-151.

—. (1994). Four senses of objectivity. In A. Megill (Ed.) *Rethinking objectivity*. Durham: Duke University Press.

Mendelson, M. (1997). Everything must be argued: rhetorical theory and pedagogical practice in Cicero's *De oratore*. *Journal of Education*, 179, 14–47.

Mercer, D. (2002). Scientific method discourses in the construction of 'EMF science': interests, resources and rhetoric in submissions to a public inquiry. *Social Studies of Science*, 32 (2), 205–234.

Merton, R. K. (1938). *Science, technology and society in seventeenth century England*. Bruges, Belgium: Saint Catherine Press.

Metcalf, R. L. (1974). A laboratory model ecosystem to evaluate compounds producing biological magnification. In W. J. Hayes Jr. (Ed.), *Essays in toxicology*. New York: Academic Press.

Meth, R. L. (1992). Marriage and family therapists working with family violence: strained bedfellows or compatible partners? A commentary on Avis, Kaufman, & Bograd. *Journal of Marriage and Family Therapy*, 18, 257–261.

Microwave News. (1990). EPA staff classifies ELF EMFs 'probable human carcinogens.' *Microwave News*, 10 (3), 1, 9–15.

—. (1991). White House's Allan Bromley in his own words. *Microwave News*, 11 (3), 13.

—. (1995a). Conflict over NCRP review of modulated RF/MW radiation. *Microwave News,* 15 (3), 1, 14–15.
—. (1995b). Draft NCRP report seeks strong action to curb EMFs. *Microwave News,* 15 (4), 11–15.
—. (1995c). From the field: clippings from all over. *Microwave News,* 15 (1), 15.
—. (1995d). NCRP to revise report on bioeffects of RF/MW radiation. *Microwave News,* 15 (5), 12–13.
—. (1996a). EPA shelves EMF-cancer report. *Microwave News,* 16 (1), 1, 7–8.
—. (1996b). NAS finds no EMF–cancer link; report stirs controversy. *Microwave News,* 16 (6), 1, 5–7.
—. (1998). NIEHS panel finds EMFs are 'possible' human carcinogens. *Microwave News,* 18 (4), 1, 4–7.
—. (1999a). Media storm over Liburdy affair. *Microwave News,* 19 (4), 8.
—. (1999c). Views on the news: wireless phones and public health: industry is in the driver's seat. *Microwave News,* 19 (6), 19.
—. (2000). Physicists: 60 Hz magnetic field effects as low as 10 mG. *Microwave News,* 20 (4), 11–12.
—. (2001). Views on the news: weapons development and public health should not mix. *Microwave News,* 21 (2), 19.
—. (2002a). Introducing brillouin precursors: microwave radiation runs deep. *Microwave News,* 22 (2), 1, 10.
—. (2002b). Views on the news: Motorola's junkyard dog. *Microwave News,* 22 (4), 19.
Midgley, M. (1980). Rival fatalisms: the hollowness of the sociobiology debate. In A. Montagu (Ed.), *Sociobiology examined.* New York: Oxford University Press, 15-38.
Mill, J. S. 1887. *The positive philosophy of Auguste Comte.* New York: Henry Holt and Company.
Miller, C. R. (1993). Rhetoric and community: the problem of the one and the many. In T. Enos & S. C. Brown (Eds.), *Defining the new rhetorics.* Newbury Park, CA: Sage.
Miller, G., & J. A. Holstein, eds. (1993). *Constructionist controversies.* New York: Aldine de Gruyner.
Miller, R. D., R. K. Adair, D. A. Bromley, D. Korn, M. G. Morgan, R. Neutra, *et al.* (1996). Unfounded fears: the great power-line cover-up exposed. *IEEE Engineering in Medicine and Biology Magazine,* 15 (2), 106-115.
Miner, R. C. (2001). Lakatos and MacIntyre on incommensurability and the rationality of theory-change. *Epistemologia,* 24, 221–36.
Mitchell, L. M., & A. Cambrosio. (1997). The invisible topography of power: electromagnetic fields, bodies and the environment. *Social Studies of Science,* 27, 221–271.

Montague, P, & M. B. Pellerano. (2001). Toxicology and environmental digital resources from and for citizen groups. *Toxicology* 157 (1–2), 77–88.

Morgan, J. L. (1973). On arguing about semantics. *Papers in Linguistics,* 1, 49–70.

Morgan, M. G. (1990). Exposé treatment confounds understanding of a serious public-health issue. Review of *Currents of Death* by Paul Brodeur [1989]. *Scientific American,* 262 (4), 118–123.

Murphy, J. J., R. A. Katula, F. I. Hill, & D. J. Ochs (2003). *A synoptic history of classical rhetoric.* 3rd ed. Mahwah, NJ: Lawrence Erlbaum Associates.

Musgrave, A. (1978). How to avoid incommensurability. *Acta Philosophica Fennica,* 30, 336–346.

Myers, G. (1990). *Writing biology: texts in the social construction of scientific knowledge.* Madison: University of Wisconsin Press.

Nair, I. (1990). Review: Currents of death and cross currents. *Physics Today,* 43 (12), 70–72.

National Academy of Science. (1998). *Teaching about evolution and the nature of science.* Washington, DC: National Academy Press.

National Council on Radiation Protection and Measurements, Committee 89–3 on Extremely Low Frequency Electric and Magnetic Fields. (1995). Draft report.

National Institute of Environmental Health Sciences. (1999). Health effects from exposure to power-line frequency electric and magnetic fields. National Institutes of Health.

Nelson, J. (1987). Political thinking and political rhetoric. In J. Nelson, A. Megill, & D. N. McCloskey (Eds.), *The rhetoric of the human sciences.* Madison: University of Wisconsin Press.

—, Megill, A., & D. N. McCloskey, eds. (1987). *The rhetoric of the human sciences.* Madison: University of Wisconsin Press.

Neville, M. C. (1972, April 21). Solute concentration gradients in frog muscles at 0° C: active transport or adsorption? *Science,* 176, 302–303.

Newman, M. C. (1998). *Fundamentals of ecotoxicology.* Chelsea, MI: Ann Arbor Press.

—, & C. H. Jagoe, eds. (1996). *Ecotoxicology: a hierarchical treatment.* Boca Raton: CRC Lewis Publishers.

Newton-Smith, W. H. (1981). *The rationality of science.* Boston: Routledge & Kegan Paul.

Nickles, T. (1984). A revolution that failed: Collins and Pinch on the paranormal. *Social Studies of Science,* 14, 297–308.

Noaksson, E., U. Tjaernlund, A. T. C. Bosveld, & L. Balk. (2001). Evidence for endocrine disruption in perch (*Perca fluviatilis*) and roach (*Rutilus rutilus*) in a remote Swedish lake in the vicinity of a public refuse dump. *Toxicology and Applied Pharmacology,* 174, 160–176.

Norris, C. (1982). *Deconstruction: theory and practice.* New York: Methuen.

Oberheim, E. (2000). Feyerabend's early philosophy. *Studies in History and Philosophy of Science,* 31 (2), 363–375.

—, & P. Hoyningen-Huene. (1997). Incommensurability, realism and meta-incommensurability. *Theoria,* 12, 447–465.

—, & P. Hoyningen-Huene. (1999). Review symposium: radical fallibilism vs conceptual analysis: the significance of Feyerabend's philosophy of science [on Preston 1997]. *Metascience,* 8 (2), 226–233.

Ogden, C. K. (1930). *Basic English: a general introduction with rules and grammar.* London: Paul Treber & Co., Ltd.

Olmsted, W. R. (1997). The uses of rhetoric: indeterminacy in legal reasoning, practical thinking, and the interpretation of literary figures. In W. Jost & M. Hyde (Eds.), *Rhetoric and hermeneutics in our time: a reader.* New Haven: Yale University Press.

Olsen, P., G. Wuertzen, E. Hansen, J. Carstensen, & E. Poulsen. (1973). Short-term peroral toxicity of the food colour Orange RN in pigs. *Toxicology,* 1, 249–260.

ORAU Panel on Health Effects of Low-Frequency Electric and Magnetic Fields. (1993, April 2). EMF and cancer: letters. *Science,* 260, 13–14.

Ortega y Gasset, José (1967). *The origin of philosophy.* (T. Talbot, Trans.). Chicago: University of Chicago Press.

Ospovat, D. (1981). *The development of Darwin's theory: natural history, natural theology, and natural selection, 1838–1859.* Cambridge, UK: Cambridge University Press.

Overington, M. A. (1977). The scientific community as audience: toward a rhetorical analysis of science. *Philosophy and Rhetoric,* 10, 143–64.

Owens, C. (1992). *Beyond recognition: representation, power and culture.* Berkeley: University of California Press.

Pagelow, M. D. (1981). *Woman-battering: victims and their experiences.* Newbury Park, CA: Sage.

Palfreman, J. (1996). Apocalypse not. *Technology Review,* 99 (3), 25–33.

Palmer, L. G., & J. Gulati. (1976, October 29). Potassium accumulation in muscle: a test of the binding hypothesis. *Science,* 194, 521–523.

Parascandola, M. (2001, November 16). Cell phone lawsuits face a scientific test. *Science,* 294, 1440–1442.

Park, R. L. (1996, November 13). Power line paranoia. *New York Times,* 23.

—. (2000). *Voodoo science: the road from foolishness to fraud.* New York: Oxford University Press.

Parks, L. G., C. S Lambright, E. F. Orlando, L. J. Guilette, Jr., G. T. Ankley, & L. E. Gray, Jr. (2001). Masculinization of female mosquitofish in kraft mill effluent-contaminated fenholloway river water is associated with androgen receptor agonist activity. *Toxicological Sciences,* 62, 257–267.

Pearce, W. B., & S. W. Littlejohn. (1997). *Moral conflict: when social worlds collide.* Thousand Oaks, CA: Sage.

—, S. W. Littlejohn, & A. Alexander. (1987). The new Christian right and the humanist response: reciprocated diatribe. *Communication Quarterly,* 35, 171–92.

Pearson, K. (1957). *The grammar of science.* New York: Meridian Books.

Pendrick, G. J. (1998). Plato and rhêtorikê. *Rheinisches Museum für Philologie,* 141, 10–23.

Pera, M. (1992). *The ambiguous frog: the Galvani-Volta controversy on animal electricity.* (J. Mandelbaum, Trans.). Princeton: Princeton University Press.

—. (1994). *The discourses of science.* (C. Botsford, Trans.). Chicago: University of Chicago Press.

Percival, W. K. (1976). The applicability of Kuhn's paradigms to the history of linguistics. *Language,* 32, 285–294.

—. (1979). The applicability of Kuhn's paradigms to social sciences. *American Sociologist,* 32, 28–31.

Perelman, C. (1982). *The realm of rhetoric.* (W. Kluback, Trans.). Notre Dame: University of Notre Dame Press.

—, & L. Olbrecht-Tyteca. (1969). *The new rhetoric: a treatise on argumentation.* (L. Wilkinson & P. Weaver, Trans.). Notre Dame, IN: University of Notre Dame Press.

Pfohl, S. (1977). The "discovery" of child abuse, *Social Problems,* 24, 310-323.

Pickering, A. (1995). *The mangle of practice.* Chicago: University of Chicago Press.

Pinch, T. J. (1995). Rhetoric and the cold fusion controversy: from the chemist's Woodstock to the physicists' Altamont. In H. Krips, J. E. McGuire & T. Melia (Eds.), *Science, reason, and rhetoric.* Pittsburgh: University of Pittsburgh Press.

Planck, M. (1950). *Scientific autobiography and other papers.* (F. Gaynor, Trans.). London: Williams and Norgate.

Pocock, J. G. A. (1973). *Language and time: essays on political thought and history.* New York: Atheneum.

Polansky, R. M. (1992). *Philosophy and knowledge: a commentary on Plato's Theaetetus.* Lewisburg, PA: Bucknell University Press.

Polanyi, M. (1958). *Personal knowledge: towards a post-critical philosophy.* Chicago: University of Chicago Press.

Pollack, G. H. (1996). Phase transitions and the molecular mechanism of contraction. *Biophysical Chemistry,* 59 (1996), 315–328.

—. (2001). *Cells, gels, and the engines of life: a new unifying approach to cell function.* Seattle, WA: Ebner & Sons.

Pool, R. (1990, October 5). Flying blind: the making of EMF policy. *Science,* 250, 23–25.

—. (1990, September 7). Is there an EMF–cancer connection? *Science,* 249, 1096–1098.

Poore, L. M., G. King, & K. Stefanik. (2001). Toxicology information resources at the Environmental Protection Agency. *Toxicology,* 157 (1–2), 11–23.

Popper, K. R. (1934). *Logik der Forschung.* Wien: Julius Springer.

—. (1957). *The poverty of historicism.* Boston: Beacon Press.

—. (1959). *The logic of scientific discovery.* New York: Basic Books.

—. (1962). *The open society and its enemies.* New York: Harper and Row.

—. (1970). Normal science and its dangers. In I. Lakatos & A. Musgrave (Eds.), *Criticism and the growth of knowledge.* Cambridge, UK: Cambridge University Press, 51-58.

Potter, J. (1996). *Representing reality: discourse, rhetoric and social construction.* London: Sage.

Poulakos, J. (2001). Sophists. In T. Sloane (Ed.), *Encyclopedia of rhetoric.* New York: Oxford University Press, 732-733.

Prasad, A. V., M. W. Miller, E. L. Carstensen, C. Cox, M. Azadniv, & A. A. Brayman. (1991). Failure to reproduce increased calcium-uptake in human-lymphocytes at purported cyclotron-resonance exposure conditions. *Radiation and Environmental Biophysics,* 30 (4), 305–320.

Prelli, L. (1989a). *A rhetoric of science: inventing scientific discourse.* Columbia: University of South Carolina Press.

—. (1989b). The rhetorical construction of scientific *ethos.* In H. W. Simons (Ed.), *Rhetoric in the human sciences.* London: Sage, 48-68

—. (1996). Empirical diversity, interdependence, and the problem of rhetorical invention and judgment: the case of wife abuse facts. *Communication Theory,* 6, 406–429.

Preston, J. (1997). *Feyerabend: philosophy, science and society.* Cambridge, UK: Polity.

—. (2002). *Paul Feyerabend.* In E. N. Zalta (Ed.), *Stanford encyclopedia of philosophy* (Spring 2005 ed.). Retrieved March 2005 from http://plato.stanford.edu/entries/feyerabend/

—, G. Munevar, & D. Lamb, eds. (2000). *The worst enemy of science? Essays in memory of Paul Feyerabend,.* New York: Oxford University Press.

Pullum, G. K. (1991). *The great Eskimo vocabulary hoax and other irreverent essays on the study of language.* Chicago: University of Chicago Press.

Putnam, H. (1975). The meaning of "meaning." In K. Gunderson (Ed.), *Language, mind and knowledge.* Minneapolis: University of Minnesota Press, 131-193.

—. (1981a). Philosophers and human understanding. In A. F. Heath (Ed.), *Scientific explanation: papers based on the Herbert Spencer Lectures given in the University of Oxford.* Oxford: Clarendon Press, 184-204.

—. (1981b). *Reason, truth and history.* Cambridge, UK: Cambridge University Press.

Quine, W. V. O. (1960). *Word and object.* New York: John Wiley and Sons,

—. (1969). *Ontological relativity and other essays.* New York: Columbia University Press.

Quintilian. (1920). *Institutio oratoria.* (H. E. Butler, Trans.). 3 vols. Cambridge, MA: Harvard University Press.

Rashdall, H. (1907). *The theory of good and evil.* 2 vols. New York: Oxford University Press.

Rauws, A. G., & M. J. van Logten. (1973). The influence of dichlorvos from strips or sprays on Cholinesterase activity in chickens. *Toxicology,* 1, 29–41.

Raz, J. (1986). *The morality of freedom.* New York: Oxford University Press.

Read, R. (n.d.). *Kuhn: a Wittgenstein of the sciences?* Retrieved 14 March 2005 from http://www.uea.ac.uk/~j339/KuhnWittgenstein.htm.

—. (1993). Organizations and modernity: continuity and discontinuity in organization theory. In J. Hassard & M. Parker (Eds.), *Post-modernism and organizations.* London: Sage.

Reifenrath, W. G., H. O. Kammen, W. G. Palmer, M. M. Major, & G. J. Leach. (2002). Percutaneous absorption of explosives and related compounds: an empirical model of bioavailability of organic nitro compounds from soil. *Toxicology and Applied Pharmacology* 182, 160–168.

Richards, I. A. (1933). *Basic rules of reason.* London: K. Paul, Trench, Trubner & Co.

—. (1936). *The philosophy of rhetoric.* New York: Oxford University Press.

—. (1943). *Basic English and its uses.* New York: W.W. Norton & Co.

Ricoeur, P. (1977). *The rule of metaphor: multi-disciplinary studies of the creation of meaning in language.* (R. Czerny, Trans.). Toronto: University of Toronto Press.

Robson, J. M. (1968). *The improvement of mankind: the social and political thought of John Stuart Mill.* Toronto: University of Toronto Press.

Rombke, J., & J. F. Moltmann. (1996). *Applied toxicology.* Boca Raton: CRC Lewis Publishers.

Rorty, R. (1979). *Philosophy and the mirror of nature.* Princeton: Princeton University Press.

—. (1989). *Contingency, irony and solidarity.* Cambridge, UK: Cambridge University Press.

Rosenau, P. M. (1992). *Postmodernism and the social sciences.* Princeton: Princeton University Press.

Rosenmeyer, T. G. (1955). Gorgias, Aeschylus, and Apate. *American Journal of Philology,* 76, 225-60.

Rouse, Joseph. (1998). Kuhn and scientific practices. *Configurations,* 6 (1), 33–50.

Rudwick, M. J. S. (1970). The strategy of Lyell's *Principles of geology. Isis,* 61, 5–33.

—. (1974). Darwin and Glen Roy: a 'great failure' in scientific method. *Studies in the History and Philosophy of Science* 5, 97-185.
—. (1985). *The great Devonian controversy: the shaping of scientific knowledge among gentlemanly specialists.* Chicago: University of Chicago Press.
—. (1990). Introduction. In C. Lyell, *Principles of geology.* Chicago: U of Chicago Press, vii-lviii.
Russman, T. A. (1987). *A prospectus for the triumph of realism.* Macon: Mercer University Press.
Ryle, G. (1954). *Dilemmas.* Cambridge, UK: Cambridge University Press.
Salomon, A. (1946, December). The religion of progress. *Social Research,* 13, 441–462.
Sanderson, J. T., J. Boerma, G. W. A. Lansbergen, & M. van den Berg. (2001). Induction and inhibition of aromatase (CYP19) activity by various classes of pesticides in H295R human adrenocortical carcinoma cells. *Journal of Toxicology and Applied Pharmacology,* 182, 44–54.
Sandler, N. D. (1993). Panic gluttons. *Technology Review,* 96 (7), 72–73.
Sankey, H. (1992). Incommensurability, translation, and understanding. *The Philosophical Quarterly,* 41, 414-26.
—. (1993). Kuhn's changing concept of incommensurability. *British Journal for the Philosophy of Science,* 44, 759–774.
—. (1994). *The incommensurability thesis.* Aldershot: Avebury.
—. (1997a). Incommensurability: the current state of play. *Theoria,* 12 (3), 425–445.
—, ed. (1997b). *Rationality, relativism and incommensurability.* Aldershot: Ashgate.
—. (1997c). Taxonomic incommensurability. In H. Sankey (Ed.), *Rationality, relativism and incommensurability.* Aldershot: Ashgate.
—. (1998). Taxonomic incommensurability. *International Studies in the Philosophy of Science,* 12, 7–16.
—. (1999). Incommensurability: an overview. *Divinatio: Studia Culturologica Series,* 10, 1–13.
—. (2000). Methodological pluralism, normative naturalism and the realist aim of science. In R. Nola & H. Sankey (Eds.), *After Popper, Kuhn and Feyerabend: recent issues in theories of scientific method.* Dordrecht: Kluwer, 211-229.
—, & P. Hoyningen-Huene. (2001). Introduction. In P. Hoyningen-Huene & H. Sankey (Eds.), *Incommensurability and related matters.* Dordrecht: Kluwer, vii-xxxiv.
Sapir, E. (1921). *Language: an introduction to the study of speech.* New York: Harcourt.
—. (1949). *Selected writings of Edward Sapir in language, culture and personality.* (D. G. Mandelbaum, ed.). Berkeley: University of California Press.

Saunders, D. G. (1986). When battered women use violence: husband-abuse or self-defense? *Violence and Victims,* 1, 47–60.

—. (1988). Wife abuse, husband abuse, or mutual combat? A feminist perspective on the empirical findings. In K. Yllo and M. Bograd (Eds.), *Feminist perspectives.* Newbury Park, CA: Sage.

Savitz, D. A. (1993). Commentary on health effects of low-frequency electric and magnetic fields. *Environmental Science and Technology,* 27 (1), 52–54.

—, H. Wachtel, F. A. Barnes, E. M. John, & J. G. Tvrdik. (1988). Case-control study of childhood cancer and exposure to 60-Hz magnetic fields. *American Journal of Epidemiology,* 128 (1), 21–38.

Schechter, S. (1982). *Women and male violence: the visions and struggles of the battered women's movement.* Boston: South End Press.

Scheffler, I. (1967). *Science and subjectivity.* Indianapolis: Bobbs-Merrill.

Schellenberger, T. E., G. W. Newell, R. F. Adams, & J. Barbaccia. (1966). Cholinesterase inhibition and toxicological evaluation of two organophosphate pesticides in Japanese quail. *Journal of Toxicology and Applied Pharmacology,* 8, 22–28.

Schiappa,E. (1991). *Protagoras and logos: a study in Greek philosophy and rhetoric.* Columbia, SC: University of South Carolina Press.

—. (1999). *The beginnings of rhetorical theory in classical Greece.* New Haven: Yale University Press.

—, and O. Swartz. (1994). Introduction. In E. Schiappa (Ed.), *Landmark essays on classical Greek rhetoric.* Mahwah, NJ: Lawrence Erlbaum Associates.

Schlag, P. (1998). *The enchantment of reason.* Durham, NC: Duke University Press.

Schneider, D. (1997). "Raymond v. Damadian: scanning the horizon." *Scientific American*, 276 (6), 32, 34.

Schrödinger, E. (1944). *What is life? The physical aspect of the living cell.* Cambridge, UK: Cambridge University Press.

Schweber, S. S. (1977). The origin of the *Origin* revisited. *Journal of the History of Biology,* 10 (2), 229-316.

Science. (1991, February 22). EPA: physicists unwelcome on EMF panel. *Science,* 251 (4996), 863.

—. (1995, August 18). Major EMF report warns of health risks. *Science,* 269, 911.

Seabury Press. (1979). *The book of common prayer.* New York: The Seabury Press.

Secord, J. A. (1986). *Controversy in Victorian geology: the Cambrian-Silurian dispute.* New Jersey: Princeton University Press.

Segerstråle, U. (2000). *Defenders of the truth: the battle for science in the sociobiology debate and beyond.* New York: Oxford University Press.

—. (1986). Colleagues in conflict: an 'in vivo' analysis of the sociobiology controversy. *Biology and Philosophy,* 1, 53–87.

Seifert, J. (1973). Troptyphan pyrrolase in rat liver after phenobarbitol administration. *Toxicology,* 1, 179–186.

Searle, J. R. (1995). *The construction of social reality.* New York: Free Press.

Shapere, D. (1989). Evolution and continuity in scientific change. *Philosophy of Science,* 56, 419–437.

—. (2001). Reasons, radical change and incommensurability in science. In P. Hoyningen-Huene & H. Sankey (Eds.), *Incommensurability and related matters.* Boston: Kluwer.

Shapin, S. (1994). *A social history of truth: civility and science in seventeenth century England.* Chicago: Chicago University Press.

Shapiro, M. J. (1988). *The politics of representation.* Madison: University of Wisconsin Press.

Sharrock, W., & R. Read. (2002). *Kuhn: philosopher of scientific revolution.* Cambridge, UK: Polity.

Shen, J. H., B. Gutendorf, H. H. Vahl, L. Shen, & J. Westendorf. (2001). Toxicological profile of pollutants in surface water from an area in Taihu Lake, Yangtze Delta. *Toxicology,* 166, 71–78.

Shore, M. L., M. H. Repacholi, B. Servantie, & P. Czerski. (1983). Review of radiofrequency and microwaves (Ehc 16)—comments. *Radiation Research,* 95 (2), 414–417.

Shweder, R. A. (1992). Post-Nietzschian anthropology: the idea of multiple objective worlds. In M. Krausz (Ed.), *New essays on relativism.* Notre Dame: Notre Dame University Press.

Simmons, L. (1994). Three kinds of incommensurability theses. *American Catholic Philosophical Quarterly,* 31, 119–131.

Simon, J. F. (2002). *What kind of nation? Thomas Jefferson, John Marshall and the epic struggle to create a United States.* New York: Simon & Schuster.

Simons, H. W. (1980). Are scientists rhetors in disguise? In E. E. White (Ed.), *Rhetoric in transition.* University Park, PA: Pennsylvania University Press.

—, ed. (1989). *Rhetoric in the human sciences.* London: Sage.

—, ed. (1990). *The rhetorical turn: invention and persuasion in the conduct of inquiry.* London: Sage.

—. (1993). The rhetoric of the scientific research report: "drug-pushing" in a medical research article. In R. H. Robert & J. M. M. Good (Eds.), *The recovery of rhetoric: persuasive discourse and disciplinarity in the human sciences.* London: Bristol Classical Press.

—. (1995). Arguing about the ethics of past actions: an analysis of a taped conversation about a taped conversation, *Argumentation,* 9, 225-250.

—. (1999). Rhetorical hermeneutics and the project of globalization. *Quarterly Journal of Speech,* 85, 86-100.

Skou, J. C. (1989). Sodium-potassium pump. In D. C. Tosteson (Ed.), *Membrane transport: people and ideas.* Bethesda, MD: American Physiological Society.

Slesin, L. (1986). Letter. *Scientific American,* 255 (6), 6.

—. (1987). Power-lines and cancer: the evidence grows. *Technology Review,* 90 (7), 52–59.

—. (1990a). Letter. *Issues in Science and Technology,* 6 (3), 17–18.

—. (1990b). Prudent avoidance, more study of EMFs. *Issues in Science and Technology,* 6 (4), 17–18.

—. (1994). Unwarranted confidence in cellular phones. *Technology Review,* 97 (1), 7.

Sloane, T. O. (1997). *On the contrary: the protocol of traditional rhetoric.* Washington, DC: Catholic University of America Press,.

Smith, B. (1921). Corax and probability. *The Quarterly Journal of Speech Education* 7, 13–42.

Smith, B. H. (1997). *Belief and resistance: dynamics of contemporary intellectual controversy.* Cambridge, MA: Harvard University Press.

Smith, J. M. (1975, August 28). Survival through suicide. [Review of E. O. Wilson's *Sociobiology* (1975a).]. *New Scientist,* 496-497.

Smith, R. P., & R. E. Gosselin. (1966). On the mechanism of sulfide inactivation by methemoglobin. *Journal of Toxicology and Applied Pharmacology,* 8, 159–172.

Smyth Jr., H. F., C. S. Weil, J. S. West, & C. P. Carpenter. (1969). An exploration of joint toxic action: twenty-seven industrial chemicals intubated in rats in all possible pairs. *Journal of Toxicology and Applied Pharmacology,* 14, 340–347.

Snelson, J. T. (1977). The importance of chlorinated hydrocarbons in world agriculture. *Ecotoxicology and Environmental Health,* 1, 17–30.

Sober, E. (1993). *Philosophy of biology.* Boulder, CO: Westview Press.

Society of Toxicology. (1989). *Resource guide to careers in toxicology.* Washington, DC: Society of Toxicology.

Sociobiology Study Group of Science for the People. (1976). Sociobiology—another biological determinism. *Bioscience* 26 (3), 182-190.

South, J.C. (2001). Online resources for news about toxicology and other environmental topics. *Toxicology,* 157 (1–2), 153–164.

Spencer, L. (1998). *Incommensurability.* Retrieved 20 February 2005 from http://www.tasc.ac.uk/depart/media/staff/ls/Modules/Theory/Incommensurability.htm

Sprat, T. (1958). *History of the Royal Society.* (J. I. Cope & H. W. Jones, eds.). St. Louis: Washington University Press. (Original edition, 1667.)

Stegmüller, W. (1976). *The structure and dynamics of theories.* Berlin: Springer-Verlag.

—. (1979). *The structuralist view of theories.* Berlin: Springer-Verlag.

Steinmetz, S. K. (1978a). The battered husband syndrome. *Victimology,* 2, 499–509.

—. (1978b). Services to battered women: our greatest need. A reply to Fields & Kirchner. *Victimology,* 3, 222–226.

Steneck, N. H. (1984). *The microwave debate.* Cambridge, MA: The MIT Press.

—. (1986). The microwave debate [letter]. *The Sciences,* 26 (4), 18.

—, H. J. Cook, A. J. Vander, & G. L. Kane. (1980, June 13). The origins of U.S. safety standards for microwave radiation. *Science,* 208 (4449), 1230–1237.

Stets, J. E., & M.A. Straus. (1990). Gender differences in reporting marital violence and its medical and psychological consequences. In M. A. Straus & R. J. Gelles (Eds.), *Physical violence in American families: risk factors and adaptations to violence in 8,145 American families.* New Brunswick, NJ: Transaction, 227-244.

Stewart, L. (1992). *The rise of public science: rhetoric, technology, and natural philosophy in Newtonian Britain, 1660–1750.* Cambridge, UK: Cambridge University Press.

Stofen, D. (1973). The maximum permissible concentrations in the U.S.S.R. for harmful substances in drinking water. *Toxicology,* 1, 187–195.

Stone, R. (1992, December 11). Polarized debate: EMFs and cancer. *Science,* 258 (5089), 1724–1725.

Stossel, T. P. (2001, July 27). Manifesto for a cytoplasmic revolution. *Science,* 293, 611.

Straus, M. A. (1973). A general systems theory approach to a theory of violence between family members. *Social Science Information,* 12 (3), 105–125.

—. (1978). Wife beating: causes, treatment, and research needs. In US Commission on Civil Rights (Ed.), *Battered women: issues of public policy* Washington, DC: United States Commission on Civil Rights, 152-170, 463-531.

—. (1979). Measuring intrafamily conflict and violence: the conflict tactics (CT) scales. *Journal of Marriage and the Family,* 41, 71–88.

—. (1990a). The conflict tactics scales and its critics: an evaluation and new data on validity and reliability. In M. A. Straus & R. J. Gelles (Eds.), *Physical violence in American families: risk factors and adaptations to violence in 8,145 American families.* New Brunswick, NJ: Transaction, 49-73.

—. (1990b). Injury and frequency of assault and the "representative sample fallacy" in measuring wife beating and child abuse. In M. A. Straus and R. J. Gelles (Eds.), *Physical violence in American families: risk factors and adaptations to violence in 8,145 American families.* New Brunswick, NJ: Transaction, 75-91.

—. (1990c). The national family violence surveys. In M. A. Straus & R. J. Gelles (Eds.), *Physical violence in American families: risk factors and adaptations to violence in 8,145 American families*. New Brunswick, NJ: Transaction, 3-16.

—. (1992). Sociological research and social policy: the case of family violence. *Sociological Forum*, 7, 211–237.

—. (1993). Physical assaults by wives: a major social problem. In R. J. Gelles & D. R. Loseke (Eds.), *Current controversies in family violence*. London: Sage, 67-87.

—. (1999). The controversy over domestic violence by women: a methodological, theoretical, and sociology of science analysis. In X. B. A. S. Oskamp (Ed.), *Violence in intimate relationships*. London: Sage, 17-44.

—, & R. J. Gelles, eds. (1990a). *Physical violence in American families: risk factors and adaptations to violence in 8,145 American families*. New Brunswick, NJ: Transaction.

—, & R. J. Gelles, eds. (1990b). How violent are American families? Estimates from the national family violence resurvey and other studies. In M. A. Straus & R. J. Gelles (Eds.), *Physical violence in American families: risk factors and adaptations to violence in 8, 145 American families*. New Brunswick, NJ: Transaction.

—, R. J. Gelles, & S. K. Steinmetz. (1980). *Behind closed doors: violence in the American family*. Garden City, NJ: Anchor Press/Doubleday.

Stuart, H. W. (1904). *The logic of self-realization*. Berkeley: University of California Press.

—. (1939). Dewey's ethical theory. In P. A. Schilpp (Ed.), *The philosophy of John Dewey*. Evanston, IL: Northwestern University Press.

Sullivan, D. (2000). Keeping the rhetoric orthodox: forum control in science. *Technical Communication Quarterly*, 9 (2), 125–146.

Sullivan, H. S. (1953). *The interpersonal theory of psychiatry*. New York: Norton.

Sulloway, F. (1982). Darwin's Conversion: the *Beagle* voyage and its aftermath. *Journal of the History of Biology*, 15, 325-396.

Tannen, D. (1998). *The argument culture: moving from debate to dialogue*. New York: Random House.

Taubes, G. (2000). The cell-phone scare. *Technology Review*, 103 (6), 117–119.

Taylor, C. A. (1991). Defining the scientific community: a rhetorical perspective on demarcation. *Communication Monographs*, 58, 402–420.

—. (1996a). *Defining science: a rhetoric of demarcation*. Madison: University of Wisconsin Press.

—. (1996b). Theorizing practice and practicing theory: toward a constructive analysis of scientific rhetoric. *Communication Theory*, 6, 374–387.

Tenforde, T. S. (1993). Commentary on health effects of low-frequency electric and magnetic fields. *Environmental science and technology* 27 (1), 56–58.

—. (1995). Lowdown on low magnetic fields [review of Bennett (1994a).]. *IEEE Spectrum,* 32 (10), 10, 12.

Tetlock, P. E., R. S. Peterson, & J. S. Lerner. (1996). Revising the value pluralism model: incorporating social content and context postulates. In C. Seligman, J. M. Olson & M. P. Zanna (Eds.), *The psychology of values.* Mahwah, NJ: Lawrence Erlbaum Associates, 25-49.

Thagard, P. (1992). *Conceptual revolutions.* Princeton: Princeton University Press.

—, & J. Zhu. (2002). Acupuncture, incommensurability, and conceptual change. In G. M. Sinatra & P. R. Pintrich (Eds.), *Intentional conceptual change.* Mahwah, NJ: Lawrence Erlbaum Associates, 79-102.

Times Literary Supplement, The. (1995, October 6). The hundred most influential books since the war. *The Times Literary Supplement,* 39.

Todorov, Tzvetan. (1998, April 27). The surrender to nature [review of E. O. Wilson (1998).]. *The New Republic,* 98, 29-33.

Toulmin, Stephen. (1970). Does the distinction between normal and revolutionary science hold water? In I. Lakatos & A. Musgrave (Eds.), *Criticism and the growth of knowledge.* Cambridge, UK: Cambridge University Press, 39-47.

—. (1972). *Human understanding.* Princeton: Princeton University Press.

—. (2001). *Return to reason.* Cambridge, MA: Harvard University Press.

Toxicology, Society of. (2005). "Toxicological sciences" Retrieved on 3 July 2005 from http://www.elsevier.com/wps/find/journaldescription.cws_home/622951/description

Toxicology and Applied Pharmacology. (2005). "Description." Retrieved on 12 March 2005 from http://www.elsevier.com.proxy.lib.uwaterloo.ca/wps/find/journaldescription.cws_home/622951/description#description

Toyoshima, C., M. Nakasako, H. Nomura, & H. Ogawa. (2000, June 8). Crystal structure of the calcium pump of the sarcoplasmic reticulum at 2.6 Å resolution. *Nature,* 405, 647–655.

Truhaut, R. (1974). La toxicologie. *Sciences,* 5, 35–49.

—. (1977a). Can permissible levels of carcinogenic compounds in the environment be envisaged? Critical remarks. *Ecotoxicology and Environmental Health,* 1, 31–38.

—. (1977b). Ecotoxicology: objectives, principles, and perspectives. *Ecotoxicology and Environmental Health,* 1, 151–174.

Truscello, M. (2001). The clothing of the American mind: the construction of scientific *ethos* in the science wars. *Rhetoric Review,* 20 (3/4), 329–350.

Turow, S. (2002, January 6). To kill or not to kill: confronting capital punishment. *The New Yorker,* 40-47.

Urban, W. M. (1916, December 7). Knowledge of value and the value-judgment. *The Journal of Philosophy, Psychology and Scientific Methods,* 13 (25), 673-687.

U.S. Environmental Protection Agency. (1990, October). Evaluation of the potential carcinogenicity of electromagnetic fields: EPA/600/6-90-005B External Review Draft.

U.S. National Research Council, Committee on the Possible Effects of Electromagnetic Fields on Biologic Systems. (1997). *Possible health effects of exposure to residential electric and magnetic fields.* Washington, DC: National Academy Press.

United Kingdom Medicines Control Agency. (2002). *About the agency: regulatory framework.* Retrieved on 7 September 2002 from http://www.mca.gov.uk/aboutagency/regframework/regframework.htm#current.

United States Department of Health and Human Services. (2002). *Food and Drug Administration.* Retrieved on 7 September 2002 from http://www.fda.gov/oc/history/default.htm.

van Eemeren, F. H., *et al.* (1998). *Fundamentals of argumentation theory.* Mahwah, NJ: Lawrence Erlbaum Associates.

Veatch, R. M., & W. E. Stempsey. (1995). Incommensurability: its implications for the patient/physician relation. *The Journal for Medicine and Philosophy,* 20, 253–269.

Vergano, D. (1999, July 2). EMF researcher made up data, ORI says. *Science,* 285, 23–25.

Vitanza, V. J. (1987). Critical sub/versions of the history of philosophical rhetoric. *Rhetoric Review,* 6, 41–66.

Voloshinov, V. N. (1973). *Marxism and the philosophy of language.* (L. M. I. R. Titunik, Trans.). Cambridge, MA: Harvard University Press.

Vygtosky, L. S. (1978). *Mind in society: the development of higher psychological processes.* (V. John-Steiner, M. Cole, S. Scribner, & E. Souberman, eds.). Cambridge, MA: Harvard University Press.

Waddell, C., ed. (2000). *And no birds sang: rhetorical analyses of* Silent spring. Carbondale, IL: Southern Illinois University Press.

Wade, N. (2002, April 30). Thrown aside, genome pioneer plots a rebound. *New York Times,* D1, D5.

Wakefield, J. (2000). *The 'Indomitable' MRI. Smithsonian* Retrieved 30 January 2002 from http://www.smithsonianmag.si.edu/smithsonian/issues00/jun00/object_jun00.html

Walker, L. E. (1979). *The battered woman.* New York: Harper & Row.

—. (1983). The battered woman syndrome study. In D. Finkelhor, G. T. Hotaling, M. A. Straus & R. Gelles (Eds.), *The dark side of families: current family violence research.* London: Sage, 31-48.

—. (1988). The battered woman syndrome. In G. T. Hotaling, D. Finkelhor, J. T. Kirkpatrick & M. A. Straus (Eds.), *Family abuse and its consequences: new directions in research.* London: Sage, 139-148.

—. (2000). *The battered woman syndrome.* 2nd ed. New York: Springer.

Wallace, A. R. (1855). On the law which regulated the introduction of new species. *Annals and Magazine of Natural History,* 16, 184–196.
Wallace, K. (1963). The substance of rhetoric: good reasons. *Quarterly Journal of Speech,* 44, 239–49.
Wartenberg, D., & M. Greenberg. (1992). Epidemiology, the press, and the EMF controversy. *Public Understanding of Science,* 1, 382–394.
Weaver, R. M. (1953). *The ethics of rhetoric.* Chicago: Henry Regnery.
—. (1970) *Language is sermonic: Richard M. Weaver on the nature of rhetoric.* (R. L. Johannesen, R. Strickland, & R. T. Eubanks, eds.). Baton Rouge: Louisiana State University Press.
Weed, D. L. (1997). Underdetermination and incommensurability in contemporary epidemiology. *Kennedy Institute of Ethics Journal,* 7, 107–127.
Weinberg, S. (1998, October 8). The revolution that didn't happen. *The New York Review of Books,* 48–52.
Weinreich, M. (1945). Der yivo un di problemen fun undzer tsayt. *Yivobleter,* 25 (1), 13.
Wertheimer, N., & E. Leeper. (1979). Electrical wiring configurations and childhood cancer. *American Journal of Epidemiology,* 109 (3), 273–284.
—. (1992). EMFs and cancer rates [letter]. *Microwave News,* 12 (4), 14.
Whately, R. (1963). *Elements of rhetoric.* (D. Ehninger, ed.) Carbondale, IL: Southern Illinois University Press.
Whelan, E. (1999, July 27). Regulatory power is the dangerous kind. *Wall Street Journal,* A22.
Whewell, W. (1840). *Philosophy of the inductive sciences.* Vol. 2. London: Barker.
—. (1984). *Selected writings on the history of science.* (Y. Elkana, ed.). Chicago: University of Chicago Press.
White, J.B. (1984). *When words lose their meaning.* Chicago: University of Chicago Press.
Whorf, B. L. (1956). *Language, thought, and reality.* (J. B. Carroll, ed.). Cambridge, MA: The MIT Press.
Wiggins, D. (1981). Deliberation and practical reason. In A. O. Rorty (Ed.), *Essays on Aristotle's ethics.* Berkeley: University of California Press, 221-240.
Wilberforce, S. (1860, July). Darwin's *Origin of species. London Quarterly Review* 108, 225-64.
Willard, C. A. (1989). *A theory of argumentation.* Tuscaloosa: University of Alabama Press.
Wilson, E. O. (1975a). *Sociobiology: the new synthesis.* Cambridge, MA: Harvard University Press.
—. (1975b, December 11). For sociobiology. *New York Times Review of Books,* 22 (20).
—. (1976). Academic vigilantism and the political significance of sociobiology. *Bioscience,* 26, 183-190.

—. (1977). Biology and the social sciences. *Daedalus,* 106 (4), 127-140.
—. (1978a). Introduction: what is sociobiology? In M. S. Gregory, A. Silvers & D. Sutch (Eds.). *Sociobiology and human nature: an interdisciplinary critique and defense.* San Francisco: Jossey-Bass.
—. (1978b). Forward. In A. L. Caplan (Ed.), *The sociobiology debate: readings on ethical and scientific issues.* New York: Harper & Row.
—. (1994). *Naturalist.* Washington DC: Island Press.
—. (1998). *Consilience: the unity of knowledge.* New York: Alfred Knopf.
Wilson, L. G. (1971). Sir Charles Lyell and the species question. *American Scientist,* 59, 43–55.
Wittgenstein, L. (1922). *Tractatus logico-philosophicus.* London, Routledge & Paul.
—. (1980). *Culture and value.* (G. H. von Wright, with H. Nyman, eds.). (P. Winch, Trans.). Chicago: University of Chicago Press.
Woolgar, S. (1983). Irony in the social study of social science. In Knorr-Cetina & Mulkay (Eds.), *Science observed: perspectives on the social study of science.* London: Sage, 239-266.
—. (1989). What is the analysis of scientific rhetoric for? *Science, Technology and Human Values,* 14, 47–49.
—, and D. Pawluch. (1985). Ontological gerrymandering: the anatomy of social problems explanations. *Social Problems,* 32, 214-27.
Worden, A. (1973). Toxicology and the environment. *Toxicology,* 1, 3–27.
World Health Organization, International Agency for Research on Cancer. (2002). *Non-ionizing radiation, part I: static and extremely low-frequency (ELF) electric and magnetic fields.* Vol. 80 of *IARC monographs on the evaluation of carcinogenic risks to humans.* Lyon: IARC Press.
Wright, A. S., D. A. A. Akintowa, & M. F. Wooder. (1977). Studies on the interactions of dieldrin with mammalial liver cells at the subcellular level. *Ecotoxicology and Environmental Health,* 1, 7–16.
Xintaras, C., B. L. Johnson, C. E. Ulrich, R. E. Terrill, & M. F. Sobecki. (1966). Application of the evoked response technique in air pollution toxicology. *Journal of Toxicology and Applied Pharmacology,* 8, 77–87.
Youmans, E. L., ed. (1873). *The culture demanded by modern life: a series of addresses and arguments on the claims of scientific education.* New York: Appleton & Co.
Young, R. W. (1979). Paradigms in geography: implications of Kuhn's interpretation to scientific inquiry. *Australian Geographical Studies,* 17, 204–209.
Young, R. (1978). English as a second language for Navajos. In M. Lourie & N. Conklin (Eds.), *A pluralistic nation: the language issue in the United States.* Rowley, MA: Newbury House, 162-172.
Yu, M. (2001). *Environmental toxicology: impacts of environmental toxicants on living systems.* Boca Raton: CRC Lewis Publishers.
Zakrzewski, S. F. (1997). *Principles of environmental toxicology.* Washington, DC: American Chemical Society ACS Monograph 190.

Index

A

D

F

I

N

O

P

S

U

V

W